W9-CQR-834

A Culture of Improvement

A Culture of Improvement

Technology and the Western Millennium

Robert Friedel

The MIT Press
Cambridge, Massachusetts
London, England

MIT Press books may be purchased at special quantity discounts for business or sales promotional use. For information, please email special_sales@mitpress.mit.edu or write to the Special Sales Department, The MIT Press, 55 Hayward Street, Cambridge, MA 02142.

This book was set in Times New Roman and Syntax on 3B2 by Asco Typesetters, Hong Kong and was printed and bound in the United States of America.

Library of Congress Cataloging-in-Publication Data

Friedel, Robert D. (Robert Douglas), 1950–
A culture of improvement : technology and the Western millennium / Robert Friedel.
p. cm.
Includes bibliographical references and index.
ISBN-13: 978-0-262-06262-6 (alk. paper)
1. Technological innovations—History. 2. Technology and civilization. 3. Civilization, Western. 4. Intellectual life. I. Title.

T15.F774 2007
303.48′3—dc22 2006050722

10 9 8 7 6 5 4 3 2 1

Contents

Preface

Like all books, this one can be read in many different ways, only some of which are readily apparent to the author. There are three particular ways that I would like to point out to the reader. The first is simply to see this as an overview of essential elements of a very large and complicated story—the development of technology in Europe and North America over the last one thousand years. This kind of overview has been done before, in both more extensive forms and briefer ones. My primary excuse for attempting yet another such survey is that those available in English are several decades old and do not take advantage of the flowering of scholarship in the history of technology over the last quarter of the twentieth century. The second way to approach this work is to see it as a proposal for how we should talk about technologies and technological change. I hesitate to suggest that this is a "model" in any meaningful sense, but instead hope that at least some readers will find the discussions of improvement and some of the historical notes on that notion useful in thinking about the motivations, patterns, and implications of technological change. Those readers averse to history that overtly propounds theory will hopefully find the approach here unobtrusive. Finally, I will point out that this work is a modest contribution to a sorely needed public conversation about the moral dimensions of technology. One of the consequences of the story told here, of the increasing size and complexity of our technologies, is that the implications of the choices these technologies pose to us—as individuals, as communities, and as a civilization—are often obscured beyond recognition. It is a function of history to highlight the choices that men and women have made in the past so that we do not neglect our own responsibilities for the choices of the future.

The writing of history is, by its very nature, a collective enterprise. We historians, after all, are ultimately dependent on those in the past that have left us evidence of their thoughts, words, and deeds. We also depend on those who intervene between ourselves and our sources—the collectors, record-keepers, clerks, bureaucrats, archivists, and librarians—gatherers and preservers of the evidence of what has gone on before us. Our dependency extends still further to include the other historians, cited and otherwise, who trod before us in fields once unknown or little explored. As someone who has spent his entire working life as a historian I've long had a sense of this dependency, but the current work has brought it home to me with greater force and urgency than ever before. Whereas most of my earlier writing depended in large degree on archival investigations, this book makes great use of the large body of scholarship, much of it appearing in the last twenty-five years, that has put the history of technology on a solid foundation. I feel humbled and privileged to have witnessed the flowering of the field and to have been able to put the products to good use.

The list of debts incurred in researching and writing this book is an unusually long one, as befits a work covering a millennium of history and depending on such a wide range of scholarship. Like almost all scholars, I have depended on support and encouragement from both institutions and individuals. The attempt to name these to make my gratitude more public and explicit will inevitably fall short of being complete, but I want to name at least the most prominent here. One institution above all must be named first, for without its material support, wonderful resources, and very special people this book would have had no chance of being written. The Dibner Institute for the History of Science and Technology, located on the campus of MIT in Cambridge, was where this book was first conceived during a fellowship more than a decade ago, and was also where much of it was written during a subsequent, longer stay, some years later. The Institute's Burndy Library, now transferred to the Huntington Library in San Marino, California, provided treasure after treasure as I surveyed the centuries. The staff members of the Library and the Institute were unfailingly helpful and gracious, and I hesitate to single out individuals from such a special group, but I would be remiss if I did not mention special thanks to Ben Weiss, Bonnie Edwards, Trudy Kontoff, Rita Dempsey, and Carla Chrisfield. The pleasure of working at the Dibner depended in great measure on those whose vision and hard work made it such a special place for scholars; Jed Buchwald and George Smith held up the highest ideals of scholarship and intellectual exchange and have thus benefited the history of science and technology for years to come. A very special thanks is owed to Evelyn Simha, who saw more in this project than I did at its earliest stages and thus gave me indispensable encouragement to see it through. She also introduced me to Larry Cohen of The MIT Press, who took up her faith in the project with enthusiasm and commitment. Evelyn and Larry were the guardian angels of this effort for more years than they could have imagined.

Several other institutions provided valuable assistance along the way. The Smithsonian Institution's National Museum of American History has been a professional home for me throughout my career, even though many years have passed since I was on the museum's payroll. Its wonderful library and other collections are key resources for any American scholar in the history of technology and at numerous points in work on this book it was a great comfort to be able to turn to the Smithsonian for help. As precious as the Smithsonian's material resources are, even more valuable to me over the years have been the personal resources. Numerous curators—many of whom I count also as colleagues and friends—guided me in unfamiliar territories, from airplanes to typefounding to computing, and I want to thank at least a few of them here: Peter Jakob, Paul Cerruzzi, Martin Collins, Helena Wright, Carlene Stephens, Hal Wallace, Terry Sharrer, Stan Nelson, Pete Daniel, Paul Forman, Paul Johnston, David Shayt, Peggy Kidwell, Deborah Warner, and, especially, Bernard Finn, whose many years of friendship and collaboration have inspired me in more ways than he could ever know.

Overseas, a host of institutions have come to my assistance in ways both large and small and I extend my thanks to them: the Science Museum in London showed me some of the varieties of technology in the past that provided key clues to the processes of change; the Deutsches Museum in Munich reinforced and broadened some of these lessons in ways that even its curators—among whom I count a number of fine friends—would find surprising. In Stockholm, the staffs of the Nobel Museum, the Tekniska Museet, and the Jernkontoret (the Swedish Iron Producers Association) were generous in their responses to my requests for help and advice.

As this book made its slow passage to completion, a number of institutions and audiences extended the kindness of hearing me out on ideas that must have seemed at the outset rather ill-formed, and I'm grateful for the opportunities they provided me to try out these notions and approaches: these included audiences at the Dibner Institute, the University of Virginia, the University of Wisconsin, Madison, the Science Museum, and the Maryland Colloquium in History of Technology. My colleagues in the History Department at the University of Maryland were unfailingly gracious when I sought out their assistance and advice, and the department's staff has been a wonderful source of steady support over the years—I'm very grateful to them all. I'm especially grateful to Jeff Coster, who pitched in with great gusto and dedication late in the day to help find the illustrations that we both hope make this work a more useful introduction to the subject.

A number of individuals read portions or all of this manuscript in the course of my work on it, and their advice and encouragement was important to me. The students in several semesters of my History of Technology survey course put up with my efforts to try out this material on them—not always without complaint, to be sure, but still managing to remind me that I was usually learning more from them than they from me. George Callcott was a kind and generous reader, always encouraging me in more ways than he probably realized. Two colleagues in particular—Ira Chinoy and Michael Schiffer—read most or all of these chapters with extraordinary care and critical acumen. My intellectual debt to them is more than I can describe, and I count myself particularly fortunate to count them as friends as well as fellow scholars.

Other scholars helped out with specific portions of the work, steering me away from error and a bit closer to truth in seas that were often quite unfamiliar to me. While all of my companion fellows helped to make my Dibner experience more enjoyable, I need to mention a few specifically for their expert guidance in some of the areas in which my own ignorance grew more and more apparent the further I studied: Pamela Long, Brett Steele, Arne Kaijser, Lindy Biggs, Mary Jo Nye, Michael Mahony, Massimo Mazzotti and Bob Seidel. Ruth Schwartz Cowan, Sheila Faith Weiss, and Paul Israel also pitched in at later stages when I found myself in unfamiliar waters and I thank them for the timely rescues.

Providing illustrations for this work turned out to be a greater challenge than I had ever imagined, and the results here, as imperfect as they may be, could not have been achieved without the gracious help of numerous individuals and institutions. The latter are largely named in the credits, but the following must be thanked for their kind assistance: Annette Ruehlmann, Peter Harrington, Philip Cronenwett, Christopher Stanwood, Helene Sjunnesson, Philip Edwards, Michael Henry, Patricia Kosco Cossard, and Susan Koechlin. Dimitri Karetnikov rendered the new illustrations in this book, and I am particularly grateful for both his skill and his patience in translating the often imperfect examples and descriptions at hand into clear and understandable images. Patti Slowiak and Rita Suffness provided invaluable assistance in preparing the index.

This book is dedicated to Rita, who made the years of my work on this book special ones indeed.

A Culture of Improvement

1 Technology and Improvement

The focus of this book is the nature of technological change: Why does technology change over time, and how? A somewhat more specific historical issue is how technological change itself has changed over the past thousand years, focusing on the region of the world loosely termed "the West." This presentation of the issue begs a couple of definitions, which are discussed here in the sketchiest of fashions: What is technology? And what is the West?

Neither of these terms is comfortably defined in rigid or even rigorous ways. This is in part because their meanings change over time and in part because their usefulness as concepts depends on a certain flexibility and looseness. By technology we typically mean the knowledge and instruments that humans use to accomplish the purposes of life. Within this obviously broad and vague realm, it is often easier to clarify what is technology by agreeing what is *not* included: processes that are completely mental or biological are excluded (this includes everything from language to sex). We also exclude that knowledge of the world (including the knowledge of ourselves and our species) that is purely in the realm of ideas and description. Likewise, technology is separate from nature—the rocks, trees, and clouds are not, in themselves, parts of the technological realm (even if, in fact, like everything else, they may be affected and even generated by that realm). It is one of the historic characteristics of technology, particularly in the last few centuries, to test these boundaries constantly, to intrude in just those places we have thought it never belonged, whether that testing be through novel instruments or through new habits of thought. Medicine, for example, would have been seen only a century ago to have been largely outside the realm of technology, whereas today it is one of the most thoroughly technological fields any of us will encounter. *Most* of the time technologies are clearly represented by artifacts, by material constructions in the form of tools, instruments, machines, structures, and the like, and these representations are the most unambiguous markers of the technological. In an age that finds itself absorbed with the idea and power of "information," however, it has to be readily acknowledged that there are less material forms of technology, even if, generally speaking, these usually have some indispensable linkage to the artifactual world. For example, a computer program is nonmaterial technology, but it is useless without a machine on which to use it. This is, in the final analysis, one of those definitions that it simply does not pay to belabor, even if we have to acknowledge the need to come to at least a general understanding of what is and what is not included in the concept of technology.

The idea of the "West," as it is meant here, has been the subject of protracted debate, especially in the last decades of the twentieth century. For our purposes, we need not be overly complicated about what is meant; we simply need acknowledge that the geography of the West changes over time. In the so-called High Middle Ages with which this work begins (the eleventh and twelfth centuries, say), the West is essentially coincident with Catholic Christendom—those parts of Europe where the spiritual authority of the Church of Rome was acknowledged. This stretched from the Atlantic coast to the plains of northeastern Europe and the mountainous regions of the Balkans, including such outlying areas as Scandinavia and Ireland. This geography does not change with the decline of Rome's authority in the period of the Protestant Reformation, but it is of course profoundly expanded with the colonial push into the larger Atlantic world, especially North America. The more arcane debates about the integrity of the "West" as a historical concept are not relevant here. The story of modern technology is largely a Western one, at least to the extent that we focus on the creation of the technologies and the technological order that is now dominant throughout the world at large.

The core of my analysis is what I call "the culture of improvement." By the culture of improvement I mean the ascendancy of values and beliefs permeating all levels of society that "things could be done better." This is not the same as a faith in progress or the belief that the necessary trajectory of history or human experience was upward, but rather the simpler daily understanding that human beings have the capacity to improve how they do things. One is not, in this view of the world, fated by God, Nature, or the limitations of humankind to carry out various tasks, pursue wants and appetites, and satisfy longings both high and low in the same old way. This is not a doctrine of explicit, stated beliefs or the product of philosophical reflection. It is instead a set of attitudes and mentalities manifested by the ways in which women and men carry out their tasks in life, whether it be tilling fields, constructing homes or temples, designing and using machinery, waging war, raising children, or making the great variety of things that give life its shape and texture, from nails to tapestries to locomotives. Measures of improvement change all the time and differ from person to person, from group to group, and the perception of improvement for one may not be shared by another. But given those caveats, the pursuit of alternative techniques, in the belief that life or work or sensibilities will be better than they were, for some if not for all, for the moment if not in the future, is an extraordinarily powerful historical force.

For this idea of improvement to be useful, a few simple characteristics should be kept in mind:

Improvement is contingent. By this I mean that there is no absolute or universal standard for what constitutes an improvement; an improvement is determined by individuals (who may or may not work in concert with others) with specific goals at specific times. What may be an improvement in carrying out a task or making something at one time for one person may be no such thing for anyone else—or even for the same person at some other time. I want to emphasize this point, for I believe that the expectation that technological activities can be judged by some kind of absolute standard, and that the history of technology is in large part the history of achievements measured by such an implicit standard—what has sometimes been referred to as "progress talk"—is one of the primary hurdles that historians have been struggling to overcome. The attention being paid to the "social construction" of technology is one sign of the recognition of this difficulty. The approach that I am proposing is different, but quite consistent with that one.

The contingent nature of improvement directs our attention to the immediate technical motivations for innovation. By this I mean simply that we need to account for technological change, not by reference to the eventual effects of these changes, but by looking at the particular improvements that individuals have in mind in pursuing a particular change. This may seem rather obvious, but in fact much of the literature of the history of technology is colored by post hoc kinds of explanations—that is, explanations that account for the emergence of a technology based on the final effects that the technology has. Looking at improvement as an act whose character is completely contingent on the immediate goals of the actors gets us away from this sort of thinking.

Improvement is possible at all levels of action. This simply draws attention to the fact that the concept of improvement applies to the simplest sort of task as much as to complicated, expensive, and important ones, and that improvement is possible for any actor. If I wish to bake a dish, build a bridge, organize an army, or raise a child, I may seek to improve my means for doing so, and I may be capable, at some level, of actually effecting improvement, at least in the short term.

One of the problems that characterize earlier broad-brushed studies of the history of technology, particularly those reaching back into the preindustrial period, has been the retrieval of the actors in the story of technological change. Concepts such as invention and revolution tend to direct our attention to relatively few individuals in giving change its shape and force. But, in fact, most technological change—improvement, if you will—comes through the small contributions of ordinary, anonymous workers and tinkerers. By focusing on improvement we are better equipped to appreciate the great range and variety of technological action and of people engaged in interesting things.

Improvement may be ephemeral, or it may be sustained. By ephemeral I mean simply that making an improvement at one time does not in itself mean anything about the future. If one simply attempts to do one task better than before or adapt a device or techniques to circumstances, the resulting improvement or adaptation very likely is irrelevant to the future or to anyone outside the immediate environment. To be sure, the person making the improvement may *learn* from the experience, and so on a personal basis "sustain" the improvement in that sense, although it is not to be assumed that such learning will necessarily result in duplication of the improvement in the future.

This notion of "improvement" is less the subject of historical explanation than it is of psychological observation. Some cognitive psychologists have long remarked on the extent to which human beings are wired, so to speak, to be dissatisfied with their current environment. The obvious display of this is in children, for whom some level of experimentation is quite natural and in some cases spectacular. In such cases, as well, the truly ephemeral nature of most experiments and improvements is also readily apparent. Given the extent to which human evolution has consistently favored mental rather than physical advantage, the potential evolutionary advantage of this tendency toward dissatisfaction and experiment is significant. There are, of course, many influences, both social and personal, at work in diminishing this proclivity toward experiment as individuals mature. In many cultures these influences can be overwhelming, and the culture comes to be seen as "tradition-bound." In the West, on the other hand, the opposite tendency is manifest, and part of the story told here is directed toward understanding how and why experimental improvement has come to seem such a normal part of life.

Much of the truly interesting part of the story of Western technology told from this perspective lies in the changing means by which improvements become less likely to be ephemeral and more likely to be sustained over time and distributed over space.

I am not the first to point out that small, gradual improvements are a great, neglected element in our understanding of technology or that European society became more receptive to technological improvement over time. I want to go beyond this, however, to suggest how a focus on improvement could give us new means for explaining not only what was distinctive about Western technological culture, but also what conditions shaped the modern experience of that culture, including its enormous capacity for expansion and dominance.[1]

To do this, I will introduce one other simple concept, which I call "capture." This is the means by which an improvement becomes not simply an ephemeral, contingent act or product, but part of a sustained series of changes. Capture may consist simply of telling other practitioners, writing down methods and discoveries, organizing distinct crafts or professions, distributing or maintaining products, or constructing legal or economic instruments. Capture does not consist simply in means of recording techniques, but also includes the processes by which a technical change is socialized, by which the case is made that an improvement is, in fact, not only better for its originator but is better, at least in some contexts, for others, and not only for the present but also for some time in the future.

Control over the processes of capture is one of the primary means by which power is exercised in the technological world. The different levels of influence that different groups and institutions have over the direction of technology and over its uses derive from a number of sources, including political power, social status, and wealth. But influence also accrues to those who record how things are made and used, who decide which new things are important and which are not, who teach new practitioners their art, who sell new customers their products, who celebrate things they judge good and denounce things they judge otherwise. As the means of capture change in history and as their significance grows, the power of some groups rises at the expense of others. These shifts in power never take place in isolation or for simple reasons—social, economic, political, and cultural factors are always present—but paying attention to the control of technological capture can tell us a great deal more about how the winners and losers are determined as technology changes.

This notion of improvement and attention paid to the great range of means for its expression and to the methods of its capture assist our history in a number of important ways. Above all, perhaps, they emphasize the great range of actors that contribute to the richness and variety of technological life in the West. Understanding the people who make and use technologies in the past, and who are most central to the processes of change, is particularly challenging due to the anonymous nature of much of technological improvement. The historian is traditionally guided and bound by the written record, and yet there was nothing at all literate about most technical activity through most of history. Of course, this is true for all but a few spheres of human action, and many fields of history have confronted this difficulty with varying degrees of success. The particular problem for technology is twofold: first, a focus on change (rather than on simply describing a state of affairs at a particular time and place) makes anonymity especially frustrating, since this makes specific statements about motives, capabilities, and the like problematic.

Second, technology lies largely in a realm that is not just nonliterate, but is what might be called "antiliterate." It not just that the practitioners of technology for most of the human experience—even in the last thousand, relatively literate, years—have themselves not been readers or writers, but the things these practitioners knew and did often resist verbalization, and the improvements they sought to make in their work or products are not easily described or accounted for in words. We are badly misled by our last couple of centuries of systematic technical analysis, engineering theory, and codified practice, a culture of measurement, delineation, and definition. These things lead us to believe that technical practice is easily incorporated and depicted within the context of literary experience. Our laws, for example, allow patents of invention upon written description of a contribution to the art, and our teaching is centered on books and formulas. But for all of human time before the last few centuries, technique was less a matter of words and recipes than of skills and feeling—what has commonly come to be called "tacit knowledge."

For much technology, as we generally conceive of it, at least, there are alternatives to documents. In particular, there may be artifacts—the tools and products of technical activity. Of course, much of what I am talking about is only marginally better represented in artifacts than it is in writing, for the technical experience lies in making and using, and improvement is often a matter of modifying action rather than things. In addition, there is a kind of prejudice toward the artifact that is just as limiting in some ways as that toward documents. Many technologies yield products that are almost as ephemeral in their existence as the actions themselves (think of food, for example, the subject of an enormous amount of daily effort throughout all of history, including our own day), but we have been blinded to this kind of production, and we tend to neglect them in our consideration of technology.

One of the useful features of looking at technological change through improvement and capture, in fact, is the distinction that this makes between essentially intellectual processes (improvement) and social ones (capture). This way of talking about technological change is not meant to be a rigid "model" or "theory," but is instead a proposed vocabulary, a way of framing questions and hypotheses about the history of technology. In the terms that I propose here, my theses are pretty simple:

1. Over the past thousand years there has developed in the West a "culture of improvement," an environment in which significant, widely shared value has come to be attached to technical improvement and conditions have been cultivated to encourage and sustain the pursuit of improvement. Related to the value attached to improvement is the widespread expectation that improvement will indeed occur in most realms of technology.

2. The rate and nature of technological change in the West has been substantially altered in the last half of this millennium through the devising of new and more effective means of capture. The extent to which technical improvement is necessarily ephemeral and transitory has diminished to an extraordinary extent, and the West's capacity for developing and harnessing technologies has consequently increased.

The real virtue of these propositions is that they raise useful questions, questions that hopefully help us make sense of this large and complex history. Here are a few examples and some hypotheses:

• How are the values attached to improvement expressed and cultivated? Particularly in the earlier parts of this story—the Middle Ages, for example—we need to seek out these values and get some understanding of how the European attitudes toward new techniques shifted. Furthermore, we need to do this less through philosophical discourse, as some scholars have recently done, and more through an examination of technological activity itself. The real significance of the culture of improvement lies in its role in the lives of the men and women who are actually making and doing things.

• What are the basic means by which improvements are captured, and how do these change over time? This focuses our attention on the nature of technical knowledge and, particularly, on the means by which technical knowledge, particularly of innovations, is transmitted. Again, for the earlier part of our story this is a question of considerable significance. And it is even more significant as we move into the early modern period, in light of our hypothesis about change. What improvements, if you will, are effected in the means of capture?

• More directed questions follow. For example, who controls the means of capture? This gets us into the large realm of the relationship between technology and power. But it is not an open-ended kind of question; it is directed toward understanding the relationship between particular people and groups—defined as our particular concerns lead us—and particular means. As these means—the institutions, tools, and processes of capturing technological improvement—change, so too do the implications of these power relationships.

When one begins to think more seriously about this mechanism of improvement and how it actually has worked over time, a number of interesting features emerge. As an illustration, let me return to the point about looking at the means by which the "capture" of technical improvement was made so much more effective and extensive in the early modern period. Few historical questions have been given more thorough and varied attention that those surrounding this transition to modernity—the conditions and features of the society, economy, and intellectual culture of Europe between, say, 1400 and 1700 that lay the foundations for the industrial world as we know it still. I have no intention of revisiting this historiography, except with a narrow, defined object, namely, trying to explain how and why, in the course of these centuries, the processes of technological change became so much more rapid. By the eighteenth century, in fact, the idea of "improvement" had become particularly fashionable and compelling in many parts of Europe. At the same time, of course, the foundations were laid for the creation of industrial technologies, both science-based and otherwise.

There are many elements that go into explaining the fundamental changes in society and economy that accompany and mark this transition to modernity. But clearly relevant is an understanding of how the forms of technical knowledge changed, and how the means for capturing improvement became more effective and widespread.

This is in itself a large topic, so let me just enumerate some of the elements that can be found:

• the appearance of written manuals and technical books—the most obvious and probably the most important

• the elaborated functions of guilds—often looked at as suppressers of innovation, but actually much more complex instruments for shaping and controlling knowledge

• the elaborated functions of other regulatory systems, such as trade regulations, sumptuary laws, and the like

• the beginnings of patents for novelty

• the emergence of artistic and graphical techniques and skills for rendering devices, constructions, and machines

• the much increased amount of travel among European craftsmen and traders

• the increased trade in manufactured goods (textiles, glass, ceramics, metalwork, firearms)

All of these—there are of course others—are social and technical instruments for embodying and carrying technical knowledge, and they all either first appeared or were significantly changed in the late Middle Ages and the Renaissance. In the centuries after about 1400, these instruments and practices began to be used much more widely among a great range of craftsmen in some parts of Western Europe, and the cumulative effect was a profound shift in the way that both large and small experiments in techniques and tools modified the means by which things were made and done. Social and economic value came to be attached to improvements and efforts to exploit them spread among practitioners. New ways of doing things came more and more to be perceived as possible sources of profit or well-being or power.

Many works have been written, most particularly by economic historians, to explore and answer the question of why the West created a modern industrial economy, and hence achieved modern standards of living, before any other part of the world. That technology is a key part of the answer to this great question is hardly disputed, but the full extent of technology's role and the causes of technology's influence are subjects of much scholarly argument. Karl Marx and Adam Smith, for example, both agreed that "machinery" was significant, but they disagreed a great deal on its relative significance. Of course, Smith was writing in the 1770s, when the creation of the industrial world was barely visible to even the most astute observer, while Marx, writing almost a century later, was reacting strongly to the character of that world as it had unfolded through an apparently unchecked pursuit of wealth on the backs of exploited workers. For Marx, therefore, changes in technology were key events in human history, profoundly altering the relations between people and the different powers available to different classes. The debates about technology's relative significance in shaping the modern world have continued to our own day, but its central role is little challenged by most historians.

The questions that occupy most attention in this connection are about why the West was able to create such a dynamic and productive technology while the rest of the world failed to do so. This book does not attempt to answer these questions, in the sense that it is not a comparative work. There is little said here about the technologies of those civilizations most appropriately compared in wealth and size to the West's, such as East Asia or Islamic Asia and North Africa. Much has been written about the roughly comparable economic conditions of these parts of the world before "the rise of the West," and a variety of explanations for the divergence have emerged, incorporating geography, culture, religion, epidemiology, politics, demography, and environment. In discussing the nature of Western technology over the last millennium, comparisons to the rest of the world are natural, but to make them useful we need a clearer sense of how technological change in the West itself changed its character and scope. We therefore focus here exclusively on the dynamics of technological change in the West and on the motives and attitudes that shaped this change.[2]

The West and its technology do not, of course, exist in isolation from the rest of the world. Over the centuries, tools and techniques have traveled wherever people have moved and traded. Technical knowledge, typically in the form of personal skills and experience, but sometimes also embodied in artifacts, depicted in images, or written down, has been as mobile as people and things. Whether such knowledge took root and had any effect in new places depended on a host of circumstances. Some societies were clearly more ready to take up novel ideas and practices, while others were hostile to them. The increasing readiness of the West to accept them and then to seek them out was both a cause and an effect of the culture of improvement, and it is part of what we wish to show and explain. The truculence of other societies, while a historical fact of considerable importance, is not part of our subject.

One concept that is very common in the discussions of economic historians and others, but that is avoided here, is the notion of technological progress. To many, economists or not, it might seem foolish to deny the fact of progress and its historical importance. There are, as the history here will illustrate abundantly, many things that we can now do that we could not do before. There are many features of life, from health care to entertainment to transportation, that few of us would wish to forgo and that are the product of technological improvement over the last several centuries. But this book is not about progress, even though readers may infer progress from the history here, by whatever values they wish to bring to the term. I do not wish to impose or assume a particular measure of progress, and it is not germane to the story here, as much as it may be expected. Improvement, as it is used here, is not progress, at least not in the large sense in which that term is typically used. As the preceding definitions attempt to emphasize, improvement refers to the perceived ends of the actors in this story, essentially short-term and local, not to more cosmic outcomes. While social and material progress, in some periods of this history at least, may indeed be loudly touted as both a goal and an achievement, this does not so much describe immediate actions as it attempts to link these actions to ostensibly larger and worthier aims, indeed aims implicitly sanctioned by God and Nature. To talk of progress is to speak in essentially teleological terms, to suggest that technology (or whatever is the subject of the term) is moving linearly toward a divine end. The reader is free to believe this or not; it is essentially irrelevant to the way the story of technology is laid out here.[3]

A number of other approaches that have been applied to our subject are missing in what follows, and some readers might find it helpful to know why. It is common, for example, to differentiate among different kinds of technical improvements, designating them as "inventions" or "innovations," sometimes as "microinventions" or "macroinventions," or some similar vocabulary. There is nothing necessarily wrong with such distinctions, but they are not used here inasmuch as they contribute little or nothing to the argument. Indeed, such categories can distract us from the more important fact that for most of history the pursuit of improvement has been carried on with no regard whatsoever for such distinctions. In the last several centuries, the emergence of legal instruments for awarding intellectual property in some kinds of improvement—most particularly patents for inventions—has made such distinctions about the extent of novelty embodied in an improvement a matter of legal and economic importance. But this need not force the historian to abandon an appreciation for the fundamental continuity from the smallest modification of a tool or process to the appearance of a great new idea with widespread implications. Of course, some improvements or inventions are much more important than others, and this work pays much more attention to these—their stories tend to be more interesting and much better known—than to others. But these distinctions should not divert us from the fact that the creativity of humans is not predicated on the importance of the outcome of their efforts.[4]

Another approach to the history of technology that has a long history as well as some current popularity is based on an analogy between technological development and Darwinian evolution. Technologies are seen as akin to species, and inventions are variations or mutations. This approach can have several virtues: it emphasizes the ubiquity of variations, both large and small; it calls attention to the importance of selection, by which societies determine which innovations are worthy of surviving and reproducing and which are not; and it can provide an escape from the teleology of most progressive accounts, the notion that technological change is inherently directed toward some final goal. The Darwinian (or, more strictly, neo-Darwinian) notion of evolution is striking for its ability to accommodate progressive change (increasing complexity, for example) while not imposing some kind of final purpose or design. This has made it attractive as a model of explanation for historians of technology. There is no fundamental inconsistency between this model and the discussion here, although there may certainly be differences in emphasis. The Darwinian analogy is, however, largely irrelevant. Much energy can be spent in developing the parallels and explaining away most (but never all) of the apparent inconsistencies between the biological and technological worlds, but no real research program seems to emerge from the technological exercise, unlike the enormously fruitful efforts that follow the work of the evolutionary biologists.[5]

What follows is a generally chronological account of European technologies beginning with the High Middle Ages, when the material basis of life was still largely as it had been for many centuries but for a few isolated areas of change. The attention is always on these areas of change, which become more and more numerous and substantial in the following centuries. The culture of improvement emerges slowly from our picture of how Europeans—and their American offshoots—responded to and cultivated these areas of change. The expressions of this culture become clearer and more insistent as the means of capturing improvements grow more effective and numerous. Once our narrative reaches the last couple of centuries, change is present everywhere, and the cases we describe are selected to illustrate both the scope of the drive for improvement and some of the profound ways in which that drive shaped society.

2 Plows and Horses

Sometime right about the year 1000, an English Benedictine monk, Aelfric of Eynsham, wrote a textbook to assist learners of Latin. In alternating lines of Latin and Old English, Aelfric's "Dialogue of the Teacher" presented a discussion between a teacher of Latin and a host of common folk, to each of whom the teacher asked the question, "How do you do your work?" The answers give us a sketchy yet meaningful snapshot of medieval labor, and, by extension, medieval technology.

After first querying a student monk, the teacher turned his attention to the workers of the field, first to the plowman, asking, "How do you do your work?"

> O, dear master, I work very hard; I go out at daybreak, drive the oxen to the field and yoke them to the plow. Never is winter weather so severe that I dare to remain at home; for I fear my master. But when the oxen are yoked to the plow and the share and coulter are fastened on, every day I must plow a full acre or more.[1]

The teacher's further questions elicited that the plowman's only helper was a boy to goad the oxen in the field, and that even after the plow was put away, he must feed and water the oxen and clean their stalls. "The labor is indeed great," the plowman exclaimed, "because I am not free." The shepherd, oxherd, hunter, fowler, and fisherman then all described their own contributions to the most basic of all human endeavors, the raising and gathering of food.

Getting food has been, before the last two centuries or so, at the heart of human life. When Aelfric's teacher asks a wise man of the village which worldly craft was most important, he instantly responds, "Agriculture." Changes in the supply of food, through climate, disease, importation, or technology, have profound implications for all of society. Typically changes have been seldom and slow, and this was no different in the European Middle Ages. But over the centuries between the demise of classical civilization and order and the rise of great cities and kingdoms—from, say, 600 to 1100—change there was, and the effect was great though hardly quick.

Aelfric's plowman hinted at one of the earliest important changes when he described how he fastened the share and coulter to his plow every morning and then proceeded to plow more than an acre in the day. A bit later in Aelfric's dialogue, the blacksmith boasted of his own importance: "Where would the farmer get his plowshare, or mend his coulter when it has lost its point, without my craft?" Such a claim would have made no sense to the farmer of classical Greece or Rome, but the dependence on iron blades for the heavy plow of Aelfric's time was a distinctive contribution of the early Middle Ages. The plow that meant such heavy labor for its tender was a novel machine, of a size and power that reshaped the European order. This plow, sometimes now supplied with wheels to help manage its bulk and allow better control over its depth, allowed the northern European farmer to attack the heavy, clay-filled, often waterlogged soils of the forested lands that had been so sparsely settled in classical times. The plowman's reference to "oxen" speaks also of the amount of energy harnessed by this tool, for teams of oxen, rather than the single animal common earlier, were now required. The full acre of the plowman's difficult day was an extraordinary achievement, and it represented the fruits of a new orientation toward the harnessing of nonhuman power for production. Even though the plowman might complain that he was not free, his place in the medieval world was far different from that of the Greek helot or Roman slave, for his Christian soul and a different set of social and economic arrangements put value on his labor that made increasing its productivity a recognized good.[2]

The heavier plow had been around for many centuries before Aelfric's time, and was even mentioned by late Roman writers such as Pliny, but its spread, probably from western Asia, was very slow. This was typical of technological novelty before the late Middle Ages, particularly in the pursuits of ordinary life (as opposed to warfare, for example). And like almost all technologies before the last four centuries, we know very little about just how the plow spread, except that its use moved, predictably, from east to west and from south to north. In the lands around the Mediterranean, the plow of classical times was a lighter wooden structure, often equipped with a metal (bronze or iron) plowshare, which cut the soil underneath and behind the point of the plow. In these dry, light soils, even that small feature might be dispensed with, and the plow was little more than an ox- or muledrawn hoe (figure 2.1). For lands with light rains and mild winters, this was fine, but north and west of the Danube and the Alps, such a plow left much to be desired, and it is to be wondered if it was used at all. Cleared lands would more likely have been worked by hand tilling, with little direct help from animals, and the vast forests natural to northern Europe remained either untouched, or perhaps cleared in small sections by fire, and the land used only so long as the ash-enriched soil yielded good crops and then abandoned for some other similarly cleared field. Such a pattern of agriculture and settlement was no basis for sustained cultural or economic life.

With the new heavy plow (figure 2.2), however, fields could be cleared, sowed, and maintained with little more difficulty than in the long-settled lands of southern

Figure 2.1
Scratch plow. The original form of the plow was simply a means to make a shallow, straight furrow in the ground.

Figure 2.2
Moldboard plow. The heavier, more complicated plow could incorporate a number of features, including moldboard, plowshare, coulter, and wheels.

Europe, while the richness of the new soils, the reliability of the rains, and the variety of crops now possible made for an extremely productive agriculture. The new tool, however, imposed new demands, technical, economic, and social. The heavy plow was a substantial piece of capital, unlike a simple hand hoe, and this had the same sorts of implications that capitalization always has—it favored the concentration of wealth and control. To boot, making full use of it required, as Aelfric's plowman pointed out, more animal power, and this had a host of implications of its own. The full importance of this was even more apparent in the centuries after 1000, when oxen began to give way in certain parts of Western Europe to horses.

The powerful, rugged farm horse was itself a product of medieval improvement, and it was part of a complex set of technical changes and capabilities. The horse-drawn world of the pre-twentieth-century West—so basic to our view of pre-motorized life—was a real novelty in the early Middle Ages. Before the ninth or tenth centuries, horses were ridden, not driven, except for very light loads or very short periods of time. The introduction of new forms of harness and foot protection transformed this animal into the single most important assist to human labor and travel. Instead of the classical throat-and-girth harness of the Greeks and Romans, there appeared from central Asia the rigid, padded horse collar. Now, when the horse pulled against a load, no longer did the load pull back against its neck and windpipe, but rather rode on the sturdy shoulders. When this innovation was combined with the iron horseshoe, the greater speed and stamina of the horse displaced oxen wherever it could be afforded. The larger importance of this lay not only in more efficient farm work, but in swifter and surer transportation between town and countryside. The farmer with horses could move products to market more frequently and at greater distances than with only oxen, and the "urban revolution" that was to transform the European economic and social landscape after the eleventh century was propelled in large part by these new horse-centered transport capabilities.

Another indicator of how compelling and important was the new horse agriculture was its the sheer cost. Unlike oxen and other cattle, horses cannot be supported exclusively on hay and pasturage; they require, particularly in northern climates where pasturing seasons are short, cropped food, such as oats and alfalfa. Unlike grass and hay, these are grown with much of the same effort and resources applied to human nourishment, and thus their acquisition represents a sacrifice, in a real sense, of human food. The importance of this in a world that usually lived at the margins of sufficient diet is hard to overstate. The increased resources that went into making the horse central to both the medieval economy and, in a separate but related development, medieval warfare, are the surest signs of the great utility the animal now assumed.

At the heart of this utility, however, remained the heavy plow. This was, in fact, an even more complicated and significant device than hinted at by Aelfric's plowman. In speaking of his morning routine, the plowman made no mention of one other feature introduced to the plow sometime before the tenth century, the moldboard. This was a flat or curved piece of wood that rode at an angle behind the coulter and share and turned the loose, freshly cut soil over to one side. The moldboard gave shape to the plow's furrow and opened up the furrow much more thoroughly, thus exposing deeper soil to air and water. Not all plows had this extra feature, particularly in Aelfric's time. It might have been part of the plowman's tool, but was not mentioned simply because, being made of wood and not precious iron, it did not have to be fastened each morning to the plow. Whatever the case, the moldboard's use appears to have spread slowly through central and northern Europe until by the thirteenth century it was common. There remained a great variety of plow types throughout Europe, defined by region, crop, economy, and probably simply the varying skills and habits of plowrights. The overall effectiveness of the medieval tool, however, when compared with its classical predecessor, was unquestionably advanced.

The implications of this were far-reaching. Early in the twentieth century, the pioneering Belgian medievalist Henri Pirenne pointed out that one of the great changes in the early Middle Ages was a shift of the center of wealth and power in Europe from the Mediterranean south to the Frankish, German, and Anglo-Saxon north. Pirenne attributed this largely to the rise of Islam in the seventh century, and its rapid spread throughout the Mediterranean world, particularly in the West, where the Muslim forces were not stopped until the battle of Poitiers, in the south of what is now France, in 732. The hero of Poitiers, Charles Martel, was arguably the first great leader of the postclassical West, the real founder of the Frankish kingdom, whose efforts would see their fullest flowering in the handiwork of his grandson, who was to earn the appellation of Charles the Great—Charlemagne. To Pirenne, "Without Mohammed, Charlemagne would have been inconceivable." Like most efforts to reduce medieval history to simple formulas, Pirenne's conclusion has been thoroughly criticized over the years, but his problem, as pointed out by American medievalist Lynn White, Jr., has remained: Why did the center of European life clearly shift from south to north in the centuries just before 1000? Amid all the complications and the struggles with contradictory and incomplete evidence, it is clear that the development and exploitation of tools to make the north more productive were central to this great historical movement.[3]

Of all the tools to make a difference in this period, none stands out more prominently than the great horsedrawn plow. The plow not only enormously expanded the amount of arable land, but it aided in changing the economic and social character of farming, and thus of the core of medieval life. The key transformation can be read in the European landscape itself, particularly when seen from the air (it is no coincidence that scholars first began to understand this right after the First World War).

High above the land, the most distinguishing feature of the countryside is the pattern of long rows in which field after field seem to be divided, as if by the design of a quiltmaker. The closer one gets to the Mediterranean the less common this pattern seems to be. It is largely the plow's work, and is dictated by the plow's size and need for power. The older scratch plow, pulled by a single animal, was easily maneuvered and turned; not so for the large plow with a team of oxen or horses. Just like a modern truck and trailer, the long train of plow and animals took much room and effort to turn. The natural thing was thus to reduce the number of turns to a minimum, and hence elongate the plow furrow as much as possible. Of course, the actual length in any field depended on the lay of the land and the allocations of farmers, but the English word *furlong*, referring simply to a furrow's length, gives us a hint of just how long these strip furrows might ordinarily be. A furlong's 220 yards (an eighth of a mile) might represent a kind of ideal furrow. A strip of land one furlong in length and a tenth that in width is, in fact, the definition of an acre—just what Aelfric's plowman told us was his day's minimum duty.

This long strip reshaped the land in other ways, not perhaps quite so apparent at a quick glance. Due to the large turning radius of the plow and team, adjacent furrows were not plowed in succession, but instead the plowing was done in a kind of moving spiral, with one edge of the strip followed by the center, followed by the next furrow in from the starting edge, and then the next one over from the center, and so on. This had the effect not only of easing the turning, but also had beneficial consequences for field drainage. This pattern meant that the moldboard, turning the soil over to one side, always turned toward the center. Thus the strip was given a camber, a bit like a road, easing excess water over to the edges. In the drier south, this would have been unhelpful, but in the much wetter north, it helped to avoid waterlogging of the fields, while retaining water in shallow ditches between strips.

The more expensive plow and team and the distribution of land in strips were part of a complex series of changes in land tenure, rights, and ownership, that eventually yielded what has come to be called "manorialism." This refers to a loosely defined pattern, with myriad variations all over Europe, in which the ownership of land was generally in the hands of a relatively small number of landlords and the working of the land was distributed among peasants who had varying degrees of freedom, rents, obligations, and opportunities. Agricultural techniques did not "create" manorialism, any more than the needs of the manors gave rise to the techniques. In a pattern of influence and reinforcement that is basic to technological change and influence throughout history, the techniques reinforced certain patterns of land use and authority that were already emerging at the expense of others. Plowmen did not simply farm their own strips, nor did they own their plows and teams. Work and tools were distributed among the workers of the manor in patterns that balanced custom, need, and opportunity. Usually a farmer would be allocated strips all over the manor, so that good land and bad would be shared evenly, and labor could be distributed efficiently, while still giving broad distribution to its seasonal fruits (and failures). This scattering of land also promoted the sharing of expensive tools and farm animals, as moving tools and teams from field to field made it sensible to work adjacent fields in order, even if they were the responsibilities of different farmers. Scattered among the tenants' fields, also, were the fields of the "demesne," which were the lands whose product belonged exclusively to the landlord. Typical rents included labor on the demesne, sometimes as much as half of all work time.

In the course of the Middle Ages there emerged a great variety of social, economic, and legal arrangements governing the relationships among farmers, landowners, and the land itself. Recall that Aelfric's plowman spoke of working in even the harshest conditions, lest he anger his master. He was, in other words, little more than a slave ("I am not free," he told the teacher), even if, in fact, he could not be bought and sold as chattel. At the other end of the spectrum of workers were free peasants, whose only obligations to landowners were rents—fixed payments that would satisfy all duties. And even from the earliest days, there were some farmer-peasants who owned their own land, owing rents to no one, but still, typically, with obligations of services or goods in return for a lord's protection or the rights of trade. The system of serfdom that was general throughout Europe in the early Middle Ages did not, in itself, encourage innovation. The owners of land saw their wealth in terms of land itself, rather than in terms of what land could produce, so techniques that increased productivity arose from tenants and farmers themselves. Manorial obligations might limit the farmer's ability to keep additional produce for himself, but new techniques still could make work simpler or easier.

Later in the Middle Ages, new conditions of land ownership and exploitation combined with new attitudes toward work and production to establish an environment much more encouraging to innovation. In the centuries before 800 or so, the status of manual labor was quite low. Work was spoken of largely as a curse or a penance, images of labor largely disappeared, and even for monks, for whom work was prescribed, it was clearly meant as a kind of self-abasement, a path to humility. This was a part of the early European inheritance from Greece and Rome as well as from the northern, "barbarian," cultures. German warriors, for example, were expected to be idle when they were not fighting, rather than be engaged in productive labor. After the eighth century, in the so-called Carolingian Renaissance and afterward, signs emerged of a change in attitudes toward work. This change was slow, but by 1000 there was a real difference in views toward work and, by extension, toward tools and techniques that either eased work or made it more fruitful. Whereas earlier the use of machines was justified only because they might allow a monk more time for the real duties of the cloister, prayer and contemplation, bit by bit other justifications came to express themselves.[4]

The idea of improvement through technology remained largely foreign to the medieval mind, but improvement itself was not. In a usage that is still common in English, "improvement" (in Latin, *meliorare* or *emeliorare*) came to be used in referring to land that has been cleared or drained and made ready for farming or other use. After 800, legal documents began to appear in which land tenure was granted provided improvement was made. The social standing or legal rights of a tenant were sometimes linked to efforts made to improve the productivity of the land or of the labor applied to it. At the same time, medieval iconography changed and pictures began appearing that reflected the importance and dignity of manual labor. The "labors of the months" came to be standard decorations in calendars (which were, of course, generally church documents), realistically showing individuals engaged in a single identifiable kind of work, generally on the farm. Poetry in the ninth century celebrated agricultural labor, and reinforced the identification of specific months with specific tasks, such as plowing in March. By the end of the tenth century, there are even references to the *artes mechanicae*, mechanical arts or crafts, which had standing comparable to the traditional *artes liberales*, or liberal arts. Whereas in the early Middle Ages, the Latin word *labor* was used in a kind of disparaging way to refer to work with little skill attached to it, such as plowing, by 1000 it and related terms come to be associated with acquisition and productivity. In the same way, though more ambiguously, *laboratores* or workers themselves began to acquire more value in the perceived scheme of things, eventually being characterized as one of the three pillars on which all medieval society depended, the others being the clergy and the warrior class.

Throughout the Middle Ages, monasteries were powerful influences on the technical and economic order. Not only were many monasteries significant landowners, but as permanent organizations, governed by established rules and hierarchies, they were also perhaps the closest medieval equivalent, in economic terms, to the modern corporation. As such, they possessed a number of the same advantages of a corporation, organizing labor, land, and capital in stable, productive systems. The analogy should not be carried too far, however, for the purposes of the monastery were religious, not commercial, but their activities often in fact resulted in the accumulation of considerable wealth. Founded in Christianity's first centuries largely as religious retreats, enabling their members to withdraw from the world and devote themselves to prayer and contemplation, the monasteries found themselves over almost a millennium of great influence (from the establishment of the first Benedictine orders in the early sixth century to the upheaval of the Reformation in the sixteenth century) often caught between the ideals of poverty and self-denial and the material success that accompanied efforts at self-sufficiency and intelligent management.

Perhaps the single most important event in the history of medieval monasticism was the founding at the end of the eleventh century of the Cistercian order. The order, like others before and after it, was formed to return the monastic life to earlier ideals of purity and self-discipline. Its influence came largely from the dynamic teaching and leadership of Bernard of Clairvaux, who founded a Cistercian house in that small French village in 1115. The order spread at a phenomenal rate over the next several decades, exceeding five hundred houses by the end of the century. Because the ideals of the order called for isolation from established cities and towns, the houses and farms tended to be in uncultivated "wastes"—forests, mountains, and swampy lands. Making these productive enough to sustain the life of the monasteries was a considerable challenge, and it was in meeting this challenge that the Cistercians extended the technical capabilities of medieval agriculture beyond anything hitherto known. At first, it was intended that the monks themselves would do most of the farming, but like other orders, the Cistercians learned that this took too much time from prayer. So another class of monk was recruited, so-called lay brothers who submitted to monastic discipline but focused on manual labor or crafts. By the end of the twelfth century, however, even this had proved inadequate to the needs of the growing establishments, and the Cistercian houses began taking on hired workers, and even accepting serfs with the donations of lands from wealthy patrons. Eventually the Cistercian houses proved even more adept at farming and commerce than their predecessors, and they became centers of great wealth, as well as of considerable technical expertise.[5]

To a large degree, the agricultural and commercial success of the Cistercians was due not to innovations, but rather to a more systematic application of the best-known techniques. Of course, success varied from house to house, but the order represented a network of expertise, exchanging personnel and experience throughout Europe, and in this way knowledge of productive ways of doing things spread much more effectively than they had in the early Middle Ages. As we shall see, this exchange of technical expertise among churchmen had spectacular results in the spread of new styles of church architecture in the High Middle Ages (after 1000), but we can infer that building was not the only subject for the exchange of technical information, expertise, and ideas. As agriculture was the primary source of sustenance and wealth for the monasteries, and the oversight of the fields was among the most important monastic occupations, no doubt the information network included the exchange of ideas and experiences for farming and husbandry. Improvements in medieval agriculture, while hardly to be found in extant written materials, were no doubt captured by this monastic interchange.

Among the most important agricultural improvements of the Middle Ages was the adoption of new crops and new patterns of crop rotation. The maintenance of soil fertility is one of the key concerns of any agricultural society, and before the last few centuries it was achieved essentially by two methods, fallow and fertilizer. Fallow simply means leaving a field uncropped for some period of time—often for more than one growing season. By leaving a field fallow, fertility is restored or maintained largely by allowing bacteria in the soil to fix atmospheric nitrogen (and possibly other trace materials). If nothing is growing, then this nitrogen is stored in the soil and is made available to future plants. The key word here is "nothing"—fallow fields have to be kept free of weeds (and otherwise desirable plants such as grass, hay, or grain would be "weeds" in the fallow), and, as any gardener knows, this takes as much or more labor than sowing and reaping. Thus fallow fields are very labor intensive, while at the same time producing nothing. This is the cruel requirement of temperate agriculture.

The earliest agricultural communities learned that animal wastes could be applied to soil to aid fertility, and the value of animals lay very much in their manure as in their work, wool, meat, skin, or milk. In southern Europe, the climate is not particularly conducive to haying or other means of sustaining animal feed, and thus husbandry is less closely associated with farming, and manuring is not so important. But in the northern, wetter and colder, areas of Europe, crops for cattle were much easier to sustain and more important, and thus cattle raising was more easily linked to grain and other food production. This fostered a much more versatile system of manuring to supplement fallowing.

The great medieval innovation was creating a system that at the same time reduced fallow and increased both the amount and the variety of crops. This innovation could not be used everywhere, but in much of the lands north of the Alps, the so-called three-course rotation that was introduced from the late eighth century onward provided agriculture—and thus economic and social life—a margin that had never before been available. In the traditional two-course farming of the classical world, agriculture centered around the sowing of grain, such as wheat, rye, or spelt (a low-value wheat) in one half of the fields in the autumn or early winter, with the other half of the fields left fallow for that year. The fallow fields would be carefully kept clear of vegetation, in preparation for their cultivation the next year, while the fields harvested in the summer would then be left fallow for a year. In the newer three-course system (figure 2.3), one-third of the fields would be planted in the traditional winter grains, and a third left fallow for that year, but the remaining third was planted in the spring with a number of possible crops, such as legumes (beans, peas, lentils), barley, alfalfa, and oats. The latter two were particularly important as feed for horses, and the first category, the legumes, was significant as broadening the nutritional sources of the common European, providing new protein sources particularly useful for peoples with little meat or fish in their diets. Perhaps Lynn White was overstating the case a bit when he enthusiastically declared that "the Middle Ages, from the tenth century onwards, was full of beans," but the three-course system undoubtedly expanded the variety of available produce from the land and provided, by offering two harvests each year, some insurance against complete crop failure.

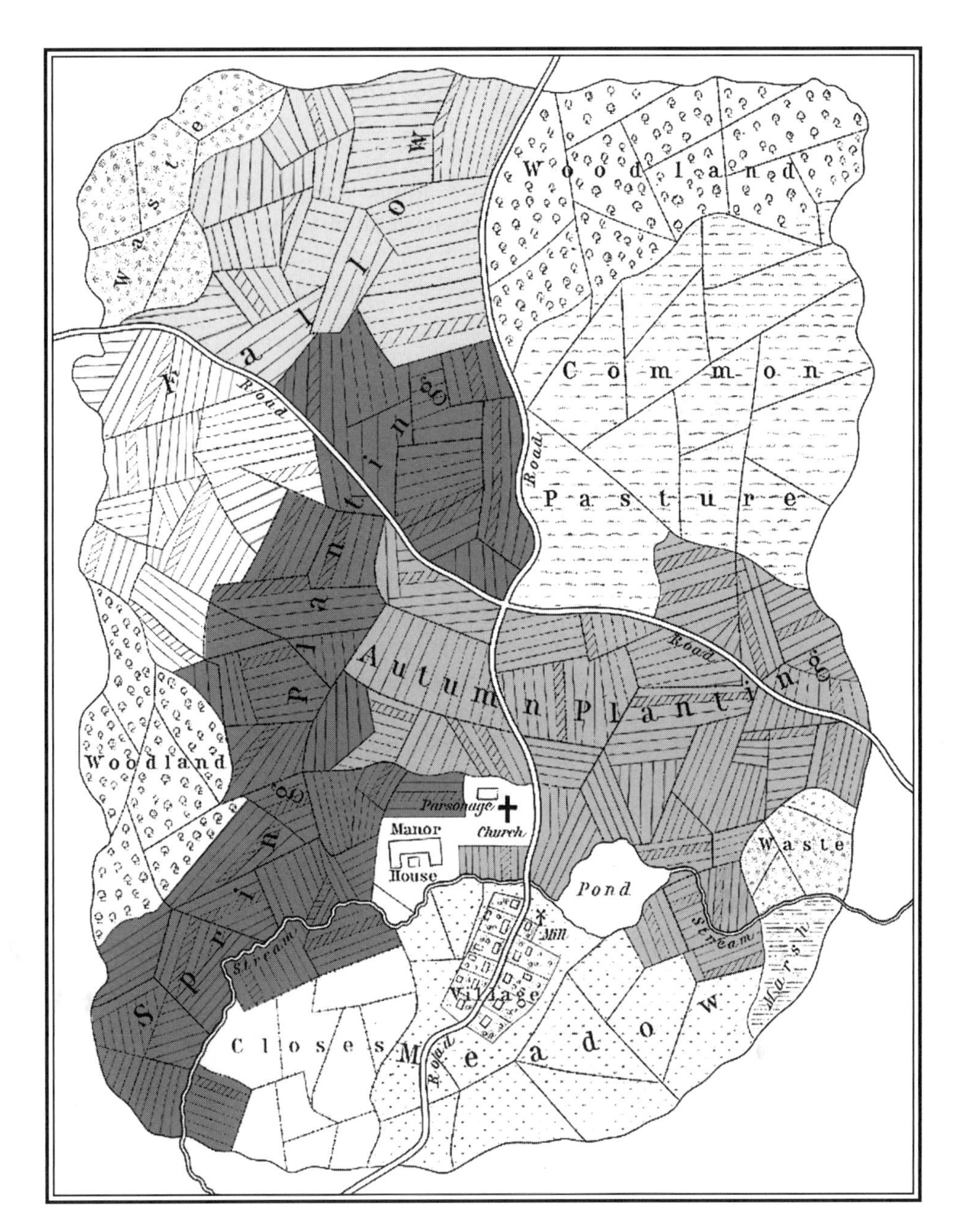

Figure 2.3
The fields of a medieval manor were typically divided into several functions, including farmed land, woodland, and pasture. In three-field rotation, the farmed lands were in turn divided into autumn plantings, spring plantings, and fallow. Adapted from William R. Shepherd, *Historical Atlas* (New York: Henry Holt & Co., 1911).

Equally significant, the additional, spring, planting worked to maintain soil fertility more effectively than simple fallowing, since most of these crops, the legumes and alfalfa especially, are nitrogen-fixing rather than nitrogen-consuming plants. That is, they tend to take nitrogen from the atmosphere rather than the soil, and through their root systems leave more nitrogen behind than they take.

In the many decades since medieval historians have studied the new tools and crop systems of the Middle Ages, there has been considerable debate about the real importance of these changes. Evidence for the actual use of specific tools and the real pattern of plantings and harvests tends to be scarce and spotty. The record must be pieced together through a patchwork of legal documents, land records, archaeological remains, and commercial paper, none of which exists in great profusion for any location. The chronology of changes—just when new tools became available or were adopted and precisely which crops were planted where—is impossible to construct fully for most of Europe. The primary result of much recent study, however, is to emphasize the extent to which European agriculture, especially as it expanded from the ninth century on, was characterized by enormous variation. Local customs, needs, and opportunities were the primary determinants of what was actually done, not theories or waves of innovation or commercial strategies. Throughout the continent, old tools and ways of working persisted for centuries alongside occasional novelties. Innovations would come, and then might possibly go, or at least be altered greatly from the originally introduced form. In this, too, medieval agriculture suggests the general shape of technological change in the West. There was enormous room for experimentation and novelty, and the institutions of society were gradually, without self-consciousness, creating the conditions under which useful improvements could be tried out, considered, and passed on.

The questions of just how much technological change there was in medieval agriculture, just how widely and rapidly it spread, and just what effect it had on life and welfare are subject to much debate and discussion among historians. The little reliable data that exist, however, tell us that the amount of land under cultivation increased considerably during most of the Middle Ages, and the population supported by agriculture increased even more. Some of the best records come from England, where the Domesday survey taken in the late 1080s provided a detailed accounting of the number and condition of thousands of estates, villages, and farms throughout the country. Of course, there is still much room for debate on how the survey's numbers should be interpreted to yield plausible figures for the entire country, since imprecision in counting was no doubt compounded by the awareness of many of the counted that the results were to be the basis for taxes and other levies. The result is a range for the English population at the end of the eleventh century of between 1.5 and 2.5 million. Two centuries later, most historians agree, the population had at least doubled, and possibly even trebled, in size. The amount of land under cultivation had also increased, as forests were cleared and marshes drained, but this increase was almost certainly no more than about 30 percent. The most likely conclusion, therefore, is that the productivity of agriculture throughout the country grew substantially in the two centuries from 1100 to 1300. The other possibility—that nutrition declined as more people attempted to live on proportionately less land—is much less likely, at least on a general basis. The prosperity of the High Middle Ages, testified to by expanding urban life and such ambitious projects as the great cathedrals, is hard to deny, just as are the relatively more difficult circumstances of the fourteenth century, when the Black Death and other difficulties beset European life and culture.[6]

At least some of the increased productivity of medieval agriculture certainly came from the intensification and better use of older methods of farming, but the new technical capabilities, combined with considerable variety in types of farming, made a great difference. Often new technologies are important simply for providing expanded choices, allowing practitioners to adapt their work to different circumstances. The new types of plow, for example, were not used everywhere, but they often made the difference in making new lands worth exploiting. The new crop rotations afforded a vast expansion of options for intensive farming. Even changes that seem very small and trivial to us could provide an important margin of success in a world in which agricultural failure, leading to famine, was never a distant possibility. The rabbit, for example, had never been a significant source of food or fur for Europeans before the Middle Ages, but the emergence of better breeds and the organization of rabbit warrens provided an extra measure of production of both meat and warmth to often marginal farmers. Most farms mixed their activities, combining grains with legumes and oats, or tending to cattle or swine in addition to sown crops, or various other combinations that depended on region, custom, and opportunity. In some areas, of course, the geography and the climate favored certain activities and crops over others, but even when there emerged dominant patterns, European agriculture was distinguished by its accommodation, in close proximity, of a wide range of cultivation and husbandry. Similarly, even when certain techniques were identifiably predominant, older ones were able to persist and newer ones were tried out.

Agricultural change necessarily has important implications for agricultural societies, and before the industrial age all but a very few societies were, at the core, agricultural. Certainly the civilization of medieval Europe centered around the cultivation of the land, the growing of crops, and the tending of herds. The greatest economic achievement of the Middle Ages was the expansion of useful lands, largely by clearing forests, rooting up brush and scrub, and draining wetlands. The work of reclaiming land from the sea for which the Dutch became so famous, for example, began in the Middle Ages, and represented simply the most spectacular and largest scale of a European-wide effort of reclamation and clearance that distinguished the twelfth and thirteenth centuries particularly. Accompanying this expansion was both a growth in population (particularly urban population) and a diversification of agricultural techniques and resources. Some scholars have remarked on how slow European farmers and landowners seemed to be to take up novel ways of doing things, with changes sometimes taking centuries to make themselves felt in some areas. This misplaces the emphasis. The important thing is that change took place at all, and that it was both widespread throughout the continent, at least from modern-day Germany westward, and that it was permanent and recognizable. Farmers have historically been conservative folk, for experiments that fail can have fatal results, not only for themselves, but for all who depend on them—which is to say, all the rest of society. Before the Middle Ages, agricultural change was much slower and was generally brought about by the need to adjust to changes in climatic, biological, or demographic conditions. In medieval Europe, however, there is much evidence that change, however slow, came more and more to be generated by the sense of new opportunity and improvement.

Historical change is never confined to a single sphere, and this was certainly true of the changes that centered on agriculture in the early and high Middle Ages. The best illustration of how the circles of change widened is the place of the horse in medieval society and culture. Horse and ox coexisted in European farming for at least a thousand years, but in the centuries on either side of 1000 the relative importance of horses in the overall scheme of life changed profoundly and permanently. This change was by no means restricted to farming itself, where the horse's increased use as a plow animal has already been described. The horse became the favored beast of carriage for a wide range of classes, and contributed to urbanization by making the transport of both persons and produce between farm and city swifter and surer. The Europeans began for the first time to breed different varieties of horses for different purposes, emphasizing a range of qualities, from the great stamina of a sturdy farm-horse, to the speed and agility of the racing thoroughbred. The importance of the horse for king and commoner alike, however, rested not simply on the animal's use for work and transport, but also, in a new and profoundly important way, for war.

The war horse has been important for at least five thousands years, but in the Middle Ages it assumed a new and important form, with implications not only for warfare but for the social and political order. While the Greeks, Romans, and German tribes used horses in battle, they were a minor part of the scheme of things, auxiliaries to tactics that were constructed around the organization of disciplined, well-armed sword, ax, and spear-wielding foot soldiers. There were during classical times tribes and armies in Asia who were famous for their adroit and effective use of horses in battle, and some famous generals, such as Alexander and Hannibal, sometimes used cavalry with extraordinary effectiveness. But fighting atop horses was a difficult and often chancy proposition due to the lack of one simple piece of equipment, the stirrup. Without some means to secure the rider to his horse's back, the mounted soldier would always live a precarious existence, and his effectiveness in battle would depend on his own strength and his horsemanship. Once, however, the stirrup is added to his equipage, not only does the horseman become much more secure in the saddle, difficult to dislodge without a direct and forceful blow, but the fighting unit of horse and man is also effectively wielded into a unified force, with the momentum and force of the horse itself added to the strength of the man atop it to deliver its blows. This does not happen until the early Middle Ages.

Exactly when and at what rate the stirrup was introduced to European warfare is a subject of much scholarly attention and debate. The argument for the central role of the stirrup in transforming medieval battle, and, indeed, in leading to profound social and political change in the early Middle Ages, was one of the great contributions of Lynn White, Jr. in his influential *Medieval Technology and Society Change* (1962). White attached his arguments for the stirrup's importance to the nineteenth-century work of Heinrich Brunner, who argued for the emergence of feudalism, the political relationship between overlord and vassal that was seen as the heart of the medieval political order, based on the increased importance of cavalry warfare in Western Europe from the eighth century onward. To White, the fact that classical horsemen lacked the stirrup and that the earliest documentary and archaeological evidence for its appearance in the West came from the seventh and eighth centuries suggested a compelling causal connection. Since White put forth this argument, much scholarship has emerged putting many crucial details in doubt, including the dates of the stirrup's introduction into Europe, of the emergence of mounted shock combat, and of the feudal systems that the new military style was said to have brought about. Simple and elegant causal connections in history rarely withstand detailed scrutiny, and the connection between the stirrup and feudalism is no exception.[7]

This should not divert attention, however, from broader connections among changes in the medieval styles of warfare, the appearance of the horse as a key animal in the European economy, and the emergence of new political, social, and economic forms. The horse became a key instrument for change in a wide range of activities, from farming to transport to battle. As already noted, the new forms of plow lent themselves to effective use of horses, which increasingly substituted for oxen, although never completely displacing them. Besides drawing the plow, the horse became a favored animal for pulling the harrow, which was drawn over the soil after sowing in order to cover over the seeds. The horse cart and the wagon made considerable inroads on the oxcart as the primary means of farm-to-town transport, and for longer travel, from city to city and region to region, the horse's advantages in speed and endurance were even more compelling. On the often miserable medieval roads, horse transport became crucial to the vigorous recovery of urban life after the eleventh century. In warfare, heavily armored mounted knights became the fiercest of battlefield weapons, even if they often did not carry the primary burden of battle. The knight and his horse, with reason, became the key symbol of the medieval warrior, and the battlefield horse became the ideal of all domesticated animals.

The place of the horse in the history of medieval improvement stems not only from the horse's contribution to key activities such as farming and warfare, but also from the horse's place in the emerging systems of technique and technical knowledge. The fact that some horses found in certain areas of the world possessed qualities distinct from those found elsewhere was widely known earlier, but in the Middle Ages we see the emergence of a more systematic categorization of horses and the development of methods for improving the categories themselves. In England, for example, there was a widely understood range of horse types, suitable by build and stamina for different tasks. The farm horse, for example, was known as an *affer*, and represented the commonest and cheapest type, costing perhaps as little as one-fiftieth as much as a good battle horse. The *sumpter* was the standard pack horse, whose importance in carrying goods with speed and reliability was key to household transport. The traveling courts of European kings, for example, depended on trains of sumpters for most of their movement. The *rouncey* was the standard riding horse, about the size of a modern pony, capable of sustaining a trot, but used in battle only for small riders, such as squires. The alternative, a bit pricier, was the *palfrey*, the preferred horse for longer journeys, trained to maintain a faster pace (an "amble"). For battle use, however, the desired horse was the *destrier*, typically ridden by only the wealthiest knights and nobles. The properly trained destrier could move steadily from a walk to a canter to a full-fledged but controlled gallop while carrying a fully armored warrior. Their expense made them precious indeed, even at the height of chivalry. *Chivalry* itself, as a code of conduct and values, represented the ascendancy of the horse and its improvement to a central place in the European scheme of things—the word, after all, comes from the Latin *caballus*, "horse," from whence also comes *chevalier* and *cavalier*, words for noble horsemen.[8]

The European Middle Ages, as commonly conceived, covers a very large span of historical time—roughly a thousand years from the fifth to the fifteenth centuries. It is a period that is hard to characterize simply, although many seem to have tried. This is in part because it is indeed a long time, filled with changes and events that are subject to generalization of only the most ragged sort. But there are other millennia in history that are not so frustrating to the historian, so why does medieval history seem so difficult sometimes? In large part, it is simply because this particular millennium is quite a bit closer to us, in time and in spirit, than, say, the prior one of Rome's ascendancy, or the ones before that for which our written records are largely those of Egyptians, Indians, Chinese, and a scattering of peoples through the Near East. This proximity evokes caution about blithe categorizations, since we are acutely aware of how impossible such generalization would be for us to accept of our own times or those of our near ancestors.

But the confusions and cautions have other sources as well. There are indeed certain important broad statements that we can make about the Europeans, and these lead us sometimes to think that similar generalizations should be forthcoming for all aspects of their lives. That this millennium, for example, was dominated by the teachings and values of Christianity is a paramount truth, with implications that reach into every aspect of life. But those implications themselves are complex and not easily characterized, changing from place to place and time to time. Similarly, the European political and social order possessed a superficial uniformity, leading past scholars to blithe generalizations about "feudalism" and "manorialism." These concepts, however, themselves resist easy and simple characterization, so much so that some scholars are willing to try to dispense with them altogether. Without advocating such a radical step, it still seems useful to emphasize variety and change in speaking of medieval structures, institutions, and ways of doing things. Change in this world did not take place quickly, for the most part, and experimentation and novelty had by no means the standing and value they were to acquire for later Europeans. But novelty was not unknown, and a sober assessment of the Middle Ages should acknowledge that, perhaps for the first time in human experience, there was emerging a society in which improvement—small but visible—was coming to be seen as part of human capabilities in the world.

3 Power

On the first of August, in the year 1086, William—the bastard son of Robert, Duke of Normandy and a tanner's daughter, Arlette, and for twenty years ruler and tyrant of England—convened an extraordinary assembly in the ancient English city of Salisbury. There, on the eve of his departure for Normandy, he brought together all of the barons, knights, and high churchmen of his kingdom for the purpose of exacting an oath of loyalty from all of them as well as money to assist his expedition. Rulers had long made a practice of getting such oaths from their subjects from time to time, but this particular meeting was extraordinary for the great range of English society thus brought together. While it had been customary for a king or chieftain to summon his great vassals and to make them swear obedience and fealty, William insisted on an oath from all who claimed rights to land in the kingdom, an oath that would acknowledge that those rights were held only at his sufferance. As the great English historian Thomas Macaulay pointed out, Norman demands on the English were as extreme as any in history: "The subjugation of a nation by a nation has seldom, even in Asia, been more complete."[1]

William and his Normans brought to England a new way of governing, one in which the power of central authority constantly challenged the ancient rights and customs that had shaped the lives of rich and poor alike. One manifestation of this new order was a new penchant for enumerating and accounting. While this was to show itself in a variety of ways through the first years of Norman rule, the greatest and most famous product was the census known by the forbidding title of the "Domesday" or "Doomsday" survey. Perhaps begun as early as 1081, the survey was essentially complete by the time that William called his subjects together at Salisbury. In all but the most remote corners of the kingdom and a few cities, a complete count had been made of the manors and other potential sources of royal revenue from which William could expect to receive rents, dues, and other feudal obligations, either directly or through his vassals. The count was intended to register every element of economic activity: plow teams (always calculated as eight-ox teams), fish weirs, tenant farmers, and one sort of machine—watermills. The Domesday book counted more than six thousand mills in more than three thousand different locations all over England. In some places there was an astonishing density of mills—as many as thirty on a single ten-mile stretch of river. By the end of the eleventh century, even this remote northwest corner of Europe had become a place where power and wealth were measured not only in land and muscle, but also in machinery.

The watermill of the Domesday was a particular sort of machine, already known to the Europeans for perhaps a thousand years. It was used almost exclusively for the grinding of grain, producing flour for bread, malt for beer, or coarse meal for porridge. Its design varied remarkably little. A wooden wheel, perhaps twice the height of a man and between one and three or four feet wide, stood vertically in a stream or under a chute. From the center of the wheel a shaft protruded, extending into the mill building. At the other end of the shaft, a set of wooden teeth engaged another set of teeth set on a shaft at right angles. Connected to this second shaft was a large millstone, thus turned by the transmission of rotary force from the waterwheel. While there might be many variations in terms of sizes of wheels, shafts, and stones, or in the internal structure of the wheel itself or of the grinding mechanism, the basic pattern of the waterwheel's construction is marked by continuity lasting almost two thousand years.

The story of change in the waterwheel in the Middle Ages is not so much the story of a novel mechanism, but rather of its intensive and extensive application. There is among historians much debate and contention over the true extent of water power's use in medieval Europe and of the precise dates of its spread, but even fairly conservative treatment of the sketchy data yields some basic impressions. The waterwheel, while known to the Romans, and even perhaps to the late Greeks before them, was not used widely in classical civilization. By the high Middle Ages, that is, by the eleventh century or so, there were parts of Europe in which exploitation of water power was commonplace. By the end of the Middle Ages, in addition, the accepted uses for water power extended beyond the milling of grain and included such activities as sawing wood, fulling and felting cloth, forging iron, and pounding a variety of substances, from dyes to rags to ore.

The meaning, too, of the spread of this machine and others like it, such as the windmill, has been a subject of historical debate. Medievalist Lynn White Jr. contended that the spread of inanimate power sources was eloquent testimony to the emergence in the West of a new attitude toward power, work, nature, and, above all, technology. There appeared in the Middle Ages, according to White, a "power consciousness" in the European mind that set the stage for the scientific and technical achievements of modern times. Other historians have demurred, pointing out the persistent limitations of the medieval view of power applications, the legal structures that seem to explain the growing use of mills more compellingly than changing mentalities, and the scant physical evidence of mills themselves or of progress in their design. But even if we do not wish to take White's claims to their dramatic extremes, even the more conservative commentators appear to recognize that over the thousand years of the medieval experience the basic elements of a power-based economy that was to so distinguish the Europeans in later centuries were being put into place. In addition, for the first time the Europeans began to show their own inventive capabilities with machinery, to harness not just water, but the wind and tides as well.[2]

The Europeans, however, did not invent the watermill. Its beginnings are obscure, like that of most technologies more than two thousand years old. The first unambiguous descriptions of watermills appeared in Latin writings of the first century BC, and there is much debate about just what may have preceded these. The vertical waterwheel of the Middle Ages bears some resemblance to a water-lifting device found in the Near East known as a *noria*, in which buckets are placed around a wheel and the wheel is turned, typically by men or animals, to bring the buckets down into a water source and then up and around to some desired height, as much on occasion as one hundred feet. Early references to wheels with buckets may be speaking of the *noria* rather than a watermill, and such references can be found in ancient India and China as well as in the Near East. Owing to the lack of archaelogical evidence as well as the ambiguity of literary references, the true origin of the vertical waterwheel is simply unknown.[3]

A bit more certain is the fact that another form of waterwheel appeared in some areas before the vertical wheel. This was the somewhat simpler and less efficient horizontal waterwheel (figure 3.1). This wheel was "horizontal" because the wheel lay on its side, with its axis sticking straight up out of the water. At the other end of the axis would be a millstone, driven directly by the waterwheel without gearing or any

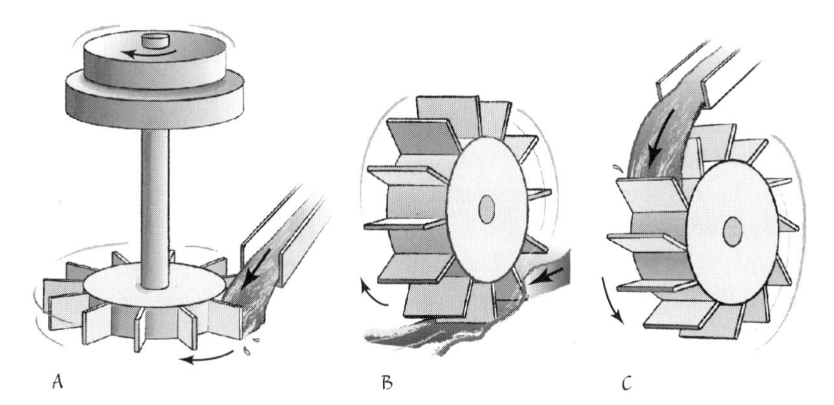

Figure 3.1
The horizontal waterwheel (A) was the simplest and least important source of water power. Much more important were the various forms of the vertical waterwheel, the most important of which were the undershot wheel (B) and the overshot wheel (C).

other complications. While this had the virtue of simplicity—one of the reasons the horizontal wheel is usually assumed to have preceded the vertical—it had the significant disadvantage of making the millstone's speed completely dependent on the direct action of the water against the wheel—no gears, and thus no gear ratios and transmission. In certain areas in which the primary water sources were rapid mountain streams (northern Greece, for example), the horizontal wheel was useful, but for most regions it simply was not versatile enough, and its poor use of the potential power of falling water was evident to even the casual observer. Some have speculated that the vertical waterwheel came by improvement of the horizontal wheel, but there is no way to substantiate this. The greater complexity and usefulness of the vertical wheel made it a very different machine.

The Roman writer Vitruvius, writing late in the first century BC, described the vertical wheel in his well-known work *On Architecture*. After describing a wheel used for lifting water, he went on: "Mill wheels are turned on the same principle, except that at one end of the axle a toothed drum is fixed. This is placed vertically on its edge and turns with the wheel. Adjoining this larger wheel there is a second toothed wheel placed horizontally by which it is gripped. Thus the teeth of the drum which is on the axle, by driving the teeth of the horizontal drum, cause the grindstones to revolve."[4] The wheel that Vitruvius described was of the form we know as the "undershot" waterwheel. This is the simplest form of vertical wheel—one in which the wheel is simply dipped into a free-flowing stream, and the velocity of the water striking the bottom paddles forces the wheel around. We know this was the object of the Roman's description because this is the one kind of wheel that can also be used for lifting water.

The Romans used another form of wheel that we know as an "overshot" waterwheel. In this, the wheel is placed underneath a source of water, and is turned by the *weight* of the water falling on the blades. Since it works by using the fall of water, it obviously cannot be used for raising water to a higher point (at least not directly), but as a source of power it is actually much more versatile. The power from the wheel can be controlled simply by managing the amount of water sent over. In the undershot wheel, such regulation is more difficult. Similarly, it is possible to build overshot wheels in a great range of sizes, if the water is available, and in areas in which the flow of water varies greatly through the seasons of the year, the overshot design makes it possible to derive some power from the wheel throughout the year, as long as some water is available. A variation of the overshot wheel, in which the water hits the wheel head-on, rather than from above, was used where larger wheels than the head might otherwise allow were wanted. The choice between undershot and overshot wheels was generally dictated by the lay of the land (hilly areas favored overshot), the supply of water, and the capital available to invest in dams, canals, and races. The Romans and medieval Europeans used both kinds with equal ease.

It is natural to think of watermills largely in terms of mechanical technology; they were, indeed, the first forms of widespread and economically important machinery. But they also represent considerable reliance on and expertise in hydraulic technology—the management of water. Improvement in waterpower technology was perhaps more often in this realm than in the purely mechanical. There were few technologies that the Romans were so adept at as hydraulic systems—the remains of their aqueducts are still marvels of ancient engineering. The primary expense of waterpower typically lies in the building of dams, canals, and races (channels to bring water to and away from the wheel), and these were clearly no problem to the Roman engineers. Indeed, from the early fourth century AD there are the remains of an astonishing mill complex that demonstrate eloquently that the potential of large scale water power was no mystery to the Romans, even if they did not exploit it widely. At Barbegal, near Arles in southern France, the Roman province of Gaul, an entire hillside was given over to no fewer than sixteen waterwheels, set out in two sets of eight descending stages. An aqueduct brought water to the lip of the hillside, and then split into two channels. The total fall of sixty-one feet was broken into eight stages, and each pair of wheels used a fall (or had a head) of about eight and a half feet. There is much that is not clear about this amazing construction—not least, just what the market for the mill's considerable capacity was, since that part of Gaul was not heavily populated. Perhaps it served as a kind of depot for troops stationed in the region. Whatever the case may be, the mills at Barbegal were apparently unique in the ancient world in their scale, and, indeed, even the Middle Ages had little to compare with them.

The Middle Ages do provide evidence for the spread of the watermill throughout the countryside and its acceptance as the primary means for grinding in all but the most tradition-bound areas (figure 3.2). The alternatives to water-powered milling were the handmill or the animal-driven mill. The handmill, which might be a simple quern, or grinding stone in a stone bowl, was the traditional means for turning wheat, rye, or barley into meal or flour. It involved laborious, numbing work, and was generally relegated to women or slaves. The mill driven by a horse, ox, or donkey was an ancient alternative, but it involved engaging an animal who might be doing other useful work in the field or in transport, and was still a capital investment of some proportions. Nonetheless, there is no reason to believe that these truly ancient means of milling were somehow unsatisfactory or inadequate to the normal needs of European society. Their large-scale displacement by an expensive machine was an event in the technological history of the West of great significance. Not only did the watermill establish new expectations of technological activity—the idea that machines using neither human nor animal power could be relied upon for important, large-scale work—but the watermill was clearly the inspiration for extending technological capabilities down avenues that would have otherwise made no sense.

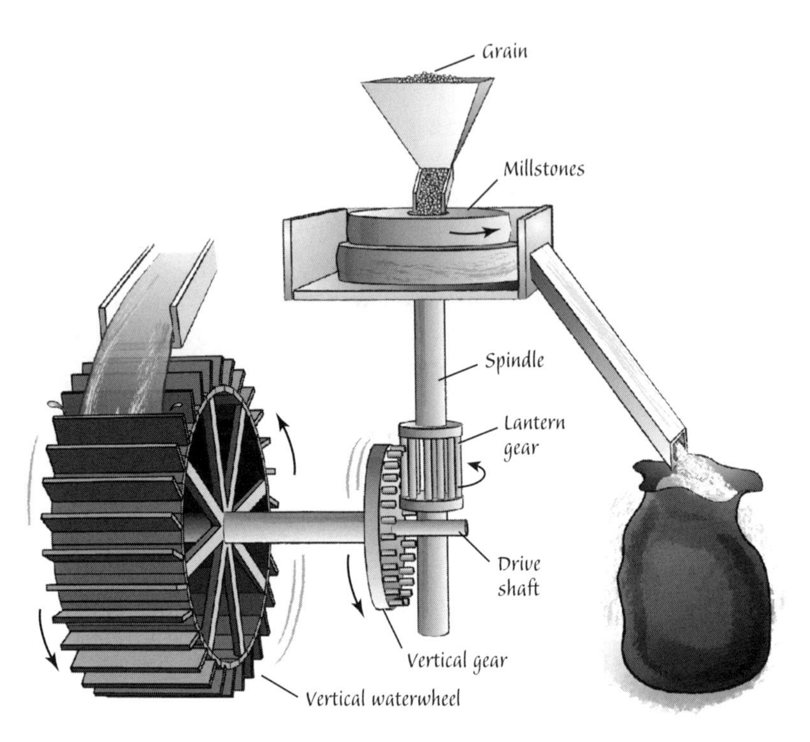

Figure 3.2
The gristmill was the most important power machine of the Middle Ages. Wooden gears translated the movement of the waterwheel to the grindstones.

Simply the effort made to extend the watermill physically required considerable technical creativity. While the basic design of the mill changed rather little from that used by the Romans, the installation of mills in many locations required much ingenuity and innovation, even if only of a local sort. In some cases, the location of a mill and adapting water supplies to power needs presents little difficulty. On a swift-flowing, reliable stream not given to regular flooding, a watermill may be built simply by putting an appropriately sized wheel directly down into the stream and attaching the milling machinery on the bank. Rarely, however, are conditions conducive to so simple a construction, and these locations are quickly occupied. If there is a fall of water in a relatively small area, then the only slightly more complicated task is to prepare a channel from the upper part of the stream, diverting some of the water at the higher level and leading it to the wheel—perhaps an overshot wheel, if there is enough "head" or difference in level between the upper stream and the wheel outtake into the lower stream. The "headrace" and "tailrace," as the upper and lower channels are called, can be simple ditches or elaborately built canals, but their design need not be a complicated matter. Nonetheless, this additional level of construction increases both the initial investment and the maintenance requirements of the mill. As the watermill spread throughout western and northern portions of Europe, the easy installation sites were taken quickly, and so the density of construction reflected in the Domesday survey represented a level of ingenuity and effort not always appreciated.

The full measure of this ingenuity and effort can be recognized most readily in the proliferation of variants of the simple mill form. The most important of these were float mills, bridge mills, and tidal mills. These were all instances in which the mill site did not allow for the simple form already described. The impetus for mill construction, however, was sufficiently great that alternatives were devised to fit the particular location. In a float mill (also called a boat mill), the mill was not placed on the river or stream bank or along a canal, but floated on a barge or boat in the stream. This was particularly useful on larger rivers where changes in river level, either seasonal or haphazard, subjected fixed wheels to problems from scarce water or, more usually and urgently, from flooding. The float mill appeared early in the Middle Ages, although records are skimpy. They were particularly important because they appeared in the heart of large towns, such as Paris, where major rivers flowed but land was scarce and streams even scarcer. Float mills, with their relatively small undershot wheels dipping moderately into the water, were not particularly powerful, and so ingenuity had to be used to enhance their output. One of the most common and effective approaches was to anchor the mill boat where the current was swiftest, either in narrow river channels or, much more importantly, near or under bridges. The large stone piers of Roman and the larger medieval bridges interfered significantly with river flow, causing currents around bridge piers to run quite swiftly. While this created real problems for river navigation and the stability of bridges, it was ideal for the floating mill. As cities grew in the later Middle Ages, the construction of mills around bridges could become quite intensive—at one point, the Seine at Paris had seventy mills in a one-mile stretch of river.

Perhaps the best illustration of the intensity of waterpower installation in a single location and the technical impetus that this could provide came elsewhere in France, in the southern city of Toulouse, on the Garonne. There, by the twelfth century, no fewer than sixty floating mills supplied the flour and meal for the town and surrounding countryside. The Garonne, however, unlike the generally placid Seine at Paris, is an unruly stream, given to periodic violent floods. Most of the float mills there were not situated around bridges, but instead were protected by three dams, which eventually allowed the mills to be made into stationary structures. The largest of these dams, the Bazacle, cut diagonally across the river for almost a quarter mile, and was constructed by ramming thousands of oak pilings, roughly twenty feet long, into two parallel rows, with the space in between filled with enormous amounts of rock, earth, and wood. It was, at the end of the twelfth century, probably the largest dam in the world.

The Bazacle was only one of three dams at Toulouse for waterpower. This level of construction and investment for the purposes of providing power (as opposed to, say, protecting property or settlements) is eloquent testimony to the growing significance of mechanical technology in the Middle Ages. The full implications of this are hinted at by a look at the financial arrangements at Toulouse. Even before the mills took on the great expense of the dam construction, their ownership was put in the form of shares, each worth one-eighth of a mill. These shares were bought and sold, just like stock, and their value fluctuated depending on the condition of the individual mill, the state of trade and agriculture, and even speculation. At first most shares were owned by the millers themselves, who had arranged for mill construction and placement and who were in charge of daily operations. But with the greatly increased expenses of the dams, and associated financial burdens and uncertainties, millers began selling their shares to wealthier townsmen, who could deal with the fluctuations in value and expenses and hold out for the longer term profits of the mills. As early as the thirteenth century, the millers were more likely to be employees of capitalists rather than owners, and in the next century the system was elaborated into a form of corporate ownership, where shares no longer corresponded to portions of individual mills, but to stock in the Société du Bazacle, for example, which owned the dam, a reservoir, related fishing rights, and several mills. The fundamental relationship between technological development and the emergence of capitalist forms of ownership and management was one of the most important products of the Middle Ages.[5]

The dams at Toulouse were particularly spectacular examples of one of the Middle Ages' most important contributions to power technology: the hydropower dam. The primary technical significance of the power dam lay in its making large and variable streams into useful sources of power. The float mill and the bridge mill were to a certain degree simply expedient ways of getting waterpower from larger rivers, but their size and power were necessarily limited and they were particularly vulnerable to floods as well debris, traffic, and other hindrances. The development of techniques for damming streams and rivers and building canals for power purposes drove European hydraulic technology down directions unknown even to the great Roman aqueduct builders. Water-supply and irrigation technology was modified to supply power to wheels. The bridge mill, in a sense, taught the lesson that human intervention in a river's flow could be put to good use by enhancing the power prospects of a river site. Dams were built for a range of purposes, sometimes to impound water where daily streamflow was unreliable or simply variable and often to create artificially sufficient head for a wheel when the natural topography did not provide it. In this latter case the dam would divert water upstream from the mill into a canal, which would lead the water at the upstream height to the mill, situated far enough downstream to yield the necessary head for the water to fall over the wheel. Power canals became prominent features of many cities throughout medieval Europe. As early as the ninth century, a canal was built across a bend of the Thames at London to provide power to three mills, and other great towns, such as Rheims, Nuremberg, and Bologna made use of canals for mills.[6]

Yet another indication of the compelling nature of watermills in medieval Europe were the attempts to create waterpower where none would seem to exist, at least at first glance. The most significant of such attempts were tidal mills, built in coastal towns where stream flow was either insufficient or nonexistent, but where the daily tidal changes in water level suggested alternatives ways of getting power from water. At first such structures were probably floating mills anchored in a harbor, where the running tide would turn a wheel in one direction coming in and the opposite going out. Such mills would have generated little power, but could have been useful where no alternatives existed. A more sophisticated (and expensive) version that was more widely used involved the construction of dams or breakwaters at the mouth of a stream or inlet. Gates could be opened to allow tidal flow in, and closed to form a reservoir that would empty out through a race to a waterwheel. The hours that such a mill could be used were obviously limited by the tidal cycle, and the power depended completely on the tidal height, but in coastal towns around Europe tidal mills began to appear by the later eleventh century, and were common in some areas through the remainder of the Middle Ages.

The use and spread of tidal mills, with all their inherent limitations, was important historically not so much for the power they produced as for the evidence they provide that the medieval Europeans had adopted a power mentality by the High Middle Ages. Historians debate over the meaning of this mentality, some agreeing with Lynn White Jr. that medieval Europe was a "power-hungry society," bent on exploiting the energy sources of nature. Others are more skeptical, siding with the British student of waterpower, Richard Holt, in the belief that the Europeans before the fifteenth century remained quite restrained in the real use they made of power sources. There can be little doubt that the skeptics are right in their claim that medieval society never lost its fundamental dependence on the muscles of humans and animals for most work and movement. Nonetheless, the expansion of machines, the spread of power devices into every corner of at least the watery parts of Europe, and the slow but still visible expansion of power machinery beyond milling into a variety of productive tasks mark in ways that are hard to deny the efforts of a society to reduce the utter dependence of the human race on muscles for the accomplishment of needed and wanted tasks.

Perhaps the most spectacular clue to this mentality's pervasiveness and significance actually lies outside the realm of waterpower itself, and comes instead from one of the greatest of the European inventions of the period, the windmill. Unlike most of the other medieval technical advances, it is very unlikely that the windmill originated anywhere other than in the European areas in which it became important—particularly in the flatter regions of northwestern Europe. The Domesday survey of England showed not only the numbers of watermills, but also their geographic distribution, and this makes it clear that there were substantial areas of the country where watermills were few and widely scattered. This had different causes in different areas, but in certain parts of the country, especially the flat and marshy lands of eastern England, the lack of streams with substantial flow or head would have made waterpower scarce. Little surprise, then, that the evidence for the first windmills comes from just these areas. Indeed, while the evidence is scant, it is possible that the windmill was an indigenous invention of this section of England. If so, the windmill has claim to be the single most important and original technical contribution of this part of Europe. While some have claimed that the mechanism of the medieval windmill was so close to that of watermills that it represented a limited technical advance, this seems to underestimate considerably both the leap of technical imagination required and the range of technical problems that had to be overcome.

While sailing ships, and thus use of the power of the wind, were ancient by medieval times, the translation of this idea into harnessing the wind for mechanical work was by no means an obvious step. It certainly does not seem to have occurred in other areas in which it was to prove of great advantage, such as waterpower-short portions of the Mediterranean world (like Greece or Spain). The first documented reference to a windmill comes from 1185, when records speak of a mill in eastern Yorkshire and other records of one in Sussex. Within hardly more than a decade, English documents speak of at least twenty further examples, and in the early thirteenth century the references to windmills, both in England and on the continent, multiplied in considerable numbers. This suggests an idea that was by no means obvious, but that, once it became known, was appreciated and exploited readily. It is true that the medieval windmill could never supplant the watermill; it was always useful as a resort for those situations where waterpower was simply not practical or sufficient. But this role was by no means trivial, and in the late Middle Ages the possibilities of the windmill for work besides grinding grain—especially for pumping water from marshy lands—were to be of surpassing importance in such areas as the Netherlands and the English fens.[7]

The technical problems that had to be solved for successful windpower exploitation were not trivial. The wind does not behave at all like a stream of water. Even when a stream's flow is uneven, the technologies for diverting and impounding water made control of water supply—and thus power input—in watermills a relatively straightforward matter. But there is no means for "impounding" the wind, and thus control technologies had to be very different. These were singularly clever examples of medieval technical improvement. The wind varies in direction and in force, and both of these variations had to be accommodated in a successful windmill. In certain areas, in fact, variations in direction were not so important, since prevailing winds were consistent enough to favor one direction over all others, but more usually efforts were made to allow the mill to be turned into the wind. The most straightforward solution was to construct the mill on a single huge wooden post, around which the mill could be pivoted to face the wind. This put a considerable burden on this post, which had to remain rigid under all conditions, or else the mill's stability was compromised. It also limited the size of windmills to that of structures supportable on a post, as well as movable. These considerations led to a uniformity of design to which a surprising number of medieval mills conformed. Typical medieval images show a wooden shed on a large post, with a pointed roof, a ladder leading up to a rear doorway, and a long pole on the other side which was used to turn the mill when the wind shifted (figure 3.3). Four vanes were standard, equipped with sails that could be reefed and furled like those of a ship.[8]

In the midst of this remarkable uniformity of design, one area that invited experimentation and improvement was the construction of foundations of the mill. The inherent problem with the postmill design was the reliance on the single large wooden post for stability. It is not surprising, therefore, that a range of methods were used to secure this post in the ground so that the mill stood with little vibration or danger of leaning or collapse. Because windmills were generally the product of flat terrain

Figure 3.3
Post mills were relatively light structures that could be turned into the wind. Mill machinery had to be relatively light.

(where water flow was slight and winds unimpeded), one common characteristic of foundations was the construction of artificial mounds, raising the windmill a bit more above the landscape and allowing for additional foundation work to secure the post. At first, the post was simply stuck into a hole in the ground, as deep as was necessary to secure it, and dirt was mounded around. Then a variety of techniques were used to add stability by using additional pieces of wood around the post, placed flat on the ground and secured to the post before the mound was made, or placed at angles up against the post, securing it further from strong winds and vibrations. As in so many other cases, it would probably be a mistake to make out a "progression" of techniques—many perceived or experimental improvements were tried, depending on local conditions, resources, and knowledge.

Toward the end of the Middle Ages, in the late fifteenth century, another kind of modification in securing the windmill appeared in the form of the tower or turret mill (figure 3.4). Unlike the post mill, the tower mill did not turn in its entirety, but in-

Figure 3.4
Tower mills were stable, permanent structures, capable of housing more substantial machinery. The top of the mill could be turned to face the wind.

stead had a movable head, to which the vanes were attached, that sat atop a permanent structure (the tower), possibly made of stone or brick. The head alone could be turned to place the vanes into the wind, while the rest of the structure, and most of the machinery, remained stationary. Problems had to be solved to make the connections between the movable vanes and axle in the turret and the stationary gears and grindstones in the tower as secure as possible. This made the mechanical connections a bit more complicated, but the turret mill could be made larger and more stable than the post mill, and so was attractive where such a larger size could be used to advantage. This was probably not important in most places, and the turret mill does not appear until relatively late, but it is just one more indication of the windmill's susceptibility to improvement, even while the machinery itself underwent little change for centuries.

Besides the new windmill, the other sign of how mills stirred European technological creativity was in the extension of applications for power. In our industrial civilization, the use of power sources for every aspect of work, transport, and daily life is so fundamental that it may said to define us as clearly as any other aspect of our technology. The Middle Ages saw the beginnings of this power mentality, through not only the extension of mills into every corner of Europe in which they could be built, but also through experimentation in other applications for power than the milling of grain. Milling was always, it must be emphasized, the overwhelmingly dominant application of waterpower and windpower, and the association of the mill with "grist" hardly even waned before the last couple of centuries. Likewise, the tasks for which waterpower came to be used on occasion were barely changed by the occasional substitution of millpower for human muscle. This power mentality, in other words, was more the emergence of a sense of potential and possibility than it was the reflection of any new economic or social reality about how work was actually done. Its importance lay not in the creation of any kind of "industrial revolution" in the Middle Ages, as some overly enthusiastic students have claimed, but rather in providing yet one more marker of the emerging culture of improvement.

The two trades in which waterpower (windpower was simply not a factor) found the most extensive application were textiles and ironmaking. Fulling was a key step in the manufacture of most cloth. It consisted of scouring and beating newly woven woolen cloth both to clean it of dirt and lanolin (the natural oils of wool) and to shrink and felt the woven fibers to make the finished product softer and suppler. Since ancient times a variety of techniques had been used for fulling, but the most common was to mix the cloth with scouring material (such as "fuller's earth") and soak it in water, while beating it with clubs or, more usually, trampling it underfoot. Such repetitive, unskilled labor invited mechanization almost as readily as grist milling, although it required devising rather different mechanisms. In a fulling mill, the axle of the turning wheel was studded with protrusions, or tappets, that engaged a

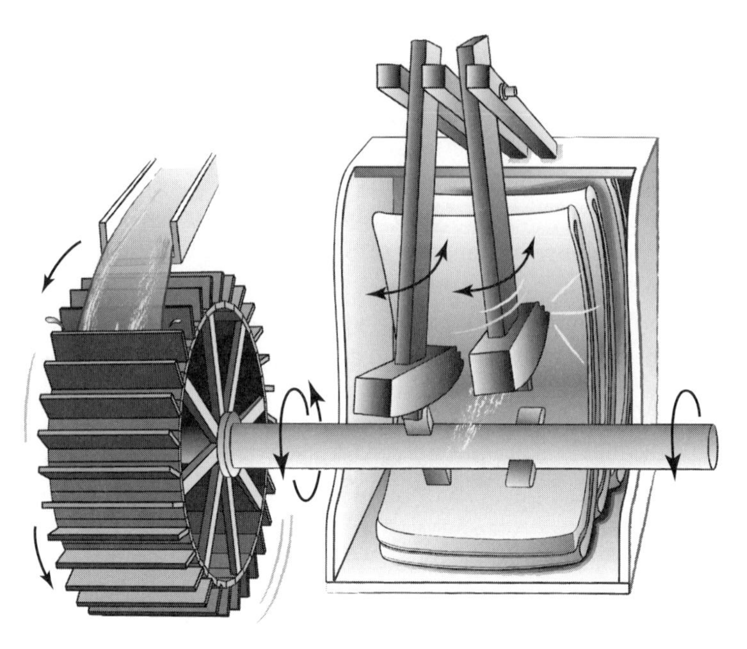

Figure 3.5
In a fulling mill, tappets on the shaft of waterwheel were used to translate rotary motion into reciprocal motion. In fulling, this mechanism was used to beat newly woven cloth to tighten and soften it. This kind of hammer mill was also used in forges, ore crushers, and other workplaces.

series of hammers or stocks along its length. The moving tappet would lift a hammer, and then slip out from underneath, allowing the hammer to fall and beat the cloth underneath (figure 3.5).

The adoption of the cam or tappet to produce reciprocating motion from a turning water wheel was an important step in the growing technical prowess of the West. This mechanism could be, and was, applied to a range of tasks, from hammers for ore crushing and beating rags for the papermills that came into Europe about the fourteenth century, to moving bellows up and down in forges. Perhaps the most important use of hammers was for beating iron, for, as has already been suggested and will be explored later, iron grew enormously in importance in the later Middle Ages, and the availability and use of waterpower turned out to be a crucial element in this development. Ironworking from ancient times had always demanded a great deal of hammerwork—the very notion of "wrought" iron, which was always the preponderantly useful form, embodies much repeated hammering, not only to give shape to the metal but also to adjust its composition and behavior—something that is done by brute physical force rather than chemical means all with way up to the nineteenth century. The rhythmic pounding of the hammers of a waterpowered forge were the basic late medieval signs of productive industry, comparable to the billowing smokestacks of the industrial age centuries later. No less important to ironmaking was the application of waterpower to the operation of bellows. Bellows themselves were ancient devices for increasing the air flow to fires, in hearths, forges, and elsewhere. The application, however, of a steady power source for bellows that could be made quite large turned out to be far more than simply a means of saving human or animal labor, but in fact transformed the very nature of ironmaking—and with it, a great range of products—by enabling the blast furnace. When enough air is provided to the heat used in smelting iron ore into metal, the fire reaches a temperature that changes the smelting reaction profoundly, speeding it up enormously and changing the product from small amounts of spongy slag-filled metal into much larger amounts of molten cast iron. This particular application of waterpower does not appear until the thirteenth century, and does not become widespread until the fourteenth century, so its implications are more appropriately discussed later.

Beyond the large and important industries of textiles and ironmaking and working, the new power technology contributed to a host of lesser tasks. New power sources did not transform this work, but they did contribute mightily to the fostering of a mindset in which one of the key avenues for improvement was the lessening of the burdens of labor by the application of powered machinery. By the end of the Middle Ages, whatever the extent to which work and manufacture was still largely a matter of human and animal muscle, simply the widespread awareness of the power possibilities gave the Europeans a profoundly different view of how technology could be altered and harnessed to serve them.

Central to this spreading awareness was the emergence of new mechanisms for harnessing power. The edge-runner mill is a nice example of these. Many substances besides grain need to be ground or crushed to be useful, but few of these substances can withstand the violence of the action of two heavy millstones. A key example is the making of oil from olives—the primary fat source for southern Europe, where butter was never important. In classical times human labor could be saved in the process of oil making by the use of human- or animal-powered edge-runner mills. In these, the work was done by a stone wheel set vertically on its edge atop the crushing surface. As the wheel ran around, it crushed the olives under its edge, yielding the desired oil. Adopting waterpower to this process was not complicated, although it appears not to have been done until the eleventh century—testifying perhaps to the importance of the emerging power mentality rather than to technical factors as the explanation for the extension of power applications. The edge-runner mill was thereafter applied to the crushing of a variety of substances, including poppy seeds, mustard seeds, sugar, tanning bark, and pigments for dyes and inks.[9]

Other mechanisms were attached to waterwheels in the course of the Middle Ages. Water-powered grindstones sharpened knives or polished plate, borers attached to waterwheels sped up the making of wooden pipe, and metals could be rolled flat or cut into strips by mills with large rollers or cutters. By the end of the Middle Ages, coins were being made by water-powered cylinders, which stamped the coin's design on heated blanks of copper, silver, or an alloy. By the sixteenth century, waterwheels were used to power fans to ventilate mine shafts and to operate a variety of pumps. Perhaps the most sophisticated application of waterpower was the sawmill. The early thirteenth century sketchbook of Villard de Honnecourt, a French builder and mechanic (we might say an "architect and engineer," but this implies professional identification that is probably misleading), included the image of a water-powered saw that used a combination of motions to move a saw back and forth against a timber while at the same time advancing the workpiece forward against the saw blade. Thus was combined application of the rotary motion of the wheel and the reciprocating motion translated to the saw by cams. Such devices for cutting stone may have appeared a couple of centuries earlier, in fact, but by Villard's time water-powered saws could be found in Switzerland as well as France, and they spread to Germany over the next century. In some regions the use of powered saws actually created environmental problems—forest depletion—as the conversion of timber into building materials was speeded up considerably. Powered sawmills were not important in England or Scandinavia until much later, so they clearly did not have universal appeal. Nonetheless, such use of waterpower made it clear that European technological ingenuity received full and robust expression in a wide range of medieval mechanisms.

The powered machines of the Middle Ages, whether moved by animals, water, or wind, were what we would call "labor-saving." This is to say that the work such machines did was work that would have been carried out by human muscle if the machines were not used. This raises the question: whose labor was being saved? The overwhelming preponderance of waterpower was for grinding grain, typically wheat, barley, or rye. Watermills converted such grain into meal or flour (flour being the more finely ground), which could then be used for cakes, bread, or ale. Without the mills, this grinding would have been done generally by hand. Handmilling is a very ancient practice, but it is a very laborious and time-consuming one. It was, in the culture of the West at least, women's work. So are we to conclude that the many thousands of mills described in King William's survey of 1086 were for the purpose of saving the labor of women? This isn't very likely—or, at least, this would be a very misleading way of interpreting the evidence. For one thing, there are alternative ways of eating grain, and in all likelihood, absent the powered mills, these would have been much more common—gruel and porridge, after all, are common elements of our images of the diet of poverty. What we find perfectly ordinary and common for the eating of oats and rice, say, earlier Europeans, at least before power and machinery changed things, would have found palatable for wheat and barley. The conversion to meal and flour, however, is not simply a matter of taste. Wheat, for example, spoils much more easily in its whole, unground state, and thus grinding becomes an important storage and preservation technique, making the food supply viable year-round rather than near the harvest. The spread of the powered mill, therefore, was probably one of the key elements in the widespread improvement of the European diet in the High Middle Ages. The growth of population that propelled European society and economy in the hundred years or so before the Black Death of the fourteenth century was the product in large measure of the technological changes in agriculture and machinery combined.

The question of "whose worked is being saved" is therefore one that cannot be simply answered. It is striking, however, to note that this first important machine in the West takes over not the tasks of men but those of women. Its value to society was clearly not seen in its saving of the women's labor, nor even in the displacement of women by men in one of the key tasks of food preservation and preparation. The value of the mill lay in the power that the technology provided to shift the opportunities for profit and production away from labor to capital. The owners of mill sites and the mills themselves became the holders of wealth and the reapers of benefit from technological opportunity. The advantages and opportunities of water-powered production, at first in the grinding of grain, and then in a host of other tasks, were often sufficiently compelling to make the saving of labor—regardless of whose labor it was—a widely recognized good. When the advantages to farmers and peasants were not so obvious, then legal structures were used to make the compulsion work anyway. But the larger lesson, just beginning to be taught by the medieval watermill, was that technology could transform human relations and human power even in the most mundane of arenas.

The Millere was a stout carl for the nones.
Ful big he was of brawn and eek of bones—
That preved wel, for overal there he cam
At wrastling he wolde have alway the ram.
He was short-shuldred, brood, a thikke knarre,
Ther was no dore that he nolde heve of harre,
Or breke it at a renning with his heed.

The miller that set out with the pilgrims in Geoffrey Chaucer's *Canterbury Tales* has stood, at least in English literature, since the late fourteenth century as the archetype of his trade. There is no reason to doubt that Chaucer drew upon the common images of his day for this, as well as for most, of his characters. The big, brutish figure who could either lift a door off its hinges or break it with his head, who, Chaucer tells us, was a good fellow at heart, even if he cheated a bit at his tolls, is, in a sense, the original machine minder and supervisor of Western literature, the source for centuries of images and caricatures. This was not the designer or builder of the mill—that was the millwright, who in this period was probably simply a more generally skilled carpenter. The miller might maintain the mill to a modest degree, although even such tasks as dressing the stones (keeping them in grinding form) might be handed over to a craftsman with more experience. The miller was also not the owner of the mill, in most cases; that would most likely be a landlord, in whose employ the miller worked and whose agent and collector of dues the miller was. Our fourteenth-century miller, therefore, begins to take on the special character of the machine-minder, a figure who henceforth stands constantly on the margins of the story of technology in the West—not designing, not building, not owning, but working and overseeing, one of the first, but hardly the last, of individuals whose very identity would be tied to a machine.

4 Buildings for God and Man

It began, in a sense, with the abbot simply wanting a bigger door and more space for the increasing crowds. But let him describe his situation:

I found myself, under the inspiration of the Divine Will and because of that inadequacy which we often saw and felt on feast days, namely the Feast of the blessed Denis, the Fair, and very many others (for the narrowness of the place forced the women to run toward the altar upon the heads of the men as upon a pavement with much anguish and noisy confusion), encouraged by the counsel of wise men and by the prayers of many monks (lest it displease God and the Holy Martyrs) to enlarge and amplify the noble church consecrated by the Hand Divine; and I set out at once to begin this very thing.[1]

So wrote Suger, abbot of the Royal Abbey of St. Denis, outside Paris, about the year 1145, describing the work he had undertaken just a few years earlier. Suger was one of the most astonishing men of the Middle Ages, a man of modest origins who rose, by dint of intelligence and determination alone, to be the most powerful man in France. Indeed, to Suger is sometimes given the credit for saving the French monarchy, more than once, when it was beset by foes from both within and outside the kingdom. This was perhaps fitting, for he spent almost his entire life, from the age of ten, in and serving the Abbey of St. Denis, the abbey dedicated to the patron saint of the kings of France and the resting place of the members of several dynasties.

But it is not for his considerable political skill, nor even for his equally impressive literary talents, for which Suger is most honored today. It is instead as the promoter of perhaps the greatest creative accomplishment of medieval Europe—Gothic architecture. Suger did not "create" the Gothic style; he did not, indeed, invent any of the key technical or architectural elements that characterized that style. These had, individually, been around for generations, and had even been used in combination with one another long before Suger's time. But the synthesis found in the Abbey Church at St. Denis was, most architectural historians would concede, the spark that would make Gothic churches the hallmark of medieval European architectural originality. This was due in part not just to Suger's active support and guidance to the unknown master builders on whom he relied, but even more to his articulation of the Gothic vision. Never before in history, probably, had the intellectual and spiritual ideals behind a technical and cultural revolution been so explicitly set forth. Certainly never in history had this happened in such a way that it sparked a passionate frenzy of ambitious building that left monuments on the landscape that continue to astonish more than eight hundred years later.

The Gothic churches that began to be built in the mid-twelfth century were singular examples of the spirit and power of the nascent culture of improvement. Suger's remarks tell us at the outset that, for all the glorious intellectual power behind the Gothic style, there was always the drive simply to improve what there was. The church that he set out to refurbish and enlarge was already, in the early twelfth century, one of the most historic in France, having been founded almost five hundred years earlier, by King Dagobert. It incorporated parts attributed to Charlemagne himself. But it could, to Suger and his contemporaries, clearly be improved. And as the Gothic spread, first in the Île de France—the French heartland that surrounds Paris—and then throughout Western and Central Europe, improvement continued to drive the church builders. This is one reason, in fact, that today Suger's church, fine as though he made it, is very much overshadowed by a host of near neighbors, just a bit younger, but enough so to have been the subject of the experiments, extensions, and artistry that make them far grander in modern eyes—Sens, Noyon, Paris (Notre Dame, to most of us), Bourges, Chartres, Reims, Amiens, and Beauvais. In short order, these churches were joined by others farther afield, in England, the German states, northern Italy, Spain, and even more distant from the Gothic birthplace.

Certain characteristics of the Gothic style made it a particularly effective platform for technical experimentation and improvement. After all, ambitious building—even driven by rivalry, as was much of the cathedral crusade—is as old as civilization itself. Pharoahs, kings, and generals had pushed their subjects to create larger and grander temples and tombs for millennia. But almost always these ambitions expressed themselves in terms simply of size or of expense (usually through precious materials or the quantity of labor required), but seldom through true technical daring. The cathedrals and churches of the twelfth and thirteenth centuries, however, represented engineering achievements of such a high order that their emulation still has not lost the capacity to impress us. This is due in large part to the values they placed on height, light, and space. Any one of these elements can be (and have been) made key elements in a building, but bringing all three together into a harmonious whole was arguably the radical center of the Gothic ambition. The combination required extraordinary daring and intelligence, and the ability to extend capabilities bit by bit through experimentation, observation, and communication. This in turn depended on the emergence of a technical elite—not one formally organized to any degree, but nonetheless an group of men who possessed specialized knowledge and experience, were capable of pushing the limits of their field vigorously and effectively, and made themselves indispensable to a wide range of communities throughout Europe.

While Suger, master of an abbey whose domain included more than a hundred villages and manors and even more churches than that, was a particularly auspicious champion of the new style, and St. Denis, ancient royal abbey and resting place of French kings, was a similarly prominent showplace, the rise and rapid spread of the Gothic are not to be attributed to a man or a single church. An array of forces were at work, from the political and economic circumstances that made even modest-sized towns sufficiently independent and wealthy to support ambitious building projects to the intellectual and philosophical currents that were gathering in twelfth-century Europe, fed by imported Arabic translations of classical philosophers and by the emergence of new institutional homes for learning, such as the cathedral school at Chartres and the faculty at Paris that was to organize itself as a university by about 1200. In addition, for at least a century before Suger's efforts, church building had been a major preoccupation throughout much of Western Europe, creating a substantial body of experience and experiment that was the indispensable underpinning of the Gothic.

The Romanesque style, from which Gothic evolved, was itself a significant achievement in Western architecture. Roman forms, most distinguished by the use of stone and of rounded arches, were never lost to the builders of the West—examples, after all, were still to be seen strewn around the former empire even centuries after the legions had departed. But the great upsurge in church building that followed the turn of the first millennium provided the impetus for a more distinctive adaptation of these forms for the sacred purposes that now motivated European builders. Perhaps as many as 1,600 churches were built in France in the eleventh century alone—the contemporary chroniclers called it a *morbus aedificandi*, or a "disease of building."[2] While at first these structures used brick or small stone, gradually masons acquired greater confidence and skill and displayed this through more ambitious stone work. Particularly significant for the development of building technology was the use of stone for roofing. The susceptibility of the traditional wooden roof to destruction by rot or, more important, fire, led builders of larger and more expensive churches to devise more permanent roofing of stone. This led to the extensive use of the barrel vault, which was simply an elongation of the semicircular arch so that it covered a space, such as the central nave of a church. A barrel vault in its simplest form, however, is quite heavy and dark, since it requires heavy supports all the way down between roof and earth. The classic response to this problem was the cross or groin vault, in which two barrel vaults are brought together at right angles (figure 4.1). For this to work, however, the roof's load must be supported by very large and strong pillars at the corners. For this reason, Romanesque churches were notable for their immensely thick walls and columns—and their dark spaces. They

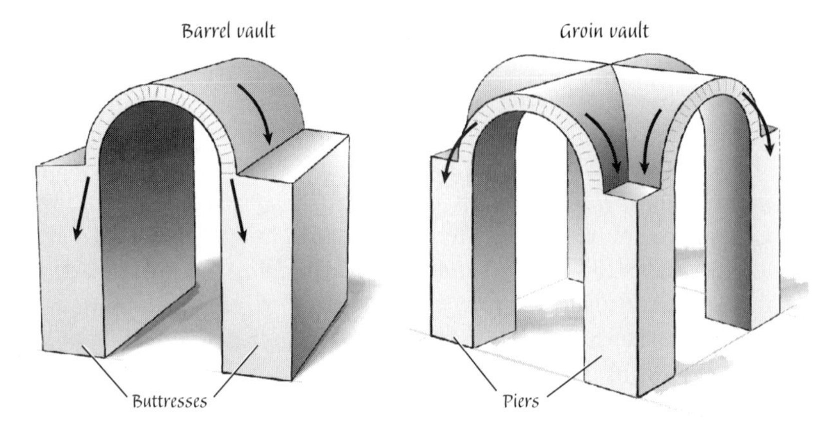

Figure 4.1
The extension of the ordinary semicircular arch created the barrel or tunnel vault. Crossing two such vaults at right angles yielded the open form of the groin vault.

were not, however, necessarily brooding or foreboding buildings; bright colors were applied to the great stone walls and columns, hangings of colorful tapestries or exuberant sculptures of stone broke up the flat spaces, and the woodwork of the interior was adorned with artistic carvings, enamels, and gilding.

The Romanesque style also lent itself to a certain amount of experimentation. On occasion, the perfect roundedness of the Roman arch was altered, largely to give more of an impression of height, with a result that was pointed (figure 4.2). The reliance on groin vaults spurred many masons to reinforce the vaulting with ribs at the intersections. The huge weights of the stone roofs and the ever-higher walls compelled the construction of buttresses at the foot of the walls, especially where vaults crossed. These were usually in the form simply of masses of stonework laid against the outside of the main walls, built to resist the large outward thrusts imposed by the heavy roof and vaults. These three elements—the pointed arch, the ribbed vault, and the external buttress—came together to define the key technical elements of the Gothic style.

It is important, however, not to confuse these technical elements with the style—and the technology—itself. Gothic building is a particularly useful reminder of the fact that technologies, no more than individuals, cannot be reduced simply to their constituent parts and still be understood. Pointed arches (or "ogives"), ribbed vaults, and buttresses had all been used long before Suger's time, sometimes even in combination with one another. Their appearance in the church at St. Denis was not in itself noteworthy (the buttressing, in particular, was very ordinary in Suger's church). What did attract attention, however, was the overall objective of the church. Here was a grand and important sanctuary that sought to proclaim its significance not so

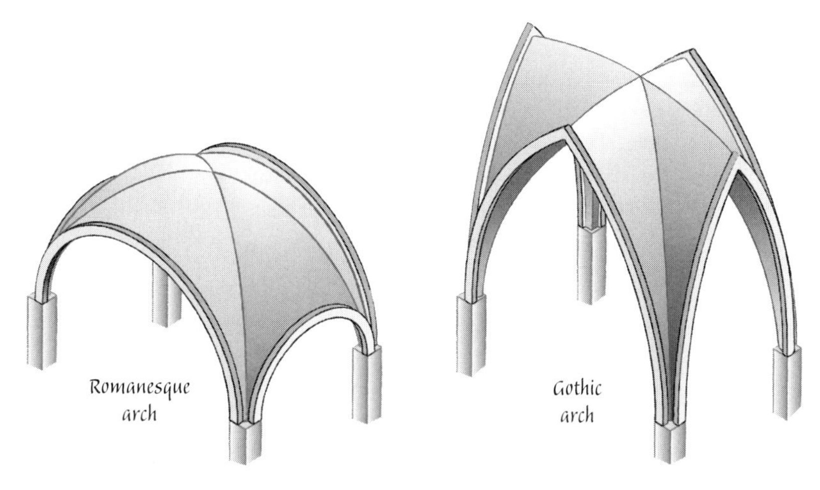

Figure 4.2
The pointed Gothic arch was derived from the semicircular Romanesque form and offered more design flexibility, particularly in the height of the arch.

much through size or expensive ornament, but through, above all, light. Visible light was, to Suger, the necessary phsical manifestation of the invisible light of divine grace. On the doors of his church, he had inscribed an invitation which included the following:

Bright is the noble work; but, being nobly bright, the work
Should brighten the minds so that they may travel, through the true lights,
To the True Light where Christ is the true door . . .[3]

The effect of Suger's belief that light embodied the divine, and that a proper church should allow this divine radiance to permeate its space, was a monumental departure from the older heritage of church construction.

What makes the place of light in the Gothic so important for our history is the extraordinary technical demand this places on the design of large structures. Before the industrial age there was only one material that could be used to make buildings of an impressive size—stone, either natural or its artificial equivalent, brick. Stone, however, is heavy, and as walls of stone rise ever higher, the weight of the walls themselves become enormous burdens to the structure, burdens that can be supported only by more stone below, in the form of walls many feet thick. The thicker the walls, however, the less opportunity light has to penetrate the structure. This, in the simplest terms, is the challege that was met and triumphantly overcome by the Gothic builders. They did so, not by the use of alternative materials or by theory and understanding of structures, but by experimentation, failure, observation, and constant striving for improvement.

— 289 —
[CAT]
2
Z
X
O
K
C
I
V
H
G
E
L
D
J
S
P
R
A
F
N
T
M
A'
B
10m
5
H.J

While light was the key to the Gothic spirit, only slightly less important were height and space. In terms of continuing improvement, in fact, these were arguably more significant, for they provided the most readily visible measures of achievement. The cathedral building that started in the Île de France in the mid-twelfth century and spread from there to possess all of Western Europe's most prosperous and vibrant corners was in a sense a great competition, in which towns and cities, bishops, abbots, and secular leaders, landowners and merchants all saw themselves in a struggle to prove their piety and worth through grander churches. The amount of light that a church admitted was a matter of judgment, aesthetics, setting, and even climate, but the heights of the towers and walls and the capacities of the buildings themselves were measurable and communicable. These thus provided the readiest benchmarks of "success" in the ambitions of promoters, patrons, designers, builders, and the entire community. It was thus these elements of height and space that were constantly pushed in the century or so of the great crusade, the core targets of the efforts for improvement—and the leading forces in failure.

Ordinarily, height, space, and light all compete against one another in a structure. Just as light is diminished by the thick walls needed for a high building, so does the effort to enclose a large open space, and thus support a large roof, work against height. The higher a wall is, the more difficult it becomes to support the load of a heavy span. This is due not only to the increasing weight of the span and the wall itself, but also to the increasing lateral (sideways) forces of wind. As a wall rises higher, it presents more surface to the wind and it exposes itself to greater forces from higher wind speeds. Before the efforts of the medieval church builders, this effect was rarely important in building, but height of the new cathedrals, combined with the north European climate, made such considerations quite serious. There were several technical solutions to these new challenges, but the most important were flying buttresses and new vault designs.

Buttresses are simply masses that are placed at or near the base of a structure to support forces from the walls and roof. The more forces these impose, the larger the buttresses must be. Masses of stone, however, placed on the walls of a cathedral would tend to block out light. The flying buttress is an extraordinarily creative and effective solution—move the supporting masses away from the walls, and transmit the forces through open structures connecting them. The flying buttress, often beautiful and ornamental in itself, is nonetheless the most obviously "technical" element of the Gothic—it bears its load away from the walls in the most obvious and

Figure 4.3
Transverse view of aisles and buttresses, Notre Dame Cathedral, Paris. French architect and writer Eugène Emmanuel Viollet-le-Duc was the foremost champion of the Gothic revival in mid-nineteenth-century France, and he did his own drawings for his ten-volume *Dictionnaire Raisonné* (Paris, 1854–68), from which this is taken.

straightforward manner possible. The multistory elevation of the typical Gothic church made the flying buttress a spectacular and useful building element. The most conventional style, found at Chartres and other early exemplars, involved the construction of three levels. The first level typically rose to one-half of the total height. This *arcade* level defined the height of the cathedral's outermost walls. Above this was the *triforium*, most usually about fourteen feet high. This smallest level ordinarily was roofed over by timbers slanting from the top of the arcade walls inward to the top of the triforium, creating a kind of gallery above the side aisles of the cathedral's interior. This made the triforium the darkest portion of the structure, but it prepared the way for the uppermost level, the *clerestory*, which stretched upward as far as the masons dared and brought in most of the cathedral's light through magnificent windows. The flying buttress was designed to carry the weight the roof away from the clerestory, allowing those upper walls to contain as much glass as possible and giving the Gothic church the sense of a roof almost magically suspended in the air (figure 4.3).

The first cathedral to take full advantage of the flying buttress was Notre Dame in Paris (most French cathedrals, in fact, were dedicated to the Virgin—"Notre Dame"—but the capital's cathedral has appropriated the name in the popular mind). Begun in 1163 and taking about a century to complete, the building of Notre Dame, defined as much as any project the great age of French cathedral building. More than five hundred Gothic churches were built in France during the years of Notre Dame's construction, but the prominent location on its island in the Seine, the wealth of the diocese, the power of the surrounding city, and the ambitions of its designers to make a church fitting for the site all combined to place Notre Dame at the focus of the Gothic age. Other cathedrals may be said to embody higher spirituality (Chartres, for example), greater ambition (Beauvais, certainly), or even more exuberant artistry (Rheims or Amiens, perhaps), but it was at Notre Dame that the Gothic builders began to show the full power of their form and their technology. Here great height called for full development of the flying buttresses' capabilities—the roof of the nave of Notre Dame rises almost 110 feet above the floor. Here the great towers of the West Front, the imposing rose window between them, showing off stone tracery and stained glass in skillful glory, defined what the front of a cathedral should be (figure 4.4). The result was called by Victor Hugo, in his appreciation of the structure (a novel that is perhaps the greatest literary tribute to any building), a "vast symphony in stone." The harmony of the whole is a key element of the Gothic, and one of the reasons why Gothic building was such an important stage in the development of technology in the West.

The harmony was a great achievement, but for the subsequent development of technology it was perhaps the means by which this harmony was achieved that was most important. More than any other sustained technological effort before it, the

Figure 4.4
Notre Dame Cathedral, Paris. Library of Congress photograph.

cathedral crusade was the product of the relationship between visionary patrons and exceptionally skilled and knowledgeable craftsmen. Out of this relationship emerged a new kind of technical knowledge and a new kind of person to carry and practice it: the master mason. The cathedral masons represented the emergence for the first time of a group of individuals, known in many cases by name, who were identified by their exceptional technical skill, who wielded unparalleled power and authority over their works, who organized themselves in a fashion that became the model for professional privilege and independence, who created and shared extensive new technical knowledge by traveling and working over extended territories, and who made improvement on a large scale the primary source of their economic and social value. The ecclesiastical figures who sponsored the great churches, like our Abbot Suger, were important sources of ideas and ideals, and the Gothic crusade would not have moved without them, but it was the masons who made the ideas into technical achievement and who set important patterns for future technological change.

Other craftsmen had been recognized for their valuable knowledge and skills; the medieval millwright, for example, was indispensable to the design and construction of the power devices of the Middle Ages. But the masons responsible for building cathedrals and other large structures were different, both in the nature of their knowledge and in the authority they wielded from it. In the master masons, for example, we begin to see the divorce between technical knowledge and technical skill. The split was not substantial in this period, nor was it that widespread, but in hindsight it is readily evident. While the mason might reach his master's status through the exercise over some years of skillful stonework, the value attached to his work and the source of his status lay not in these skills, in the abilities to shape stone accurately and artistically, but in the capacity for visualizing the larger work and for organizing and supervising the myriad details that such work required. In daily life the master mason might still wield a chisel, but he was much more likely to spend more time with men than with stones and to oversee the making of scaffolding, falsework, and stonecutting forms. He was much more like a modern architect and contractor than a craftsman, but above all he was the embodiment of the most advanced technical knowledge of his day.

Master masons rose from the ranks, beginning as boy helpers and apprentices, becoming itinerant craftsmen and journeymen, rising to a post as assistant to an experienced master, where it was possible to observe the full scope of the work, and then achieving a master's post by replacing a deceased master or answering the call from a project elsewhere. This kind of occupational mobility was uncommon in the Middle Ages—the monastery represented one of the few other institutions in which it might be observed. Of course, mobility waxed and waned, depending on social and economic conditions, but the heated competition for skilled masons during the period from the mid-twelfth to the mid-thirteenth centuries, in northern France then more widely across Europe, opened the field up more than ever. Just as important, the nature of the competition, with the striving for grander, higher, more ambitious structures, fostered the identification and success of technically bold and knowledgeable master masons. It was during a period such as this that the accumulation of experience, the pursuit of savvy experiments, and the cultivation of imaginative ideas could all be combined to make an individual particularly attractive to prospective patrons. The process by which individuals came to be identified with advanced technical knowledge and capability was of enormous importance for the development of the West's taste for technology. Classical and earlier Western experience offered few if any opportunities for such identification—building and design were jobs done at the direction of patrons, and who actually did the work was rarely of interest, except within narrow notions of competency. The idea that a project provided scope for originality and improvement by individual technical minds was a novel and significant development.

Many master's names are lost to us, and are known to scholars simply as the "First Master" of Beauvais or some such—evidence not of their obscurity or lack of importance, but of the simple fact that much did not get written down in medieval Europe, and much that was did not survive. But others have come down to us by name, thanks to the careful record keeping of a diocese or the interest of a verbose and literate monk. An example of the latter, who can represent for us the character and influence of this new technical elite, was William of Sens, who was instrumental in bringing the Gothic to England. In 1170, the English church experienced one of the most traumatic events in its long history, the murder of the Archbishop of Canterbury, Thomas à Becket, in his cathedral at the hands of Henry II's swordsmen. Just four years later the eastern parts of Canterbury cathedral were destroyed in a great fire. The monks of the cathedral were understandably distraught, but they called in advisors from around England and from France (it is well to remember that in the Norman England of the twelfth century, French influence was taken for granted in most aspects of life). There was much debate about what could be saved of the ruins of the cathedral's choir, but out of the conflicting voices that of William, who had worked on the great cathedral at Sens as well as other French churches, emerged as the most authoritative. William waited to gain the monks' confidence before telling them that the old walls of the choir would need to come down, and the east end of the great church entirely replaced. He then proceeded to supervise the demolition, acquire new stone from France, design machinery for moving and lifting stone, and provide forms for the carvers to use in their work. At each stage he took the opportunity to import the new techniques and styles from France. Within less than three years, columns and walls of the choir was sufficiently complete that work could begin on the vaulting. It was at this point that William was injured: scaffolding gave way beneath him as he was supervising the centerings and he fell fifty feet, with stone and timber piling on top of him. Severely injured, he nonetheless was able to continue his supervision in the coming months from his bed, using a trusted young monk to oversee the work directly.

A closer look at the work of William and other master masons gives some idea of how technical knowledge was coming to be captured and communicated during the cathedral crusade. A monk of Canterbury, Gervase, chronicled the reconstruction of his cathedral, telling us about William of Sens and his successor, known as William the Englishman (a name suggestive of the extent of French influence). We learn that one of the first tasks that William of Sens took up was to spend three weeks "making the molds in the tracing house." These were then given to two masons and an apprentice to use in cutting needed stones. The master, in other words, would convey the needed components of the building by directly tracing the patterns of the desired stones on to boards. A carpenter would then cut and fit the boards together into forms for the masons' use. There are numerous references to "tracing houses," buildings put up on the cathedral site to house the work of drawing plans or sketches and tracing needed patterns, perhaps on parchment or directly on the boards to be cut into molds. There are very few surviving drawings from medieval building, and these are not working drawings in any modern sense but means for conveying a general idea of what the mason or patron were visualizing. In the course of the Middle Ages, architectural rendering did improve, in the sense of becoming more skillful, consistent, and informative, but drawings never reached the point where they could effectively guide the work alone. Models, written descriptions, and references to earlier buildings were all used on occasion to communicate both the designer's overall vision and some elements of the detailed structure. But the limitations of these media meant that the master mason could never be far from the work itself, and his supervision, either direct or through trusted assistants, was indispensable to success.[4]

The chronicle of Gervase also provides us with some understanding of how aware contemporaries were of seeing novelty presented before their eyes. In describing the new choir of William of Sens, Gervase remarked on the difference between old and new:

> The pillars of the old and new work are alike in form and thickness but different in length. For the new pillars were elongated by about twelve feet. In the old capitals the work was plain, in the new ones exquisite in sculpture. No marble columns were there, but here are innumerable ones. There, in the circuit around the choir, the vaults were plain, but here they are arch-ribbed and have keystones. There there was a ceiling of wood decorated with excellent painting, but here is a vault beautifully constructed of stone. The new work is higher than the old.[5]

We could hardly have asked for a more evocative description of how the new Gothic style would appear to those accustomed to the Romanesque, nor for a better sense of the combination of artistic and technical admiration that the new style evoked.

Such was the scope of the Gothic effort in its first century that technical improvements manifested themselves everywhere. One of the great features of the new style was how it combined reliance on some basic architectural elements and fundamental aesthetic principles with enormous freedom for creativity and experimentation. Each cathedral was seen as a source of knowledge about structure, proportions, materials, forms, and the like, and masons traveled or were sent on journeys of many miles for the express purpose of studying other works, both finished and under way. It can be argued, in fact, that the cathedral projects, lasting as most of them did for many decades, constituted the first technical academies in the Western world. While, like most trades, masons protected their "mysteries" from prying outsiders, they also created in this period the most extensive organized "brotherhood" of technical practitioners before the modern era. By the end of the twelfth century, the tracing houses of the early projects had given way in many places to more substantial facilities, masons' lodges. The lodge came to represent the authority of masons over their work, their independence from their patrons, their acquisition and ownership of knowledge and tools, their ability to provide assistance and training for members of the brotherhood, local and foreign, and their distinctive status within the heirarchy of medieval society. As time went on, and the independence and status of masonry came to be recognized by custom and necessity, the lodges developed into schools, repositories for drawings and records, and places of important social interaction among members of a privileged class. The lodges were new mechanisms for capturing improvement, and their effectiveness at this is one explanation for the longevity and extent of the technical creativity that characterized cathedral and church construction in the High Middle Ages.

The lodges also exercised a certain amount of power. During the most feverish decades of cathedral building, during the decades on either side of 1200, there never ceased to be a shortage of qualified masons. This allowed the craftsmen to appropriate to themselves power that had never before been accessible to technical men. This showed itself in the authority given to masters over their works, and in the collective capacity of masons to govern their own conduct and lives. In the year 1230, for example, when the bishops of France decided that masons had become too unruly and disobedient, they decreed that masons throughout the kingdom would have to shave their beards and shorten their hair, in the fashion of the monks alongside whom they often worked. The masons resisted. They stopped work throughout France (a national habit of great antiquity, it would seem) and informed the church authorities they had no intention of following such orders. As the Church, in turn, made clear that it wished to be taken seriously on the matter, the masons upped the ante and threatened to demolish every religious structure in France. In an astonishing testimony to the ascendancy of urban, secular power over ancient sacred authority, the bishops backed down, and the masons resumed both their work and their acquisition of independence and control.[6]

This independence both fostered and was supported by continuing experimentation and innovation. The expense and magnitude of cathedral projects, however, posed some fundamental problems for experimenters. This was long before there existed any body of theoretical knowledge that set abstract boundaries around possible structural approaches or could guide some kind of calculation of the results to be expected from some untried technique. As in all other technical areas, innovation required empirical observation of the results of novel materials or designs. But the great expense, long time frame, and sheer size of cathedrals made haphazard experimentation impossible and even controlled experiment overly risky. How then could the master builders pursue improvement, as they most emphatically and spectacularly did? The simplest answer is that they took advantage of the long, slow construction process, and discovered what to watch. In particular, the lime-based mortar that was used between the cut stones of a cathedral was sensitive to the forces of tension (pulling) that were the most important to manage with high, heavy structures. Cracks in this mortar would develop over time as portions of a building went up, and if these were observed in one part of a structure, subsequent portions could be modified (typically by adding or enlarging pillars or buttresses) to prevent stresses that could not be handled. Gothic churches were generally built, after the lower sections were completed, by completing one bay or segment at a time, all the way to the roof level. If cracks in mortar were observed in the first bays, then it was generally sufficient to modify and strengthen subsequent bays. This is a perfect example of learning through failure, illustrating that this can be done without resorting to catastrophic failure, but by adjusting techniques from stage to stage.[7]

An example of the kind of innovation that might emerge from such observational methods can be seen in the changes in the styles of Gothic vaulting. In the decades after St. Denis's construction, the most common kind of vaulting for the bays of a large Gothic church was "sexpartite," in which the ribs of the vaulting, which carried the roof's weight, were designed to cross diagonally over two bays at once, creating six spandrels or triangles between the arches and the ribs (hence the name). Since the ribs of a vault are difficult and dangerous to construct (William of Sens was injured while overseeing just this stage), it made sense to try to cover two bays at a time. This was also seen as a way of holding the series of ribs together visually. But as walls reached ever higher, a difficulty was encountered, as the forces from sexpartite vaulting pushed equally in a direction parallel to the walls (toward the yet-to-be-constructed bays) as they did outward, perpendicular to the walls (toward the buttresses). Since one of the goals of the Gothic style was to reduce the size of internal buttresses or piers, a means was needed to reduce these inward forces. The solution was "quadripartite" vaulting, in which the ribs crossed only one bay, creating four spandrels rather than six. Since the width of a bay was only a fraction of the width of the roof itself, this meant that more of the load was directed outwards, where it could be supported by the external or flying buttresses (figure 4.5). This kind of innovation was almost certainly the result of careful observation of the cracks in supporting piers during construction, an illustration of improvement at work even under the constraints of large-scale building.[8]

The innovative forces unleashed by the Gothic crusade reached (one might say overreached) their literal and figurative height in the cathedral of St. Pierre, in the town of Beauvais, about twenty-five miles north of Paris. Construction began in 1225, after the Gothic fever had been burning for a long lifetime, and the grand achievements of the le de France were scattered all around. Just thirty or so miles farther to the north, the great cathedral of Amiens was rising in that town, and was, after only five years of construction, demonstrating the full glory of the High Gothic, especially in its verticality and light. The ambitious bishops of Beauvais, driven not

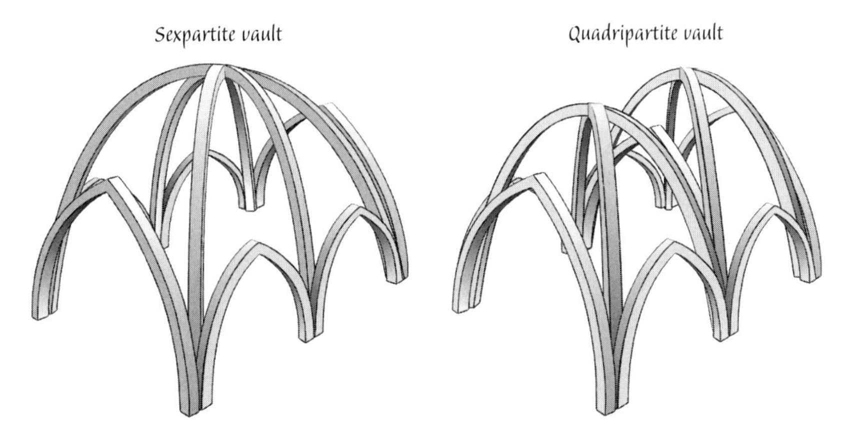

Figure 4.5
The ribs of the Gothic vault created sectors, typically six (sexpartite) or four (quadripartite), depending on the choices of the architect and the mason.

only by the works of rival neighbors but also by the political storms of the early thirteenth century, in which they were caught in the midst of battles for power, influence, and authority with the secular nobility, the increasingly rich townsmen, and, above all, the kings of France, sought to proclaim their power and wealth through the greatest cathedral of all. This might seem obvious overambition for a town the modest size of Beauvais, but in the early thirteenth century it was in fact a significant center of the increasingly prosperous textile trade of northern France. The construction of dozens of fulling mills in the late twelfth century, for example, testified to the industry's growth, and the diocese and bishop of Beauvais were intimate sharers in this wealth. The town's cloth hall, for example, belonged to the bishop, and his rights included the imposition of taxes on the trade in cloth, wine, and fish, as well as many other important commodities.[9]

In 1225, the cathedral at Beauvais was destroyed by fire. As in so many earlier cases, this precipitated an ambitious rebuilding effort that resulted in the adoption and extension of the Gothic style and technology. The bishop at the time, Miles (or Milon) of Nanteuil, was very much in the thick of political struggles with the crown and the town's bourgeois. The new cathedral of St. Pierre was a statement by Bishop Miles not only of the relative importance of his town, but also of the importance of the Church in a world in which its ancient privileges were under sustained attack. The result was astonishing ambition; indeed, so ambitious was the plan of Beauvais that it was never brought anywhere close to completion. At Beauvais today we find a beautiful and impressive cathedral, but one without most of the parts of a finished church: there is no nave, which ordinarily constitutes the primary body of a church. The benchmark for the ambitions at Beauvais was obviously the rising cathedral at

Figure 4.6
Beauvais Cathedral was particularly notable for the height of its nave and for the collapse of its central tower. The transept was never built. Courtesy Visual Resources Collection, School of Architecture, University of Maryland; Juan Bonta photograph.

Amiens. In that church, the reach for highest heaven was pushed farther than ever before. Not only were the finished vaults an astonishing 140 feet from floor to arch top, but the proportions of height to width were almost 3:1. These same proportions were adopted at Beauvais, but with a wider span, and thus even greater height. The choir at Beauvais reached 157 feet (48 m), the highest ever built (figure 4.6).

It is not only the achievement of Beauvais that calls it to our attention, however, but the failure as well. The financial strains of the grand project, despite Beauvais's prosperity, were great and slowed the building. After twenty years, the lower portions of the choir was largely complete, but nothing of the high vaulting. The strains of battles between the bishops and the crown, as well as the drain of wealth from efforts to finance the Sixth Crusade, reduced resources considerably. Nonetheless, when Bishop Robert became bishop in 1238, he combined his efforts to patch up relations with the king with a push to advance the cathedral, at least to the point that the lower choir was complete by the time, ten years later, he went off with

King Louis IX on crusade, never to return to France. His successor, William of Grez, gave the building his utmost attention for almost the next twenty years. Under his patronage, the great vaults of the choir were constructed and the transepts were largely completed. Finally, by the 1270s, the full extent of Beauvais's ambition could be seen in the finished choir vaults and the beginnings of the transepts (the north and south arms of the cathedral cross). All would seem to have been well, although the slowness of the cathedral's progress was no doubt frustrating—by this time, for example, the only slightly less ambitious cathedral at Amiens, started just five years before Beauvais, was essentially complete in its full glory. But at Beauvais, disaster struck in the evening of November 29, 1284. The high vaults of the choir came crashing down to earth, destroying much of the glorious structure around. The ambitions of the bishops and builders of Beauvais had reached beyond the limits of their technology.[10]

There is still much debate about the actual causes of Beauvais's collapse. Modern engineering investigations have directed attention to intermediate pillars or buttresses, and the accumulated effect of more than a decade of fluctuating wind stresses, particularly on the mortaring between the stones. Master masons from around France were called in to inspect the damage and recommend repairs, but they had no analytical tools with which to propose more than guesses about the failure's causes. The rebuilding that ensued brought the vaults back up to their designed height, but at the expense of adding numerous additional piers around the choir, thus breaking up the formerly open spaces. The result that can be seen today is still magnificent, and Beauvais is still a triumph of architecture and engineering. But it also in many ways marks the limits of the Gothic ambition and techniques; no Gothic church ever attempted such heights again, and few buildings reached so high for centuries yet to come. The end of the thirteenth century brought with it the end of northern France's great prosperity and relative peace; the Hundred Years War, the visitations of the Black Death, and breakdown of political and religious authority made the fourteenth century one of retreat and survival in much of Europe, especially northern France. The wealth that had been available for the great cathedrals was gone in a cloud of war, plague, and disorder. Beauvais was by no means the last Gothic cathedral, but it marks the end of the great crusade of improvement.

Spectacular failure is an integral part of the history of ambitious building, even up to our own day. Despite advanced mathematical theories, new techniques and models, and a deep (though by no means complete) understanding of materials, structures, and forces, constructions still fall down. Now, however, the learning that comes from failure is systematic, testable, and codifiable, at least to a point. The response to the collapse of Beauvais's vaults was essentially not to try anything so foolhardy again. Today the more likely response would be to study the failure inside and out until causes could be specified and the paths to ambitions simply recharted. This

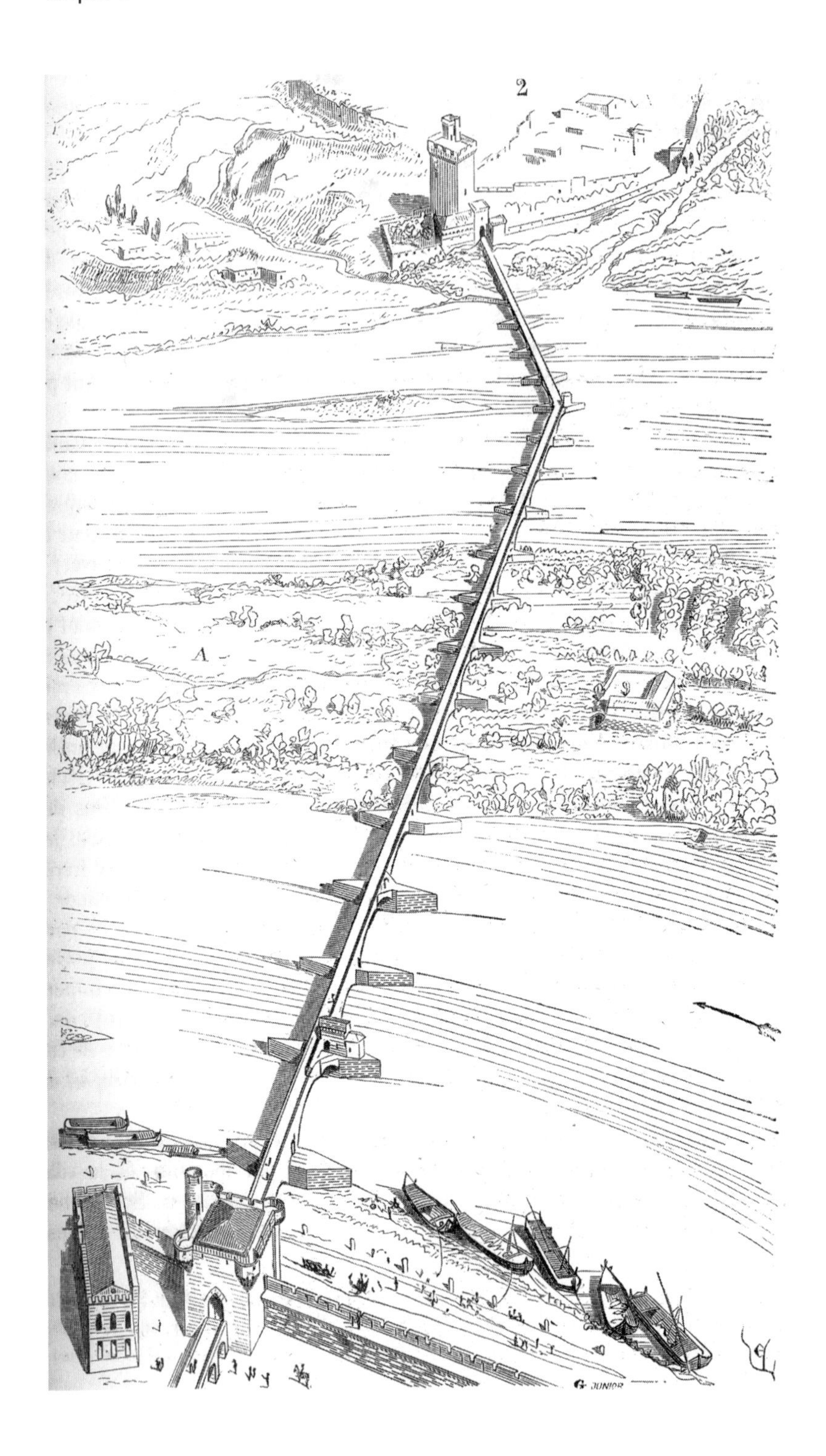
2
A

is one of the fundamental ways in which the course of improvement has changed in the last seven centuries. As if to mark just how difficult it would be to effect this change, the cathedral at Beauvais provided yet one more lesson in ambition, ignorance, and failure. By the early sixteenth century, the French kingdom and economy were well recovered from years of decline and chaos, although an unsettled period of religious strife lay ahead. The chapter (governing body) of the cathedral at Beauvais sought finally to move its work forward. First the transepts were completed, and then they sought to construct an ambitious tower and spire that would accentuate the impressive height of the building. After some discussion, a grand stone tower in the late Gothic style was commissioned. Begun in 1564 and completed five years later, the tower rose 291 feet and this was topped by a wooden steeple that brought the total height to 438 feet above the floor. So high was the tower and spire at Beauvais, the rooftops and church towers of Paris could be seen in the distance. Inspired by this achievement, building commenced on the first bays of the nave, but within a few years problems were observed: the central piers, which took most of the tower's weight, were shifting and even cracking. Experts were called in and recommendations made, but soon before the new work was to begin, on the last day of April 1573, the tower and steeple collapsed, barely missing the celebrants and congregation of the Ascension Day mass. This last failure, apparently the result of simply asking the piers to take on too much weight, ended at last the ambitions of Beauvais. A couple of decades later the chapter decided that, after more than 375 years, the cathedral of St. Pierre was finished. A wall was placed across the west end, where the nave would have met the transepts, and the Gothic age was well and truly ended.[11]

For all but a very few medieval towns that possessed them, the cathedral was by far the largest and most imposing and complex structure, both within the town and for many miles around. But cathedrals were by no means the only important construction projects of the medieval Europeans. The revival of urban life, in particular, led to extensive building throughout the wealthier parts of Europe, from the trading and textile centers of northern Italy to the new towns growing up along the Baltic and North Sea coasts. Guild and market halls, parish churches in great profusion, even workshops and storehouses provided work for the masons, both master and common. The other class of large-scale works, besides the cathedrals, that were of continuing importance and that saw considerable advancement in design and techniques, was military structures, particularly castles and other fortifications. Such buildings posed different problems from the great churches, and evoked different approaches. One consideration was that they were seen as the project of weeks and

Figure 4.7
Pont d'Avignon, the bridge over the Rhône River at Avignon built in the late twelfth century. From E. E. Viollet-le-Duc, *Dictionnaire Raisonné de l'Architecture* (Paris, 1854–68).

months, not years and decades, and thus could require intensive works. The construction of Beaumaris Castle, on the northern coast of Wales, at the end of the thirteenth century, for example, involved at one point four hundred masons, two hundred quarrymen, thirty smiths and carpenters, and about two thousand other workers, in addition to extensive transport equipment. Works on this scale had, of course, been seen long before, but they became almost routine at certain times and places in medieval Europe, particularly when rulers were consolidating new conquests or claims. This provided another important opportunity for experimentation and development by the community of masons, although the technical daring was of a much lower order than in the cathedrals. The fractious military history of these centuries provided plenty of opportunity for innovation in construction, and the basic form of the medieval castle changed greatly in the twelfth and thirteenth centuries, influenced particularly by the encounters with eastern practices during the Crusades.[12]

One other form of medieval construction deserves some mention, for its contribution both to commerce and defense and to technical expertise—bridge building. The Romans had been famous for their roads and bridges, and up at least to the eighteenth century these continued to provide much of the transport infrastructure of portions of the former empire. Smaller bridges were usually built of wood, and were not expected to withstand floods or the vicissitudes of time, fire, and rot. In times of peaceful trade, military necessity, or political convenience, such bridges would be maintained and rebuilt as needed and they posed few technical challenges or opportunities (such does not, by the way, have to be the case with wooden bridges, as numerous non-European developments testify, from ancient Asian suspension structures to American railway trestles of the nineteenth century). Major river crossings, however, were invariably made of stone, and were hence important and expensive construction projects.

One of the most famous, which can be used here as a general example, was the Pont d'Avignon, which carried traffic over the Rhône in a series of elliptical arches, each spanning about one hundred feet, which an overall length of almost three thousand feet (figure 4.7). The construction of the bridge has been surrounded by legend, largely involving the inspiration of the young St. Bénézet, who was said to have been divinely inspired to travel to Avignon for the express purpose of building a span over the Rhône. The construction began in 1177, and took about eleven years to complete, by which time Bénézet was deceased. In his wake there emerged a brotherhood with the responsibility for completing and maintaining the bridge.

The actual role of this "Brotherhood of the Bridge" (*Frères Pontifes*) is somewhat cloudy, and despite stories that this brotherhood was an extended network of bridge-builders, they were more likely simply another charitable order attaching itself to a particularly strategic location for the collection of alms or dues. Similar brotherhoods at other important crossings, such as at Lyon or at the Pont-Saint-Esprit in Languedoc, had differing levels of responsibility for their bridges, but it does not seem likely the brotherhoods were ever true technical communities. Like other medieval constructions, the great bridges were likely built by traveling masons, who could apply their skills and knowledge of materials, forms, and techniques to the task at hand. There were probably not enough bridge projects, however, for specialist knowledge to develop. Bridges like that at Avignon, however, do testify to the capacity of the medieval builders to build on the Roman legacy when given the opportunity. The itinerant medieval builders, of bridges, fortresses, and cathedrals, constituted one of the world's first independent technical elites, and thus one of the earliest means by which Europeans began to capture the improvements that emerged from a competitive, creative environment.[13]

5 Transforming Matter

The first detailed technical treatise in the West from the hand of a craftsman was written by a German monk who styled himself, in the Byzantine manner, as Theophilus Presbyter (which is to say, Theophilus the preacher):

> Another method of hardening is also carried out in the following way for those tools with which glass and the softer stones are cut. Take a three-year-old goat and tie it up indoors for three days without food; on the fourth day give it fern to eat and nothing else. When it has eaten this for two days, on the following night shut it up in a very large jar perforated at the bottom, and under the holes put another vessel, intact, in which you can collect its urine. When enough of this has been collected in this way during two or three nights, let the goat out and harden your tools in this urine.
>
> Tools are also made harder by hardening them in the urine of a small red-headed boy than by doing so in plain water.[1]

Theophilus was, in fact, most likely Roger of Helmarshausen, who lived in the early twelfth century and who was credited with the fashioning of some beautiful church items, including a portable altar that can still be seen in the German town of Paderborn. His treatise, *De diversis artibus*, or simply "On Various Crafts," described in a remarkably matter-of-fact manner the "art of the painter," "the art of the worker in glass," and, in the section of the book most clearly reflecting the writer's own skills and interests, "the art of the metalworker."

Theophilus addressed his work to "all who wish to avoid and subdue sloth of mind and wandering of the spirit by useful occupation of the hands and delightful contemplation of new things." He was clearly aware of how unusual it was to commit the knowledge of his crafts to writing, but he reminded his readers of the "inheritance that God bestowed upon man." "Whoever devotes care and attention to the task," he promised, "can acquire, as by hereditary right, the capacity for the whole range of art and skill." The kinds of crafts with which he was concerned, however, were limited to the creation of fine and beautiful things, suitable for the decoration of churches and the worship of God—paintings for the walls of a chapel, glass for the great windows of a cathedral or for fine goblets, gold or silver chalices, or bells and organ pipes for sacred music. The craftsmanship that Theophilus exalted was technique as worship, and the manufacture of workaday things was of little interest. But he still had no difficulty in acknowledging improvement through these secular concerns, as "human ingenuity, in its varied activities in pursuit of gain and pleasure," developed over time so that it might serve the devotion of God.[2]

Most of what Theophilus wrote would be judged today as sound and useful technical description and instruction. His discussions of pigments for painting and techniques for using them, for example, would allow a modern student to do a passable job of replicating the medieval materials, and provide such subtle suggestions as how to differentiate the beards of adolescents from those of old men, or what differences need to be observed for the paints used to depict a robe on a ceiling as opposed to those on a wall. Nonetheless, as the opening passage reminds us, Theophilus worked in a world in which the qualities and behavior of different materials, crucial as they were to success in making things, were understood only at the level of direct observation, and were not part of a "scientific" worldview. Observation showed that an iron or steel tool, such as that used for engraving precious metals, could be given a more durable point if they were quenched—that is, dipped in a liquid while red hot—in cool water. Clearly, some experimentation led to the use of urine rather than water, and the modern metallurgist may indeed recommend some organic matter in the cooling liquid. But the virtues of a three-year-old goat, not to mention those of a small redheaded boy, are completely outside the rationale of the modern mind. They are, nonetheless, the products of improvement, in their own way. Just as significantly, Theophilus's careful record suggests how the combination of a monk's literacy and a craftsman's diligence could extend the possibilities of capturing improvement in important new directions.

Every technology begins with material transformation. At its simplest, this is little more than cutting down a tree or adding water to dirt to make mud. Over some thousands of years, the transformations have become more complex, more subtle, more dramatic. For all but the last century or two, this increase in mastery over matter has come from craftsmen's observations alone, without theory or philosophy. While there is an extent to which this is largely true of all crafts and techniques, it is particularly the case with material transformations, for with buildings and machines it is possible to discern a logic of mechanism or structure that can be worked out by eye and feel. This is less true with the making of materials, and so pre-modern change is slow indeed in this area. The techniques that Theophilus described in the early twelfth century—contemporaneous with the first stirrings of the Gothic campaign—were little altered during the millennium preceding. Over the next few medieval centuries, however, profound changes indeed were to emerge, with equally profound effects on everything from the texture of daily life to the powers of princes and the wealth of nations.

Three important new materials appeared to the Europeans in the High and Late Middle Ages (from, say, the thirteenth through the fifteenth centuries): paper, cast iron, and gunpowder. It is no coincidence that each of these were known to, and probably invented by, the Chinese centuries before they made their way to the West. In the years in which the West was defined by the Roman Empire and the gradual emergence of a Christian civilization—the first millennium of the Christian calendar—the artistry and technical mastery of the Chinese Empire far surpassed anything the Europeans ever knew. Why this should be so, and the arguably much more interesting question of why it should cease to be over the next several centuries, are fascinating and important historical issues, but they are beyond our scope here. Suffice it simply to say that the list of medieval novelties that originated in the Far East is an impressive one, in both scope and significance. From pasta to pants, Asian imports gradually reshaped European life and habits, although quickly taking on native identifications. Much of the impetus for this came from the Crusades and the more sustained contact with the nearer Eastern civilizations frequently described as "Arab," but which were in fact much more complex ethnically, incorporating peoples extending a wide swath of the world, from India through Persia and Anatolia and across North Africa. Theophilus himself spoke of the sources of new knowledge, from Russia to the Arab lands, and celebrated the European capacity for assimilation and learning.

More important than the Crusades for sustaining the exchange of ideas with the East was the development of commerce in the Mediterranean. Whatever the changing geopolitical complexion of the region, the profitability of trade kept traffic going through the centuries. The most profitable commodity was without question spices, for pepper, cinnamon, and other substances were irreplaceable to the Europeans as the means for preserving foodstuffs and, just as important, making those foods that were not well preserved palatable. Oranges, apricots, rice, perfumes, and medicaments added value and variety to the Eastern trade. The very large return for relatively small amounts made these commodities the ideal basis for great profits, and Italian merchants, particularly from Venice and Genoa, great and sometimes hostile rivals throughout the Middle Ages, became the first exemplars of private capitalist wealth in European history.[3]

The growing importance of trade from the twelfth and thirteenth centuries, both within Europe and with the regions to the south and east, fostered the development of techniques in native crafts as well as the introduction of new products from abroad. Woolen cloth, in particular, became a commodity of great significance, and a craft that was formerly located in homes or small workshops of women began to take on a modern commercial character, concentrating in towns, and losing its formerly entirely female cast. Indeed, throughout the Middle Ages, the making of textiles constituted the single most important European manufacture. Flanders (roughly

Figure 5.1
The simple spinning wheel allowed a spinner to keep a spindle steadily turning, either by hand or by a foot treadle.

northernmost France and most of Belgium), in particular, became a region of cloth-making towns. The industry began there due to large stocks of sheep feeding on the rich meadows of the north European plain, but these soon became inadequate to supply the growing industry, so wool from Britain and elsewhere was imported to supplement supplies. This, in turn, further encouraged the development of mercantile structures—permanent exchanges (rather than seasonal fairs), joint stock enterprises, and other protocapitalist instruments.

Alongside these organizational changes were technical ones, affecting spinning, weaving, and finishing. The spinning wheel was a medieval invention, appearing probably sometime in the thirteenth century. Its primary value lay in increasing the productivity of the spinster, who formerly had to twist fibers into threads by a hand-held drop spindle (essentially a top, with thread being twisted off one end) and a distaff (a rod held in the left hand, typically, for winding the spun thread). While this method had the virtue of being cheap, simple, and portable and could thus be used by anyone (just about always women) almost anywhere, it was slow and tedious work. Using a wheel and pulley to turn a horizontal spindle with one hand while pulling and then winding up thread with the other speeded up the process as much as threefold (figure 5.1). Later in the Middle Ages, a flyer and bobbin were added to the wheel, which complicated the device somewhat but made it possible to twist and wind threads in one, continuous action (figure 5.2). Still later, a foot-operated treadle

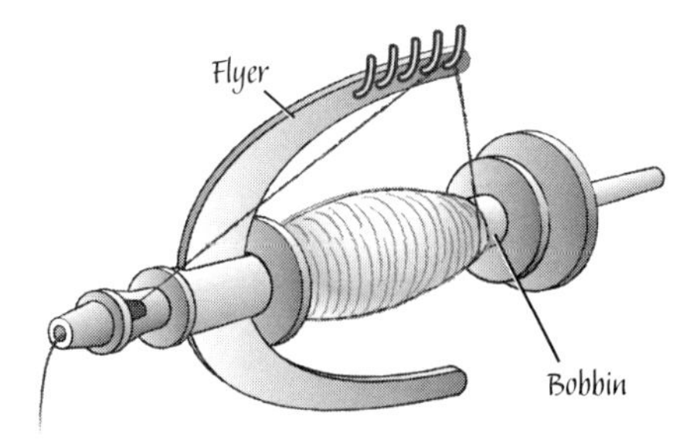

Figure 5.2
The flyer and bobbin were put in a spinning wheel in place of the spindle. Different sized pulley wheels turned the two elements at different speeds, imparting twist to the yarn while it was being wound on to the bobbin.

was added by some users to free the hands from turning the wheel. The use of these devices was sometimes resisted, out of concern either for maintaining quality or for maintaining employment, but by the end of the Middle Ages wheel spinning was the norm throughout the West.[4]

The new wheel increased the productivity of spinning, but it changed little the social character of what was even then the most feminine of occupations. The same was not true for the technical changes in weaving, for the introduction of new forms of looms contributed to significant realignment of the occupational structure of weaving. Before the Middle Ages, the loom was a rectangular wooden frame, usually vertical, on which the warp threads would be made to hang down (typically with weights at their ends). Alternating threads were fastened to a stick called a heddle. When this was moved to one side, a space was created with the alternate threads on either side. Through this space, called the shed, the weft threads were pulled. When the heddle was moved to the other side, the warp threads switched sides, the weft thread was pushed more tightly in, and a new shed created for another weft thread, from the other direction, to be pulled in. This manner of weaving remained unchanged for thousands of years, although there were plenty of variations in particular weaves and styles (by varying the numbers and proportions of threads in warp and weft, for example). Like spinning, this was an activity traditionally identified with women's work.

Sometime in the thirteenth or fourteenth centuries a rather different instrument emerged, using the same basic principles of warp, weft, heddle, and shed, but on a larger, more complicated device: a horizontal loom with a treadle (figure 5.3). The treadle operated the heddle up and down, allowing the weaver to use both hands for throwing the weft threads, wound in a shuttle, across the shed. The horizontal loom could also be made much wider—a broadloom—for weaving large pieces of cloth. This would typically require two individuals' working on the same cloth, one on

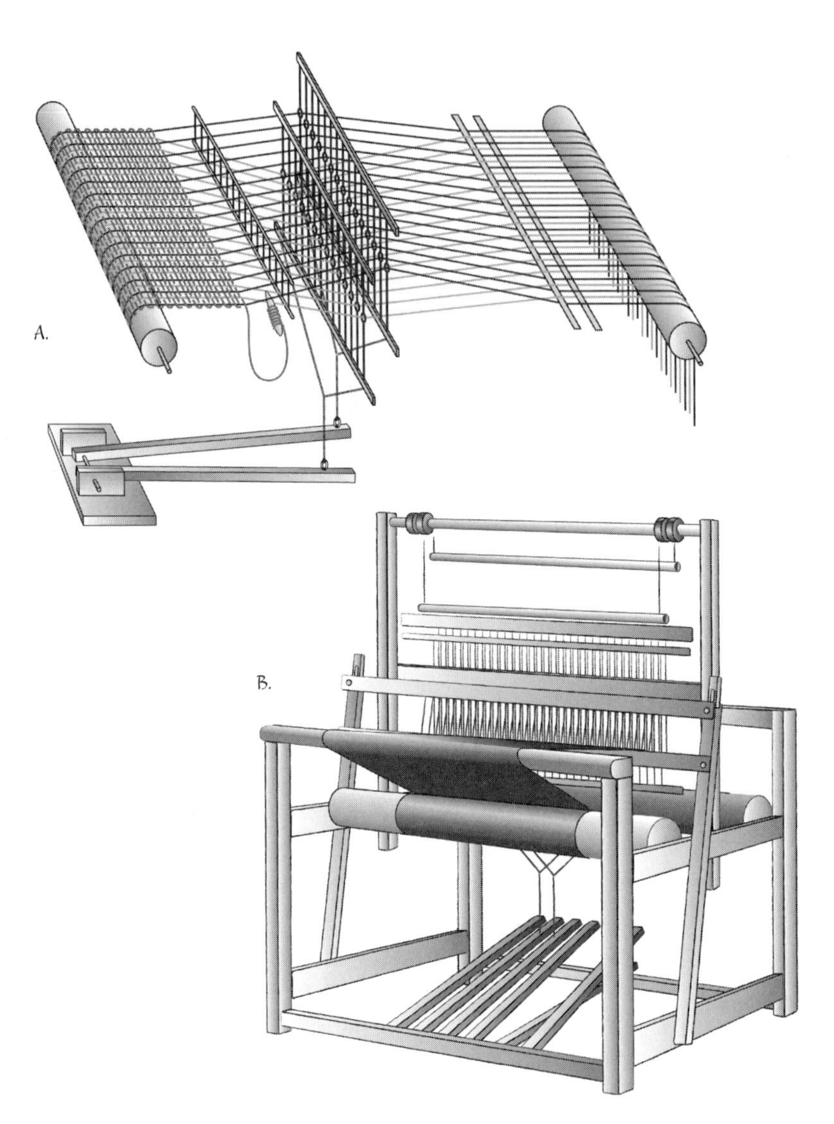

Figure 5.3
While the principle of weaving by introducing a thread alternately above and below a set of cross threads (the "warp") is quite ancient, the medieval loom became increasingly complex. Foot treadles (A) were used to speed up the process of raising and lowering the threads of the warp; multiple sets of treadles (B) were used to make different patterns of weaves.

each side throwing the shuttle back and forth at each other. Both of these devices, much more complicated and expensive, but with much higher productivity with less skill, became the tools of men rather than women. Full-sized broadcloth was about six feet wide, and a full bolt some thirty yards long. This required some two to three thousand (or more) warp threads (going the length of the bolt). This was a much larger product than the older looms produced, much more clearly a market commodity.

Alongside the increased productivity of spinsters and the enlarged scale of weavers' work was the application of waterpower to the most tedious of finishing processes, fulling. Substituting water-driven hammers for the wet feet of fullers trampling on freshly woven woolens further increased the scale of textile production. The spread of mechanized fulling also had the effect of diminishing the Flemish domination of textile production, as the areas better endowed with water power, such as northern Italy and parts of England, became more competitive in both cost and quality. These developments further encouraged an economic revolution that had already begun in the textile industry—the organization of the industry into a complex of merchants, factors, specialized workers, outworkers, and guilds—which became the prototype for early modern capitalism.[5]

One of the characteristics of the textile trade was the variety of products and techniques that it fostered. Within the textile regions like Flanders and, later, northern Italy, individual towns came to be identified with particular products and styles. Particular weaves, colors, or finishes of fabric were associated with specific towns or workshops. Native dyestuffs, such as woad or madder (sources of blue and red, respectively), came to be supplemented by imported, exotic substances, such as vermilion, kermes, or saffron (producing brighter reds, purples, and oranges). Woolens themselves varied in quality and character (there were more than fifty grades of wool in England alone by the fifteenth century), and other fibers, such as linen, cotton, and silk, were combined with wool to produce even more variety.[6] The commercial growth that characterized the textile industry in Europe after the twelfth century further promoted the differentiation of products and experimentation with techniques. While the guild controls over textile manufacture sometimes seemed to diminish experimentation on a local basis, intense competition among towns and regions fostered variation and improvement.

An obvious source of variation, particularly in the later Middle Ages, was the growing use of other fibers. Cotton and silk had been known for centuries, but they have no native European sources and thus were always thought of as costly imports. With the growth of trading networks, on the one hand, and of capitalist organization, on the other, a vigorous expansion of the manufacture of these materials from imported raw stock became a mainstay of the most successful European economies. Italy, in particular, was able to take advantage of its Mediterranean location, the military and commercial domination of the relevant sea lanes by Venice and Genoa, and the commercial orientation of the key city states (in addition to the two powerful ports there were Lucca, Bologna, and, later, Florence) to establish a dominant position in both silk and cotton manufacture and trade.

Linen, on the other hand, is made from flax, a strawlike plant that grows readily throughout much of Western Europe. Its working, however, is somewhat labor-intensive, as getting usable fibers from the plant requires a series of processes (retting, breaking, scutching, and hackling), that take both time and hard work. Nonetheless, linen was an important fabric to the Europeans, being their only readily available native lightweight alternative to woolens. The manufacture of linen grew in the High and Late Middle Ages, encouraged, like so much else, by urban prosperity, commercial networks, and population growth. Linens, too, experienced change in tools and processes, specialization of styles and weaves, and the expansion of uses. This was particularly important as Europe entered the extended period of cooling that some have called the "Little Ice Age" toward the end of the Middle Ages (the beginning has been put in the range from the early thirteenth century to as late as the mid-fifteenth century). Increased clothing options, particularly for underclothing, were important (other technological responses included more fireplaces and smaller rooms in large houses), so the consumption of linens increased measurably from the fourteenth century on.

The expansion of the linen trade fostered, indirectly but crucially, one of the key materials innovations of the late Middle Ages—the development of papermaking. Without linen, quite simply, there is no paper, since before the nineteenth century linen rags provided the only raw material that could be successfully converted into paper (the Chinese used silk or mulberry bark, but these were not readily available materials for the Europeans). The craft itself began probably in the first or second century of the Christian era in China, and it made its way gradually across the southern tier of Eurasia toward Europe, finally entering by way of Islamic Spain in the mid-twelfth century. By the late thirteenth century the Italians were making paper, and from thence the craft moved slowly across the Alps, not reaching Britain until the very end of the fifteenth century.[7]

The Europeans very quickly changed papermaking technology, particularly by mechanizing the most laborious steps. Papermaking began with wetting and rotting piles of rags for several weeks or even months. This fermented pile was then washed to reduce a yellowish hue cast by the fermenting process, and put into water and beaten to a pulp. In Asia this beating was done by hand, but the Europeans adapted the technology of water-powered fulling to the pulping process and thus reduced the labor needed for pulping as well as speeding it up. The first appearance of papermaking in Europe, in Valencia, Spain, in the mid-twelfth century, was marked by the appearance also of a water-powered pulping mill, in which heavy stampers were moved up and down by the tappets on the axle of a watermill. Other adaptations soon followed—rigid, metal-covered moulds for the sheets, variations on the stamping and pulping processes, and more extensive use of waterpower. The amount of paper used by the Europeans before the introduction of printing in the mid-fifteenth century was modest, but the material did become increasingly more readily available, encouraging experimentation with uses. In the period just before Gutenberg, paper was used not only for manuscript volumes but also for engravings, playing cards, handbills, and the like. Communications were encouraged and the recording of all sorts of ideas and information promoted by the availability of the new medium.

The alacrity with which the Europeans adopted waterpower for papermaking suggests the wider and continuing implications of the power revolution that spread after the tenth century through those sections of Europe endowed with flowing streams and rivers. In the case of papermaking, power made it possible to reduce the labor requirements for the material, in exactly the same way it had done for cloth finishing and grain milling. In other cases, however, the implications of power were even more profound—most significantly in the making and spread of a new form of iron.

Very broadly speaking, there are three forms of iron that have been useful historically: wrought iron, cast iron, and steel. These categories are not rigid, and they merge into one another to some degree, depending on composition, treatment, and quality, but for the purposes of general discussion it is adequate to differentiate these three and to understand their distinct histories. For most of history, everywhere in the world in which iron has been used, wrought iron (sometimes called "malleable" iron) has been what people referred to when they spoke of iron. This is almost pure iron, with small amounts of other materials, depending on the quality and location of the source. As the name suggests, this is iron that can be shaped relatively easily, usually by hammering. It is thus not particularly hard, but it is tough and ductile. For thousands of years this iron has been made into useful implements and weapons, for it is more durable than any of the other metals available to premodern peoples.[8]

Useful as it is, however, wrought iron does not hold an edge very well without further treatment. Much of the great value given to the skills of the smith in traditional cultures lies in this fact, for the hardening of an iron edge was a craft requiring considerable skill and diligence. By heating an iron edge or surface in a charcoal fire and with repeated hammering, the smith can create a shallow surface of steel on the iron. Steel is the form of iron that most effectively combines toughness with hardness. This quality is largely (but not exclusively) imparted to it by the integration of a small amount, from 0.5 to 1.5 percent, roughly, of carbon into the microstructure of the iron. From the earliest manufacture of iron some 3,000 years ago, ironworkers learned that careful working at a forge with iron from the hearth could produce a strong, durable edge, and this fostered the identification of forged iron with weaponry. Quenching the hot hammered edge in water or some other liquid will help maintain its hardness—this is the technique for which Theophilus offered advice.

Another source of iron's great and ancient value lies in its ubiquity. It is one of the most common metals in the earth, although it very rarely occurs in metallic form and its truly workable ores are found only in specific regions. These, however, are still more widespread than those for any other metal, and they tend to be accessible rather than deep in the earth. To extract the metal from its ores is not a complex process, but neither is it an obvious one. For this reason, iron was the last of the great traditional metals to be exploited, after the precious metals, gold and silver, and copper and useful copper alloys such as bronze and brass. When the ores of copper and its primary alloying metals, tin and zinc, are roasted in a properly constructed fire, they will yield liquid metal that separates out readily from its source. When iron ore is roasted in a traditional hearth, however, it remains solid, and the metal is typically in a spongy mass, mixed with considerable amounts of slag—glassy by-products of silicates and other materials that must then be physically separated from the iron, typically by heating and hammering. The traditional "bloomery" hearth was very simple. It consisted of a shallow pit in which layers of broken chunks of iron ore alternated with layers of charcoal. The charcoal was set afire, and a primitive hand-operated bellows was used to foster the burn, which continued for some days. When this had run its course, there was dug out from the pit a "bloom," a heterogeneous mass of spongy iron and glassy slag, small enough to be handled by one man using a pair of tongs.[9] This bloom then had to be reheated and hammered at the forge for some time to beat out the slag and air, consolidating the iron into bars or other useful shapes.

Work with iron was, like almost all crafts, based purely on experience and the skills and techniques handed down through generations of practitioners. Of all the useful metals before the nineteenth century, iron was the most variable in its qualities and forms. While impurities might affect copper or other metals, they were usually readily eliminated or at least controlled. Because traditional iron was never liquefied, however, and because iron more readily combines with nonmetallic substances such as silica, sulfur or phosphorus, the constitution and behavior of iron from the hearth varied a great deal, depending on the source of the ores, the fuels used to smelt it, even the materials used to line the hearth or furnace. Likewise, the techniques used to make iron workable and to give the final product desirable qualities had many variants and were constantly the source of experimentation and change.

By the High Middle Ages—the twelfth and thirteenth centuries—the broad utility of iron was widely recognized. In the mid-thirteenth century, for example, a wide-ranging treatise by Bartholomew the Englishman declared iron the most useful of all metals: "Without iron, the commonalty be not sure against enemies; without dread of iron the common right is not governed; with iron innocent men are defended; and foolhardiness of wicked men is chastised with dread of iron. And well-nigh no handiwork is wrought without iron: no field is eared without iron, neither tilling craft used, nor building builded without iron."[10] While the actual quantities of iron used by medieval workers was small by later standards, the uses were critical in providing the hard and durable pieces that even wooden machinery or weaponry required in places. In buildings, iron nails, spikes, or tie bars could add an important element of strength and reliability; in machines or mills, iron sleeves or bearings made mechanisms durable and less prone to failure; iron horseshoes gave cavalry an extra edge on even the roughest fields; iron knives and sickles were indispensable to the harvest, and iron fittings for carts and harnesses. Due to the wide range in quality of iron and iron products, there was a thriving continentwide trade in both the material and in iron goods. Where quality demands were high, as in weaponry or building materials, certain regions, such as Spain and Sweden, became the most desired sources, and economic power and wealth flowed back to these edges of the European world in consequence. The economic expansion of the period leading up to the fourteenth century was accompanied by a dramatic rise in iron production and consumption.

The association of iron with weapons and the waging of war was already an ancient one by the Middle Ages. While in fact there was a range of weaponry that was important to the medieval warrior, including longbows, crossbows, lances, and pikes, it is no accident that the sword became the true symbol of battle and power. In addition to the iron and steel blades of swords and smaller knives, in the course of the Middle Ages the dependence of the warrior on iron increased, as the styles of battle and the growing resources available to support warfare led to a steady increase in the size and weight of armor. Whereas the early medieval soldier was largely protected by thick pieces of leather and a shield made mostly of leather and wood, by the fourteenth and fifteenth centuries the mounted warrior was encased in an almost impenetrable shell of iron armor, and when it could be afforded, the iron protection was extended to the horse as well. A complete suit of armor in the late Middle Ages might consist of as much as a hundred pounds of iron plate, and plate was supplemented by chain mail, drawing even further on the skills of the smith/armorer. The fully armored knight and horse was a formidable but ponderous weapon, representing the heaviest and most expensive mobile implement of land warfare until the introduction of the tank in the early twentieth century. The medieval military uses of iron, which extended well beyond the arms and armor of the knight to include arrow and spear points, spurs and stirrups, and key parts of crossbows and siege machinery, made particular demands on the most precious forms of the metal, especially hardened steel. These demands increased even in the period of economic decline that marked the fourteenth century and the coming of the Black Death to Europe.[11]

The combination of military uses, spreading dependence of peaceful arts from agriculture to architecture, and the commercial expansion of urban Europe after the early thirteenth century increased enormously the demand for iron. Pressures were put on both the sources of ore and fuel and on production capacities, and the responses to these fostered technological developments that were to have enormous long-range consequences. Older sources of easily worked ores were not adequate to meet demand, so ironworking spread into new areas, and new ores, more difficult to smelt, began to be exploited. Many of these ores did not work well in the older bloomery hearths, and so experimentation with furnace design and construction was encouraged. Out of this emerged furnaces that were higher, larger, and more complicated (figure 5.4). These larger furnaces were particularly dependent on larger and steadier sources of air supply, and so water-powered bellows, originally seen as convenient adjuncts to the traditional furnace, became more and more necessary. This, in turn, drove furnace location down from the barren hillsides or the boggy swamps where ores were found to river and stream beds where waterpower was more reliable. Even more importantly, however, the larger furnaces, with their steady air supplies, were hotter and burned fuel at a faster rate than older designs. The combination of more fuel and higher temperatures changes significantly the chemical reaction taking place within the iron-charcoal mixture. Specifically, much more of the charcoal—almost pure carbon—is combined chemically with the deoxidized iron, and this chemical combination has a significantly lower melting point than pure wrought iron (about 1200°C as opposed to nearly 1550°C). This lower melting point, combined with the higher heat in the furnace, means that the resulting iron comes out, not as a spongy bloom, but as a continuous supply of molten iron. This molten metal is vastly different in its character and applications than the traditional product of the hearth, and it was the recognition of its possible uses as well as of the key techniques of its manufacture that brought about a tranformation in late medieval metals technology.[12]

The new furnace, called "blast furnace" due to the key role of the more forceful air supply, and the new iron, called "cast iron" for the most radical change in its working qualities, engendered profound change both because there became available a very new material and because the new techniques yielded their product in vastly greater quantity and speed than traditional ironmaking. This transformation, however, like most technological changes, both small and large, was not the product of conscious effort. It was instead the unsought result of efforts to improve the productivity of furnaces. The blast furnace and cast iron were known to Chinese ironmakers as much as two thousand years before they appeared in the West in the late Middle Ages (probably the late fifteenth century). While it is possible, particularly given the increased trade between China and Europe in this period, that Chinese techniques

Figure 5.4
Late medieval ironmaking was transformed by the blast furnace's rapid production of molten iron. This late-sixteenth-century illustration of iron works suggests the increased importance of machinery as well as hotter furnaces. From G. Agricola, *De Re Metallica* (Basel, 1556). Burndy Library.

made their way westward, most scholars believe the European blast furnace was an independent development.

The blast furnace markedly increased the speed of iron extraction, but it was by no means an obvious improvement to all observers. Cast iron behaves very differently from wrought iron. It is, particularly, much more brittle than traditional iron. When cast iron is struck with a hammer it is more likely to chip or shatter than to be beaten into shape. Any smith observing this would initially reject the material as essentially useless. While Chinese ironmakers made many useful implements directly by casting the carbonized iron, European smiths were reluctant to do so, perhaps because, by the fifteenth century, they were working with markets and communities familiar enough with wrought iron that the more fragile nature of the newer iron would have been seen by consumer and craftsman alike as unacceptable. While in the coming centuries molded iron did indeed find important outlets, especially for armaments, the primary use of cast iron was always as an intermediate material in the making of wrought iron. Cast iron would be drawn from the blast furnace and led into standard molds—usually in a bed of sand or loam with bar or ingot shapes that branched off a longer channel. The pattern of ingots and channel was familiarly likened to piglets feeding off a sow, hence the common term "pig iron" for this unrefined material. The pig iron then had to be processed in a "finery," which consisted of a forge with hammers, which would gradually beat out the carbon from the pig by exposing the surface to an oxidizing flame and thus burning off surface carbon, and then hammering to expose more carbonized surfaces. This was a very labor-intensive process, and it is testimony to the radical economies of the blast furnace that it eventually won out over the older bloomery hearth with its direct production of wrought iron.

The full importance of the blast furnace and the proliferation of iron came through linkage with another, equally portentous, technological innovation, also pioneered by the Chinese—gunpowder. Indeed, the end of the Middle Ages as a coherent, definable episode in the history of Western culture is perhaps most clearly marked by the conjoined introduction of these two technologies. But where the blast furnace and cast iron most probably arose in Europe independent of external sources, gunpowder was more like paper in its gradual move westward from China until it made its way into Europe, most likely sometime in the mid-thirteenth century. It is not true that gunpowder was only appreciated by the Chinese for firecrackers and other celebratory uses—they recognized soon after its discovery, probably three centuries or so before its European introduction, that the explosive could be used for bombs or in metal tubes to shoot projectile weapons. The Chinese, and the Islamic peoples who first took their knowledge westward, explored uses that included "fire lances" (probably a kind of flame thrower), bombs (both explosive and incendiary), and cannon. The originators of gunpowder, however, always thought of the material as more than powder for guns, and they made use of a wide range of compositions, with purposes that ranged from medicinal to noisemaking.[13]

The Europeans, on the other hand, focused from the beginning on gunpowder's utility for weapons. Their version of gunpowder was somewhat more narrowly prescribed—a mixture of charcoal, sulfur, and saltpeter (potassium nitrate), in proportions with roughly equal amounts of charcoal and sulfur, and two to four times that amount of saltpeter (the Chinese experimented with smaller amounts of saltpeter, which mixture would burn but not explode). The first European description of gunpowder came from Roger Bacon, an English friar who gloried in experimenting with all sorts of materials and phenomena. Bacon wrote, in the 1260s, "We have an example of these things . . . in that children's toy which is made in many parts of the world: i.e., a device no bigger than one's thumb. From the violence of that salt called saltpeter together with sulfur and willow charcoal, combined into a powder, so horrible a sound is made by the bursting of a thing so small, no more than a bit of parchment containing it, that we find the ear assaulted by a noise exceeding the roar of strong thunder, and a flash brighter than the most brilliant lightning."[14]

Within a generation, Western discussions of gunpowder were much more specific and more oriented toward that usable in guns. The preferred mixture was about two-thirds saltpeter, which made for a mixture that was certainly suitable for explosives, but which would also have been quite expensive, for the West had, in the first century or so of gunpowder, no native source of saltpeter. In spite of this difficulty, guns appeared soon after gunpowder. The record is very sketchy, but certainly by the early fourteenth century there were to be found guns and orders for ammunition. The first ammunition appears to have been arrows, and these first firearms were perceived simply as additions to the host of machines already available for hurling arrows either over great distances or in substantial numbers. The early guns and the gunpowder used in them would not have been able to shoot heavier munitions. Even though it was almost a century before firearms actually made a difference on the battlefield, they spread quickly among armies that could afford them, perhaps because the noise they made was so frightening.

Whatever the real effectiveness of early gunpowder weapons in killing people or knocking down fortifications, it is clear that they became popular tools of warfare, and the making of both guns and powder became technologies of the utmost importance. They were also technologies rather different in character from most crafts that had preceded them, and these differences, in hindsight, provide a glimpse at the crucial change in direction on which the West was about to embark. For one thing, both gun and munitions manufacture were novel skills, drawing to only a limited extent on the practices of preceding generations of craftsmen. These skills brought forth with a special emphasis the importance of new knowledge, of arts unknown to the ancients. For another thing, the making of guns and of gunpowder were skills that demanded, at least in the first couple of centuries, extensive experimentation. The promise of power through effective firearms was so alluring, and yet the problems in the performance of actual armaments so apparent, that the possibilities of improvement were widely evident and evoked much experimentation.

Gunpowder is a simple composition, on the one hand—just three simple ingredients in proportions that are not that mysterious—but it is also a very complex and subtle substance, whose principles of action are not even fully understood today. There were two obvious sources of difficulty in the first century or so of its use: the preparation of saltpeter and the milling and preparation of an effective mixture. Saltpeter is ordinarily and ideally potassium nitrate (other nitrates may be referred to, inaccurately, as saltpeter), a substance that appears very seldom on earth as a recoverable mineral, and almost nowhere in Europe. It is instead most often the product of bacterial action on organic waste materials, a process in which certain specific bacteria break down ammonia-containing substances and produce soluble salts as a result. This particular organic action is extremely common, but potassium nitrate is only a minor product among much else. The problem lies typically not with finding sources of organic by-products—we are talking, after all, of manure, night soil, and other stores of animal and human wastes—but with refining the products of this widespread action, eliminating unwanted substances and maximizing the return on the desired product. This turns out to be a complex and difficult-to-control process, and thus provided much room for experiment and improvement. It also pushed into prominence the skills and ingenuity of a new class of practitioners, who were highly valued for their knowledge of substances and of their abilities to manipulate them.

Not only did the extraction of a key ingredient make gunpowder manufacture an elusive and expensive art, but the compounding and final preparation of the material turned out to be another subtle and variable craft. Gunpowder is not a chemical combination of sulfur, charcoal, and saltpeter, but a physical one. Its action depends on the combined responses of these three ingredients to fire, and the burning of the mixture is a complex affair. Unlike modern explosives, gunpowder does not combust rapidly or turn into gaseous intermediate products before explosion, but the particles of powder actually burn from the outside in. This means that the shape and sizes of the particles are crucial variables in its action. Another key variable is the amount of moisture absorbed by the particles, for the three ingredients, saltpeter especially, all have some tendency to take on water. After the first century or so of powder making and use, the Europeans widely adopted the technique of "corning," in which the final mixture of ingredients was made into a paste by addition of a liquid, formed into balls, dried, and then either used as such or pulverized further immediately before firing. The size of the balls ("corns"), the composition of the wetting agent, and the subsequent treatment of the dried product were all the subject of extensive experiment and variations in final practice.[15]

Just as the making of gunpowder called into existence a range of skills and variations unknown before, so too did the design and manufacture of firearms themselves. From the fourteenth century onward, there emerged a range of weapons, varying in size, power, and reliability. They also widely varied, significantly, in cost. Indeed, one of the historical implications of gunpowder weapons was the gradual but inexorable tightening of the relationship between the capacity to wage war and money. Warfare, either defensive or offensive, of course always required resources, but the circumstances and technical requirements of the new weaponry and the times in which they appeared redefined the requirement in fundamental ways. No longer were land and manpower the measure of warmaking capabilities, but the money to buy arms and ammunition came to define military possibilities. The results of this in the next few centuries were to be profound, as nations such as the Netherlands were able to turn their mercantile wealth into military power and the moneyed bourgeois acquired political power. The great cost of cannon, in particular, startled observers of the fifteenth and sixteenth centuries, and this, along with the growing cost and complexity of naval ships, redrew the lines of power in the West, both among nations and between classes.

While the use of gunpowder in metal tubes or chambers for propelling projectiles was not an elusive idea, the manufacture of effective devices to make this into a truly useful way of fighting or threatening fortifications required important technological changes. There had never been need before, after all, to put explosive mixtures into barrels and then ignite them without destroying the barrel. The challenge evoked a number of solutions, but economic as well as technical considerations kept the problem alive for centuries. The first cannon were cast of bronze, in a poignant historical irony, by bellfounders who had learned their art for the cathedral crusade. Bronze, however, was simply too easily worn and too expensive to be widely popular in such applications. The first serious alternative was the use of strips of wrought iron, bound together as in a barrel, by iron hoops. While such guns had many defects, they were much cheaper and could be made in increasingly large sizes, for shooting shot of increasingly large weights. In 1377, the Duke of Burgundy used cannon in a siege that shot 200-pound stone balls, and he ordered another to be made that could throw a 450-pound ball. The high cost of powder in this early period discouraged attempts at much larger munitions, but as powder became dramatically cheaper in the fifteenth century (with the development of native saltpeter manufacture), the Europeans strove for ever-bigger gunnery. By the early fifteenth century, guns that could hurl stones of more than 500 pounds were not uncommon, and an Austrian piece was built to throw more than 1,500 pounds.[16]

Just how effective these large guns were is not clear, although it would appear that they did in fact make the difference in some sieges. On the battlefield, however, they were so ponderous and difficult to use that their value was truly problematic. The re-

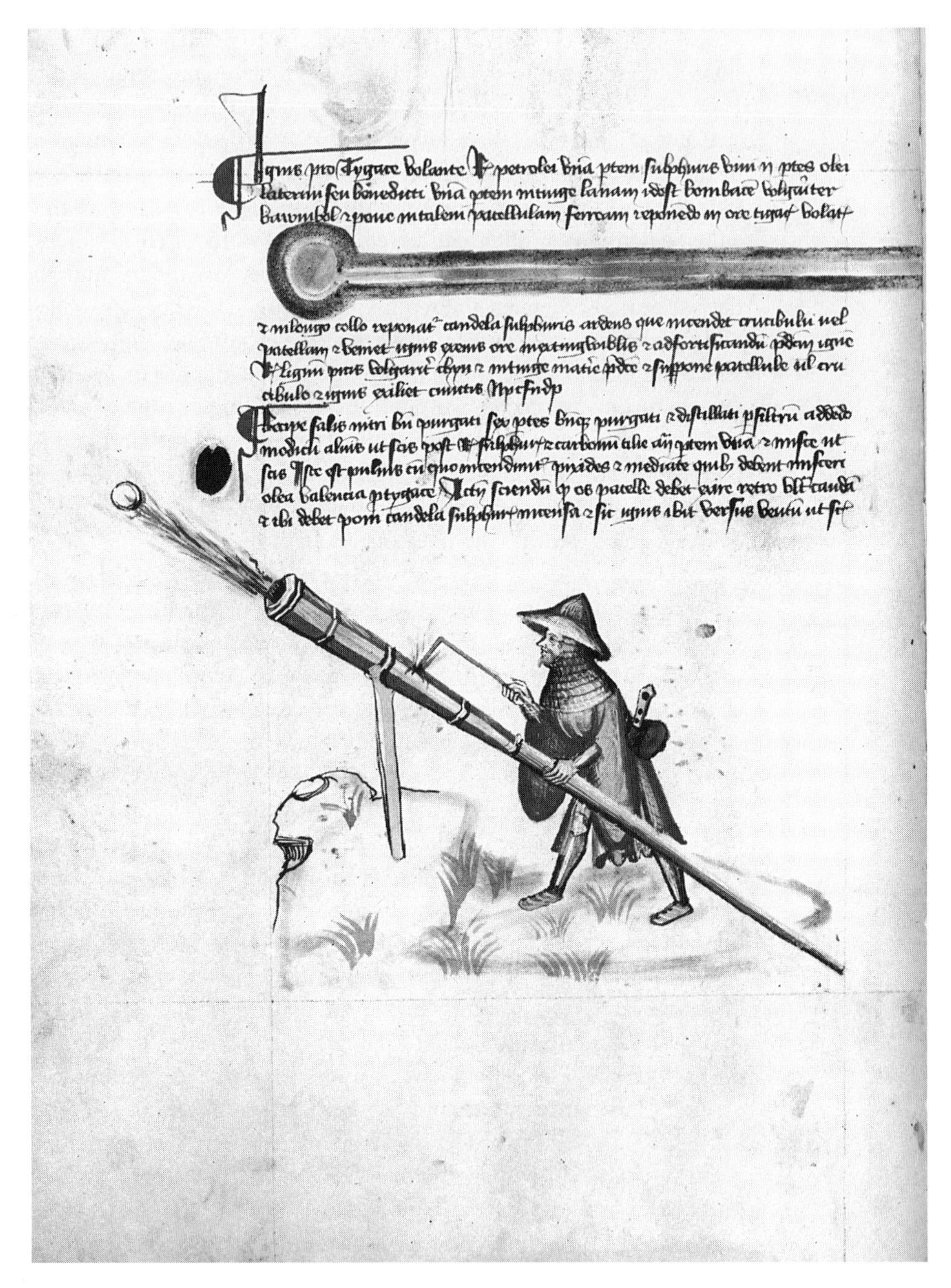

Figure 5.5
A small culverin, one of the first steps in the emergence of shoulder guns. Conrad Kyeser provided the earliest illustrations of such gunnery in his *Bellifortis* of 1405. From Conrad Kyeser, *Bellifortis*, facsimile publication (Düsseldorf: Georg-Agricola-Gesellschaft zur Förderung der Geschichte der Naturwissenschaft und der Technik, 1967).

sponse, of course, was to design and put into service a range of smaller firearms. Cannon that used projectiles of between two and thirty-five pounds began to be used, and iron and lead shot began to replace stone. When such devices were applied in large numbers, they could indeed turn the tide of battles, although through the fifteenth century the primary value of gunpowder weaponry remained as siege devices. Cannon were joined in the fifteenth century first by culverins (figure 5.5), which were small-bore devices that could be held up by one or two men, and then by the arquebus, which in its standard form had a barrel of about forty inches and a very small bore, of little more than a half-inch. In the fifteenth century such weapons were typically made of cast bronze, and bullets of iron or lead began to appear. The small arms put an even greater premium on carefully made gunpowder, and thus advantages began to accrue even further to the technically skilled and precise. After the mid-fifteenth century, cast iron weaponry began to displace bronze and wrought iron, largely as the economic demands of ever-larger arsenals combined with increasing ability and confidence in the design and fabrication of cast-iron gunnery.[17]

While there is considerable debate over the extent and nature of a "military revolution" at the end of the Middle Ages, the larger importance of gunpowder and the metalworking technology that it encouraged lay not so much with particular military capabilities but with the perception that military power now lay with technological prowess, rather than traditional sources like manpower, native courage, and the justice of one's cause. Good weapons had always been acknowledged as important advantages in warfare—magic swords were the stuff of legend for thousands of years—and the value of a good smith for an army was similarly the stuff of myth as well as practical generalship. But in the older order, there was no question that the prowess of the warrior was the primary cause of battlefield success. In the new, post-medieval world, the makers of weapons and of munitions were as important as the users. Just as significantly, the capacity of these makers to experiment and improve was a widely appreciated virtue, one that came to be seen as part of the military order.

6 Light and Time

Friar Giordano di Rivalto was a popular preacher from the northern Italian city of Pisa at the beginning of the fourteenth century. His lectures were written down by fascinated listeners, and they became part of the flowering of Italian language and literature that made that century a period of the most remarkable cultural transformations in the West. On the morning of February 23, 1305, Giordano gave an early Lenten sermon in the church of Santa Maria Novella, in the nearby city of Florence. It is in the record of this sermon that we find one of the earliest celebrations of invention in Western literature, attached to the specific reference to a recent and notable invention that was indeed perhaps of great value to the forty-five-year-old preacher: spectacles.

> All the arts are found by man. Man doesn't make them, but he finds them; and yet they haven't all been found. The finding of arts will never come to an end. . . . Many of them have been found, but any day a new one could be found and new ones are always being found. It is not yet twenty years since there was found the art of making eyeglasses which make for good vision, one of the best arts and most necessary that the world has. . . . I have seen the man who first invented and created it and I have talked to him.[1]

Giordano's sermon has been used as a primary source to date the invention of spectacles to the year 1285 or 1286 by a Pisan monk, it has been said, Alessandro Spina. When the evidence for this is examined closely, it fails to convince, and so eyeglasses join almost every other human invention, great and remarkable as well as small and trivial, in the anonymity of premodern technological history. What is noteworthy, in fact, is not that we really don't know who invented eyeglasses, but that the claims that we do have been taken so seriously. We are, at the outset of the fourteenth century, indeed at the verge of a turning point in history, and in particular, in the history of things, techniques, and useful knowledge, for it is at this point, as Giordano's sermon unambiguously tells us, that the men and women of the West begin to notice and appreciate the novel improvements appearing in material life.

The invention that Giordano so appreciated probably was indeed a product of north Italian glassmakers, from Pisa or Venice most likely, and does first appear at the end of the thirteenth century. Since antiquity, it had been known that certain curved pieces of crystal, glass, or even bowls of water could magnify images. Indeed, references to the use of these magnifying glasses for reading can be found throughout the medieval period. Roger Bacon wrote in the 1260s, "If anyone examines letters or small objects through the medium of a crystal or glass if it be shaped like the lesser segment of a sphere, with all the convex side towards the eye, he will see the letters far better and larger. Such an instrument is useful to all persons."[2]

The Italian innovation was, of course, to make the lenses quite small and to mount them, paired, in such a fashion as they might be "worn" and thus become part of a person's general equipment for seeing the world, or, at least, for seeing what was being read. In addition, the first spectacle maker had to devise lenses with a very different curvature that those in a simple magnifying glass (presumably, a concave-convex lens). By the middle of the fourteenth century, spectacles were moderately common, at least among the well-to-do, they begin to appear in paintings, and references to orders and sales may be found. By the fifteenth century, the city of Florence had emerged as a center of the optical trade. While at first rock crystal (polished quartz) was far preferred over glass, since truly transparent glass, without color or flaw, was extremely difficult to make, clear glass came to be more readily available in the fifteenth century, and the price of spectacles consequently began to fall. Venetian glassmakers were admonished not to try to sell ordinary glass lenses as though they were crystal, although later it was apparent that good glass lenses were appropriate products, as long as they were properly labeled. The spread of spectacles was no small technical achievement, for it went hand in hand with an increasingly literate culture, particularly in government and commerce. The means to combat the nearly universal tendency of the human eye to lose with age the ability to focus on nearby objects or words—presbyopia—with technological means extended the useful working lives of thousands and then millions in the generations after the friar of Pisa celebrated the discovery.[3]

The larger meaning of spectacles and their appearance at this time is worth contemplating, for they represented, it may be argued, the first invention that was clearly perceived and appreciated as a novel contribution to the quality of individual life. It is one of the fundamental characteristics of technology in our time that it is seen as being clothed in the promise of individual betterment, through comfort, health, amusement, information, or enlightenment. But, in fact, very few specific devices introduced into life in the past—certainly before about five hundred years ago—came into the world with such claims or possibilities. The devices, machines, buildings, and materials that we have spoken of up to now were in the aid of work, worship, and warfare, but they never were directed toward the improvement in the condition of individual life (except, of course, to make labor less onerous). Spectacles were addressed to alleviating a problem that besets almost all humans in their advancing years. To don a pair of eyeglasses was to cheat old age, to recognize the capacity of a novel technology to rewrite the rules of the human condition by prolonging the full participation of men and women in life—especially literate life—into and beyond middle age. Eyeglasses thus become a commodity in the fourteenth and fifteenth centuries with special meaning for the perception of technological improvement itself. In addition, the usefulness of lenses became more and more evident in the next centuries, as they extended the capacity for observation of the world beyond the capabilities of the healthiest human into telescopic and microscopic realms that were to reshape the very image of the universe itself.[4]

The material that went into these early spectacles was in one sense quite ancient, and yet in another, important one, a product of several centuries of attempted improvements. Glass had been known to the ancient Egyptians at least a couple of thousand years before, and by the time the ancient Romans were using the material, its manufacture was well understood and glass articles were important features of Mediterranean commerce. But the glass of the Egyptians, Romans, and everyone else in earlier times was seldom truly clear. The consistent manufacture of glass that was colorless and without blemish was a late medieval invention of considerable difficulty and merit. Earlier in the Middle Ages, much attention had been lavished on developing the art of colored glass—nearly a thousand years later the evidence still dazzles us in cathedral windows. But clear glass requires a different kind of skill and facility with materials. It requires, in particular, a purity of ingredients that very few chemical processes carried out on any scale before modern times could muster.

The remarkable nature of glass as a material was recognized by medieval and Renaissance writers, and with good reason. Glass is the one nonmetal available to preindustrial craftsmen that can be made hard and sturdy and yet can be readily worked (and reworked) by a vast range of techniques, from molding to grinding. Glass can be worked in every state from liquid to soft plastic to rock hard, and can then be reshaped if necessary with no loss of material. The fact that glass can be colored in a remarkable range of shades and transparencies added to the fascination that it held for early writers. Glass was essentially a mixture of two basic ingredients: sand, as a source of silicon dioxide (silica)—the whiter or clearer, the better—and ashes from specific plants. These ashes, we know and Renaissance workers discovered, are the source of key mineral salts that act as "fluxes," which lower the melting point of the silica in the sand to temperatures available to early furnaces. The key fluxes found in ancient glass were oxides or carbonates of sodium or potassium, but because the ashes were often used directly (and even when not were never reduced to pure salts) the glass composition always contained other matter, such as compounds of iron, aluminum, manganese, or magnesium. The final color of the glass depended on how much of these other materials were in the glass, as well as on impurities typically found in the sand. But it is important to remember that these factors were unknown, in these terms at least, by the early glassmakers. They only knew that particular sorts of ashes, mixed with sand or pebbles from a particular source, heated for a certain length of time, in a furnace of a particular type and a clay pot of a certain make, would usually yield a glass of such-and-such a color. Theophilus, in his twelfth-century discussion of glassmaking, for example, instructs that beechwood ashes must be used, and that different colors, from yellow to purple, could be produced by different numbers of hours heating the melt. Glass was, in other words, a perfect example of a product whose qualities depended utterly on a combination of skill, technique, and materials.[5]

Unlike many other important products, the raw materials that go into glass are not difficult to find. Theoretically, anywhere in which one can find sand of reasonable purity and plants for useful ashes, along with wood to fuel this heat-intensive process, is a candidate for glass manufacture. In history, however, glassmaking has been highly localized, depending on high quality sources of sand and, even more importantly, pure and readily available alkali sources (the word "alkali" comes from the Arabic *al-qili*, "ashes"). The location of materials was, however, not as important as the location of skill. In the late Middle Ages, European glassmaking came to be closely identified with one place in particular, Venice. Venice was located near good sources of useful sands and on trade routes for the Near Eastern sources of high-quality soda ash. The manufacture of good flux material, not unlike the later development of making saltpeter, required careful craftsmanship as well as good starting materials, and for centuries the regions at the far eastern end of the Mediterranean were recognized as the sources of the best soda. With its strategic location at the head of the Adriatic Sea and a seafaring and trading tradition that went back to the beginnings of the state, Venice was ideally located for the glass trade. When that was combined with a long history of commercial and cultural ties with the Eastern Roman, or Byzantine, Empire, which gave Venice good access to a long and distinguished tradition of making luxury glass, the conditions were perfect for the emergence of an industrial preeminence that was to last for centuries, and which established a modern pattern for dominating a product through technical excellence, a tradition of innovation, and marketing savvy (and sometimes ruthlessness).

From the fourteenth through the sixteenth centuries, Venice exercised an extraordinary influence on European life and commerce, in large part through its command of particular technologies and products. The Renaissance was a period of widespread awakening in many aspects of European culture, and there were many centers of power, wealth, and influence, from a newly emerged Swedish state in the north to the extraordinarily wealthy imperial powers of Spain and Portugal in the southwest. To focus on Venice is not to say that this small Adriatic republic dominated the European economy or even the Italian peninsula—this was, after all, the period in which German mining, Flemish textiles, Swedish iron, and a variety of other trades and products made their influence felt far beyond their origins. In Italy itself, cities such as Florence, Genoa, and Milan were economic powers quite independent of Venice. But the Venetian story is nonetheless worth a closer look, for there one can see the emerging capability of technology to shape the dimensions of power and wealth, through a range of products from glass to ships.

On November 8, 1291, the Great Council of Venice decreed that all glass furnaces in the city must be moved to the island of Murano, well removed from the Rialto and the center of Venice. There were already glassmakers in Murano, and it is not clear why all production was moved there—possibly the danger of fire from the furnaces led city leaders to move a dangerous trade from the city. While this had been a common practice of civic protection from centuries, it is perhaps more likely in this case that it was part of a larger plan to organize city industries. Other trades were moved and concentrated in other parts of Venice. It was, in other words, a kind of industrial and urban planning. Already the glass trade was under the jurisdiction of an officially sanctioned guild, which set out dozens of regulations and standards, governing everything from who could be admitted to how disputes were to be settled and when furnaces could be operated (an annual vacation of several months was required for glassworkers). One problem addressed at this early date, and which concerned Venetians for centuries into the future, was that of workers migrating out of the city to practice their trade elsewhere. This was strictly illegal, but the records show that it could not be effectively prevented. At the least, the vacation—say, from mid-August to mid-January—gave workers both the opportunity and the incentive to travel abroad to work elsewhere. The value attached to the technical skills and knowledge of the glassworkers was highlighted by the efforts to keep them at home.[6]

While the Venetian glass industry was an important source of wealth and trade as early as the mid-thirteenth century, it was in the fifteenth and sixteenth centuries that the trade truly flourished. In particular, during this period the rising demand for luxury goods, not only in the thriving cities of Renaissance Italy, but across much of Europe, combined to provide the incentive and the resources to elaborate glassmaking technology to an extraordinary degree. The same thing happened, it should be added, in other cities in other arts and crafts, and so the Renaissance still appears to us, some five centuries later, as a period of exceptional innovation and artistry. In glassmaking, the innovations included new glass compositions, as well as new artistic effects.

The most important product was a clear glass known as *cristallo*. This was a colorless glass, free from defects, that could be used as the basis of more complex luxury effects. The term *cristallo* at first probably referred not to glass at all, but to rock crystal, or clear quartz, and the use of the word for a new kind of glass signified the importance of the clarity and transparency of the material—qualities that we take largely for granted in modern glass. *Cristallo* is a remarkable example of the significance and mechanism of improvement, for its composition was superficially no different from that of earlier glass, but it was the product of gradual refinement of old techniques. In particular, *cristallo* was made by very careful selection and treatment of raw materials. Instead of sand, the Venetian glassmakers insisted on white river pebbles (*cogoli*), found in only a couple of river beds in northern Italy and shipped to Venice. Instead of raw ashes, they refined the ashes by a laborious and time-consuming process (rather like the making of saltpeter) that yielded a pure alkaline flux (in this case, sodium hydroxide or carbonate). Finally, the clarity of the glass was further enhanced by the careful addition of small amounts of manganese oxide, which was known as a "decolorizer" for fine glass. Manganese, we now know, tends to react with any iron in the silica-alkali mixture, reducing the iron's tendency to turn the glass blue or green. From *cristallo* there could be made a host of other luxury products, often featuring filigree, jewel-like coloring, or other decorative effects—the kinds of things that Venetian glass is still famous for today. It was also the perfect glass for lenses.[7]

Glass was by no means the only important manufacture in Venice, nor the sole trade in which innovation was seen as important. Another luxury item that is also still associated with the city is silk, and this too was a trade in which the Venetians sought to carve out a reputation and a market, in large part through technical excellence. As has already been remarked, the manufacture and trade of textiles constituted the single most important industry in medieval Europe. Specialization developed early, and, along with it, a sophisticated system of marketing that depended on a transportation and communications network that not only extended throughout western Europe but also had important linkages far to the East. In the case of silk, of course, these linkages in fact extended as far as China. It was, after all, from Venice that the Polo brothers set out along the silk road in the mid-thirteenth century for the great adventure that culminated in Marco Polo's fabulous reports of the Chinese empire. By the High Middle Ages the Europeans ceased being totally dependent on eastern sources for their silk cloth (although the raw materials might have to be imported). As often happened, the craft was initially fostered in Europe by the expulsion of craftsmen deemed ethnically undesirable by rulers. Some of these, primarily from Greece, settled in Italy before the thirteenth century, and by the fourteenth century silk manufacture was established in several Italian cities, including Venice. It is important to remember that the term "silk" actually covers a wide range of cloths, from the purest and finest brocade (which typically included threads of gold and silver) to mixed cloths, in which silk was combined with linen, wool, or cotton. Indeed, the creation and control of varieties of silk was an important aspect of the technical innovation in the trade. Different mixtures of different qualities of thread, woven in different fashions, with varying levels of fineness or care, yielded a wide range of products. Unlike the glass trade, in which the Venetians long held a special and privileged place, the silk trade was widespread throughout Italy and other Mediterranean areas (and, later, into northern Europe as well), and so special care had to be taken to differentiate the Venetian product from others. This effort, overseen by guild and state authorities, fostered innovation, both in different mixtures and grades of cloth and in techniques for winding, spinning, weaving, dyeing, and other processes. Here, too, the experimental spirit was encouraged by commercial growth, increased flow of information, the potential of competition, and a receptive attitude on the part of merchants, craftsmen, and the state.[8]

The significance of Venetian technical prowess lay in large part in the republic's capacity to take full advantage of it by controlling trade. At just the point, in the late thirteenth and early fourteenth centuries, that Venetian craftsmen began to acquire significant reputations, the technology of ships and shipbuilding began a long transformation, one that was eventually to give the Europeans a mobility unrivaled in the rest of the world. Before the fourteenth century, the typical Mediterranean ship was a "round ship," being no more than about three times as long as it was broad, had two triangular sails, on two masts, and two or more decks, with a displacement of perhaps two hundred tons (about the size, for comparison, of the *Mayflower*, of three centuries into the future). These ships were steered by side-oars—essentially a long oar stuck out from one side of the rear of the ship, operated just as one would steer a canoe today. While the triangular (*lateen*) sails allowed the ship to sail to some degree into the wind, this capacity was quite limited and so most ships sailed almost exclusively to the windward. Sailing vessels were supplemented, particularly for naval action, by galleys—much longer ships (with a length, say, eight times the breadth) propelled largely by oarsmen (figure 6.1). Galleys could be relatively fast (at least for short periods of time) and maneuverable, but they had little cargo capacity (with only one, very low, deck) and were obviously very expensive to operate. They could provide the crucial element in naval engagements, however, and so were indispensable in the picture of naval power. In Venice, the oarsmen were not slaves, but were typically citizens fulfilling their military obligations.[9]

At the end of the thirteenth century, the Venetians devised a new kind of ship, the great galley, which extended their shipping capabilities. When this ship was combined with new policies that directed almost all shipments to be sent out in protected convoys, the security and power of Venetian commerce exceeded that of any other Mediterranean state. At the same time, the Venetians experimented with modified round ships, based on northern styles that used square sails. The resulting ship, the "cog," gave merchants an effective cargo-carrying vessel that required far less manpower than the galleys. The fleets of Venice were the basis for the greatest commercial power the European world had yet seen. Due to the larger size and weight of the great galleys, they came less and less to rely on oarsmen, and instead used the sails

Figure 6.1
The galley was the most important kind of ship for medieval and Renaissance Venice. The ship shown here is a war galley smaller than some commercial vessels.

that had originally just supplemented rowing in favorable winds. It was not the size, but the complexity, stability, and versatility of the great galleys that made them formidable weapons of both trade and naval power. With a crew of about two hundred men, who could man the oars, wield weapons, or handle the sails and lines, a galley was a complex and versatile machine. Building and equipping both galleys and cogs became a task of the Venetian state, and in creating the capacity for this, Venice pioneered an entirely novel concept of making things—one that borrowed from the complex and multistage task of erecting a building but extended it to fabricating machinery and systems. In addition, the ships of the Venetian Republic presaged the further development of vessels that resulted, at the end of the Renaissance, in the caravel and other modern forms that gave the Europeans the unique capacity to embark safely on voyages over the oceans of the world (figure 6.2).

Two important innovations were introduced in the thirteenth century that spurred the development of the new ships, innovations that enhanced the steering and navigation of ships to such an extent that the larger investment represented by the new designs made good economic sense. The side-oars that had been used for steering were replaced, gradually, by the sternpost rudder, an innovation from northern Europe that made it much easier to steer a ship, particularly in heavy seas. The rudder

Figure 6.2
The caravel emerged through combining northern and southern ship forms and was the most important vessel in the European voyages of exploration.

protruded directly from the center of the stern, was hinged and connected to steering gear that allowed a single man to keep the ship on course without undue effort. Even more dramatic was the introduction of the magnetic compass, which had been known for centuries to the Chinese. The north-pointing compass made it possible for the first time for a navigator to determine and maintain bearing, even in the cloudiest weather. The popularizing of the compass and the devising of appropriate mounts that would allow the device to hang freely aboard a pitching vessel took place in the thirteenth century, and by the fourteenth century navigational charts of the Mediterranean were in widespread use, fostering an expansion of shipping through all seasons of the year, even those in which the clouds obscured the sky for days or weeks at a time.

At the beginning of the twelfth century, the Venetian state established a place for outfitting ships of the fleet, called the Arsenale, from the Arabic for "place of manufacture." For two centuries, this was a modest establishment, with a small number of workers who were largely occupied with inspecting and reconditioning ships after they returned from voyages. But in the early fourteenth century the Arsenale was transformed by the republic's efforts to equip itself with the new great galleys and cogs, and the size of the shipyard grew four-fold by 1325. This expansion was observed with wonder by none other than Dante Alighieri, who found in the place some inspiration for his vision of the *Inferno*:

As in the Arsenal of the Venetians
Boils in winter the tenacious pitch
To smear their unsound vessels ov'er again,
For sail they cannot; and instead thereof
One makes his vessel new, and one recaulks
The ribs of that which many a voyage has made;
One hammers at the prow, one at the stern,
This one makes oars and that one cordage twists
Another mends the mainsail and the mizzen. . . .[10]

Over the next couple of centuries, the Arsenale became not only the primary guarantor of Venice's naval power, but more importantly the primary shipyard for its merchant fleet. The number of workers at the Arsenale varied considerably from year to year as the needs of the state fluctuated, but it would not have been unusual for several hundred carpenters, sawyers, caulkers, and other to have been all at work at the same time in the early fifteenth century.

By the sixteenth century, the total number of craftsmen on occasion reached into the thousands, and the Venetian Arsenale became an industrial wonder renowned throughout Europe (figure 6.3). While, of course, large building projects had brought together masses of workers ever since the construction of the pyramids of Egypt or the Great Wall of China, the Arsenale represented not construction but industry—the sustained production of very complex artifacts by coordinated, economically driven, effort. Galleys and cogs were put on the ways of the shipyard sometimes by the dozen, and the management of manufacture became a central concern of the state. Indeed, the complexities and economic demands of the Arsenale promoted the development of modern accounting techniques, which are among the most valuable and enduring of Renaissance innovations. New manufacturing techniques, such as the use of standardized parts, appeared as the Arsenale adopted what amounted to an assembly-line approach on those occasions when the state ordered great numbers of new galleys in the mid-sixteenth century. The division of labor, through the assignment of specialized tasks to craftsmen (as Dante observed), came to be widespread, although in these preindustrial days work discipline was never of the rigid form it was to take in nineteenth-century factories, and workers' rights were broadly protected by state statutes and custom.[11]

The involvement of the Venetian state in the building not only of naval vessels but also of the city's merchant fleet was typical of the active role played by the government in commerce and technology. This was not exceptional in medieval Europe—

Figure 6.3
Water Entrance to the Arsenal by Giovanni Antonio Canale (Canaletto), painted about 1732. For several centuries the great Venetian shipyards were key attractions of the city. By kind permission of His Grace the Duke of Bedford and the Trustees of the Bedford Estates.

governments and rulers everywhere had a hand in fostering trade, promoting manufactures and other sources of wealth, and in regulating work and production. Especially for urban crafts, oversight by state authorities was widely accepted. Sometimes this was done, as in the Venetian glass trade, through the structure of guilds and associated regulations, and sometimes it was done through controls on buying and selling or on labor practices. Little has been said by historians about the implications of this for technological innovation. It has often been assumed that the hand of state regulation lay heavily on would-be innovators, but this is perhaps more a prejudice of thinking heavily influenced by the early industrial preaching of such thinkers as Adam Smith, who were so adamant in their insistence that little good could come from state intervention in the market. In the medieval and early modern period, at least, the evidence would suggest that innovation was encouraged and promoted by at least some governments—particularly those, such as at Venice, who long made it their concern that the state thrive on reputable craftsmanship and effective trade and transport.

Not only the state could be a constructive and encouraging influence for technological improvement, but so also could be that other great center of medieval power, the Church. Its generally positive role in promoting advances in architecture and in the application of waterpower and windpower has already been noted. In the late Middle Ages there appeared, almost certainly first in a church or monastery, one of the most significant of all Western contributions to technology, the mechanical clock. The clock appears at just about the same time as spectacles, but we do not have the good fortune of specific references to its invention, so its first appearance is a matter of considerable debate and uncertainty among scholars. There is widespread agreement, however, that the first devices that incorporate the key mechanisms of the clock appeared in a religious setting. This is in part because the slight evidence and remains that we do have point to its use in churches, and in part because monasteries and cathedral chapters are about the only institutions of premodern Europe that show any sustained interest in timekeeping. While keeping track of the hours of the day was an ancient practice, it was one that depended almost solely on astronomical (largely solar) observation or on crude instruments such as water or sand clocks. The times of work, meeting, and play were not kept by instruments but by the rhythms of nature—the rising of the sun, the stirring of cattle, the shortening and lengthening of shadows, the grumble of the stomach. This was, in fact, quite adequate for the purposes of daily life. Work was done to fill the hours between sunrise and sunset, to accomplish the tasks of each day, with a pace determined by need, opportunity, and nature's own demands. Keeping hours, much less minutes, had little or no place in such a world. The only exception that we can find—and even that is largely by inference—is in religious institutions, where by the Middle Ages the rules of the monastic orders required a schedule of worship that went outside the bounds of the easily measured day. Following ancient Jewish tradition, the groups of pious men who gathered together to follow the rule of St. Benedict and other disciplines adopted extensive requirements for worship at intervals during day and night. One of the key duties in the monastery was keeping track of the praying ("canonical") hours—traditionally signaled to the monks, from the High Middle Ages, by bells. It was almost certainly as a bell-ringing machine to assist in these duties that the first mechanical clocks appeared, probably in the late thirteenth century.[12]

We really know nothing about these early devices; the first ones that we have any record or remains of come from the mid-fourteenth century, by which time the art of clockmaking had clearly made substantial advances. These clocks, such as the tower clock of Norwich Cathedral, in eastern England, or the astronomical device of Richard of Wallingford, built for a church in St. Albans over the course of almost thirty years, were instruments that went well beyond bell-ringing for monks or even timekeeping itself. They incorporated astronomical dials and, in some cases, complicated machinery to keep track not only of the passage of the sun (which is what any ordinary clock is doing) but other heavenly bodies, such as the moon or the planets. It seems quite reasonable, therefore, to conclude that simpler mechanical timekeepers preceded these devices by at least a generation or two, perhaps back into the mid-thirteenth century. Part of the complication in dating the clock is an accident of terminology: the common Latin term used for the mechanical clock, *horologium*, is the same word that was applied to older timekeepers, such as water and sand clocks, and thus documentary allusions alone are ambiguous. In addition, these latter devices could, by the thirteenth century, be connected to clever mechanisms in order to ring bells at an appropriate point, obscuring yet further the meaning of early documents that refer to "alarms" or bells. But by the early fourteenth century there are enough references to suggest that mechanical timekeepers were no longer strange novelties. Dante, once again, provides a fine example when he speaks in the *Paradiso* of

> . . . the wheels in the works of horologes
> Revolve so that the first to the beholder
> Motionless seems, and the last one to fly. . . .[13]

It is hard to imagine a more evocative image of the train of gears in a clockwork mechanism.

The great medieval contribution was making a self-acting, self-regulating machine to keep time. At the heart of the early clock lay a particularly clever and momentous mechanism—an escapement—that arguably marks the single most important mechanical invention to emerge from the Middle Ages. Every clock must have some means of dividing time into regular, measurable intervals. Until the medieval clock, this could only be done by grains of sand or drops of water. Making a device that would take a continuous mechanical force and break it up in this way was an astonishing achievement. The first escapement was probably the verge-and-foliot (figure 6.4), in which the steady force provided by a falling weight and a pulley turned a toothed wheel (the crown wheel), which engaged a vertical rod with two protruding pallets (the verge). The pallets were placed on the verge so that when the toothed wheel engaged one (say, at the top of the wheel), the other pallet (at the bottom) was pushed out of the way. When the force applied to the wheel pushed the top pallet out of the way, the verge turned to put the bottom pallet into position to stop the wheel's progress again, but only after the desired interval of time. This interval was governed by the inertia provided by a crossbar (the foliot) at the top of the verge. Weights on either end of the foliot could be moved to adjust this inertia, and thus the length of the interval. In this way, the continuous force of the falling weight was carefully measured out, with the back-and-forth motion of the verge-and-foliot halting the continuous progress of the crown wheel at just the regular interval desired. By the mid-fourteenth century important variations had already appeared, such as the crown wheel used in the very elaborate "astrarium" of Giovanni de' Dondi, whose device is the first for which we have a complete description.[14]

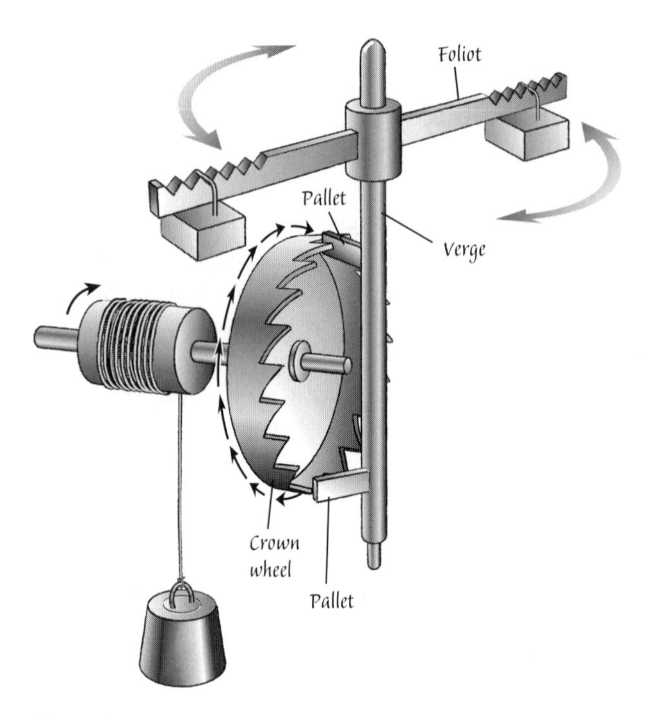

Figure 6.4
The verge and foliot escapement was the fundamental mechanism of the first mechanical clocks in the West. The regular engagements of the pallets on the verge with the crown wheel provided the beat of the clock. The rate could be adjusted by moving the weights on the foliot arms.

De' Dondi's construction was much more than a clock—indeed, its keeping the time of day was almost incidental to the amazing array of astronomical functions it incorporated (figure 6.5). It kept track of almost a dozen motions or events, including the positions of the five known planets, the sun, and the moon as well as the fixed and the movable feasts of the church. Its designer and builder was a physician in the city of Padua, not far from Venice. Giovanni de' Dondi was, in fact, a professor at the famous university in that city, and was known for his mastery of rhetoric, philosophy, and astronomy. It is clear that he was also, like his physician father, Jacopo, before him, a master mechanic. He built his astrarium working over about sixteen years, from 1348 to 1364, largely as a contribution to science, a kind of teaching device or a means of appreciating the divine order of the regular motions of the universe. He actually scorned its timekeeping capacity, remarking in his instructions that this was simply an "ordinary clock," and thus needed no detailed description or explanation. De' Dondi's creation became famous in his lifetime, eventually ending up in the hands of the rulers of Milan (who invited de' Dondi to come to teach at the University of Pavia), and eliciting comment and written appreciations by a

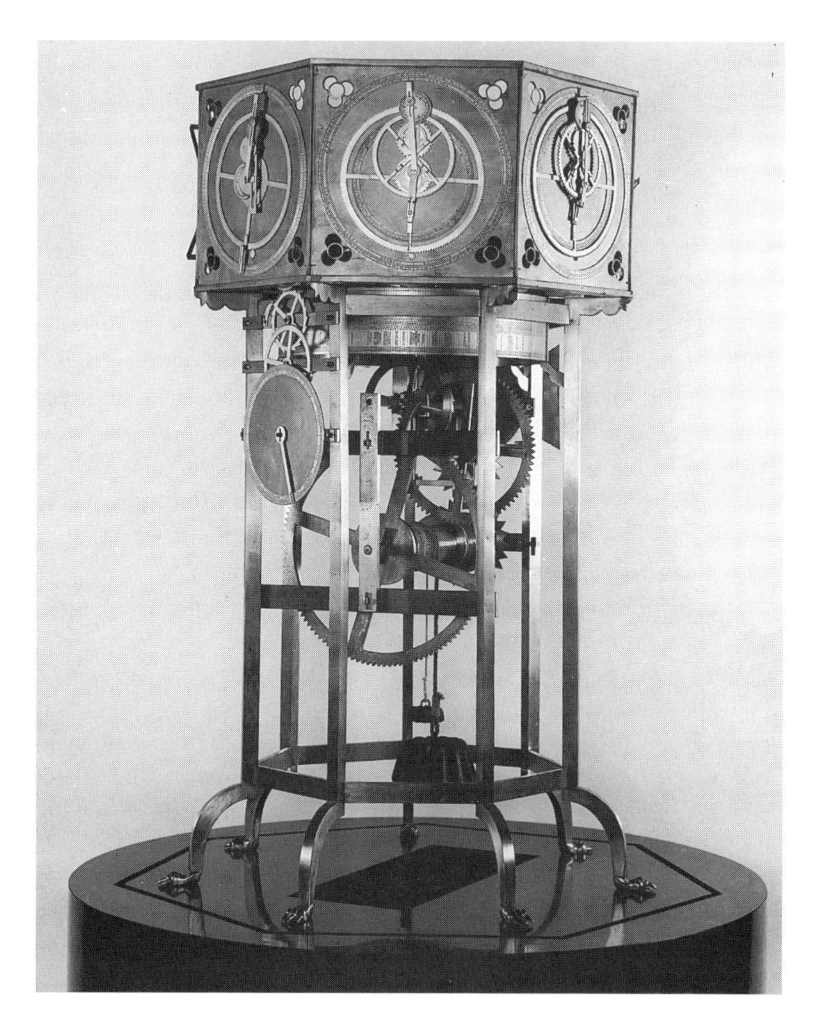

Figure 6.5
Giovanni de Dondi's *Astrarium* (replica, National Museum of American History, Smithsonian Institution). De Dondi's detailed drawings of his machine make it possible to reproduce it with some accuracy. Smithsonian Institution photograph.

number of observers. The historical significance of the astrarium lies largely, in fact, simply in our knowing so much about it. The records made of the device, both at the time of its construction and in subsequent writings, constitute a remarkable change in the way in which technological knowledge was held in learned esteem. These records also, by their thoroughness and accuracy, tell us a great deal about emerging notions of technical communication, through careful, proportioned drawing as well as detailed textual appreciation—a novel means for capturing improvement that was to be of enormous significance in the future. The clock itself has not survived, but the descriptions of its complex workings, its gearing—including elliptical gears to handle the irregular motions of Mercury and the Moon—and its extremely high level of workmanship have allowed modern scholars to replicate it. It stands as remarkable testimony to the emerging value attached to mechanical genius in the West at the end of the Middle Ages. In the next two centuries, there was to be an outpouring of contrivances—not only clocks but also automata, toys, and scientific instruments that proclaimed the new mechanical philosophy well before the philosophers themselves grasped it in the seventeenth century.[15]

After de Dondi's astrarium was acquired by Duke Gian Galeazzo Visconti, it rested in a prominent place in the ducal library in the castle of Pavia for more than a hundred years. There it continued to attract attention, and records show that the dukes had to call on skilled clockmakers from time to time to maintain or restore the instrument. One of the visitors to the library, about a century after the astrarium was placed there, was an ambitious and talented artist and engineer (as we would have called him) from the small town of Vinci, Leonardo di Ser Piero. In 1490, Leonardo da Vinci, as we know him, was working for Lodovico Sforza, regent and later Duke of Milan. At Pavia he saw the famous astrarium and sketched at least one of the complex dial and gear mechanisms, that for the movements of Venus. This drawing is to be found today as just one of the hundreds of images that Leonardo committed to his notebooks. Collectively, these images represent one of the most astonishing records of artistic and technical genius in history. But beyond their testimony to Leonardo's distinctive talents and fertile imagination, they also are important reminders that at this time and place—northern Italy in the fifteenth century—there occurred an extraordinary, perhaps unique, joining of art and engineering that signified the new status and power of technology in the West. The artist-engineers of the Renaissance had, it may be argued, only modest influence on technical practice. Leonardo's great notebooks, after all, were just that, notebooks sketched largely for his own intellectual purposes and not public documents. But the presence and practice of the artist-engineers among the military, political, cultural, and economic ferment of the age were eloquent and often very public, indicators of the new place of technological creativity in the European world.[16]

Figure 6.6
The Duomo (Cathedral), Florence, with Brunelleschi's octagonal dome. Library of Congress photograph.

When Leonardo first arrived in Florence in 1469, at the age of sixteen, he entered the workshop of Andrea Verrochio, in order primarily to apprentice as a painter and sculptor. At this time, Verrochio was assisting in putting the final touches on the single greatest work of architecture that the city was ever to see, the great *duomo*, or cathedral. The cathedral church of Santa Maria del Fiore was, in the tradition of the great Gothic churches of northern Europe, not only a great work of art, but also a triumph of engineering (figure 6.6). Its primary distinction was its great dome, designed by and constructed under the supervision of the first of the heroic artist-engineers, Filippo Brunelleschi. When the city of Florence sought to finish the cathedral that had been started at the end of the thirteenth century, a council of advisors was called together to solicit and make proposals. The call for this council explicitly sought the advice of *ingegneri*, or "engineers," a term hitherto limited to military builders. Brunelleschi proposed and won backing for a structure of astonishing originality and ambition. Only two large domes had been used before in the West—the Pantheon in ancient Rome and the great church of Hagia Sophia in Constantinople. There was, therefore, no contemporary body of knowledge and practice for Brunelleschi to draw upon for either design or construction. In addition, he chose to accept two additional difficulties—to adopt the octagonal plan already in the cathedral's design for the shape of the dome itself, and to build the structure without wooden centering (thus saving considerable time and materials, if it could be done). Not only did he, with the assistance of a sometime rival, Lorenzo Ghiberti, complete a dome that still astonishes by its height and girth (the rise, not counting the lantern on top, is 105 feet, 3 inches—compare to Rome's St. Peter's, for example, at 91 feet, 10 inches; the mean diameter at the dome's base is 143 feet 6 inches, compared to, say, the U.S. Capitol dome at 95 feet or St. Peter's at 134 feet), but he also did so in a scant fifteen years. There is no more fitting symbol of the Renaissance's capacity for heroic engineering than Brunelleschi's accomplishment.[17]

A key to the success of Brunelleschi's ambitious project, widely recognized at the time, was the use of large and capable lifting machinery. Some of this machinery was still in use when Leonardo arrived, well after the dome itself was completed (in 1436), and he made careful sketches. Others, such as Lorenzo Ghiberti's nephew, Bonaccorso, and an engineer from Siena, Mariano di Iacopo (known as Taccola), also made careful drawings of these devices. Together, these renderings show not only the sophistication of Brunelleschi's own mechanisms, but, just as important, the new capabilities for capturing technical improvements by drawings. Brunelleschi was himself well aware of what these new capabilities might mean; he thus sent his own drawings of the parts of his giant crane to different shops, so that the secret of the entire device might remain safely his. For others, however, these drawings became important products in their own right, prepared in careful manuscripts, finely illuminated, and dedicated to wealthy patrons or others who might be of assistance to their artist-authors. Taccola, of Siena, was a contemporary of Brunelleschi and even reported in one of his books a dialogue with the Florentine about the difficulty an inventor had in getting credit for his own work. While Taccola did not display the extraordinary range or the deeper artistry of Leonardo, his work, in Latin, showed the deep learning of some of the artist-engineers, as well as his own willingness to speculate about the possibilities of novel machines and devices. His work, and that of his younger Sienese compatriot, Francesco di Giorgio, a contemporary of Leonardo, also shows that the artistic and technical skill of Leonardo, while unmatched in scope and depth, was to be found in some measure in other men of the day.[18]

Leonardo da Vinci himself has come to represent the paragon of "genius." The very notion of "Renaissance man," while not exclusive to Leonardo, is certainly most aptly applied to him. This is due not only to the deeply humanistic message of his art, but also to the great range of interests we find recorded in the notebooks. It is estimated that less than one-third of his written notes and drawings have survived, so that the true scope is only hinted at in the remainder. These include not only drawings of machinery (figure 6.7), but many depictions of fortifications and other military works, plans for hydraulic projects, such as canals and water-supply systems, studies of optics and other aspects of light and shadow, famous anatomical drawings as well as depictions of a great range of birds and beasts, from embryo to old age, geological and astronomical drawings, and theoretical studies of everything from mechanics to urban plans. His famous depictions of flying machines represent the fullness of his imagination, but they also represent the extraordinary extent to which he made every effort to attach his speculations to what he could truly see and understand about the world, both natural and artificial. Many drawings remind us that Leonardo was, above all, a practicing engineer, employed by a series of powerful men who needed his talents to devise and oversee practical works, from waterworks to fortifications. His painting and sculpture tended, except toward the end of his life, to be done on borrowed time, when he could find the prized commissions or patronage. The artistry, however, was by no means irrelevant to his technological work. This was clearly an age in which the boundaries that we take for granted that separate the possible realms of human creativity simply did not hold much meaning for those with skill and imagination. Just as Leonardo sought to understand the workings of the human body by comparing them with machinery, so it made perfectly good sense for him to attempt to improve his machines by comprehending the patterns and mechanisms of nature. The ability to delineate this knowledge by line and pen was integral to Leonardo's notions of both understanding and creation.

In the West, after the fifteenth century, technical understanding was no longer confined to what could be described in words or executed with the hands. The capacity to put down on paper pictures that would enable the skilled artisan to replicate another's work or reported improvement enlarged enormously not only technical communication but also the tools of the technological imagination. Brunelleschi himself is given credit for introducing geometric perspective drawing ("linear perspective"), although none of his own drawings survive. By the time of Leonardo's birth in 1453, the rules proposed by Brunelleschi had been widely disseminated, and artists and engineers alike came to rely on them for effectively conveying information. For the first time in history, the preservation of technical knowledge in its fullness would not depend on the survival of the artifact or the always shaky capacity of language to capture form, structure, or function in words.[19]

The additional value now placed on technical information is also evident from one other contribution of the fifteenth century, one that takes us back to Venice. On March 19, 1474, the Venetian Senate passed the first general law explicitly directed at promoting technological novelty:

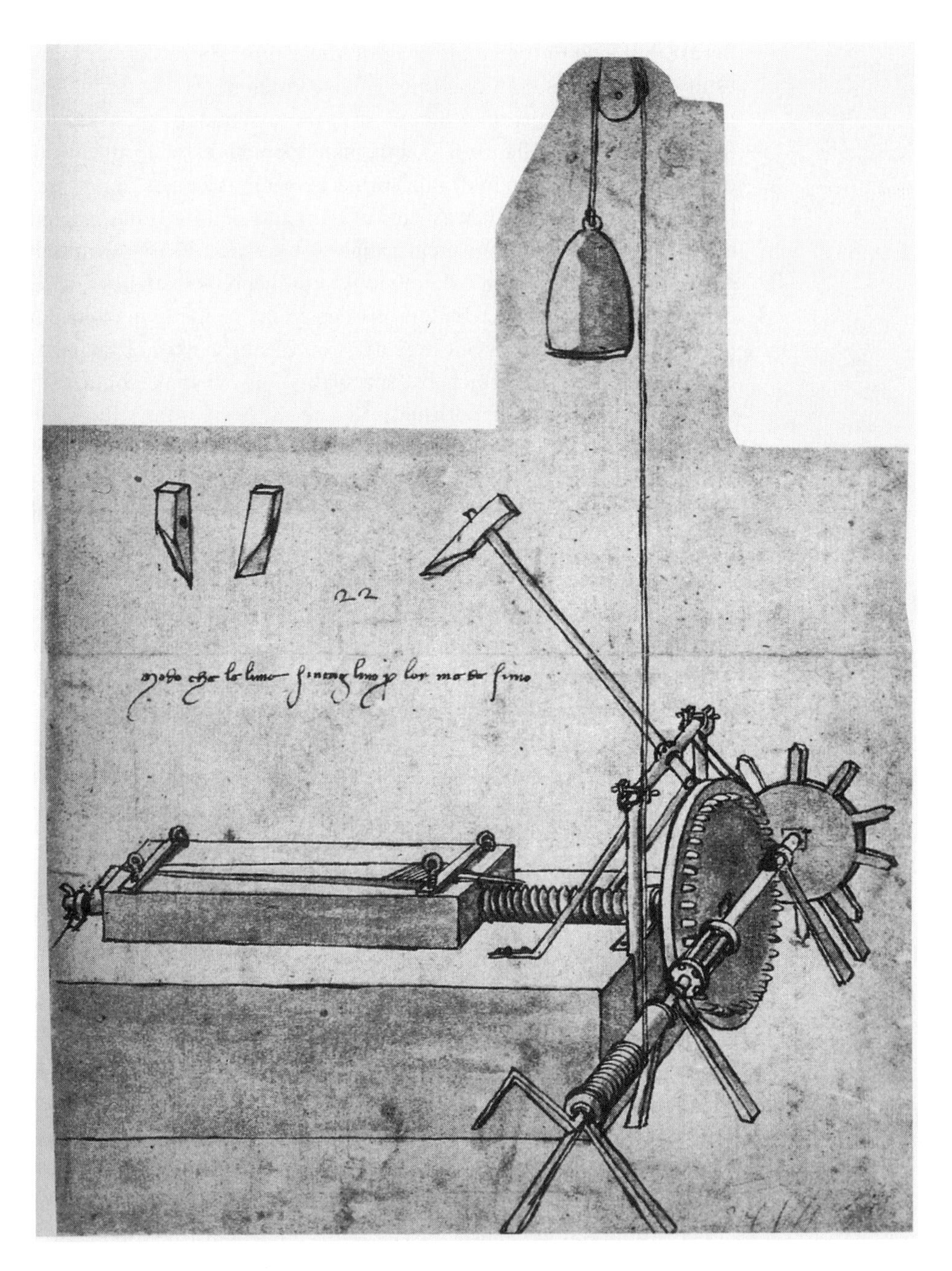

Figure 6.7
Leonardo Da Vinci's drawings of machinery were never published, but his notebooks have become some of the most valuable sketches ever made. This is a mechanism for cutting files. Burndy Library.

> There are men in this city, and also there come other persons every day from different places by reason of its greatness and goodness, who have very clever minds, capable of devising and inventing all manner of ingenious contrivances. And should it be legislated that the works and contrivances invented by them could not be copied and made by others so that they are deprived of their honor, men of such kind would exert their minds, invent and make things that would be of no small utility and benefit to our State. Therefore the decision has been made that, by authority of this Council, any person of this city who makes any new and ingenious contrivance not made heretofore in our Dominion, shall, as soon as it is perfected so that it can be used and exercised, give notice to the same to the office of the Provveditori di Comun, it being forbidden up to ten years for any other person in any territory and place of ours to make a contrivance in the form and resemblance thereof without the consent and license of the author.[20]

This "first patent law" was not, in fact, the first instance of official legal encouragement of invention, nor even of granting the privileges of a patent to an inventor. This had long been done in numerous cities and states, by many rulers and authorities. In Venice itself, records show that 150 years earlier, the Grand Council gave support and privileges to a German millwright for "ingenious" millwork, and other grants were made later for dredging machinery. Brunelleschi convinced the commune of Florence in 1421 to grant him exclusive privileges for a new boat he devised for hauling the heavy loads of stone on the Arno that his building required. The grant reported that the inventor claimed his ship would make hauling more economical, but that he "refuses to make such a machine available to the public in order that the fruit of his genius and skill may not be reaped by another without his will and consent, and that, if he enjoyed some prerogative concerning this, he would open up what he is hiding and would disclose it to all." Here is to be found not only the essential logic of a modern patent, from the inventor's point of view, but also the necessary requirement from the public policy perspective, which is that, in return for the privileges, the inventor must disclose fully the essential details of the claimed improvement.[21]

What was distinctive about the law of 1474 was its statement of a general policy, supported by a clear procedure. Up to this time, anyone could petition a ruler or state authority to grant an exclusive privilege, for making a product, for selling a commodity, or for using a technique. Such privileges were generally viewed as prerogatives of the state, and they might, in fact, be used as patronage or favors or as sources of state income (monopolies were ancient sources of revenue and equally ancient causes for grievances by common folk). From the point of view of public policy, such grants of privilege were seen as appropriate inducements for craftsmen to introduce new products or trades, thus reducing the state's dependence on outside sources. Over the centuries, for example, glassmakers from Murano were lured to other towns, and even as far away as England, by the promise that they would be given exclusive right to make their product in their new home, at least for a limited period of time. The privilege was a reward for bringing something new and useful into the realm, but it was not, usually, a reward for creativity, much less a "right" to the benefits of intellectual property. In the fifteenth and sixteenth centuries, however, the attitude began to emerge that originality deserved special attention and encouragement from the state. In the seventeenth century, this notion merged into ideas of property and natural rights that gave birth to the modern concepts of the rights of authors and inventors to benefit from their creations.

7 Types of Change

> . . . we should notice the force, effect, and consequences of inventions, which are nowhere more conspicuous than in those three which were unknown to the ancients; namely printing, gunpowder, and the compass. For these three have changed the appearance and state of the whole world: first in literature, then in warfare, and lastly in navigation; and innumerable changes have been thence derived, so that no empire, sect, or star, appears to have exercised a greater power and influence on human affairs than these mechanical discoveries.[1]

This oft-quoted passage from Francis Bacon's *Novum Organum* of 1620 is a bit of a shocker to the student of Renaissance history. The shock comes from the modernity not only of Bacon's message but also of his style. While Bacon's great plea for new ways of acquiring knowledge and understanding was nonetheless written in scholar's Latin, it was delivered with a tone and authority that made his rejection of ancient ways unmistakable. Furthermore, even the passage of almost four centuries and a much deeper knowledge of the changes of the late Middle Ages and the Renaissance has not altered the deep insight of Bacon's choices for agents of change. Gunpowder and the compass, as we have seen, were indeed key inventions for bringing the medieval era to a close. These were not, however, themselves European discoveries, but rather much appreciated borrowings from the East. Bacon's third great agent, on the other hand, and the most recent one of his list—printing—was, despite Asian antecedents, a distinctly European creation, and arguably the greatest of them all.

Equipped with their new ships and navigation skills and instruments, and armed with their new weapons of iron and fire, the Europeans were able to make their influence felt all over the globe and were themselves forced to redefine the sources and forms of temporal power. The new technology of printing, however, marked a fundamental break in European culture itself, a shift sometimes described as the transition from "scribal culture" to "print culture." The world before printing was one in which the whole pattern of learning, communicating, and storing information was defined by what could be written down or drawn or spoken in a singular and immediate fashion. The world after printing was one in which the repeatable message or lesson was the one that carried authority and influence, in which there was an explosion of words and pictures spread through all Europe in the space of barely two generations. The two cultures—scribal and print—did, in fact, coexist for that period of time, but it was a time in which there developed, for the first time in history, a widespread consciousness of change and discontinuity.

The invention of printing or, more precisely, of printing with movable type, marked another transition in the history of technology—the movement from the anonymous inventions that characterized everything significant up to that point to the identifiable (although still contestable) inventions that we associate with our modern awareness of technological change. But it is important to remember that this was a transitional invention, not a modern one, and thus, once again, we simply cannot say for certain when or where printing with movable type first appeared. We can, however, come very close.

Sometime in the years just before 1450, there appeared in the German city of Mainz a shop from which there came printed items—calendars or religious tracts, and then, eventually, a complete Bible. Tradition, backed by incomplete records, ascribes the work of this shop primarily to a gold- and silversmith, Johannes Gutenberg. Printing, however, was a complicated invention—one of the first complex systems, in fact, that had to be all worked out in some detail before making any technical sense. It is not too likely that such an invention sprang from the work of a single man, perhaps not even a single shop. It will not do, therefore, to worry too much about the Gutenberg tradition. Instead, it is important to understand just what the invention consisted of, and what this communicates about the new Renaissance way of looking at technology and its improvement.

The technique of reproducible images or lettering is as old as writing itself—the oldest examples of writing that we have are symbols impressed repeatedly into clay. That the same form—a shaped stick, a seal, a thumb—could be used to produce the same "message" in different pieces of clay was clearly apparent from the first. More important, the possibilities of reproduction in this fashion were exploited early, most obviously in coinage. Coins generally represent a combination of image and words that is stamped into a piece of metal (usually) to mark that metal's value and usage. For thousands of years this kind of reproducible image, in metal or clay, joined by more evanescent versions, such as wax seals, represented the functions and limits of reproducibility. In the Middle Ages there began to appear a few techniques and products that suggested how reproducibility could be expanded. The best example was the making of printed cloth. Actually, the use of wooden blocks with a cut design on the face to stamp patterns in plaster or to apply pigment to cloth appears to have been known to the Romans, and with the expansion of the textile trade in medieval Europe, it's no surprise that this means of decorating fancy cloth should become widely known.[2]

There is quite some distance, however, between using large blocks of wood to apply designs to cloth and the use of hundreds of small pieces of cast metal to print text. The older technique would seem to have few implications for the revolution to come, but the linkage was supplied by a medieval development already mentioned—paper. By the early fifteenth century paper was widely available, although still by no means cheap. The spread of paper throughout Western Europe is closely associated with the spread, first, of more readily available manuscript books and pamphlets, and then with the reproduction of woodcuts, playing cards, and other forms of popular imagery. Papermaking is an excellent example of what can be called an "enabling technology." The alternatives to paper—parchment and vellum, primarily—were both too expensive and technically unsuited for making rapid reproductions. Printing was by no means "caused" by paper, but it would have been unthinkable without it.

"Xylography," the printing of images and words with wooden blocks, became an important trade by the early fifteenth century. Woodcuts, at first most typically of religious images but later of a wide range of pictures and some text, even became cheap enough to be afforded by illiterate laborers. Printed playing cards, which had been seen by travelers to Asia, were another popular form of woodcut, and Venice, once again, used its strategic place between East and West to establish itself as the early center of the trade. In this case, however, the trade soon migrated northwards, establishing itself in German cities, such as Nuremberg and Augsburg. It may have been in these places that another key element of the printing system was devised—an oil-based ink, using lampblack or powdered charcoal as pigment. The inks used by scribes, typically water-based with a gum binder, simply would not have worked with metal type. In addition, the woodcut shops of Germany probably also devised the standard method of applying the linseed oil–based ink to the printing surface, using a pad of stuffed leather that had been dabbed into the thick ink.

By the middle of the fifteenth century, therefore, there were a number of the key elements of the printing system in place. Another element that is hard to trace specifically but which was known in different forms rather widely was the press. The use of a screw to facilitate bringing together two flat wooden surfaces had been common for many years, found in everything from wealthy households, where they were used for keeping linens smooth, to papermills, where stacks of wet paper had to be tightly pressed before drying, to oil factories, where olives were pressed, or vineyards, where presses had been used for grapes for a thousand years. To be sure, with the advent of printshops, the design of the press underwent considerable modification, primarily to speed up the work, but the principle of the screwpress was no mystery to any observant mechanic.

The part of printing that posed the greatest difficulty—for the fifteenth century mechanic as well as for the modern historian—was the type. Carving an image into a block of wood and then spreading ink on the carved surface to use it for printing was not a complicated idea, but it was a laborious technique, dependent for its success on the skill of the individual cutter and requiring much time for its execution. It also produced a printing surface that not only had limited capacity for detail but was easily worn or broken. Xylography was fine for pictures, and it was, in fact, an important source of published images for centuries to come, but it was never seen as an appropriate technique for reproducing more than very small amounts of text—captions or religious instructions that might be attached to an image, but little more. Finding a workable alternative to the woodblock required a series of inventions and improvements that not only produced a printing system that transformed the literate culture of Europe, but also represented perhaps better than any other single technical development the emerging capacity of the Europeans for manipulating materials and machines in complex and useful ways.

The printing that emerged in the fifteenth century was what is called "relief printing," which means that it relied on a raised printing surface, with the image defined by forming a flat surface on which the ink is placed for transfer to paper, surrounded by lower, non-inked space defining the boundary of the printed image. To use this technique for a text page consisting of hundreds of individual letters, each rather small in size, necessitated a practical means of making letters with rigidly standard heights and depths, while at the same time allowing for considerable variation in widths (that is, providing for a **W** that takes up as much space as several **i**'s). It is clear from the earliest examples of printing that the creators of the technology did not feel free to depart from the norms established by manuscript practice, so not only did letters have to vary in width, but accommodation had to be made for evenly justified margins, columns, enlarged initial capitals, hyphenation, diacritical marks, and even ligatures (that is, letters joined together, such as **fl**). It has been estimated that a single type size in the earliest printshop would have required about 150 different matrices to produce. Variety and quantity within the framework of rigid standardization thus characterized the making of type from the very beginning. Little wonder, then, that this is sometimes called the beginnings of mass production by interchangeable parts—the core means by which the great quantity of manufactured things is still made.

The controversy over how much credit Johannes Gutenberg should get for the invention of typography will never be completely settled, but one of the key elements in Gutenburg's background that favors his claim is his experience in precious metals. Probably only a goldsmith could have come up with the means for accurately shaping large numbers of very small pieces of metal, through the use of punches, matrices, and molds. We don't know precisely how type was made in the Mainz workshop of 1450, but the method that does appear in the fifteenth century involved shaping a letter on a positive punch (that is, with a relief surface) of a hard material (eventually

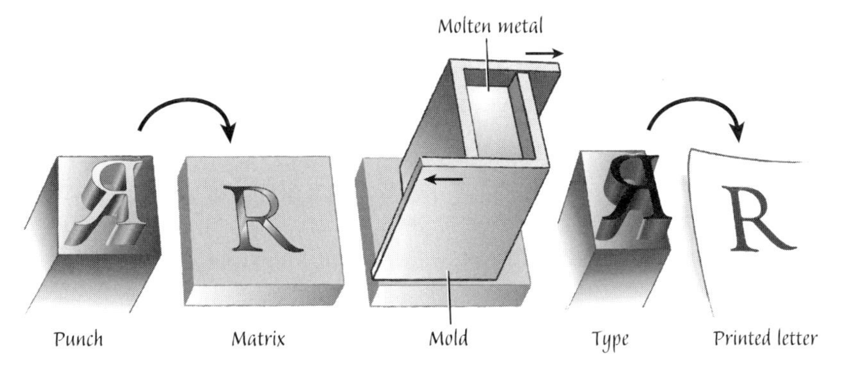

Figure 7.1
Gutenberg's method of molding type of standard size and variable width was key to his system of printing. The paired L-shaped halves of the mold would slide against one another to vary the width of the type piece while keeping the height constant. The actual molds were typically held in one hand and had additional features to assure a uniform height for each piece of type.

steel), which was used to make a matrix (negative impression) in a slightly softer material (eventually copper or a copper alloy). This matrix was then placed at the bottom of a mold, the sides of which were two L-shaped pieces of metal (figure 7.1). The L-shape allowed the two parts of the mold to slide past one another, making a rectangular form of precisely even height, but variable width. Molten metal—a soft lead-tin alloy, generally—was then poured into the type mold, and upon hardening it was removed when the two halves of the mold were separated. A notch in the sides of the mold eased the process of making sure the metal type was the standard depth (from face to bottom) for the composing form. The use of punches, of a range of metals with distinctly different properties, and of carefully designed molds all suggest the key role of the goldsmith's skills in developing the technology of type founding. It is one of the elements of the increased technological consciousness of the Renaissance that a metal worker should put his hand to the making of books—a craft that would seem otherwise to be far removed from his competence or concerns.

Making the type was the most difficult puzzle to be solved by the first printers, but certainly not the last one. The entire practice of organizing type into "cases" and of composing type into pages—the craft of the compositor—had to be invented from scratch. Making up a high-quality page from the hundreds of small pieces of type was not a simple task to define and perfect. Even the simplest thing, such as how to effectively sort and make available the type pieces, had a significant effect on the efficiency and economy of the trade. The forms for composed pages, the means of dividing and folding sheets into quarto, octavo, and smaller sizes (practical printing involved printing large sheets on which as many as eight or sixteen pages might be set, each oriented in the correct fashion so that when the sheet was folded properly, a "signature" of pages in the right sequence would result), and even techniques for printing in more than one color (the forty-two-line Bible of 1452–56, the earliest full-sized book, followed established scribal practice and used both black and colored inks) all emerged from the Mainz printshop.

Johannes Gutenberg was originally supported in his Mainz shop by Johannes Fust (described in some sources as a fellow goldsmith, in others as a lawyer), who was primarily a financial backer. By the mid-1450s, the two partners had a falling out, Gutenberg was forced from the shop (no printed work, in fact, has his name on it), and Fust recruited a young man, Peter Schoeffer, later his son-in-law, to take up the trade. In Schoeffer's hands, the craft of printing took on most of the elements that it retained with only minor change for the next three centuries. In the shop of Fust and Schoeffer, the practice of making books began slowly to diverge from scribal norms, adjusting to the demands of the technology (so that, for example, the number of ligatures and other marginally useful characters was cut considerably). Most remarkably, perhaps, the craft began to spread out from Mainz (despite its originators' early efforts) with extraordinary speed. Before the 1450s were over, a printer in Strasbourg was producing a Bible, and new shops opened in Mainz itself. The 1460s and 1470s saw an astonishing spread of the new and complex technology, best illustrated by a partial list: Rome (1463), Basel (1466), Venice (1469), Paris (1470), Utrecht (1470), Milan, Naples, and Florence (1471), Lyon (1473), Budapest (1473), Cracow (1474), Bruges (1474), Westminster [London] (1476), and so on, until by 1480 it is estimated that there were presses in more than 110 towns and cities throughout Europe.[3]

> Thus end I this book, which I have translated after mine Author as nigh as God hath given me cunning, to whom be given the laud and praising. And forasmuch as in the writing of the same my pen is worn, my hand weary and not steadfast, mine eyne [eye] dimmed with overmuch looking on the white paper, and my courage not so prone and ready to labour as it hath been, and that age creepeth on me daily and feebleth all the body, and also because I have promised to divers gentlemen and my friends to address to them as hastily as I might this said book, therefore I have practised and learned at my great charge and dispence to ordain this said book in print, after the manner and form as ye may here see, and is not written with pen and ink as other books be, to the end that every man might have them at once.[4]

So wrote William Caxton at the end of his first book, the first one printed by an Englishman, *The Recuyell of the Histories of Troye* (Bruges, 1474). Caxton was about fifty years old when he joined with an accomplished scribe, Colard Mansion, to learn the printer's trade, buy the needed equipment in Germany, and establish a printshop in the Flemish city of Bruges, where he had lived for more than three decades as a merchants' representative, well-connected with high-placed figures, both English and continental. The modern reader can sympathize with Caxton as easily as one five hundred years ago when he writes of weariness of hand and eye with copying his new translation of the popular French version of the history of Troy and the Trojan war. But William Caxton was apparently of a particular type—not so unfamiliar today—a person who sought respite from his labors not by passing them on to another but by learning the high technology of his day. He probably encountered the new art of printing in the shop of one Ulrich Zell, in Cologne. He asked Zell to print his new translation, and probably learned the trade as a helper filling his own order. He brought his new skills back to Bruges, recruited Mansion, hunted up the odds and ends of the printer's trade, and began to turn out books of his own. Caxton and Mansion were not, in fact, very good printers. Their type, in particular, was of rather poor quality, but this was perhaps to be expected in a shop where, unlike that of Gutenberg and Fust, there was no metalsmith to oversee the typefounding. Other improvements which had already appeared in printing's first couple of decades were either unknown or ignored by Caxton, but the partners still met with some success. A couple of years after the enterprise's beginning, Caxton decided to move back to his native England and to take with him the pieces of his shop.

In 1476, the press, type, and even loads of paper (there was as yet no papermaking in England) were all shipped across the English Channel and on to a small shop, rented by Caxton for ten shillings, in the shadow of Westminster Abbey, known as the "Sign of the Red Pale." The task of setting up the press and its accompanying equipment was sufficiently complicated that it was not until the next year that the first book issued from the first English press, *The Dictes and Sayings of the Philosophers*. This was another popular French work, translated by an English nobleman. One of the key elements of Caxton's success was his lifetime of connections to ennobled and wealthy friends, from whom he could solicit orders, patronage, and protection. The following year Caxton published the first printed contribution to English literature, Geoffrey Chaucer's *Canterbury Tales*. His first edition of Chaucer was, in fact, from a poor original, but Caxton proved himself a scrupulous scholar as well as an ambitious businessman, and so corrected this in another edition six years later. His careful scholarship, however, did not translate into careful proofreading, and so his books are often filled with typographical errors. He also remained technically backward, not justifying the margins of his type in his first books and using no or only very poor illustrations. But his contributions to printing—which surely would have made its way to England soon in any case, just as it had to every country in Western Europe—were not really technical, but political and intellectual. His good social standing and his intelligent choice of works and editorial oversight gave the start of the new technology in England a status and intellectual quality that served it in good stead in the tumultuous political and religious climate of the coming decades.[5]

The story of William Caxton serves to emphasize both the attractiveness of the new technology and its accessibility. Caxton was, after all, well into middle age (by the standards of his day, nearly an old man in fact) when he took up his new trade, and he had never worked with his hands. But with the assistance of an experienced calligrapher (no mechanic himself), he was able to take up the novel craft by purchasing the needed equipment and promoting his product to his high-placed acquaintances. Caxton was not only England's first printer, but at the Sign of the Red Pale he was also publisher and bookseller, as well as translator, editor, and author. His career emphasizes the extent to which this technology differed from almost all preceding it by its literate, intellectual nature. It was not always so easy to enter as Caxton found it, for in many places governments were leery of the power of books, pamphlets, and handbills made common and readily available throughout society. But the limitations that were placed on presses tended to come after the fact—only when the danger they might pose to established authority and customs became manifest. The trade and technology were themselves open, and enterprising individuals in cities and towns throughout Europe took advantage.

The spread of printing in the late fifteenth and early sixteenth centuries coincided with a large scale expansion of the European economy, and with the spread of European economic activity well beyond the continent itself into every corner of the globe. This, after all, was the age of Christopher Columbus, Vasco da Gama, and a host of other explorers and adventurers whose voyages remade the relations between Europeans and the rest of the world. The great explorations were much aided by printing, as well as by the continuing improvement of maps and navigational charts. The presses of Europe publicized the news of a New World (Columbus's report to Ferdinand and Isabella, the *Epistola*, was translated into Latin and published within months of his return), and stoked the fires of ambition in princes, prelates, merchants, and common folk. The very naming of America was in many ways the result of printing, as Amerigo Vespucci was able to have his book, *Mundus Novus* (*The New World*), published in nine different cities between 1503 and 1508, and a map using the term "America" first appeared in 1507. The other great revolution associated with the first century of printing was the Protestant Reformation. Of the first products of Gutenberg's printshop, indulgences—certificates offered for sale by the Catholic authorities to raise money by promising heavenly dispensations—were perhaps the most profitable. There is irony, therefore, in the fact that the printing press became the single most important instrument of those who would reform the church or, failing that, split it. A key tenant of the Protestants was expanded direct access to scripture—that is, to the Bible—by worshippers. And not only was the Bible printed in great numbers in the first half of the sixteenth century, and in vernacular tongues, but sermons, theological tracts, handbills, religious images, and posters also appeared, exhorting the faithful to follow one form of worship or the other and laying out in a flood of print the arguments for new or old religious truths.[6]

What printing did for exploration and religion, it did for every other element of Western culture: it captured information and made its dissemination surer and easier.

When Bacon celebrated printing's contribution to literature, he was using that term much more broadly than we would—he was, in fact, referring to what we call "information" in the broadest sense. While the copying of important works of philosophy, theology, and the like was an ancient and useful tradition, and books had already established their importance in European culture by the late Middle Ages, the limitations of scribal culture were enormous. This was compounded when nontextual information was important, as in geography or technology. While scribes can copy books with care and reasonable accuracy, the copying of maps or technical diagrams is far more difficult and prone to error. Before the late fifteenth century, someone interested in the latest and best map or atlas, for example, would have to travel to a specialized library, such as that of Klosterneuberg, near Vienna, where the monks specialized in gathering together the best geographic information and making as good copies as possible. But most of the information stayed, for the most part, in the monastery, not because of secrecy but because transferring the information was simply too difficult and unreliable. Printing changed all that.[7]

It is easy to understand how the spread of information, both textual and pictorial, through printing began to change the accessibility of technical knowledge in the fifteenth and sixteenth centuries. But the course of this change was not in fact smooth, nor was its effect that obvious. Up until this point, the knowledge of how to make and do things was only rarely gained from books, and even then usually by persons curious about technical matters but not actually concerned with carrying out a craft or managing a construction. Perhaps a writer such as Theophilus was in fact addressing himself to fellow monks who might wish to color glass or work gold, but most technical writings that existed were descriptions of ancient architecture or methods of waging war, less likely to be read by masons or soldiers and more likely by possible patrons or rulers. The first printed technical works fit solidly into this tradition. Sometime around 1455, the ruler of the Italian city of Rimini, Sigismondo Malatesta, bid his secretary, Roberto Valturio, write a work on Roman warfare. Valturio was not himself a soldier or engineer, so he approached the subject as a historian, commissioning illustrations of Roman war engines. Mixed in with these, however, were obviously modern touches, such as proposed mountings for cannon or wind-propelled vehicles. Valturio's *De Re Militari* (*On Military Subjects*) was published in Verona in 1472, thus becoming the first engineering text ever printed (figure 7.2). It can be argued that Valturio's book, which was very popular, actually set back the prospects for serious technical publishing—establishing a model for technical treatises with very old-fashioned illustrations (not at all of the quality of, say, Taccola or Leonardo) and no claim to truly applicable knowledge.[8]

It was not until several decades into the sixteenth century that technical literature began to reflect the new skills and new interests of the artist-engineers of the previous generation. The work of Leonardo, Taccola, and Francesco di Giorgio Martini were

Figure 7.2
From Roberto Valturio, *De Re Militari* (1472 edition). Valturio illustrated both practical machines and, as here, more imaginative ones. Anne S. K. Brown Military Collection, Brown University Library.

not printed in their lifetimes, but they became well known among careful students of architecture and machinery. About 1475, for example, Francesco di Giorgio completed a seven-volume treatise on architecture (*Trattato di Architectura*), which was largely his effort to elucidate and expand on the work of the great Roman architectural writer, Vitruvius. He was one of the first serious students of Vitruvius who had a good command of the new pictorial skills that had been developed by Brunelleschi, Taccola, and others, using linear perspective and a range of artistic techniques to give drawings a dimensionality that they had never had before. The last book of his treatise went well beyond Vitruvius in its attention given to machinery. It depicted dozens of types of mills, as well as pile-drivers, winches and cranes, mechanical cars, a great variety of pumps, and many other devices, both imaginative and mundane. It did so in a style that incorporated a host of key expository techniques, such as exploded views, cutaways, careful attention to scale, and other elements of technical drawing that we now take for granted. Francesco di Giorgio probably did not originate any one of these techniques, but he successfully brought them together in a style that became well known, even though they existed only in manuscript copies. He also made clear how useful careful drawings could be, correcting mistakes in others' drawings on the basis, not of models or working machines, but simply through visual constructions that were so good that they made clear what could and could not work.[9]

The artist engineers of the fifteenth century succeeded in setting a new higher standard for technical illustration. In manuscript form, the works of men like Taccola and Francesco di Giorgio circulated among patrons and others with the good fortune to have access to such expensive works. A new kind of literature, directed toward capturing and conveying the latest in technical information in a range of fields, from architecture to metallurgy, emerged out of the scribal culture of the fifteenth century. It became appropriate and even fashionable in certain literate and well-to-do circles to become educated in technical subjects. It should be noted, however, that such education was not directed, for the most part, toward practitioners, toward individuals who would actually learn how to do things from books. It was, instead, directed toward the potential patrons or customers of such individuals. Taccola and Francesco di Giorgio and others prepared and distributed their works in large part as advertisements of what they themselves could do, not as instructions to show others how things might be done. They succeeded, at least to the extent that by the end of the fifteenth century, technical knowledge of the sort they displayed, or of the kind demonstrated by a Leonardo da Vinci, whose own writings and drawings received very little circulation, became noteworthy and appreciated additions to the expanding body of learning that characterized the Renaissance.[10]

It was not a quick or easy task to bring the new standards of technical illustration into print, but in the first half of the sixteenth century there began to appear new kinds of technical books that took advantage of the new techniques of perspective drawing, copperplate engraving, and cutaway, transparent, and exploded views to convey information about technology, both old and new, at a level of detail and precision that had simply been unknown in the past. This was revolutionary not only in terms of the ease with which technical information became available, but also in terms of the quality of the information itself. Not only did the artistic skills exemplified by the best of the artist-engineers of the fifteenth century influence wider practice in the sixteenth century, but the very idea that accurate, reproducible technical information should be disseminated was also a radical innovation. In the past, the value of technical knowledge was seen as directly related to how few people possessed it, and thus even when descriptions of devices or processes were prepared and distributed, they often left out key elements. Like the work of Taccola, they were advertisements for why one needed to retain the technician's services, not means for substituting for his presence by capturing his knowledge on paper. But in the late Renaissance, the process of capture and dissemination itself acquired great value and prestige. Authors appeared who sought to gain their reputation and fortune by the publication of all the knowledge they had gained of their art or by showing precisely what they had accomplished or thought they could accomplish. This approach to the diffusion of knowledge did not take hold suddenly or universally, but the examples that did appear were impossible to resist or ignore. Since books could now be made by the hundreds and even thousands, they themselves—and the information they contained and conveyed—could be sources of wealth and reputation. Another consequence of the new means for capturing and spreading technical knowledge was the increasing tendency to link technological invention to human progress. The sentiments that Francis Bacon gave such emphatic voice to in the early seventeenth century were being heard with great frequency beginning a century earlier.[11]

Indeed, the appearance in the sixteenth century of books of and about machinery stands as eloquent and evocative testimony to an emerging enthusiasm about machines and their possible uses. In the later decades of the century there were published a number of examples of exuberant displays of machine design, in what came to be referred to as "theaters of machines," or simply "machine books." These were prepared by engineers and mechanics who desired to show off their ingenuity and the potential of the machinery of the day. Their works borrowed heavily from such unpublished predecessors as Francesco di Giorgio and Leonardo, but they were distributed far more widely and ranged in subjects from the prosaic to the wildly speculative. The two most important authors in this genre were Jacques Besson and Agostino Ramelli. Besson was a French Protestant who found his aspirations to be a practicing engineer frustrated by the religious wars of the day, but who was still able to compile a wide range of mechanical images from older engineers (like Francesco di Giorgio) and have them engraved (figure 7.3). His ambitions for a grand book of

Figure 7.3
From Jacques Besson, *Theatrum instrumentorum et machinarum* (1582). Smithsonian Institution.

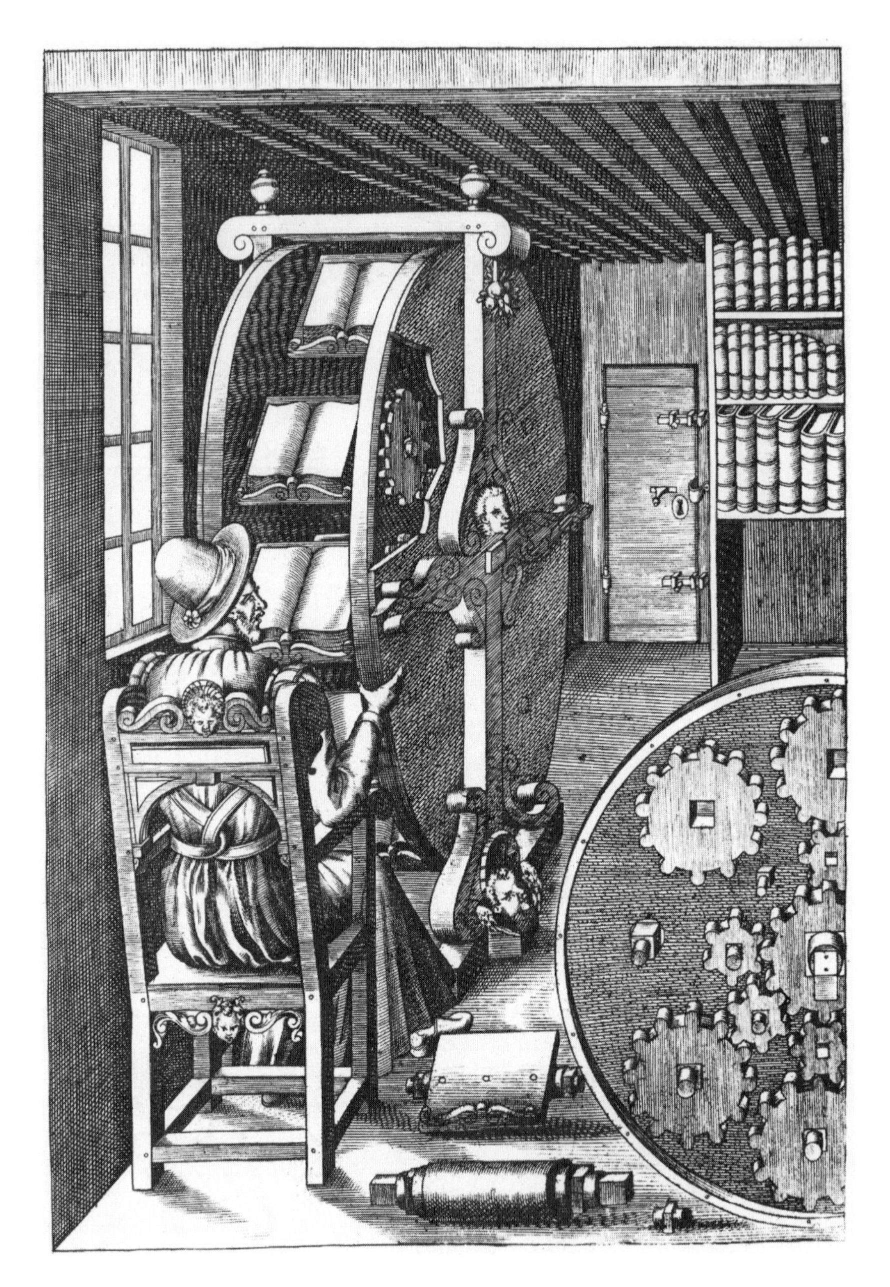

Figure 7.4
From Agostino Ramelli, *Le diverse et artificiose machine* (1588). Ramelli illustrated both well-known devices and fanciful inventions, like this machine for keeping books upright while reading. Burndy Library.

Figure 7.5
From Agostino Ramelli, *Le diverse et artificiose machine* (1588). The Archimedean screw was an ancient pump form; Ramelli often made use of cutaway drawings to extend the technical information in his images. Burndy Library.

these with learned commentary were thwarted by the need to flee France in 1572, but before he did so he was able to publish a modest edition. He died in London the following year, but shortly thereafter a French publisher came out with an enlarged edition, the *Théatre des instruments mathématiques et méchaniques*, which became widely popular.

This may have inspired an Italian soldier and engineer working in France, Agostino Ramelli, to put together his own compilation of machines, which was much more comprehensive in its coverage of different styles and applications. Ramelli's *Le Diverse et Artificiose Machine* was published in Paris in 1588, and since he had support from wealthy patrons, including King Henri III, he was able to include a large number of engravings of the highest quality (figure 7.4 and 7.5). The books of Besson, Ramelli, and a number of later authors in the late sixteenth and early seventeenth centuries depicted with care and precision a great range of mechanisms, both real and imagined, that defined the technological capabilities of Renaissance Europe.[12]

8 Earth, Fire, Water, and Air

Theophrastus Bombastus von Hohenheim was the son of a physician, raised in the Alps of Switzerland and Austria, and himself trained as a physician in the first years of the sixteenth century. His father became the municipal physician in the mining town of Villach, and thus Theophrastus grew up surrounded by medicine and miners. Unwilling like his father to settle down, however, the son became an itinerant healer, traveling widely throughout Europe, and as far afield as the Greek islands, Turkey, and Egypt. He wrote widely, often angrily railing against the tradition-bound medical practices of his day and seeking both to resurrect the highest standards of ancient medicine and to create novel paths to healing and to knowledge in general. He had a very high opinion of himself, and a rather low one of most everyone else. To signify this, he began to write and publish under the name "Paracelsus," which was his way of claiming to surpass in knowledge and skill the ancient Roman encyclopaedist, Cornelius Aulus Celsus, whose *De medicina* was one of the first medical texts printed (1478). Here is a representative passage:

> The arts are all within man.... Therefore learn to recognize alchemy, that she alone doth by means of fire change the impure into the pure. Also that not all fires will burn, but every true fire that will remain fire. Thus there are alchemists of wood, such as carpenters who prepare the wood that it may become a house; also the wood-carvers who make of the wood something quite alien to it, thus is a picture formed from it.... Bread is created and bestowed on us by God; but not as it cometh from the baker; but the three vulcans, the cultivator, the miller, and the baker make of it bread.[1]

To those who would understand why the sixteenth century was such a period of radical and dynamic change for the peoples of Western Europe, Paracelsus is as good a figure to start with as any, for he combined an almost furious rejection of tradition with an extraordinary allegiance to ancient forms of knowledge and practice. This is perhaps best brought out in his references to alchemy, the body of learning that sought to bring together what was known and believed about the transformations of matter and to organize this knowledge in a philosophical and spiritual corpus that would reflect the harmonies and conflicts of the world. Because alchemy dealt with such substances as sulfur and mercury, such tools as alembics and stills, and such processes as transformation and reaction, it is easy to think of alchemy as a primitive chemistry, but this characterization, convenient as it is, is badly misleading. Alchemy, with deep roots in Eastern mystical traditions and dogmas, was as much a belief system as it was a body of knowledge or practices. It revolved around beliefs in connectedness, hence the joining of kinds of matter, positions of heavenly bodies, and moral and spiritual well-being. For Paracelsus, it was an important adjunct to medicine, which was always his first concern. He believed strongly in the healing power of the right substances used in the right combinations (what came to be known as "iatrochemistry"), and made a point of learning all he could of the folk remedies, herbal lore, and other medicinal knowledge that was rejected by the university doctors. He also believed in making knowledge widely available. He wrote in German, not Latin, lectured widely, and found audiences throughout central Europe. When, however, he accepted an appointment in the Swiss city of Basel and began to lecture at the university there, his unrestrained style and iconoclastic message quickly got him in trouble with authorities, and he had to flee the city.[2]

The influence of Paracelsus on technology was small—he is interesting, not as a contributor to technical matters, but as a useful symbol of the ferment of change, particularly in attitudes toward knowledge. Even while authorities reviled him, and he ended his life a lonely and bitter man, the challenges that he raised to orthodox thinking—about medicine, but also about all our knowledge of nature and of what we can do with it—were taking hold in a variety of ways, many of which could not be appreciated until well after his death. Paracelsus himself noted that he was called "the Luther of the physicians," but he never attracted the popular following of the German reformer. His work, however, did serve to demonstrate effectively that old assumptions about what humans could know and do, just like old assumptions about how they must worship and believe, were crumbling in the sixteenth century.

Paracelsus's origins in the mining towns of south central Europe were particularly fitting sources for an iconoclastic seeker of new knowledge. In the mines here—stretching from the Alps eastward into southern Germany, Bohemia (today's Czech Republic), Silesia (in modern Poland) and surrounding territories—a new technological and economic order was taking shape from the mid-fifteenth to the mid-sixteenth centuries. This new order was not founded on significant inventions or novel technologies, but on a much increased scale of technology-based enterprise, a new scale that, more than any other thing, rewrote the rules that defined who was rich, who had power, and who worked for whom. As we have seen, mining and metallurgy became increasingly important in the High and Late Middle Ages, particularly as uses for iron and for copper alloys multiplied and as techniques for their preparation reduced their cost. But up until the late fifteenth century or so, the scale of metal extraction and working was a small, local one. The coming of the blast furnace increased considerably the production of iron, but reliance was still largely on readily available ores found near the surface. Deep mining was reserved for precious metals, particularly silver, and then only in a very few locations.

The commercial expansion of Western Europe in the twelfth and thirteenth centuries—the same period that supported the great cathedral crusade—fostered the growth of mining and metalworking wherever the resources were available. Ironmaking tended to be widely distributed, since iron ores—at least surface ores—are found scattered throughout Western and Central Europe. Other metallic ores, such as silver, lead, copper, and tin, are much more localized, as a fact of geology, and hence their working was concentrated in particular districts. In the ancient Roman empire, Spain had been the most important source for metals, but the Spanish deposits grew less important in the Middle Ages and those of Central Europe more so. The demand for metals, as for so much else, declined with the demographic disasters of the fourteenth century. Even though the Black Death and other plagues were less severe in a number of the key mining areas, such as Bohemia and the northern parts of Bavaria, the overall decline of the European economy took a considerable toll on all metal working. The never-ending warfare of the fourteenth and early fifteenth centuries did serve to support the demand for iron, but this same warfare still made prosperity elusive even for the iron districts.[3]

The fortunes of Europe in general and of miners in particular began to change after the mid-fifteenth century. Economic recovery across the continent—sustained even though the plagues had not completely died away (and would not until the eighteenth or, in some areas, the nineteenth century)—fed a growing demand for metals, both for commodities and for money. The need for more coinage, in particular, lent a kind of urgency to the expansion of mining after about 1450. Gold and silver had always, of course, been of great value, but in periods of economic expansion a shortage of precious metals—the only widely recognized basis for currency—could choke commerce completely. The easily available sources of these metals in Western and Central Europe were exhausted, however, and thus a concerted drive began for deeper and more elaborate mines. The miners of Bohemia, Hungary, Saxony, and surrounding territories were up to the task, not by devising significantly new techniques, but by applying well-known ones on a scale hitherto unthinkable. In the Middle Ages, demand for metals was modest enough and the potential of unexploited surface sources substantial enough that the costs and effort of investing in deep digging, tunneling, and drainage systems could be avoided, even though the Romans had shown the technical possibilities. New demands and new concentrations of wealth made investment in mine works a profitable enterprise after the mid-fifteenth century, and a metals boom took over the ore-bearing areas of Central Europe. Entire new towns were built around older mines that could be developed further as well as around newly discovered or, at least, newly profitable sources of silver, copper, and iron.

The hunger for precious metals drove the boom and the improvement of techniques. The quest for silver, for example, resulted in the elaboration of important methods for extracting the metal from ores in which it was combined with lead or copper. These methods, generally called "cupellation," consisted of heating the combined metals in an oxidizing furnace until the more easily oxidized base metal separated from the precious one, often being absorbed into the porous clay of the container, or "cupel." The idea had been known for centuries, but successful practices and tools had to be further developed before it became an important contribution to silver supplies. The new success of cupellation further encouraged the exploitation of lead and copper—metals with a wide range of applications, particularly when alloyed with zinc or tin. Alloys had been known since the beginnings of real metalworking, but the Renaissance Europeans took the art of alloying to new lengths, albeit still very much based on empirical experience and observation. The new skills and uses of alloys, along with the new importance of extraction methods like cupellation and the assaying techniques required for judging the results, were significant sources for the wider interest in alchemy. The transformation of ordinary looking rocks into shining, valuable metals was no more or less miraculous or improbable in the alchemist's eyes than any of the great range of changes and feats that were sought after. If lead can, in a sense, indeed be changed into silver (at least in part), then why should gold be any less possible, even if a bit more elusive?

The combination of the mystical and the practical was very characteristic of the discussions of both nature and of technology in the early sixteenth century. In mining and metalworking this can be seen in the first books printed on the subject. A work in German, the *Bergbüchlein* (a "small book on ores"), first appeared about 1500, and a work about assaying, the *Probierbüchlein*, appeared about twenty years later. The two books were reprinted often in the course of the sixteenth century (and were sometimes printed together, beginning in 1533), for they represented easily accessible works on a subject of profound importance, particularly to inhabitants of German-speaking Europe. To the modern eye, these works are filled with peculiar and wrong-headed statements, such as "Gold . . . is made from the very finest sulphur —so thoroughly purified and refined in the earth through the influence of Heaven, especially the Sun, that no fattiness is retained in it that might be consumed or burnt by fire . . . and from the most persistent quicksilver [mercury]." But to the Renaissance reader such statements were perfectly consistent with the established view of things. In any case, such observations did not much detract from the books' value, for it was for applicable information and directions rather than explanations that the books were appreciated. The useful information in the *Bergbüchlein* and the *Probierbüchlein* was not extensive or, in most cases, very detailed. It is hard to see how, for example, someone could take the information as a guide for actual mining or assaying. But as a means for appreciating the work of the miners, craftsmen, and assayers and for getting an idea of what questions to ask, what problems to look for, or how to judge whether someone really knew what they were doing, they were probably quite useful. While the market or audience for early technical works is very elusive to the historian, it would seem most likely that these books were of greatest value to patrons, customers, and investors, on the one hand, and to practitioners already skilled in at least some portion of the art who wished to extend their familiarity, if not their actual capabilities.[4]

The small German works are important simply for being the first indicators that there was a market for printed information on mining and metals, but they contributed little to greater understanding of these subjects. This is very different from the next work on the topic to appear in print, the *Pirotechnia* of an Italian architect, engineer, and metal worker, Vannoccio Biringuccio, which was published in Venice in 1540, soon after the author's death (figure 8.1). Not only did Biringuccio know what he was talking about, having traveled widely observing metal works of all kinds and having been in charge of significant works himself, but he was a literate, careful, and insightful writer, who wished to provide reliable and thorough guidance to those interested in any metallurgical topic. Like most of the good early technical writers, Biringuccio made no claims to innovation, but simply to comprehensive knowledge of his subject and the desire to dispel misunderstanding. In ten chapters he discussed the ores of metals and other minerals, techniques of smelting, assaying, and alloying, and how castings and other articles could be fashioned. His last chapter was about gunpowder and the making of firearms and other weapons, remarking, "I do not want to omit anything in which fire or its operation has a part," but making a point of ending, not with weapons but with fireworks, and finally with an almost poetic reference to "the fire that is more powerful than all other fires," which is love.[5]

The most important thing about Biringuccio's book is not what it tells us about sixteenth-century mining and metallurgical practices, but the message it delivers about the new attitude toward technical knowledge. Biringuccio made every effort to understand and record the best practices in the entire range of his subject, both those in which he was directly engaged and those that he only knew secondhand. Unlike the earlier, modest German works, his *Pirotechnia* was a thorough and intelligent discussion of all that the metal worker should understand. Unlike the works of the alchemists or of Paracelsus, he insisted in grounding his discussion exclusively in useful and observable practice. While he spoke throughout the work of the contributions of the alchemists to metallurgical practice, it is obvious that he judged the worth of these solely on the basis of the results, and not on the spiritual or mystical qualities that the alchemists always insisted on touting. The *Pirotechnia* was written in Italian, thus following the pattern of the earlier much more modest German

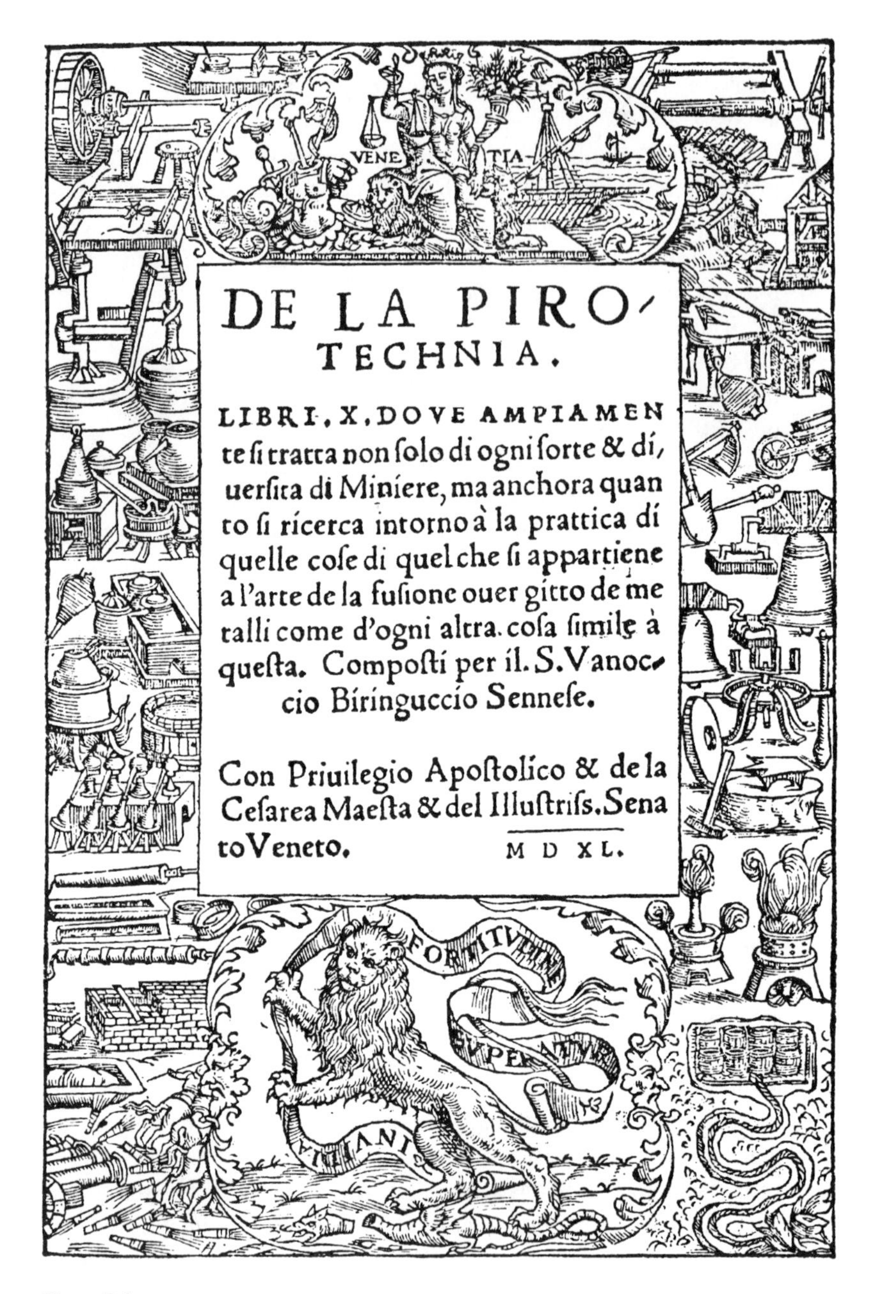

DE LA PIRO-
TECHNIA.

LIBRI. X. DOVE AMPIAMEN
te ſi tratta non ſolo di ogni ſorte & di,
uerſita di Miniere, ma anchora quan
to ſi ricerca intorno à la prattica di
quelle coſe di quel che ſi appartiene
a l'arte de la fuſione ouer gitto de me
talli come d'ogni altra coſa ſimile à
queſta. Compoſti per il. S. Vanoc-
cio Biringuccio Senneſe.

Con Priuilegio Apoſtolico & de la
Ceſarea Maeſta & del Illuſtriſs. Sena
to Veneto. M D XL.

Figure 8.1
The title page of Vannoccio Biringuccio's *De La Pirotechnia* (1540) combined symbolic images with depictions of laboratory and workshop apparatus. Burndy Library.

works, and marking it even further as the product of a practitioner, not an outside observer. The modest illustrations in the work made it even more accessible to a larger audience, although the use of the vernacular may, in the long run, have reduced Biringuccio's extended influence outside of Italy—the only translation for several centuries (except for modest excerpts) was into French, meaning, for example, that only the best educated German readers were able to appreciate the work.[6]

Among those well-educated German readers was Georg Bauer, a physician in the mining town of Chemnitz, in the German state of Saxony. Bauer, in the fashion of many educated men of his day, adopted a Latinized version of his name (which means "peasant" or "farmer" in German), and so his writings have come down to us under the name of Georgius Agricola. Despite his name, Agricola was clearly no peasant; he attended the University of Leipzig, taught Greek and Latin for several years, and then, at the age of thirty, studied medicine and a host of other subjects at some of the best universities in Italy. During this period, in the 1520s, Agricola came to be well acquainted with the humanist scholarship that was sweeping over Europe, associated with such noted figures as Erasmus of Rotterdam. When he returned to Germany, he began practicing medicine in the boom mining town of Joachimstal, in Bohemia (the town was only eleven years old when Agricola arrived), and there he made a study of the most exciting technology of his day—mining. It was with the encouragement of Erasmus and other humanists that he decided to put some of his observations about the subject into print, publishing a dialogue under the title of *Bermannus* in 1530. *Bermannus* was primarily a loose discourse comparing contemporary German knowledge of minerals with the standards of the ancients, but it was recognized as having some technical value. Biringuccio himself acknowledged learning some things about silver ores from the work. The reception of the book must have encouraged the humanist-physician, for he published another small work, on weights and measures, only a few years later. As he built his career and family, however, there came a pause, until the mid-1540s, when he began turning out a series of works, ranging over such subjects as fossils, underground animals, minerals, and the like. All of this, however, was simply preparatory to his creation of one of the most important contributions ever made to technical literature, his *De Re Metallica* (*On Metal Subjects*), which was so ambitious that it took three years to print, not appearing until 1556, a year after the author's death.[7]

More than any other single work, *De Re Metallica* signaled the Europeans' new capabilities and attitudes toward technology and technological knowledge. The comprehensiveness of Agricola's effort, the skill with which the book and its magnificent illustrations were put together, the rapid recognition throughout Western Europe of the work's importance all marked a departure from past practices and ways of thinking. This departure itself was not, of course, a sudden phenomenon. As the last chapter made clear, by the fifteenth century the Europeans were embarked on the path of open technical knowledge and a wider dissemination of the most useful practices and of novel approaches. The artist engineers of the Renaissance demonstrated both the new capabilities for rendering knowledge and speculation about technology and the new audiences for this knowledge. Agricola himself acknowledged an ancient model, the *De Re Rustica* of a first-century Roman writer, Moderatus Columella, which had been printed in 1472. *De Re Metallica* was thus simply the most spectacular event in a sequence of developments that had been going on for a century or so, and would continue well into the future—the creation and distribution of authoritative, accessible, skilled descriptions of improved ways of doing things. These developments were critical changes in the processes by which Europeans captured improvement and made technical knowledge widely accessible.[8]

De Re Metallica was a large volume of over five hundred pages, covering every topic related to mining and metals. The most striking thing about the work, however, was the large number of carefully rendered woodcuts, for the author was acutely aware of the limitations of mere verbal description: "with regard to the veins, tools, vessels, sluices, machines, and furnaces, I have not only described them, but have also hired illustrators to delineate their forms, lest descriptions which are conveyed by words should either not be understood by men of our own times, or should cause difficulty to posterity." Posterity has been very grateful indeed to Agricola, for his renderings provide us with the first truly large-scale depiction of the many facets of a complex technology. In addition, Agricola's attitude to the sources of his information was one of great care. He went on to explain, "I have omitted all those things which I have not myself seen, or have not read or heard of from persons upon whom I can rely. That which I have neither seen, nor carefully considered after reading or hearing of, I have not written about." He acknowledged the contributions of Biringuccio, and even of the earlier German books, but pointed out the limitations of those works in their coverage of many subjects. He also mentioned the writings of the alchemists, pointing out their preoccupation with gold and silver, but also mentioning that none seem to have been able to get rich themselves from the secrets they claimed to possess, and that some of them seemed to be downright frauds. "No authors," he concluded, "have written on this art in its entirety," and thus he felt the need for his own effort, written in Latin, "since foreign nations and races do not understand our tongue." There follows this introduction twelve books that cover everything from a general defense of mining and metalworking to the location of ores, the digging and maintenance of mines, assaying, smelting, working in all metals, and finally information about making nonmetallic products of the earth, such as salt, alum, sulfur, and glass.[9]

As an educated and widely traveled physician—a scholar, really—living in the lively technical environment of a new mining town, Agricola clearly saw himself as a privileged observer of one of the most important activities of his time. He saw nothing strange about a humanist and scholar paying such careful and detailed attention to a complex technical subject. He, in the scholarly fashion of his day, went to lengths to connect his efforts to a fuller appreciation of the ancients and of classical learning, but his embracing of the practical knowledge of the miner and the metalworker (and of other craftsmen, from millwrights to charcoal burners) was complete and unreserved. In addition, Agricola was acutely sensitive, as few technical writers before (or since, it might be added), to the social context of the technology he described. He described the legal traditions that governed mining, the organization of supervision and work, and the qualities of managers and foremen. This social dimension was made vivid by the illustrations, which were not lifeless drawings of shafts, seams, and machines, but lively, sometimes poignant, depictions of men (and just a few women) actually engaged in life and labor (figures 8.2 and 8.3).

Particularly striking in Agricola's illustrations is the depiction of machinery. There are machines, some quite complex, for just about every task, from bringing dirt or ore from a mine shaft, to pumping and ventilating, to grinding or stamping ores, to extracting the heavy products from smelting furnaces. The most spectacular machines were the pumps and windlasses, driven by a range of power sources, from human and animal to water and wind. As the mines of central Europe grew ever deeper, the problem of pumping groundwater out was one that occupied many, and the range of solutions depicted in *De Re Metallica* spoke eloquently both of the great concern this problem caused and the ingenuity that sixteenth-century technology was able to muster. The physician Agricola was also particularly sensitive to another need of deep mines—ventilating systems to keep the air fresh deep underground. This challenge was also met by a range of powered and hand-operated devices. The machines were not the only extensive capital investments represented by the mines. Considerable works were required for carrying ore, breaking and washing it, and preparing it for smelting. Agricola's pictures provide the most vivid indications we have of the new capitalist scale of these works—mines were clearly no longer places where a few dozen men, at most, worked, often for their own shares of the product, but were large enterprises with hundreds of wage laborers, organized into shifts that sometimes kept the works going around the clock.

It appears that Agricola completed *De Re Metallica* in 1550 (the dedication is dated that year), but the woodcuts required several more years of work. Everything was sent to Basel for printing by the middle of 1553, but it still required several more years to get off the press. In the meantime, the impending work came to the notice of important people—the Elector Augustus of Saxony, for example, Agricola's ruler, wrote requesting that a translation into German be prepared even before the Latin work was out. This was done, although quite badly. An Italian edition came out in 1563, from the Swiss publisher of the original Latin one. There were no more translations made for several centuries—none into English until mining engineer (and

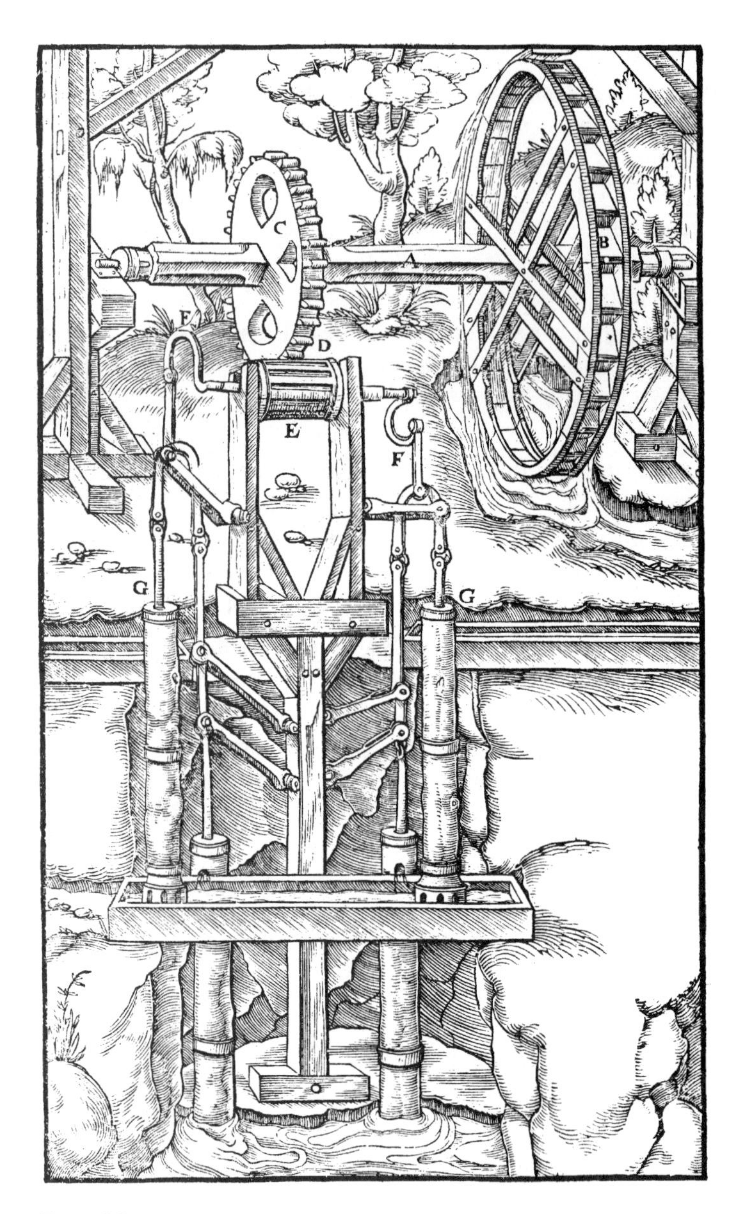

Figure 8.2
From Georgius Agricola, *De Re Metallica* (1556). Mine drainage was a key concern of almost any mining engineer, and pumps are among the most common machines illustrated by Agricola. Burndy Library.

Figure 8.3
From Georgius Agricola, *De Re Metallica* (1556). Agricola took pains to illustrate the parts of unfamiliar machinery, as in this equipment for washing ores.

later president of the United States) Herbert Hoover undertook the task with his wife, and issued a very fine edition in 1912. Nonetheless, the book exercised enormous influence, not only on mining and metallurgical knowledge—Agricola became the primary source to cite for many decades to come—but also on technical matters in general, for the work set an impressively high standard for what a comprehensive and useful treatment of an important technology should look like. It also spread widely the high standards of illustration that had been pioneered by the artist-engineers, and showed that these were good not simply for speculative and elaborate efforts, but for depicting the great range of tools, furnaces, and even work sites that every technology, from the mundane to the exotic, depended upon.

It is somewhat ironic that Agricola's great work appeared just as the mining boom, and the European economy in general, were declining. There were a variety of reasons for the economic troubles that set in after the mid-sixteenth century, but one that directly affected the mining areas was the influx of gold and, especially, silver from the New World. While the riches of the Far East that Columbus and his backers had sought turned out to be much farther away than suspected, the conquests of Cortés, Pizarro, and others did give the Spanish access to enormous treasures to plunder from the Aztec and Inca empires. The result was to enrich, at least momentarily, the Spanish monarchy, and to create enormous inflation in the European economy, as the influx of precious metals devalued not only the coinage but also all the metals throughout the continent. Another cause of economic distress was the intensification of warfare, as the religious disputes of the Reformation began to flare out in more and more intensive armed struggles, moving around the continent, from the German states, into France, and on into protracted warfare in the Low Countries. The upswing in military activities did keep the armaments and gunpowder makers busy, and the iron founders became crucial strategic resources, but the upheaval of what was in essence civil war was so materially destructive that investments and improvements were constantly at risk.[10]

In spite of the problems that set many industries back in the later sixteenth century (except, notably, in England, where isolation from the religious wars along with the vigorous administration of the Tudors provided some stability in this period), technological improvement was in fact to some degree encouraged. The economic growth of the earlier decades, along with the inflation in prices, created a great deal of wealth on the continent, and tended to increase the concentration of wealth in the cities. The prosperous urban classes, for their part, increased the demand for manufactured goods, such as textiles, glass, and ceramics. The variety of textiles, for example—already considerable—grew even more during this period as the various textile regions—Flanders, northern France, northern Italy, and the like—increasingly competed against one another for the trade. The competition encouraged the adoption of more sophisticated machinery, and it was in this period that we encounter the first efforts to make machines to harness power sources for such basic tasks as spinning or for multiplying the number of spindles tended by a single worker. Glass became a widely available commodity for the first time, moving outside the realm of luxury goods that the Venetians had so effectively dominated during the late medieval period and inserting itself in a variety of uses in urban households. The number and size of glass windows, for example, increased measurably during the late Renaissance, particularly in northern Europe. The sixteenth-century urban dwelling was a much airier and better-lit structure than its medieval counterpart. Mirrors, too, become common household articles, as depicted so well by some of the earliest of the great Netherlands and Flemish painters. Indeed, the spectacular rise of fine painting and drawing in northern Europe from the early fifteenth century into the seventeenth century, from Jan van Eyck to Rembrandt van Rijn and Jan Vermeer, would have been inconceivable without the enormous increase in window lighting that readily available clear glass now made possible.

The rise of European ceramics was a remarkable story in itself, covering the centuries from the High Middle Ages to the first decades of the Industrial Revolution. Before about the thirteenth century, all fine glazed pottery was imported from Asian sources. Middle Eastern potters mastered the technique of using tin based glazes on fired clay to make a fine, glassy finish to their wares, and these became among the most prized of the Islamic imports into medieval Europe. Beginning in the thirteenth century, Italian potters discovered the secret of tin glazes themselves, and over the next several centuries continued to improve their techniques and expand the product. New colors, such as cobalt blue and orange derived from iron, and new techniques and kilns gave the Italians the capacity to surpass the foreign producers they sought to emulate, and their "majolica" became a prized product throughout Europe. By the fifteenth century, the market began to expand sufficiently that other countries sought to lure Italian potters to set up works elsewhere. In addition, fine pottery came to be one of the first widely recognized markers of "good taste," particularly among the expanding urban classes. Here was a commodity that was not so expensive, piece-by-piece, as to be out of reach of an upwardly mobile artisan or merchant, but which had all the attributes of artistry and fine workmanship typically associated with precious metalwork or luxury glass. It was also a product that leant itself readily to technical experimentation and aspirations. Rare pieces of fine Chinese porcelain made their way into Renaissance Europe and they were recognized for their exceptional quality. They set a high standard toward which the European potters continued to work for centuries, until finally in the eighteenth century they achieved mastery of the material. Ceramics represented a technology that encouraged continual innovation in design, technique, materials, and marketing.[11]

Another, rather different, sort of technology also fostered experimentation and improvement in the sixteenth century, and prepared the way for the extension of mechanical capabilities that gave the Europeans unmatched capabilities—clockmaking. At first, as we have seen, clocks were large and complex devices, the subject of much curiosity and craftsmanship, but with very limited distribution or application. By the sixteenth century, however, this had changed. For a variety of reasons—technical, social, and economic—clocks had become important articles of commerce and urban life. The early clocks had, in fact, been problematic timekeepers, and the difficulties of designing and regulating accurate escapements were considerable. And since notions of time were originally totally bound to observations of the sun and the stars, with their seasonal variations, the artificial hours of the mechanical clock were generally seen as inferior substitutes for the natural hours of the heavens. The introduction of public clocks into town life was therefore not a straightforward matter of providing a useful means of measuring time, but rather was just as much a slow process of spreading fashions and status symbols. From the fifteenth century onward, the possession of a public clock came to be closely associated with prosperity and forward-looking civic enterprise. Only against this background did "clock hours" begin to displace "natural hours," and the rhythms of town life come more and more to be regulated by clockwork.[12]

By the middle of the fifteenth century, European clockmakers had learned how to replace successfully the weight drive of the first clocks by coiled springs. This apparently simple alternative energy source actually placed considerable demands on mechanical ingenuity. Unlike a falling weight, which exerts exactly the same amount of force on the escapement at the finish as it does at the start, a spring exerts less and less force as it unwinds. To correct for this, clockmakers used a device known as the "fusee," which was a conical element placed between the unwinding spring and the primary clockwork (figure 8.4). A cord (later a chain) ran from the mainspring and wrapped around the fusee, starting at the narrower end and running to the larger diameter as the spring ran down, and thus provided an equal force against the main wheel. This extremely clever device was adopted by clockmakers for a range of products, from exquisite and ornate table clocks to, by the sixteenth century, watches so small they could be worn in a ring or as a locket. Beyond astonishing pieces of artistry and craftsmanship, the clock and watchmakers of the sixteenth century were also able to make the timepiece into an article of commerce. In cities like Augsburg and Nuremberg, the clockmakers began to establish their trade, not as simply a source of luxury goods but as an important article for the expanding urban bourgeoisie. The implications were far-reaching, for the smaller timepiece not only made the consciousness of time and of its measurement a hallmark of postmedieval thinking, but it also lay the foundations for new levels of careful and precise craftsmanship, not solely in the service of art and fine effects, but for the design and construction of machines and measuring instruments. In the centuries to come, the capacity for ever

Figure 8.4
Spring-driven clockwork; note the conical fusee in the upper center. Smithsonian Institution photograph.

more precise measurement, not only of time, but of every other conceivable quantity in nature, was to be a foundation of a new scientific worldview.[13]

Glass windows, mirrors, fine pottery, and clocks were only a few of the articles that the sixteenth century saw making their way through the upper and middle classes of society. Stoves, often with fine tiles, cast metal implements for the kitchen, carriages and coaches, even furniture and table settings of a recognizably modern form all began to reshape the material life of the Europeans. The interiors that are revealed in the paintings of sixteenth century artists like the elder Bruegel or the younger Holbein reveal spaces and everyday articles that are not that strange to the modern eye. In addition, elements of a modern sensibility begin to emerge from the postmedieval mindset; an appreciation for novel and clever things that occasionally flashes in the medieval light now shows itself more and more often, and in the habits and thoughts of ordinary men and women, at least in the towns and in some favored spots. No place better exemplifies these new ways of thinking and feeling than the little corner of northwestern Europe that with historic suddenness flowered into an exceptional brightness, the Netherlands.

The rise of the Netherlands had significant technological underpinnings, as well as technological implications. While during the Middle Ages Flanders represented one of the most prosperous and economically important areas of Europe, famous for textiles, ironwork, and other trades, the area to the north, near the delta of the Rhine River, was of much less significance. The low marshy land was settled in early medieval times, providing pockets of very rich and productive soil as well as extensive supplies of peat, a valuable fuel. But settlement was possible only in small areas, for the North Sea and the great rivers from the south combined to make extensive flooding common and destructive. During the great agricultural expansion of the High Middle Ages, there was sufficient incentive to encourage the building of protective works, such as dikes, to reduce the damage from floods and to carve out additional productive land. The geology of the region, along with the extensive cutting of peat from the surface, combined to contribute to extensive subsidence of much of the land, requiring more protective works, but also encouraging the building of drainage canals and sluices. By the late Middle Ages, the Netherlands had already emerged as a largely artificial environment, the product of an extensive network of waterworks, administered by an elaborate and powerful system of regional drainage authorities. Agriculture was particularly productive on the resulting lands, the population quite densely settled, and growth tightly constrained by geography and geology.[14]

The constraints were overcome by pumps, and the pumps were driven by windmills. When windmills were devised by Europeans in the early Middle Ages, they were largely used as gristmills, just like their water-driven counterparts. But the needs of the Netherlands for constant, reliable, substantial drainage works promoted the extension of power sources. Previously drainage had depended solely on ditches and dikes, with water sometimes moved up by hand-operated buckets. By the early fifteenth century, records show that windmills were beginning to be applied to the task of pumping, and in the course of the sixteenth century the *wipmolen* became a common feature of the Dutch landscape. This kind of mill was a relatively lightweight structure, unlike stone tower mills, but stronger than the traditional post mill, since it placed more of the works at the base of the mill. The actual water lifting was generally done by a scoopwheel, which operated somewhat like a waterwheel in reverse (an alternative was an Archimedean screw pump). The wipmolen was soon joined by the smock mill—so called because the outer structure flared out like a smock. This was a larger mill that could be used to drain another novel Dutch innovation, the *polder*. Polders were plots of land enclosed by earthen embankments. Such areas could then be drained by the mills through steady pumping, and thus reclaimed for extended agriculture or settlement. By the fifteenth century, almost all of the low-lying land of the Netherlands had been made into polders, and new ones were created as portions of the sea were pushed back. The reliance on polders and mills made the Dutch the European masters of drainage technology, and Dutch engineers went on to provide their services wherever low-lying land was to be claimed from the water.

The enormous prosperity that made the Dutch the greatest economic power in Europe by the seventeenth century had many foundations. Dutch entrepreneurs and shippers showed themselves to be men of great vision, fortitude, and energy in carving out trade routes and monopolies, making the Dutch the heirs in the sixteenth and seventeenth centuries to the economic power that had been associated with Venice. But now the Atlantic and seas beyond were where the action and wealth were, and Venice's location in the middle of the Mediterranean was a hindrance, not an advantage. The Dutch also proved to be particularly good at adapting lucrative crafts—not only ceramics but also a host of others, such as printing, fine textiles, clockmaking and watchmaking, and the like. In this regard, they took advantage of the woes and mistakes of other regions, particularly those generated by the religious strife in the German states and France. The St. Bartholomew's Day massacres in August 1572 drove thousands of Protestants from France, and despite the Edict of Nantes twenty-five years later (that gave the French Protestants ostensible rights), the Dutch (as well as the English and the Swiss) found permanent advantage in their relative toleration of dissident religions. The religious wars and the consequent movement of great numbers of persecuted—who tended to be overwhelmingly urban and thus often artisanal—were important forces in shifting technological knowledge and skills around Europe, concentrating them in areas that had earlier possessed no particular technical distinction.

The Netherlands itself, of course, did not escape the religious wars of the sixteenth century. But there the conflict was translated from a civil war into a protracted, and eventually triumphant, war of national liberation. The Spanish authority over the Low Countries clung tightly, but was eventually worn out by the persistence and resourcefulness of the Dutch Protestant rebels. On the heels of the defeat of the Spanish in the first decades of the seventeenth century, the Dutch state and economy entered a period of unparalleled power and influence. In this period, the material condition of the entire population began to take on some of the flush characteristics not experienced by any other peoples until well into the Industrial Age. And with the new prosperity came new patterns of production, consumption, and desire. In particular, it is in the Dutch Golden Age that we can see the emergence of new values attached to novelty and change—values that were to spread throughout the West in the next two centuries and remake the relationship between people and things.[15]

Visitors to the Netherlands in the early sixteenth century were generally struck by not only the general prosperity of the republic, but also by the new habits and tastes of the dominant urban middle class. These were displayed most prominently by a new level of consumption that spread well beyond the sorts of people for whom goods like glass, ceramics, and fine art were customary. The flourishing of Dutch painting, for example, produced not only art of the highest quality—the works of a Rembrandt or Vermeer—but also astonishing quantities of art, distributed throughout all but the lowest levels of society. Paintings could be found in any decent urban household, and often in great numbers. A weaver in the city of Leiden, for example, was reported to have sixty-four paintings in his house in 1643, and this was not an exceptional case. For men and women who lived by the work of their hands, to possess such things was a new kind of society indeed. Particularly striking to many visitors was the Dutch preoccupation with novelty—with the possession of the newest and, often, the most unusual thing. The most famous manifestation of this taste and its hold on Dutch culture was the "tulip craze" that swept up the Netherlands in the 1630s. The tulip had been introduced to the country only a couple of generations earlier, and it became a popular garden item quickly. In the early seventeenth century, however, a fashion developed for particular varieties of tulips, in particular, ones with variegated coloring (typically striped red and white or blue and white). These particular kinds were not possible to breed, but were the result of infected bulbs. Their rarity and distinctive appearance led to a mania and, in fact, a financial speculative bubble which led to extraordinary prices (at its height, one bulb of the "Semper Augustus" variety could fetch the cost of a house). While the tulip mania became an oft-told cautionary tale about fashions and speculative investments, it spoke of a culture in which special value was placed on novelty itself. This value was to be an important influence in the fostering of more rapid technological change in the centuries to come.[16]

The early modern acceptance of novelty was by no means universal or complete. Indeed, as in every time, including our own, there were many forces resisting the unfamiliar and the innovative. This conservatism is in fact a reasonable and unavoidable aspect of the human condition, for an uncritical or too-ready acceptance of the new is a wasteful and dangerous abandonment of the constructive and useful elements of human experience. Sometimes the forces of conservatism were readily apparent and imposing—the Reformation answered by the Counter-Reformation. Just as often, however, the forces arrayed against change were more subtle and less directed. They were simply in the habits of people and institutions, going about their business in familiar ways, not looking for them to be disturbed or disrupted. In the sixteenth century, however, we begin to see the first suggestions of an alternative cultural response to novelty, one in which opposition and stagnation were displaced by acceptance and movement. This was typically to be seen in only a small portion of the population, in people stimulated by new knowledge and ideas or moved by new opportunities and discoveries or inspired by new capabilities or new problems. On occasion, as in the flowering of the Dutch Republic, the quest for novelty could become a widely recognized value of the culture, but this was rare. More often, the new possessed its visible appeal only to a relative few, but these tended to be articulate and ambitious individuals, capable of giving their values lasting expression. From the discovery of a "New World" at the beginning of the sixteenth century to the flood of books and pamphlets addressing novel sciences, beliefs, techniques, ideas, products, or practices that marked the literary output at the turn of the seventeenth century, the possibilities of the "new" ("*novum*," "*novelle*," "*neu*") began to take command of the European imagination.

9 Improving Knowledge

Daughter, for those that has been brought up to a trade,
When they are marry'd, what use can be made
Of that employ, when as they have a Family,
To guide and govern as it ought to be?
Then if that Calling, and work, it be done,
All things beside that to Ruin must run . . .
Maids by their trades to such a pass do bring,
That they can neither brew, bake, wash, nor wring,
Nor any work that's tending to good housewifry;
This amongst many too often I see;
Nay, their young children must pack forth to nurse,
All is not got that is put in the Purse.[1]

Among the many anonymous ballads that made up an important part of the popular culture of seventeenth-century England can be found these lines, from a song entitled "The Good Wife's Fore-cast: or, the Kind and Loving Mother's Counsel to her Daughter after Marriage." They captured a dilemma that might be faced by a young woman, most likely of the commercial or better-off artisanal classes. The attraction of learning a trade, of acquiring skills that might provide some measure of independence in life, was traded off against the high likelihood that such skills would be of little real value after marriage. The married state was, for almost all women except those disadvantaged by poverty or disability, the normal and expected state of life in adulthood. "Good housewifry" would then be the primary measure of a woman's value within the family and within society at large. This constituted, of course, a wide range of skills and duties, the final nature of which would depend on the economic status and material well-being of the household.

The changing experiences and status of women in Europe from the Middle Ages into the early modern period have been the subject of considerable discussion and debate among scholars. Dependable written evidence is slender, variations from region to region and even town to town make generalizations difficult, and contemporary perceptions or agendas often color even the most scholarly discussions. To understand the place of technology in the life of men and women, however, requires some sense of the sorts of work, the kinds of skills and technical knowledge, and the products most associated with different classes, backgrounds, and sexes. The very definition of technology, in fact, or "the arts" in the parlance of the preindustrial period of which we are now speaking, depends on an understanding of how different roles in making and doing things have been associated at different times with different categories of people.

As suggested by "The Good Wife's Fore-cast," a woman's economic and social role in preindustrial Europe was traditionally defined largely by her marital status, and the "normal" status for an adult woman was that of a wife. As such, her responsibilities were expected to be centered around the home and, except for the most well-off, the care of children. "Housewifery" was the primary "calling" of most women, and their economic contribution, both to family and to society, was often foremost in that role. But it would be a mistake to think of this as either trivial or outside the realm of technology. Such a perception is, of course, a common one—our seventeenth-century ballad itself makes the distinction between "a trade" and family life. The ballad's references, however, to brewing, baking, and washing hint at the kinds of production incorporated within even the most circumscribed notion of home-based work. Just because these activities were typically centered in the home and because they generally resulted in products for consumption in the household and in the near-term, rather than for market exchange and possible long-term use, they have often been considered outside the realm of technology itself. Logically, however, this makes no sense. Food preparation, for example, is a core concern of many technologies, from the grinding of grain to the building of ovens, and tools and devices are as crucial to its efficacy as they are to the making of any other product. The activity is no less technological for being directed toward home consumption rather than the market. Market quantities, of course, may spur the development of a wider range of tools or techniques, and this is in fact what happens from the eighteenth century onward.

Food preparation is perhaps the best example of an activity that has over the centuries generated enormous variety in techniques and materials, a variety that does not allow for the construction of narrow categories of "best practice" or "progress." The bread baked in Sweden is traditionally different from that made in Greece, say, but there is no reason to say that one is "better" than the other, except as an individual may simply like one more than another. Change in food making practices over time is typically not seen as technological change, largely because such changes are seen as shifts in taste or style or are purely part of local traditions. This is too limited a perception, however, and it contributes to the idea that women and women's work lie largely outside the domain of technological improvement. It is perfectly reasonable to surmise that preparers are at least as likely to experiment with their creations and methods as any other craftspeople. We, however, have no record of this activity in earlier times, nor has such activity been acknowledged as examples of technological creativity. This is of course not surprising, and it would perhaps not be worth even noting in the context of the history of technology, except that this is very much the kind of thinking that has led to technology being viewed as an almost purely masculine domain.

In the sixteenth and seventeenth centuries, some of the ancient patterns of distributing work and responsibility for making key commodities began to change visibly. In the making of food and textiles, in particular, some of the roles of men and women, in at least some areas of Europe, were changing in ways that suggest shifts in power, authority, and learning. It is no coincidence, of course, that these two areas were the two most important areas of traditional commodity production that had been focused in the home and been largely dominated by women. By the beginning of industrialization in the eighteenth century, they were still largely domestic and female domains, but the earlier changes laid the groundwork for the factory-based changes to come, changes that were to, among other results, displace entirely the female dominance of production (although not the female identification of home-based labor associated with them). The preindustrial changes were only in very minor ways changes in technology, as typically construed, but they signaled important trends in the ways in which the technological world, on the eve of important transformations, was viewed as relating to the respective roles and worlds of women and men.

Until the late fourteenth century, ale was the basic brewed beverage in England, and this was made almost universally by women—"brewsters." Ale is made by mixing together water, yeast, and barley malt, allowing the mixture to ferment so that some portion of the malt sugar is turned into alcohol. It is a fairly sweet drink, and it keeps for only a few days. Beer was an alternative beverage, brewed for some time on the continent but not in the British Isles in the Middle Ages. Hops are added to the ale brew to make beer, and this imparts a somewhat more bitter taste and enhances considerably the keeping abilities of the beverage. Hops make it much easier to convert more of the malt sugar into alcohol, so that to produce a drink of a given strength, the addition of hops both reduced the amount of malt required (by as much a one-half) and speeded up the fermenting process. Increased alcohol content and other factors meant that beer could be made in large batches, casked, shipped over considerable distances, and kept as long as a year. Despite these material advantages, beer displaced ale in England only slowly, beginning in the late fourteenth century in southern and eastern towns, probably imported by Dutch and German merchants. By the 1430s, however, beer was being brewed in a number of these towns, and the production was overseen by Dutch or German brewers. Whereas ale was identified as a country drink, beer was an urban one, and the marketing of beer was a part of town commerce. Because of the more complicated techniques for beer making, as well as the keeping qualities of the brew, the scale of beer making could be advantageously increased much more readily than that of ale, which always was made in small batches. This scaling up, in turn, meant that capital could be put into beer brewing to increase profitability. Due to continental traditions and the larger scale of making and marketing of beer, its manufacture began as and remained a male province. As beer displaced ale as the beverage of choice for much of the population after the mid-fifteenth century, brewing disappeared as a female craft. In the words of Judith Bennett: "Instead of a modest trade that a singlewoman or widow might turn to for support or that a wife might pursue as a by-industry, beer brewing worked on a grand scale, with larger capital costs, higher risks, and expanded markets. . . . On the scale of commercialization that beerbrewing made possible, women quite simply could not compete with men: they lacked the necessary capital, they lacked ready access to distant markets, and they lacked managerial authority." It was through patterns of this kind that women in early modern Europe began to lose their place in formerly important crafts, and to be distanced from the paths of technical improvement.[2]

Women's loss of standing in traditional crafts was typically the result of shifting economic, especially market, factors along with the persistence and even intensification of long-standing prejudices and assumptions about female activity, property, and family responsibilities. Technology was, however, often a key factor, usually because of new tools and techniques that were less available to women. Another example from this period was midwifery. Not surprisingly, in traditional societies, women had typically controlled all the activities surrounding childbirth, include those of assisting delivery. The midwife was a woman, generally trained through practical experience by her mother or mistress, who oversaw the health of expectant mothers and, most prominently, presided over the delivery of children. A midwife could take care of routine difficulties, although she did not generally practice any surgery and so would not have gone so far as to perform Caesarean sections. She would also see to the health of a newborn and instruct a new mother in caring and feeding techniques. While midwives did not generally have formal education, they nonetheless had some status attached to their skills and the great value of their services. In the late Middle Ages, physicians came more and more to be educated men; few women were allowed to attend the new university medical schools (there were, to be sure, some famous exceptions, but they became rarer as time went on). In the late seventeenth century, educated physicians began gradually to intrude into the domain of midwives. To an extent, this seems to have been encouraged by higher status that was attached to being attended to by an educated man. Such services would have been much more costly than those of a midwife, but this emphasized the high status of the practice. With the introduction of obstetrical forceps in the early eighteenth century, the process of assisting childbirth was more easily characterized as a largely mechanical affair, and the new device was almost exclusively the tool of men, who came to dominate obstetrics, to the almost complete elimination of midwives except among the poor. It is important to note that the obstetrical forceps did not *cause* the displacement of women, as the substitution of beer for ale a couple of centuries earlier had not caused it. But new tools, techniques, and products, in the context of economic pressures, widespread notions of status and domestic responsibility, and changing patterns of work, family life, and trade tended to diminish the access of women either to new technological possibilities or even to their traditional areas of expertise.[3]

For all of European history, the realm of production most prominently female was that of the making of cloth. As we have seen, textiles were perhaps the most important urban commodity in the Middle Ages; the wealth of towns and even entire regions rested largely on fabric production. Even by the late Middle Ages, the variety of available fabric materials, weaves, and styles was sufficiently broad that regions, towns, and even small villages were able to specialize, and thus carve out for themselves areas of expertise and reputation that could be the foundation for local prosperity and identity. As the economic importance of the textile trade grew, women's dominance of the trade diminished, first in the distribution of products (where their role was probably always circumscribed, particularly as markets became larger or more diffuse), and then in various areas of production itself. The causes for this are the subject of some controversy and were clearly complicated. The role of technology was substantial, but it was not simple. In some areas, such as weaving, the development of new tools and techniques came to be associated with the growing role of men. The introduction and spread of the broadloom or the ribbon frame in the late Middle Ages, for example, allowed a single weaver to handle larger quantities of more complicated products, and these tools came to be seen as the province of men. But spinning also gained in efficiency in this period, with the development of flywheels, pedals, and the flyer-and-bobbin, but these innovations (quite possibly the work of women) did nothing to diminish the relationship between spinning and female craftsmanship. One reason that has been suggested for this is that the new looms increased the output of weavers to such an extent that the demand for spun thread and yarn was kept high, and pressures to keep costs down encouraged the continued reliance of lower-waged women for spinning.[4]

The role of guilds and other trade organizations in the restriction of women's work and thus women's relations with technology and technological change is a subject of some controversy among scholars. In the High Middle Ages, when guilds were first beginning to assume economic importance in the cities, there were many prominent instances of guilds open to women and, in a few cases, mixed guilds, in which women and men were, at least ostensibly, treated equally. Some earlier scholars took these facts to suggest a kind of "golden age" for women workers and artisans, but later students of the subject have generally downplayed the overall significance of the open and female guilds, suggesting that the guild system was, by and large, always a means for asserting power over trades by privileged groups, by class, gender, and kinship. In the later Middle Ages, as guilds in some towns and cities came to assume greater economic and even political power, they tended to guard their prerogatives more jealously and to limit their membership more closely. In times of economic stress, such as many regions experienced in the fifteenth and sixteenth centuries, pressures for restricting entry into crafts or control over trades by less privileged groups—such as women in general—increased. Even in earlier periods, participation by women in skilled crafts was often limited to daughters and wives, who were seen as necessary family helpers for craftsmen. After a master's death, it was not uncommon in many places for a widow to continue a trade and even to take on apprentices. But after the fifteenth century, the open training of women became much less common, and was increasingly forbidden by statute or guild regulations or charters. Under these circumstances, the perception of "skill" came more and more to be attached to masculinity, and the particular talents associated with women and women's work, such as dexterity, deftness, and patience, were seen as merely "attributes" and not skills. That is, they were perceived as innate qualities of femininity, not learned or subject to improvement. By such mentalities, women were increasingly, by the sixteenth and seventeenth centuries, seen as outside the domain of technical improvement.[5]

The consequences of this exclusion were made all the more significant by the fact this same period saw the emergence of new patterns of thought about skill, knowledge, and innovation in technology. With the continued spread of printing and the refinement and proliferation of means for recording and disseminating technical information, as well as institutional, economic, and social changes in the way in which this information was valued and used, the sixteenth and seventeenth centuries were a period of enormous changes in the status and meaning of technology in European life. Yet one other key factor was instrumental in these changes—the radical innovations in thinking about how humans comprehend, describe, and explain the natural world that popularly are termed the "Scientific Revolution." The relationship between science and technology in this period has been the subject of debate among scholars for several generations, a debate occasionally heated by ideological fervor. As long ago as the 1930s, Marxist scholars sought to demonstrate that the impetus for changes in scientific thought was provided by the growth of capitalism and increasing needs for mechanical invention. Other scholars responded with heated defenses of the philosophical and intellectual foundations for new ideas in science, and the emerging discipline of the history of science in the mid-twentieth century was profoundly shaped by the issues highlighted in this debate. At its heart lay the question of the extent to which significant change in science, particularly in the most basic sciences such as physics and astronomy, was generated within the scientific enterprise, through new ideas and discoveries, and to what extent such change was due to external forces, such as the material needs of society or the state.

Among the issues this question raised in turn was that of the fundamental relationship between science and technology. So astonishing and significant seemed the apparent gifts of twentieth-century science to technology, from electronic communications to nuclear weapons to genetic manipulation, that it became difficult for many to imagine that technology of any but the most primitive kind could exist outside the context of scientific knowledge. From this point of view, technology was, in essence, applied science. For the period before the Industrial Revolution, however, this notion clearly made little sense. Nonetheless, scholars sought to link changes in science with those in technology, making the great changes of the eighteenth century and afterwards at least indirectly the results of the Scientific Revolution. From both sides, therefore, the linkage between science and technology was asserted, the Marxists maintaining that the revolution in science was largely the result of economic forces urging the enlargement of technical capabilities, and others promoting the notion that the intellectual and philosophical upheaval of the late sixteenth and the seventeenth centuries provided the indispensable wherewithal for rapid and extensive technological change in the years following. Neither of these perspectives appreciated sufficiently the independence of these two spheres of human action. The effort to describe and comprehend nature is nearly as ancient as the effort to provide the material means for existence. The systems of thought produced by the first have naturally guided efforts to improve the latter, just as observations of the workaday world are indispensable elements in the formation of scientific ideas. The connection between science and technology is thus ancient and unavoidable, but it is not a connection of dependency. The relationship does, however, begin to change in the period we are now talking about.[6]

It is natural to tie these changing relationships to the transformations of scientific thought that began in the mid-sixteenth century, but, while these intellectual changes did indeed profoundly reshape the relations between science and all other realms of human endeavor, including technology, they were not themselves the sources of either key technological changes or of new ways of thinking about technology. Rather, the new sciences directed attention to problems that had technical manifestations, such as the principles behind the operation of machines or the action of forces in nature or the behavior of different materials under stress. The enlarged understanding of these problems did indeed eventually have important implications for the development of new mechanisms and materials, but it was a long time before scientific theory was generally a significant aid in solving technical problems.

The origins of the Scientific Revolution lie comfortably in the Middle Ages, when questions about the order of nature were raised by churchmen, seeking for the most part to reconcile the teachings of the Bible with the impressive and compellingly logical corpus of Aristotle and other classical writers, in many cases known not from direct transmission but by way of Arabic scholars and commentators (who added their own glosses to the ancient texts). Whatever the origins, however, the beginning of revolutionary change is conveniently and popularly marked by the publication in 1543 of two books which served to announce, at least to a limited Latin-literate audience, that traditional ways of describing and depicting the world were not satisfactory. The seventy-year-old Nicolas Copernicus, a Polish mathematician, lay dying of a stroke when, legend says, a copy of his *De revolutionibus orbium coelestium* (Concerning the revolutions of the celestial spheres) was thrust into his hands. His book was to set the small world of astronomers and mathematicians afire with a somewhat esoteric, but eventually shattering, controversy about how the earth and the primary heavenly bodies—the sun, the moon, and the five known planets—moved in relation to one another and to the stars. Equally shattering, in its own way, was the publication of a brash young (he was twenty-eight years old) physician from Brussels, Andreas Vesalius, *De humani corporis fabrica* (Concerning the fabric of the human body). With this lavishly illustrated work (figure 9.1), Vesalius upset ancient notions of human anatomy, and, just as important, of how properly to extend our understanding of the body and its workings.[7]

Both of these books were constructed as direct challenges to generally accepted and long-taught ways of looking at the world, but the fundamental nature of the Copernican challenge is easier to grasp. The elements of astronomy in the centuries before telescopes were essentially a kind of cosmic geometry, involving the mathematics that would make sense of the observed positions of those heavenly bodies that clearly moved in relation to each other (the moon and the planets) as well as the earth and the fixed stars. The authoritative statement of this geometry was that put together from Aristotle and other older sources by Claudius Ptolemy in the second century and known in the West through its Arabic compilation, the *Almagest*. It had long been recognized by the scholars that could follow the subject that Ptolemy's system of the world, which had the stars, planets, sun, and moon orbiting around a stationary earth in a complex system of movements depicted by sets of perfect circles and uniform, unvarying motion, required all sorts of compromises with notions of regularity and simplicity in order to describe plausibly what was observed in the heavens. Copernicus's alternative, in which the earth was displaced from the center of the world system by a stationary sun, did not really "work" any better than the traditional geocentric model, but it was much simpler to describe, and the compromises were, at least for its champions, less irksome and offensive to common sense. The argument in the *De revolutionibus* was largely a complicated mathematical one, sufficiently complex that the more timid could maintain they were simply entertaining geometrical hypotheses and not upsetting ways of describing the world order.

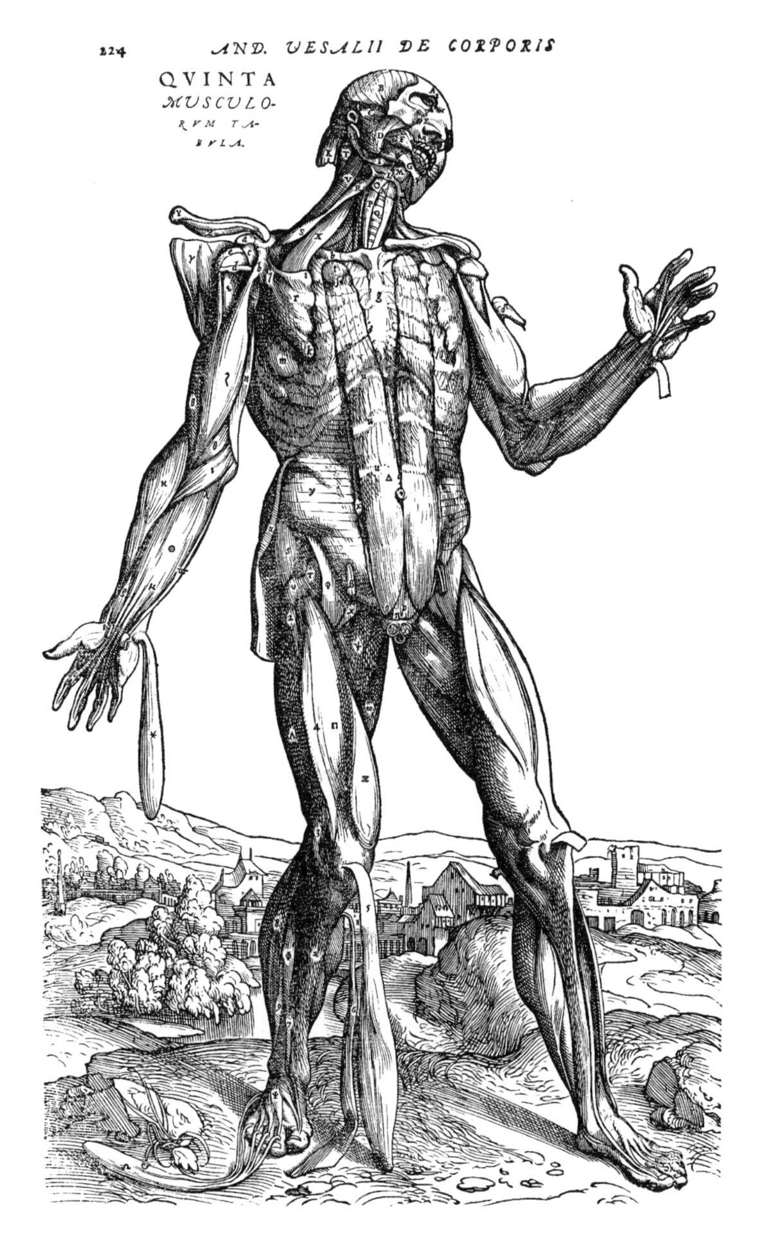

Figure 9.1
From A. Vesalius, *De Humani Corporis Fabrica* (1543). Vesalius's illustrations, with their references to parts and their elaborate natural settings, were in the same tradition as those found in Agricola's work. Burndy Library.

Nonetheless, it was widely recognized among the educated of Europe that the Copernican alternative did indeed have at its core a fundamental challenge to tradition.

The essence of this challenge lay not so much in the features of Copernicus's "heliocentric" system, but in the basis for his belief in the new system. Copernicus believed his system was true, not because it solved some problem that could not be solved otherwise, but because it made more *mathematical* sense than the Ptolemaic system. There were plenty of reasonable objections to the idea that the earth was in constant motion in space, spinning around on its axis once a day and revolving in a huge circle around the sun once a year, and Copernicus was intimately aware of all of them. But they were to be pushed aside, he insisted, because the mathematics demanded it. So dependent is modern science on this notion of mathematical simplicity and consistency that it is easy to lose sight of what a revolutionary notion it was to give it primacy over "common sense," but this is exactly the implication of Copernicanism. For the remainder of the sixteenth century, the debate over astronomy was largely a polite and esoteric one, but additional elements began to accumulate at century's end. The Danish astronomer Tycho Brahe was able to persuade his patrons to help him build the largest and best astronomical sighting instruments ever made, and with these he finally equipped the scholarly community with measurements much more precise than those available to medieval astronomers (or to Copernicus). Tycho's philosophical objections to the Copernican model led him to propose an alternative (in 1588), in which the moon and the sun described orbits around a stationary earth, and the planets were given orbits around the moving sun. This fooled no one, however, for it was clear that this was simply the Copernican model described from the vantage point of earth, and Tycho's contrivance was yet additional testimony to the fact that the newly precise observations made the tricks and devices necessary to "save the phenomena" of the Ptolemaic system even more unaesthetic than before. Tycho's improved observations also allowed him to cast grave doubt on another key element of the Aristotelian cosmology. A bright new star—a supernova in modern terms—appeared in the sky in 1572–73, and three comets were carefully observed between 1577 and 1585, and all of these new heavenly bodies were, to Tycho and some others, clearly beyond the orbit of the moon. But the classical view held that the heavens were perfect and unchanging (the one necessarily implied the other), and these new phenomena (previously always dismissed as atmospheric or "sublunary") could not be fit into this picture.

These sorts of observations, much more than anything available to Copernicus himself, constituted "anomalies" in the picture of the world of the kind that philosophers of science, like Thomas Kuhn, have suggested tend to lead to revolutions in science. The precision of Tycho's measurements of the relative positions of the planets also provided anomalies, at least to those willing to trust them and to work out their implications. By far the most important of these men was Tycho's onetime assistant Johannes Kepler. Kepler was as brilliant and iconoclastic a mathematician as Tycho had been a careful and innovative observer. Unlike his mentor, he felt not in the least beholden to Aristotelian orthodoxy, and the truth of the Copernican model seemed self-evident in its relative simplicity and elegance. Kepler struggled with the desire to find the harmony that he was sure—as a matter of faith alone—was hidden in the cosmos and the numbers that described it. In 1596 he published his *Mysterium Cosmographium*, in which he sought to describe the harmony of the spheres in terms of geometric perfect solids—the orbits of the planets around the sun were to be described and accounted for in terms of spheres nested within the perfect Euclidean solids (the five solid figures that have identical faces on all sides). It was this work that brought him to Tycho's attention, and the appointment as his assistant at the court of the emperor in Prague. Kepler was himself appointed imperial mathematician the following year, 1601, after Tycho's death, and he set about trying to make Tycho's new numbers fit into the Copernican model he so believed in. This turned out to be exceedingly difficult, however, for the better the data, the more compromises all the known models required. Finally, in a radical tribute to the new authority of numbers, Kepler abandoned many of the philosophical suppositions formerly underlying not only Aristotle and Ptolemy, but Copernicus as well. The key results were published in his *Astronomia Nova* of 1609, in which he propounded the first two of his laws of planetary motion, describing orbits not as circles but as ellipses (with the sun at one focus), and planetary motion not as uniform in speed, but as speeding up and slowing down in the course of the orbit, following a precise mathematical pattern depending on distance from the sun ("equal areas are swept out in equal times"). At this point, and arguably not before, the new astronomy became truly "revolutionary."

The man who did more than any other both to make the new astronomy truly radical and to extend the revolution to other fundamental questions of science was Galileo Galilei, professor of mathematics at the University of Padua. Galileo's interest in astronomy was actually secondary to a long-standing concern with how things move—what we would call "kinematics," or, more broadly, mechanics. He had slowly been persuaded, however, by the Copernican arguments for a new system of the world, and he began to make some observations and arguments of his own, helping to establish, for example, that the nova of 1604 had been truly a new star in the heavens. What caused Galileo to focus more of his attention to astronomy, however, were not new ideas but a new instrument. At the same time that Kepler was putting out his "New Astronomy," Galileo learned of an instrument made in the Netherlands that allowed one to see distant things as though they were close by. Within a few weeks of getting this report, he had built for himself a small telescope and began experimenting. By the end of 1609, he had a larger instrument and had begun a series of careful observations of the heavens, with results that were to cause a public furor.

Unlike the esoteric ideas of Kepler, the new phenomena that Galileo described were accessible to all: mountains on the face of the moon, a proliferation of hitherto unseen stars, and, most stunning, a set of new bodies in the sky that were clearly attached in some way to the planet Jupiter—in fact, four moons that seemed to circle that planet on their own. Their implications were self-evident and a look through a telescope would show just what he was talking about. Galileo quickly compiled his telescopic observations and his interpretations and published them in 1610 as the *Siderius Nuncius*, or "starry messenger." The book was a sensation.

Galileo's book and its reception marks more than any other single event the transition of the Scientific Revolution from a debate among mathematicians and astronomers about esoteric problems in calculating and modeling to a broader, more public challenge to the traditional philosophies of nature. Only a month after the *Siderius Nuncius* appeared, Kepler himself wrote a letter, soon published, supporting Galileo's conclusions. Before the year was out, observers throughout Europe reported confirmations of his discoveries and even began adding to them, making the first record of sunspots, for example. Galileo himself was able to take advantage of his new fame by getting a lifetime appointment as professor at the University of Pisa and "Philosopher and Mathematician to the Grand Duke" of Tuscany. His fame also, however, brought him and his ideas to the notice of the Inquisition, the office of the Catholic Church concerned with maintaining the purity of Christian doctrine. Despite questions raised by some zealots, the Inquisition gave Galileo little trouble for many years, and some highly placed churchmen clearly enjoyed discussing Galileo's ideas with him and became patrons. One of these, Maffeo Barberini, became Pope Urban VIII in 1623, but this did not prevent increasing pressure on Galileo to make clear that his apparent disagreement with traditional interpretation of Scripture concerning the order of the cosmos was a purely theoretical one, and not a challenge to the Church's authority. Discussion of the Copernican system as an alternative view of the world was expressly forbidden. This did not prevent Galileo from putting together his view of the key arguments about astronomy in a work he called *Dialogue Concerning the Two Chief World Systems*. He attempted to mollify Church concerns after completing the book, and so almost two years passed before it was finally printed in 1632 (in Italian, not Latin, it should be noted). Despite this, the Inquisition brought Galileo up on charges of making the Copernican case in his book too strong, and he was forced, in the spring of 1633, to stand trial in Rome and to abjure his errors. Despite this capitulation, he was still placed under house arrest, and his movements were tightly restricted by the Inquisition for the rest of his life.

In spite of his tribulations, Galileo exerted enormous influence on the practice and precepts of science, in his own day as well as thereafter. There had been previously men who gained great reputations as philosophers, but no one before whose fame rested on his mastery of mathematics and nature. Legend quickly surrounded Galileo and his work; stories of dropping weights from the tower of Pisa or observing swinging chandeliers from the cathedral ceiling became the stuff of myth. He published most of his work not in Latin, as almost all his predecessors did, but in Italian, where it was accessible to literate merchants and shipbuilders as well as to churchmen and doctors. This, perhaps as much as the substance of his ideas, was a source of the consternation that Galileo caused in the Italian Catholic establishment. Making a novel philosophy of nature, characterized by mathematical argument and carefully designed experiments (some physical and some mental), the subject of popular comment and discussion was as radical a challenge to the established order as the European elite had ever confronted, in some ways even more disturbing than the Protestant Reformation, which was, more often than not, bent to the expediencies of rulers rather than liberating to the minds of worshipers. Galileo's "method" has often been too facilely described, and it is worth remembering that he was in no way a "modern" scientist. He desired to transcend his place as a mathematician and to claim the title of "philosopher," and he was thus easily pulled into debates about philosophical terminology and argument, and he could not accept some ideas that struck him as unaesthetic, such as Kepler's laws of planetary motion. But the core of his methodological contribution was clear to all who followed his arguments: nature was to be described in ways that invited measurement and that always provided an avenue for experimental falsification. The experiments that he began in 1604, rolling smooth balls down inclined planes, which led in a few yeas to the fundamental law of falling bodies (that free fall consisted of uniform acceleration), are still models of experimental design and argument. He boldly set out new areas of systematic investigation, such as how to determine the strength of a beam or the path taken by a projectile in flight, and with astonishing speed his ideas transformed the terms and shape of debate and investigation among European philosophers of nature. In 1638 he published the last of his great contributions, the *Discourses on Two New Sciences*, reporting his findings on the sciences of motion and of the strength of materials. Within a decade or so of his death in 1642, European science was essentially Galilean in many respects, particularly in the place accorded mathematical reasoning as the final arbiter of debate.[8]

From the earliest years of his career, Galileo readily took up technical problems. His lifelong model was the Greek mathematician Archimedes, and like Archimedes, Galileo did not hesitate, when asked, to apply his skills to practical questions. Records show that even after he became professor of mathematics at Padua, he supplemented his salary by giving private lessons on a range of practical subjects, including fortifications, surveying, and optics. He also devised new instruments for solving mathematical and drawing problems and even received a patent from the Venetian Senate in 1594 for a novel horse-drawn pump. About this time he also began work on a treatise on machinery, although this was not published until the 1630s, after his reputation was secure. His writings, such as the *Dialogue* and the *Discourses*, were replete with references to practical work and problems. He began the *Discourses*, for example, with observations made of workers in the Venetian Arsenale and descriptions of some of the problems they had to solve. His discussion of the strength of materials (one of the "new sciences" of his title) was also clearly shaped by technological considerations. In the creation of a new way of talking about nature, in other words, Galileo readily used the problems and experience of the technical world. When it is remembered that Galileo's approach often required discussing "ideal" situations, such as frictionless surfaces or motion with no air resistance, the implications for the ready incorporation of the workaday world into the problems he posed and the methods he advocated for solving them become significant indeed. Already, at the birth of what was to become modern mechanics, we can see the intimate connection between the ascendancy of mathematics in natural philosophy and its emergence as the key to new technological prowess.[9]

It would be a mistake, however, to see this connection as readily apparent or transforming seventeenth century technology. In a few limited areas, such as gunnery and navigation, mathematical skills and techniques were widely recognized as important practical tools. But the idea that technical improvement itself could be guided or, at least, made swifter or surer, by the systematic application of either mathematics or natural laws was still in the future. What Galileo, his "school," and those who followed his precepts represented was the laying of a foundation of new attitudes and assumptions about the relationship between a precise understanding of nature, an understanding most appropriately couched in the language of number and universal law, and the identification and exploitation of possibilities for technological change.

For the English-speaking world, the new attitudes had an eloquent spokesman, who perceived the possibilities of change long before there were many changes themselves to behold. Francis Bacon, whose astuteness has already been drawn upon here, gave his voice and his name to a truly new mindset, in which the mastery of the laws of nature held the keys to power, over things as well as men. This preoccupation with the relationship between power and knowledge was appropriate for Bacon, for he was a politician as much as a philosopher, and his life was driven by the ambition for high office. The son of an official of the court of Queen Elizabeth, Bacon did not inherit office but worked his way up, largely on the basis of legal skills and much political wrangling. Like many an ambitious soul in Elizabeth's day, Bacon found the road to power tortuous, but once James I succeeded the Virgin Queen in 1601, his climb was steady and strong, and he reached the highest of political offices in 1618. His ascendancy was brief, however, for he was tried and convicted of bribery in 1621 and lost all his offices and barely escaped imprisonment. The dates of his political rise are noteworthy, for these were the very years in which Bacon put together his reflections on the paths to learning and the place that knowledge had in society at large. His great argument for the reorganization of the sciences, the *Novum Organum* (the *organum* or *organon*, "tool" in Greek, was a term applied to a key compilation of Aristotle), appeared in 1620 and his *New Atlantis* in 1626, the year of his death. Here, for the first time, we can read a systematic discussion of the best methods for people to gain knowledge about the world, and an explicit recognition of the implications of this for their power to change the world through technology.

The application of new knowledge was not the central focus of Bacon's philosophy—the creation of a new method for the improvement of understanding was at the core—but at every step he reminded his readers of the relationship between knowledge and power. The *Novum Organum* consists largely of aphorisms, by which Bacon sought to make the case for his new method and its uses. At the outset, there can be no doubt why he thought this was so important; the first three aphorisms are these:

I.
MAN, being the servant and interpreter of Nature, can do and understand so much and so much only as he has observed in fact or in thought of the course of nature: beyond this he neither knows anything nor can do anything.

II.
Neither the naked hand nor the understanding left to itself can effect much. It is by instruments and helps that the work is done, which are as much wanted for the understanding as for the hand. And as the instruments of the hand either give motion or guide it, so the instruments of the mind supply either suggestions for the understanding or cautions.

III.
Human knowledge and human power meet in one; for where the cause is not known the effect cannot be produced. Nature to be conquered [*vincitur*] must be obeyed; and that which in contemplation is as the cause is in operation as the rule.

This last theme was echoed later, in one of Bacon's most famous passages. Shortly after he praised the importance of printing, gunpowder, and the compass (as quoted at the beginning of chapter 7), he remarked, "the empire of man over things is founded on the arts and sciences alone, for nature is only to be commanded [*imperatur*] by obeying her" (aphorism 129). The meaning of this sentiment is perhaps not as obvious as it could be, but the theme of submitting to nature is one that Bacon returns to repeatedly. It is closely tied to his argument that human beings are fully equipped to comprehend nature, but they must first be aware of the hurdles they face in doing so. Bacon describes these obstacles as "idols," and enumerates four. The idols of the tribe are the weaknesses of human nature and the senses, as they are shared by every person, and thus limit what can be observed and known. The idols of the cave are the limitations of each individual, due to their talents, education, and experiences. The intercourse that humans have with one another is limited by the idols of the market, by which Bacon largely means language and the difficulties that words pose for deeper understanding and proper communication. Finally, idols of the theater are the "various dogmas of peculiar systems of philosophy," and other "principles and axioms in science, which by tradition, credulity, and negligence have come to be received" (aphorism 44). For Bacon, the advancement of knowledge began with recognition and then avoidance of these idols.

The particulars of Bacon's "method" are not very relevant here. The so-called Baconian or inductive method of science, simplistically characterized as the gathering of facts without prejudice until general patterns emerge, was never historically important, if it was even strictly possible. Facts are never observed or collected without some guidance or assumptions, for this is to make the human mind a mere indiscriminate sponge, which could neither organize nor learn nor even discern such "facts." In other words, Bacon's "idols" are not only inescapable, they are necessary to provide meaning to the sensory data the world of nature provides. That this is so was quickly recognized by philosophers as well as students of nature, but they still honor Bacon and his proposed improvements to learning for making one of the first serious examinations of the difficulties that the scientific enterprise always must confront, while imparting an infectious optimism about its eventual progress.

It is this optimism that is perhaps the most important legacy of Bacon's works on science and technology. He was not the first to tout the significance of key discoveries, as he did in singling out printing, gunpowder, and the compass, nor to proclaim stridently the superiority of modern times over the supposed golden ages of the ancients. Nor, of course, was he the first to make explicit the problem of hewing too closely to ancient—that is, mostly, Aristotelian—dogmas about nature or to suggest that such relatively recent discoveries such as those of the overseas explorers or of the post-Vesalian anatomists and physicians marked the current age as one of unparalleled discovery and innovation. But no one better than Francis Bacon was able to suggest the endless possibilities that their age heralded, the extent to which the world was not at the apex of discovery and learning, but rather at the beginning of an upward journey that had no earthly end. It was very much a part of Baconian optimism that he attempted to delineate the vehicles that seemed to him to make such a journey inevitable. The most important of these was indeed the knowledge of nature. The sentiment that nature was to be conquered (*vincitur*) or ruled (*imperatur*) by knowing its secrets and by following its own laws was moved by the belief that this was indeed the human destiny and duty. Improvement was to be boundless, as long as it was guided by a respect for the material world and fueled by the endless quest for understanding of that world.

Bacon did not merely, however, proclaim optimism about progress and the philosophy that he believed facilitated it, but, as behooved a politician, he thought hard about how society should foster these things. In his *New Atlantis*, published after his death, he set down an institutional foundation for promoting both the expansion of knowledge through the application of the inductive method and the application of new knowledge to fashioning improvement in the arts. He described in his fanciful utopia an imaginary society somewhere beyond the charted seas, an institution called "Solomon's House" that existed to allow the Baconian method full reign, to focus energies on the gathering of facts about nature, the collection of different methods and crafts from different parts of the world, and the cataloging of new knowledge to make it accessible and useful to all. It is hard to assess the real influence of the *New Atlantis*, but certainly Bacon's model of a society dedicated to the expansion of natural knowledge and of the exploration of means for exploiting such knowledge was one that found important devotees in the second half of the seventeenth century. Within a generation of his death, Bacon and his ideals, both of an unobstructed search for the truth of nature and of a concerted effort to use this truth to bend nature to human will, achieved powerful affirmation in English society, eventually leading to the formation of the Royal Society of London in 1662.

The Royal Society, which actually began meeting in 1660, soon after the end of the great civil war that had torn English society in the middle years of the century, was the formal continuation of a group of gentlemen who could trace their earliest meetings to discuss philosophical and scientific questions back to the 1640s. They first met at Gresham College, in London, which had been established a few decades before as an alternative to the tradition-bound universities at Oxford and Cambridge. In this environment, the ideals propounded by Bacon became key sources of guidance. The organization, for example, undertook to fulfill his instructions to compile histories of the trades and crafts ("artificial history," as opposed to "natural history"), although the effort was never pursued systematically (this particular notion would have much more influence among the French *philosophes* a century later; see chapter 13). Since the organization was very much a club of well-off gentlemen, few if any of whom had ever worked with their hands except as enthusiastic amateurs, this high regard given to the knowledge of crafts and craftsmen was remarkable testimony to the extent to which Bacon's creed of the intellectual and social worth of technical knowledge had found disciples.

Another way in which the Royal Society represented not only full acceptance of Bacon's call for systematic action to advance learning but also an appreciation of the great change in the sources of knowledge that the new science heralded was the central place given to experiment in its meetings. Soon after the Society's formation, the organization appointed its first salaried officer, Robert Hooke, as "Curator of Experiments." Hooke's job was to prepare one or two experimental demonstrations for each meeting and to keep the society's collection of instruments. This insistence on public and repeatable experiments was one of the most distinctive features of the new society, for other groups had been formed in the past for meeting and debating philosophical ideas, but the focus on experiments and instruments in the Royal

Society made it resemble Bacon's Solomon's House more than any prior institution, in spirit if not actually in operation (figure 9.2).[10]

So much is experiment a part of the modern image of science that we have difficulty comprehending the extent to which its place in "improving natural knowledge" (to quote the stated purpose of the Royal Society, from the charter of 1663) was a matter of some controversy. Before the seventeenth century, in the spirit of Aristotle and most other philosophers, the facts of nature came from experience, as opposed to experiment. That is, one used ordinary and common observations of the world to make plausible statements about how and why things happen. It is my common experience, for example, that if I push something harder, it will go faster than if I push it softly. When I lift a heavy object, it pushes harder on my hand than a lighter object. It therefore follows that a heavy object will fall more rapidly than a light one. It does not occur to me, as a traditional philosopher, to devise some kind of experiment to test this logical, indeed self-evident, conclusion. Furthermore, even if I did do an experiment, that would prove nothing if the experiment's results contradict everyday experience. Why, after all, should a contrived situation (an experiment—such as dropping weights from a high tower) be a better source of truth than the ordinary experience, repeated many times, of everyday life? Even Galileo, it should be pointed out, would have been hard-pressed to answer this question (and, hence, the experiments he did carry out—which did not include dropping weights from high places—he tended to repeat many times), But Bacon argued that sometimes it was necessary to seek out critical experiments that would point the way to unseen truths. He called these "instances of the fingerpost," with reference to a road sign that marked a division of ways; a carefully designed experiment would, if properly performed, point the right direction. It was largely in this spirit that the Royal Society relied on public experiment, so much so that it took as its motto "*nullius in verba*," or "On no one's word (alone)." A number of scholars have pointed out that this attitude was much encouraged by the Protestant establishment of seventeenth-century England, with its reliance on personal witness rather than received orthodoxy as the basis for belief.[11]

The Royal Society, for all the attention given to Bacon's linkage between knowledge and power, and for all the inspiration it received from his admonition to study and extend knowledge of the useful arts, was not a force for technological change, either in Restoration England or thereafter. After some bright and promising early years, the society's reputation went into steep decline until, toward the end of the century, the ascendancy of Isaac Newton, whose synthesis of the mechanical and mathematical picture of the world was the crowning achievement of seventeenth-century English science, resurrected its standing. Newton was elected president of the Royal Society in 1703, and he remained the most powerful and prestigious figure in English, if not in fact in all European, science until his death in 1727. The corner-

Figure 9.2
From Robert Hooke, *Micrographia* (1665): a fly (flea). Library of Congress photograph.

stone of Newtonian physics, the *Principia Mathematica Philosophiae Naturalis*, or *Mathematical Principles of Natural Philosophy*, was published in 1683, and it demonstrated most decisively the power of the new science—building on the astronomy of Copernicus and Kepler and the mechanics of Galileo—to describe the key phenomena of the natural world in spare, direct mathematical language. Armed with new mathematical tools, most importantly what we call calculus, Newton and his followers could tie together disparate facts and observations. The central intellectual triumph was the law of universal gravitation, by which the Newtonians could ascribe everything from the tides of the earth's seas to the elliptical orbits of the planets to, indeed, the falling of an apple to the ground, to one single ever-present force, whose action could be described and calculated in the most precise way. The specifics of the *Principia* and of the mathematical tools that it employed were nearly impossible for any but trained mathematicians to follow, but the implications in terms of unifying principles for the description of physical phenomena were rapidly and widely appreciated. The fact of Newton's enormous reputation was itself eloquent testimony to the long-lasting influence of Bacon and his disciples, for everyone could appreciate the significance of a man who apparently made the command of nature that much greater through mathematical genius.

The Newtonian synthesis was also a triumph of what became known as the "mechanical philosophy," and this too was to have implications for how people, especially in England, viewed the possibilities of technological change. The mechanical philosophy held that all phenomena in the world could be and ought to be described purely in terms of matter and motion. In its clearest form, as propounded by such champions as Robert Boyle in England and, most importantly, René Descartes in France (who disagreed strongly with some of Newton's approaches), all matter in the world could in turn be described as collections of indivisible particles, or, most commonly, corpuscles. These corpuscles—not to be confused with the "atoms" of modern science, although in a crude way comparable—might possess distinct size, shape, or form. All the qualities of things in the world could, ideally at least, be ascribed to the particular combinations and movements of these corpuscles, and the goal of natural philosophy was to break phenomena down to the point that they could be described purely in these terms. On the continent, the Cartesians took this approach as the beginning of speculative systems, seeking to resurrect a philosophical structure that would be as comprehensive as that of Aristotle, once complete. Descartes became famous also for his effort to describe living things in mechanical terms as well. In Britain, under the influence of Bacon, Puritanism, and other philosophical tendencies, speculative systems of this sort were shunned in favor of more "Baconian," or empirical, forms of explanation. But regardless of these differences, the importance of mechanism was paramount to all followers of the new science. The reigning metaphor for the universe and its working was the clock, and an understanding of the clockwork of the world was understood as the core goal of natural science. When Newton's work appeared, universal acceptance was not immediate, but within a generation the Newtonian framework was ascendant throughout European science.[12]

The triumph of the mechanical philosophy and of the clockwork picture of the world made the machine the central image of the modern worldview. This displaced an older conception of nature that was essentially organic. Some scholars have seen this as a kind of masculinization of the world picture, and others have adopted the notion of the famous German sociologist Max Weber in terming this the "disenchantment of the world." Either way, the change could not help but to make mechanical technology seem somehow to be the key to understanding the universe. In addition, in making the clock the central metaphor for how the world works, there emerged a technological goal that had before been only implicit, and that was the making of the perfect clock. Clockwork represented machinery that could always be improved, and for which there existed, at least in theory, an indisputable measure of perfection. If the universe was indeed clockwork designed and put into motion by God, then human technological striving had a self-evident goal: to emulate this perfect clock. It was no coincidence that the rise of philosophical mechanism was accompanied by striking improvements in the accuracy of clocks. In 1656–57, Christian Huygens in the Netherlands introduced the pendulum clock, rediscovering what Galileo had observed decades earlier, but had not applied. The improvement was astounding—daily variations were measured in seconds, while the best older clocks had daily variances of at least a quarter-hour. The actual uses for this sort of accuracy were rather minor; astronomers could take advantage and navigators were inspired (but not immediately aided) by the thought that an accurate clock aboard ship would solve the ancient problem of determining longitude at sea. But for the world at large, the improvement in clock accuracy was less a practical advance than it was an inspiration of mechanical marvels to come.[13]

10 Improvers and Engineers

Mr. Edward Darcy was a groom in the household of Queen Elizabeth I in the last years of the sixteenth century. We do not know much about him or his duties, but we can infer that he carried them out commendably, for he was able to ask of his queen (and employer) a favor common in that day, which was the grant of a monopoly to allow him exclusive rights to some product or service. And so it was that Edward Darcy was granted the sole right, "in payment of his long and acceptable services to the Crown," to make or import playing cards into England. For this he paid an annual "royalty" of £100 per year. Shortly after receiving this grant, however, it came to Darcy's attention that a London haberdasher, one Thomas Allin (or Allein), was selling playing cards on his own. He took Allin to a court of common law to enforce his royal grant of monopoly. He lost, and in that losing he won for himself immortality, in the form of the court's decision in the case of Darcy v. Allin, also known to historians as "the case of monopolies." The court ruled in 1602 that such grants of exclusive rights by the Crown violated common law—that is, generally accepted notions of the rights and privileges of ruler and subjects. Such grants offended the court's sense of what was fair and equitable, and it warned against making them, with one exception:

> ...Where any man does by his own charge and industry or by his own wit or invention doth bring any new trade into the realm, or any engine tending to the furtherance of a trade that never was used before, —and that for the good of the realm—that in such cases the king may grant to him a monopoly patent for some reasonable time until the subjects may learn the same in consideration of the good that he doth bring by his invention to the Commonwealth. Otherwise not.[1]

Just as the Venetian Senate had declared more than a century earlier that it would grant exclusive privileges to someone with a new "art" for the state, so now did an English law court confirm the wider appeal of this notion. The politics of royal privilege, however, were much messier than a simple law court ruling could manage, particularly in the reign of James I, when king and parliament were in a constant tangle

about power and money. Feeling quite provoked by James, in 1623–24 the English parliament enacted the "Statute of Monopolies" to codify the court's decision, expressly forbidding the Crown from making exclusive grants for manufacturing, selling, or importing things, but with the now generally recognized exception, in which grants could be made for fourteen years or less for "the sole working or making of any manner of new manufacture within this realm to the true and first inventor and inventors." The fourteen-year term was twice the standard seven years of a craft apprenticeship, and the logic was that the period of the monopoly privilege should be used to teach others the new art, and that they should then be free to make full use of their knowledge. As clear as the new statute might have seemed, the Crown continued to seek the expansion of royal privilege, in this as well as other matters, and Parliament continued to resist until Civil War was the inevitable result. By the end of the seventeenth century, however, the idea of the patent for invention had entrenched itself in English law and custom, and the quest for patents and their related privileges became one of the characteristic ambitions of the age.[2]

In the century after England's "Glorious Revolution" of 1688 (in which the powers and prerogatives of Parliament over the Crown came to be more firmly entrenched, and Britain's slow march to constitutional monarchy quickened its pace), the elements of a true patent "system" were gradually but steadily worked out. The emergence of the patent for invention as a legal and economic instrument, approaching its modern form and use, marked important and permanent changes in the very conceptualization of invention and improvement. In particular, the source of novel things came to be perceived more and more as human ingenuity and enterprise and less and less the work of divine Providence, revealing hidden treasures and possibilities. In addition, the need to work out the legal elements of a patent system required more carefully defined notions of what constituted invention and the place of invention in the social and economic order. Daniel Defoe, for example, one of the most prolific and acute commentators on the English scene, set out the distinction between invention and improvement, "For I do not call that a real invention which has something before done like it, I account that more properly an improvement." He went on elsewhere to declare that "it is a kind of proverb attending the character of Englishmen, that they are better to improve than to invent, better to advance upon the designs and plans which other people have laid down, than to form schemes and designs of their own and which is still more, the thing seems to be really true in fact, and the observation very just."[3]

As the eighteenth century went on, however, Englishmen began to see invention as indeed a characteristic of the nation, and one of crucial importance to economic prosperity and the general welfare. At the same time, the distinction between invention and improvement, so clearly stated by Defoe, came to appear more and more significant to those looking for the sources of national wealth and prosperity. In the mid-seventeenth century, despite the example of Bacon and his praise of the great inventions of the era, most commentators put their faith for advancement in the improvements wrought by skilled craftsmen, incremental steps in tools, methods, and materials that would maintain and increase the quality of products (generally of much greater concern than quantity). Invention—the development of something quite new and unexpected—was looked upon as much with suspicion as admiration. Inventions could be "fabrications," or "deceptions," meant to imitate a more genuine article or to lead one astray. The display of ingenuity was admired, but usually in the context of a familiar art or craft. If it embodied something too novel, on the other hand, it was more likely to be looked upon with suspicion. This traditional attitude, however, gave way in the eighteenth century to the notion that invention was the source of new wealth and prosperity, the valued display of human genius. Economic writers were actually rather slow to take up this theme. Adam Smith, for example, in his *Wealth of Nations*, praised the virtues of the division of labor as a means of rationalizing and improving production, but he neglected the development of novel tools or techniques as another source of new wealth. Others, however, began to praise innovation as the source of much promise for the future. This was both a cause and an effect of the growing value attached to patents. Since it was agreed that incremental improvements or simply greater displays of skill were not patentable, the promoters of patents focused attention on inventions that were clearly novel, and so emphasized the distinction between inventions and improvements (in the manner of Defoe, above). This distinction, not a particularly important one before the eighteenth century, became more and widely perceived as crucial, not only in law (where patent rights depended on it) but in the wider perception of technical and economic progress.[4]

Accompanying this change in attitudes was another one, just as important for the future of technology. Before the mid-eighteenth century or so, the social desirability of an invention (and thus, to some extent, its patentability) was judged on the extent to which it encouraged employment. The importation of a new trade into the country, or the introduction of a new machine or tool, was valued in large part by the extent to which the innovation provided work for Englishmen (and would reduce imports). The flip side of this notion was that inventions or improvements that reduced, or threatened to reduce, employment were inherently undesirable. "Labor-saving," in other words, was seen as destructive to the social order. This notion was not simply the reflection of narrow parochial interests, but was a widespread attitude about what constituted a healthy and prosperous economy. In the course of the eighteenth century, this attitude gradually changed, but not easily nor universally. A shortage of productive work, especially for the poor, long remained a feared evil, and the slowly growing number of writers promoting mechanization and the rational organization of labor had constantly to assure their readers that the long-term benefits of producing more goods per worker would outweigh the short-term difficulties of unemployment. It was not until well into the second half of the eighteenth century that this point of view became the more common, and the controversy continued well into the nineteenth century.[5]

The virtues of "improvement," however, were widely touted in seventeenth- and eighteenth-century Britain. In no other period or place, with the possible exception of early nineteenth century America, did improvement become such a watchword of society. For most Englishmen in the seventeenth century, the term improvement had a rather specific connotation, and that applied to land. Such a usage was actually quite ancient and was initially connected with legal rights—a person's property rights in land, according to long-held legal theory, rested fundamentally on the improvements that he brought to the land. Royal grants of land, going at least as far back as Charlemagne's day, were often given provided that the recipient "improve" the land, which generally meant cutting trees and planting crops or raising livestock (see chapter 2). In the late seventeenth century, the English political philosopher John Locke made this notion of improvement the foundation of his conception of property and of why private property was sanctioned by God. On a more practical level, writers throughout the century, particularly in Britain, described the means and virtues of agricultural improvement, as a route to increased wealth (on the part of the landed, to be sure) and greater prosperity for society. To a degree, the gospel of improvement was less a technical and economic one than it was political and social—it was used to justify the abrogation of ancient rights, long held by tenants and others occupying the land, in order for landowners to increase profits, often by enclosing formerly common fields and then using them either for grazing or for intensive, market-oriented cropping. Nonetheless, this possibly hypocritical use of "improvement" would have been ineffective if it had not been done in an environment in which the larger values of improvement, particularly for the greater good of the community and the realm, were not generally accepted.[6]

There were some technological elements to the campaign for agricultural improvement, and the productivity of British agriculture rose so significantly from the late seventeenth century to the nineteenth century that historians have written of an "agricultural revolution." The elements of this revolution are matters of debate, but largely in terms of what particular items contributed most to the success of British farming. Probably the single most important factor was simply the expansion of the land put under cultivation. In the course of the seventeenth and eighteenth centuries, millions of acres, formerly used for pasturage, common grazing, or simply waste lands, were cleared, fenced, and put under systematic cropping. Sometimes this involved technological innovations, as when large sections of low-lying land in eastern England were drained and thus made available for farming. The draining of fens and other marshy areas had been going on since the Middle Ages, but in the seventeenth century, Dutch engineers crossed the North Sea to eastern England and proceeded to apply and adapt their successful systems of dikes, ditches, canals, pumps, and windmills to the fertile but waterlogged areas of Lincolnshire, Cambridgeshire, Suffolk, and other eastern counties. The Duke of Bedford, who began pushing for drainage schemes in the 1620s, was the prototypical "gentleman-improver," and the creation of the Bedford Level and other drainage areas was one of the first large-scale and sustained engineering works undertaken in England. The drainage projects were not simply a matter of building dikes and canals, but were long-term commitments to the maintenance and administration of extensive constructions. In earlier cultures, going back at least as far as the ancient Assyrians and Egyptians, the maintenance of hydraulic works for agriculture was one of the key technical tasks of the state. The seventeenth- and eighteenth-century English projects showed how this could be translated into the workings of a capitalist economy, largely through the active entrepreneurship of self-styled improvers.

The passion for improvement also led to extensive experimentation with different crops and breeds. Different patterns of rotation could maintain production while preserving soil fertility, and novel crops could provide better fodder for cattle or new sources of protein for the peasant diet. The most common form of new rotation patterns was "convertible husbandry," which involved more intensive use of fields for grazing cattle in-between crops. New fodder, such as clover and "artificial grasses" (that is, grasses purposely sown for pasturage), allowed much more successful raising of cattle, both for dairying and for meat production. The most celebrated new crop was the turnip, which came to be grown widely (as animal feed) in eastern England in the mid-seventeenth century. A few new tools, such as the seed-drill and the horse-hoe, were experimented with in the seventeenth century and were sometimes vigorously promoted by such "improvers" as Jethro Tull (figure 10.1), but their contribution to agricultural change in preindustrial England, where labor was still rather cheap, was very modest indeed. While experiments with new breeds of cattle and sheep were an ancient tradition, these became more systematic and widely noticed during this period. Significant changes did not appear until the middle of the eighteenth century, when the work of Robert Bakewell, who bred successful longhorn cattle and new, more profitable, breeds of sheep, received widespread attention. Bakewell's methods of selective breeding, what was sometimes called "artificial selection," influenced all later breeders, as well as naturalists in the next century (Charles Darwin's idea of "natural selection" was an adaptation). The new crops, breeds, and tools were less key innovations in themselves, as they were visible, public symbols of an attitude toward change. Agriculture, that most basic and widespread of all human endeavors—and arguably one of the most difficult to change—was seen as amenable to improvement in an infinite variety of ways, and the benefits of improvement were perhaps most readily and widely appreciated in that realm.[7]

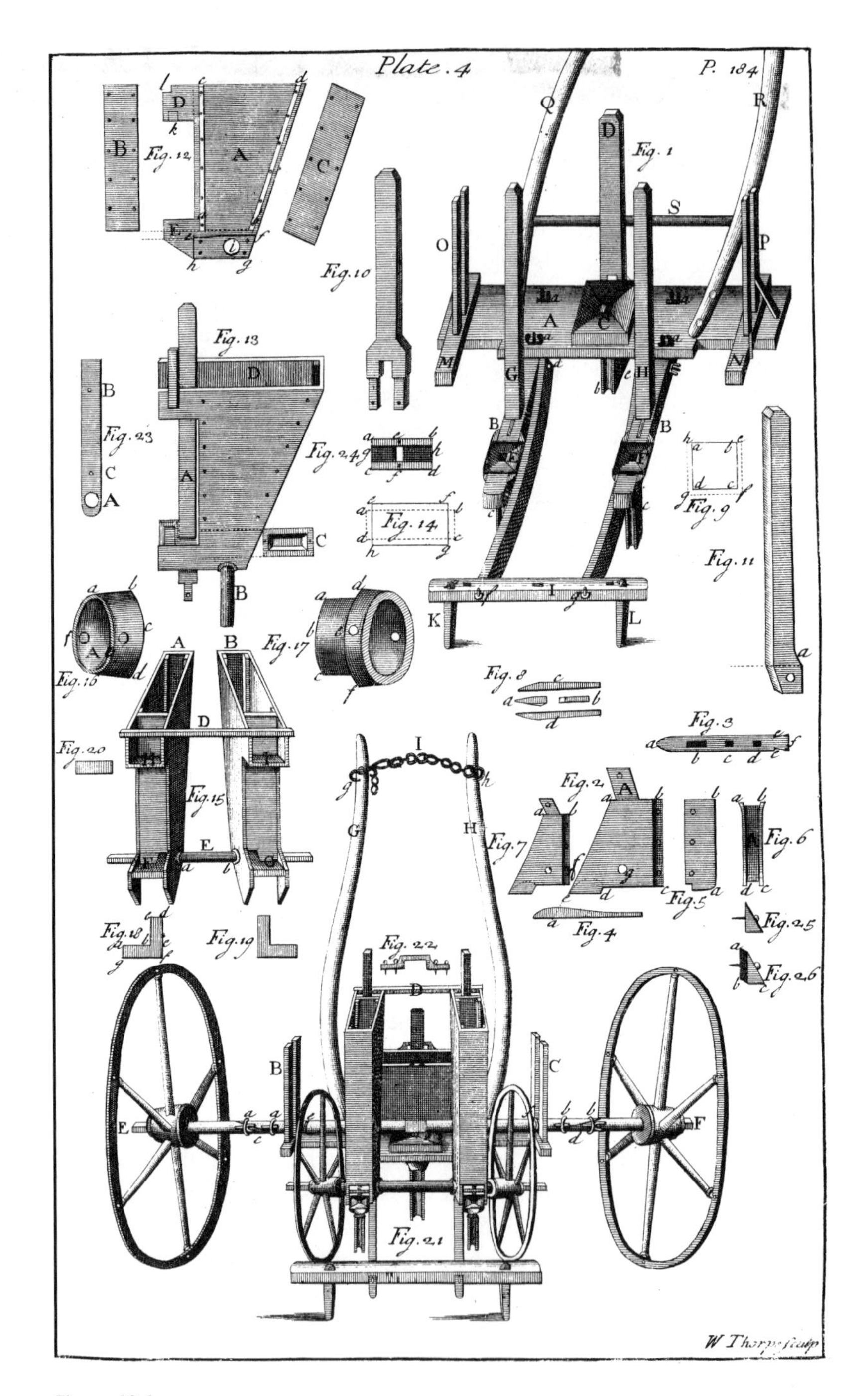

Figure 10.1
The seed drill, from Jethro Tull, *The Horse-Hoeing Husbandry* (1733), was an effort to rationalize through machinery the ancient practice of sowing seed. Rare Book Collection, Special Collections, National Agricultural Library, Beltsville, Maryland.

The campaign for agricultural improvement revealed some of the important new mechanisms for fostering technical improvement more widely. A large number of publications, touting new crops, new methods of fertilizing, new tools and methods, or simply promoting the extension of improved land, made their way into print in the middle of the seventeenth century and thereafter. Walter Blith's *The English Improver: or, A New Survey of Husbandry*, for example, appeared in several editions after its first appearance in 1649 (the 1652 edition was actually called *The English Improver Improved*). Other works promoted specific techniques, such as the wider use of clover or treatments for animal diseases (an example of the latter was Gabriel Plattes's *Practicall Husbandry Improved: or A Discovery of infinite Treasure* of 1656). While, as we have seen, a tradition of technical writing was well established by the seventeenth century, these agricultural works were notable in their advocacy of "improvement" explicitly and in their addressing an audience of gentleman farmers who might make direct use of them. The tradition of agricultural writing grew and diversified in the course of the eighteenth century, and was joined late in the century by important surveys of agricultural practice by agriculturalists, who traveled around different regions to collect and compare techniques. In the middle and late eighteenth century, Arthur Young and William Marshall, in particular, published a host of studies to highlight the best practices of different regions and to inform a wider audience about systematic agriculture and improvement. Agricultural societies, generally groups of gentlemen farmers who may have gathered as much for social intercourse as for learning and debating new farming methods, sprang up throughout England. Their real effect on agricultural practices was probably quite small, but they represented early models of the kinds of voluntary associations that were to reshape technical communications in the coming centuries, on both the local and national levels. As exclusively male organizations, they also heralded the increasing disadvantages that women were to face in full participation in technical discussions, even in areas, such as dairying, they had formerly dominated. Through such means, agriculture, that most basic and hidebound of human activities, provided models for a new world of technological change.[8]

Another source of models for change emerging in this period—and another widely heralded venue for "improvement"—was a growing program of what was to be called "internal improvements," or the construction of a transportation network. The same cohort of gentlemen entrepreneurs that identified themselves (and their interests) with agricultural improvement sought to promote the building of turnpikes and canals and the improvement of rivers and harbors. A number of factors spurred the great rise of investment in the British transportation system in the seventeenth and eighteenth centuries: more intensive agriculture depended on more accessible markets, the example of Dutch and other trading centers led to efforts at emulation, increased mercantile wealth sought appropriate long-term investments as the source of profits, and the growing commercial influence in British governing circles, particularly after the Civil War, eased the legal and financial hurdles that had long dogged large-scale projects. The seventeenth century also saw the development of important new financial mechanisms and instruments that the British often borrowed from Continental pioneers, but which they brought to new levels of efficiency and sophistication. Banks, credit systems, trading exchanges, and joint stock companies, all of which existed rudimentarily earlier, grew into something approaching their modern form in the seventeenth and eighteenth centuries. These were indispensable tools for focusing entrepreneurial energies on transport improvements. And the improvements, in turn, were indispensable for the rapid commercial growth of the eighteenth and early nineteenth centuries which supported the transformation of manufacturing and consumption.

Until well into the nineteenth century, water transport possessed enormous advantages over land travel in terms of economy, safety, and reliability. All bulk trade, in particular, such as grain, wood, or coal, depended on water transport, either on rivers or by the coastal trade. For centuries, in Britain as in the rest of Europe, trade was fundamentally dependent on natural waterways and harbors, with only modest works being done to promote safety or to clear obstructions. In the seventeenth century, however, this began to change, not only in Britain but elsewhere in Europe. From a few, tentative canal projects in the wealthier areas of Renaissance Italy and the Low Countries to very ambitious cross-country efforts in France and England, the water transport system of Europe received growing attention during this period. Different areas brought different resources and different styles of organization to the effort, and the contrasts reveal a great deal about the different paths that technological improvement took in the preindustrial period. The same contrasts, particularly in the different roles of the state and private enterprise, can be seen in the more modest, yet still important, efforts to improve road systems as well as in the different levels of support that governments gave to emerging industries.

In Britain, the most important transport network was the coastal trade. The advantages of being an island, in which a very large proportion of the population lived within just a few miles of the coast or of a river leading to the sea, were considerable for internal communications. But as populations grew from the late sixteenth century onward, particularly in cities and towns, and as agricultural production became more and more market oriented, the older systems of transport strained to keep up. It became evident to promoters that there were profits to be made from providing attractive alternatives, and so the concept of improvement was explicitly extended to transport works. At first these took the form of river improvements, which largely meant the clearing of obstructions, dredging silted areas, and either demolishing or moving mills and other structures that interfered with navigation. In more ambitious efforts, channels were widened and eventually canals were constructed to bypass loops or immovable obstructions. In the course of the seventeenth century, these efforts created new inland port towns, such as Bedford and St. Ives, in eastern England. These places became important sources for the shipment of grain from the newly improved farms into London and other growing cities.[9]

In the eighteenth century, the improvement of waterways in Britain became even more ambitious, and a period commenced that was later seen as a kind of "canal age." The beginning of this is typically given as the construction in the 1760s of the Duke of Bridgewater's Canal, which connected the Duke's coal mines to the growing city of Manchester about ten miles away. While the canal was not a long one, it was still ambitious engineering, featuring underground portions near the coal mines and, most famously, an aqueduct, two hundred yards long and thirty-nine feet high, carrying the canal over the River Irwell. The great profitability of the work (it halved the cost of coal in Manchester), financed entirely by the Duke of Bridgewater, sparked a canal boom which lasted, off and on, until the middle of the next century, by which time Britain boasted of 4,250 miles of canals. Many of these canals were simple constructions, with little technical or even economic merit, but some of them were pioneering pieces of engineering, with aqueducts, ambitious cuts, series of locks and dams, and other fundamental features of hydraulic engineering. The canals served as an indispensable training ground for Britain's civil engineers, who went on to make great names for themselves in the building of harbors, lighthouses, roads, bridges, and, above all, railways. The canals also served as the model for financing of much of British civil works, but not in the fashion of the Duke of Bridgewater's Canal, with its sole source of noble funding, but rather in the form of joint stock companies, which turned out to be the tools not only of ambitious improvers but also of unscrupulous stock manipulators. Even with the enormous amount of waste and loss that accompanied unwise speculation, the private financing of the British canal system probably produced the most useful canal network in the world.[10]

The British model was not followed on the Continent. Indeed, in many ways continental—primarily French—engineers preceded and superceded the British canal efforts, but in a very different style of both engineering and enterprise. The pioneering canal builders in Europe were the Dutch, who excelled in all things hydraulic, out of necessity, and the Italians, whose wealth and ambition led them to stretch the limits of the possible before most other Europeans. The construction of Milan cathedral in the late twelfth century led to the building of a thirty-one-mile-long canal to carry marble down from the nearby hills. This functioned, however, as largely a one-way passage, and it was not until the fourteenth century that fully functioning pound locks (which held water in a basin between two gates to change the level of a barge, going either up or down) were being built on European canals (the Chinese Grand Canal used them a century earlier). In the seventeenth century, canals began to be political tools, sponsored by newly emerging monarchies to bind together formerly disparate regions. In Prussia, the rulers of Brandenburg and, after 1701, Prussia, promoted canals to link the Elbe, Oder, and Weser rivers, not incidentally bypassing smaller states that had exacted sometimes exorbitant tolls on their small stretches of river. The most spectacular example of this sort of effort, however, was in France, where kings beginning with Henri IV sponsored canal projects of breathtaking scale, seeking to link the farther regions of what was by far the largest European state. The Briare Canal, completed in 1642, linked the Loire and Seine rivers by using forty locks to rise up 128 feet on one side of the summit level and then drop 266 feet down to the other side. Even more ambitious was the Canal du Midi, also known as the Languedoc Canal, which stretched 144 miles to link the western river port of Toulouse (and thus the Atlantic port of Bordeaux) with the Mediterranean Sea on the east. Its construction stretched from 1666 to 1681 and required 119 locks, three major aqueducts, and even a 180-yard-long tunnel. Supplying water to the three mile stretch at the summit (620 feet above the Mediterranean level) was itself a major accomplishment of water diversion and storage. Just as significantly, the Canal du Midi provided the model for civil engineering works in France—which were largely state efforts, from financing to engineering to maintenance. This in turn provided the impetus for the French to create the first systematic effort to organize and train what we today would call civil engineers.[11]

The great canal projects of the seventeenth and eighteenth centuries were the first large-scale engineering works of modern European history. They were also the first examples of important engineering that engaged a wide range of practitioners, from promoters to engineer/designers to builders/contractors to philosophers/scientists. The problems of the more ambitious works, particularly on the Continent, were of such magnitude that advice and assistance was sought from all quarters, and by the late seventeenth century this included the great scientific academies that had sprung up in the major European states. These, such as the Académie Royal des Sciences established in France in 1666, were state-sponsored groups of mathematicians, natural philosophers, naturalists, and physicians whose work was meant to be at the disposal of the state. While inspired to some degree by the same Baconian model that led to the Royal Society in London, the continental academies tended to be more organized, even, perhaps, more "professional," although that term is somewhat misleading in its seventeenth century context. Their role was much more explicitly advisory for the state, particularly in France. For the building of the Canal du Midi, for example, the king's chief minister, J. B. Colbert, not only turned to the Académie des Sciences, but, realizing its limitations for the task at hand, founded the Académie Royal d'Architecture.

Even more significant for the great projects than the academies, however, was the appearance of a new class of practitioners, the first individuals that were to be identified as "engineers," rather than architects, masons, or builders. While the term engineer, in one form or another, had been used since the Middle Ages, it always applied to the designers and builders of "engines" of war, which meant siege devices, catapults, and the like. Such individuals also typically were in charge of building fortifications, and might in fact have a wide range of skills, but their military orientation was foremost, and nonmilitary efforts would identify them more as architects. In the mid-seventeenth century, however, the crossover from military engineering to what came to be called civil engineering became more pronounced, with important implications for the organization of engineering knowledge and practice. This process was accompanied by the emergence of individuals whose skills were highly valued by promoters and rulers and who were able to parlay these into fame, wealth, and status.

The appearance of the professional engineer was most evident in France, where state sponsorship of large projects like the great canals and road and bridge construction placed a high value on individuals who could organize such efforts. One of the first to make a name for himself in this way was Sébastien de Vauban (1633–1707), who began as a military engineer in the French army, learning his craft in the field and distinguishing himself quickly as a designer of fortifications. In his long career, Vauban demonstrated the utility of a skilled and imaginative engineer, not only for military works, but also for some of the ambitious civil works of the late seventeenth century. He designed the great aqueducts of the Canal du Midi and at his death was working on plans for an extensive canal work in French Flanders. Roads and bridges turned out to be even more important proving grounds for the new profession, as demonstrated by the remarkable careers of the Jacques Gabriels, father and son. The elder Gabriel (1630–1686) was one of France's most successful architects, thriving with the ambitious projects of Louis XIV. His most important engineering work, begun just before his death, was the Pont Royal in Paris, where he oversaw the construction of the bridge designed by the architect Jules Hardouin Mansart. This bridged the Seine to connect the Tuileries Palace with the opposite bank of the river, and it demonstrated the grace and utility of elliptical arches. Gabriel also was the first since the classical Romans to use hydraulic cement ("Pozzalana," from the area in Italy from which it was derived) for bridge construction. His son, Jacques (II) Gabriel, was an even more daring engineer, overseeing the building of thousand-foot-long bridge over the Loire River at Blois, featuring eleven elliptical arches and a central span of eighty-seven feet. In 1716, the younger Gabriel became the first chief engineer of the newly organized Corps des Ingénieurs des Ponts et Chaussées, which gave France the first permanent agency for the designing and building of civil works.[12]

While water transport long remained the key means for moving both goods and people, often overland travel could not be avoided, and as urban populations and trade increased in the sixteenth and seventeenth centuries, the unsatisfactory condition of European roads became a matter of serious concern both to governments and to commercial interests. As happens so often in history, concerted efforts to rectify matters were initially the product of military concerns. The widespread warfare of this period, particularly on the Continent, spurred states that could afford it to pay extra attention to the improvement of roads, and for the first time since the end of the Roman Empire more than a millennium earlier, the road network of Europe began slowly to improve. Until the mid-eighteenth century, the only useful model for the road builders was that of the Romans, although economic necessity often required the development of alternatives to the expensive Roman road model, based on carefully shaped and fitted paving stones. The French were once again the most systematic in their attack on the problem of road improvement. A member of the Corps des Ponts et Chaussées, Henri Gautier, published influential treatises on roads and on bridges at the turn of the eighteenth century, and in the next generation Bernard Forest de Belidor published the first works that attempted to set down the mathematical and physical foundations for engineering construction (*La Science des Ingénieurs*, 1729, and *Architecture hydraulique*, 1737–53). This new emphasis on the mathematics and theories of engineering led to the founding in 1747 of the world's first engineering school, the École des Ponts et Chaussées, which was put on a professional footing by an accomplished bridge engineer, Jean Rodolphe Perronet. By the middle of the eighteenth century, the French had not only created the first organization of professional engineers, but had also started to send engineering education down a novel path, oriented toward mathematical and physical knowledge rather than simply empirical rules and practical experience.[13]

The British experience in road building was very different from the French, and in those differences, just as in the contrasting styles of canal building, could be seen distinctive approaches to technological improvement generally, as well as to the development of professional engineering. In Britain the building and maintenance of roads was exclusively a local, "parish," responsibility, and was generally based on the annual duty of each man in a parish to give six days of labor towards the upkeep of local roads (wealthier men had to supply a team of horses or oxen and two laborers). This system naturally led to widespread variation in the quality and capacities of roads in different regions, and furthermore encouraged neglect of through roads at the expense of work that was only of local benefit. As long as water carriage could handle all but minor traffic, this system was workable, but in the seventeenth century, growing population and trade along with new fashions of travel put unsupportable strains on the ancient road system. The stagecoach, first for public service and then increasingly, from the 1650s or so, for wealthier individuals, displaced travel by wagon or horseback. These coaches, much more comfortable and accommodating than the older means of travel, were also much heavier and more damaging to roads. In the countryside around London this posed particularly difficult problems, for a band of heavy clay soil, some fifty miles in width in places, separates London from much of the north and west of the country, and this clay soil was easily turned into an impenetrable mire through much of the year (England is not, after all, a dry country). After the middle of the seventeenth century, new means—more economic and political than technical—were devised to improve roads. The most important of these were the turnpikes, authorized by legislation that allowed local commissions or private companies to put gates on roads and charge tolls in return for substantial improvement of their condition and upkeep.[14]

It was not really until the mid-eighteenth century, however, that British internal transport began the steady improvement that helped to sustain domestic commerce during the years of expanded production that followed. The canal boom, as we have seen, did not begin in Britain until the 1760s, and it was not really until the 1780s and 1790s that the new canals and navigations began to make important contributions. Similarly, while the turnpike effort accelerated from the late seventeenth century on, new techniques were required before significant improvement could really by marked in overland travel. The building of the late-eighteenth-century canals and roads was also marked by the emergence of the British style of civil engineer, which contrasted markedly with the products of the highly organized French system. The chief engineer of the Duke of Bridgewater's Canal, for example, was originally a wheelwright's apprentice of very humble background. James Brindley had established himself as a successful and resourceful millwright when the Duke approached him to take change of the works for the new canal. It is not clear why Brindley, whose only construction experience up to that time would seem to have been in mills, received this charge, but his hiring was evidence of the scarcity of obvious engineering experience in England. Brindley's success, however, was readily parlayed into other engineering assignments, including the Grand Trunk Canal, which eventually connected the Mersey with the Trent and the Severn, thus connecting Manchester with Birmingham and Bristol and linking the key industrial areas of midland England. Brindley became the first of Britain's heroic engineers. The drive toward improvements in roads and other forms of transport provided the stage on which the hero civil engineers of the nineteenth century won their reputations.[15]

Increased trade was an indispensable stimulant of the road and canal projects of the seventeenth and eighteenth centuries, particularly in Britain, where military needs were less significant than on the continent. Not only was the amount of trade, fostered by growing populations, especially in cities and larger towns, increasing significantly in this period, but the kinds of commodities that needed to be shipped longer distances spurred serious investment in improved transport. Of particular importance was the increasing trade in iron and coal. The iron industries of Europe grew significantly in the seventeenth century, encouraged not only by increased commerce generally but also by large-scale military consumption. Cast-iron cannon almost entirely replaced the much more expensive bronze cannon of earlier armies, and other military requirements, including naval usage, gave European ironfounders unprecedented demand for their products. Combatants typically were not able to fulfill their own requirements for iron, and so depended on imports, and this dependence in turn made reliable transport, overland as well as by water, particularly important. It also brought to the forefront of commerce some areas of Europe long at the periphery, such as Sweden and Russia. Sweden, in particular, became a factor in European technology and commerce for the first time in history, exploiting vast resources of timber, copper, and iron, and in the process extending its technological capabilities in impressive ways.

Since the Middle Ages, Swedish metalworkers had gained a reputation for their production of copper and iron. The copper was originally more important, since iron was widely distributed throughout Europe but copper was much scarcer and more valuable. In Sweden, the great mine at Falun, worked since the early Middle Ages, became famous throughout Europe, for both its scale and the quality of its output (figure 10.2). But in the seventeenth century, iron became Sweden's key export, and some parts of Europe, including the Netherlands and Britain, came to rely on Swedish imports for a significant proportion of their iron. The success of the Swedish iron industry began with the high quality of the country's ores combined with the readily available fuel represented by huge forests. The upheavals of the early seventeenth century, such as the Thirty Years War, also gave the industry a boost, as the Dutch, in particular, found it increasingly difficult to acquire German iron and so had to go across the North Sea for their needs. The English market followed, in large part because native ironmakers simply could not keep up with demand, but also because the low-phosphorus iron of Sweden was far easier to work than most English iron. A growing scarcity of wood for charcoal—an indispensable fuel for ironmakers—also placed considerable constraints on British industries. In Sweden, the expanding overseas markets combined with strong interest from an increasingly powerful monarchy and vigorous participation in the trade by a growing merchant class to encourage investment and experimentation. In the course of the seventeenth and eighteenth centuries there were no key technical innovations in Swedish ironmaking technology, but instead a tradition of improvement that continued to yield useful results. Blast furnaces grew larger, processing facilities became better organized, and systems for quality control and inspection were put into place. The larger blast furnaces were, in turn, made possible by better mill construction and even improved bellows. The growing industry thus provided the country the foundation for the development of other technologies in the years ahead.[16]

The Swedish example is worth exploring just a bit further, for it provided an instance of the kind of individual who was to become an important feature of European technological development after the seventeenth century. Christopher Polhem came from a comfortable background, but personal circumstances forced him to

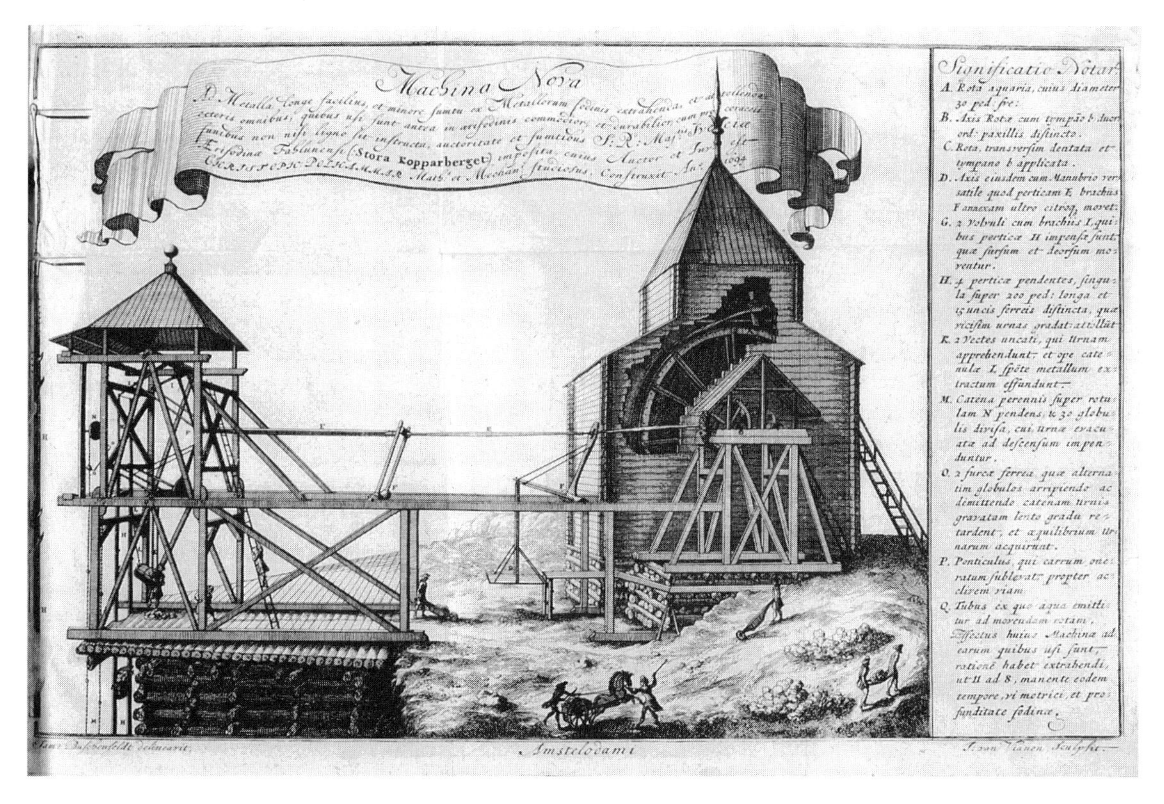

Figure 10.2

The Falun mine works, from Olof Naucler, *Delineato magnae fodinae cuprimontanae* (Uppsala: Johan Werner, 1702–3), were the most extensive works in early modern Sweden. From the Roy G. Neville Historical Chemical Library, a collection in the Othmer Library, Chemical Heritage Foundation; photograph by Douglas A. Lockard.

learn mechanical crafts, such as forging and carpentry, at a young age and to seek practical employment. Nonetheless, he perceived the limitations of these skills, and so enrolled as a student at the University of Uppsala in 1687. There he helped support himself by repairing clocks, and he acquired such a reputation for his work that he was called to Stockholm by the king, who gave him a salary to encourage further mechanical work. In this role, he began designing machinery for some of the key mines of the country. After spending several years abroad studying mining and manufacturing sites throughout Europe, Polhem was appointed engineer for the Royal Bureau of Mines. He proceeded to do far more than this title might suggest, however, designing not only mine hoists and pumps but also a host of specialized machines for the mass production of metal goods, in both copper and iron. He rapidly became an inventor of machinery, as well as a promoter of mechanical technology more generally—his notebooks, essays, drafts, and other papers totaled more than twenty thousand pages at his death in 1751. Perhaps his most famous legacy, however, was the collection of some eighty carefully constructed wooden models of mechanisms, made for instructional purposes. These became a key part of the "Royal Chamber of Models," which was one of the true treasures of Sweden in the late eighteenth century (figure 10.3). The more than two hundred models of the collection in the 1770s (it grew to more than 350 by the nineteenth century) were intended to show the possibilities of technology, not simply the devices that were already known to be useful and practical. The models also represented a recognition of the difficult ground occupied by technological improvement—territory that was not always susceptible to successful verbal description, but which needed to be represented in three-dimensional, often operable, form. These models were the uniquely Swedish contribution to the Enlightenment effort to catalogue and disseminate all knowledge about useful things.[17]

The rise of Swedish iron did not go unchallenged. In the eighteenth century, the same factors that had given the Swedes an edge in European markets helped the Russians to overtake them. The difficulties of transport in Russia were addressed by canals, river improvements, and the construction of port facilities in the new Baltic city of St. Petersburg. Just as in Sweden, ironmakers from established centers of the trade, such as southern Germany and southern Flanders (latter-day Belgium), were brought into the country to set up blast furnaces and other works with the latest technology. In England, the response to Swedish imports took a different form, since charcoal scarcities continued to be a growing problem, and rising fuel costs made native iron uncompetitive with imports. As early as the first decades of the seventeenth century, some British ironmakers had experimented with a charcoal substitute, coal, but with little success. In the early eighteenth century, however, an iron founder in the western English town of Coalbrookdale managed to succeed where others had failed, and in so doing provided an important new direction for a key industry. Abraham Darby had gotten his beginning making mills for malt (the form of barley that went into ale and beer), but he set up a foundry in Bristol for the casting of iron pots and experimented with improved ways of making thin castings in sand molds. His focus on cast iron, as opposed to the wrought iron that most ironworkers used, encouraged him to operate his own furnaces, and in 1708 he moved up the Severn River from Bristol to the small rural village of Coalbrookdale and took over an old blast furnace. As the name of the town suggests, this was a place with easily accessible coal, and the possibilities of adapting it for ironmaking were obviously very appealing to Darby.[18]

Ironmaking was a largely rural craft, slow to change, but with a tradition of experimentation and improvement. Since the introduction of the blast furnace into late medieval Europe, makers of iron had devised basic rules for the shape and

Figure 10.3
Christopher Polhem's Mechanical Alphabet was a set of models that sought to illustrate the "vocabulary" from which all machines were constructed. From the collections of the Tekniska Museet, Stockholm.

proportions of furnaces, the form and size of bellows, the preparation of fuel and ores, and all of the other key elements for the reduction of iron ore to useful metal. Within the framework of these rules, however, there was considerable room for experimenting with the optimum sizes of furnaces, the millwork needed to operate bellows and hammers, the treatment of pig iron to make wrought iron, and other processes. Abraham Darby had some additional incentives for experimentation, for his cast pots required a particularly liquid cast iron from the furnace, so that the walls of his pots could be made fairly thin. His experience with malt mills and malting probably introduced him to the usefulness of mineral coal as an alternative, often cheaper, fuel, and, more to the point, to the usefulness of coke—coal that had been roasted (much in the manner wood was roasted to make charcoal) to remove undesirable materials, such as sulfur. It was largely the sulfur that occurs in almost all coal that made it unacceptable for iron smelting, since sulfur will readily combine with the reducing metal, yielding a brittle, largely useless product. Coke was desirable for malting to prevent the product from absorbing the offensive sulfur odor (it was used by bakers for the same reason). Darby experimented with the alternative fuel soon after setting up his Coalbrookdale furnace, and by 1711 was regularly using coke in his blast furnaces. His success was due in part to some good fortune in the type of coal in the Coalbrookdale seams, lower in sulfur than other coal, and in part to Darby's great care in selecting the coal and overseeing the coking process. Very quickly, however, the new process proved a profitable success, but it was a success extremely difficult to export.[19]

It was not until the 1750s that other British ironmakers adopted coke for their furnaces. Up to that time, the difference in price between charcoal and coke for most furnaces was simply not sufficient to make the change economical. The Darby coke furnaces made a type of iron that was different from charcoal iron. The higher heats produced in the coke furnaces produced a more liquid iron, which was excellent for Darby's cast products, but contained additional amounts of silica, making it more difficult to convert into wrought iron. In the 1750s, however, the increasing price for charcoal outstripped the price of coke by such a margin than ironfounders throughout Britain began to adopt the coke furnace. Within thirty years, coke iron output was more than double that of traditional furnaces, and by the 1790s only 10 percent of British iron was smelted with charcoal. In this same forty-year period, total iron output increased threefold, and Swedish imports ceased to be a major share of the British market (although Russian imports grew in importance). The causes for this great growth of British iron making and using went well beyond the contributions of Darby and coke smelting, and we shall return to them shortly. But along with iron came an even more dramatic increase in another heavy commodity, coal.[20]

From the mid-sixteenth century to the late seventeenth century, a period of about 130 years, the annual production of coal in Britain increased fourteen-fold. Over the next century, to the late eighteenth century, there was another threefold increase (to a bit over ten million tons). The earlier, particularly dramatic, increase preceded the changes in iron technology that made coal so valuable for that trade, and even the eighteenth century increase owed little to ironmaking. The increase in coal production and use was instead the product of a growing economy along with continuing shortages in wood, charcoal, and other fuels, and an increasing capability to mine and ship coal. Ironmaking was unusual in its stringent demands on the quality of fuel; most industrial processes, the making of everything from salt to glass, from bricks to soap, simply required a good steady source of heat and made no demands on the chemical composition of the fuel. While mineral coal from readily available surface deposits had been used in Britain since before the Middle Ages, it was not until the mid-sixteenth century, when timber prices began to climb rapidly due to building and other uses, that coal began to be shipped from its sources around the mouth of the Tyne River (Newcastle) and elsewhere to population centers such as London. By the seventeenth century, the new fuel, typically called "sea coal" in London because it invariably came by coastal vessels from the north, was an indispensable part of city life, used for domestic heating as well as industrial processes. It also rapidly became the source of the first widespread complaints about urban air quality, so much so that legislation, ineffectually to be sure, attempted to control the areas in which coal could be burned. The availability of coal gave the British a flexibility in the growth of their cities and towns and the expansion of an enormous range of industries that was crucial to national prosperity even before the industrialization of the late eighteenth and the nineteenth centuries. The readily available, shallow, and easily shipped deposits of the Newcastle district coal fields, however, rapidly proved inadequate in the face of the enormous growth of demand in the seventeenth century. Deeper mines, more distant from the sea or natural waterways, had to be developed through the century, and became indispensable in the course of the eighteenth century. Coal thus placed great demands on the transport system of Britain, and on the development of mining techniques and mechanisms never before required. This latter demand, in particular, helped propel British technology into dramatic new directions, with implications for the future that we are still living with.[21]

11 Raising Fire

"To the Gentlemen Adventurers in the Mines of England,"

I am very sensible a great many among you do as yet look upon my Invention of raising Water by the impellent force of Fire, a useless sort of a Project, that never can answer my Designs or Pretensions, and that it is altogether impossible that such an Engine as this can be wrought under-ground, and succeed in the Raising of Water, and Draining your Mines, so as to deserve any Incouragement from you. I am not very fond of lying under the Scandal of a bare Projector, and therefore present you here with a Draught of my Machine, and lay before you the Uses of it, and leave it to your Consideration, whether it be worth your while to make use of it or no.

So begins a little book, *The Miner's Friend*, by one "Thos. Savery, Gent.," printed in London in 1702. Here Savery, about whom we know next to nothing, attempted to persuade the promoters of mines, and whomever else would listen, that he had solved a problem that had been plaguing English miners for several generations, and which had recently come to be seen as acute. The skepticism that he took note of in his first lines owed to the fact that such widely noted problems seemed always to turn up a host of crackpot solutions, which usually succeeded in doing no more than bilking trusting investors of their money. Not only had a wide range of novel pump designs been proposed by the end of the seventeenth century, but even the idea of using fire in some way to provide power had been broached by several experimenters. Savery's device, "an Engine to raise Water by Fire," was simply another in a line of proposed experimental devices, but the first that was actually put to public use. It is thus a little arbitrary to call it, as so many history books do, "the first steam engine," but it is still a convenient place to begin a story that is so central to understanding technological change in the eighteenth century.[1]

The Miner's Friend was not, in fact, Savery's first announcement of his invention. Clearly concerned by the poor reputation that earlier efforts had given such enterprises, he took careful steps to publicize his creation in prominent and proper ways. He had prepared the way for a patent application by demonstrating the principle behind his device at the royal court in 1698, and the patent followed shortly. The next year, he managed to get an invitation to demonstrate the machine at a regular meeting of the Royal Society, which then published the first description and drawing of it in its *Philosophical Transactions* in June, 1699, simply noting that "Mr. Savery... Entertain'd the Royal Society with shewing a small model of his Engine for raising Water by the help of Fire, which he set to Work before them; the Experiment succeeded according to Expectation, and to their Satisfaction." Later Savery was to make much of this event, including the announced "Satisfaction" of the fellows present. While we know very little about the man—not even why he was addressed as "Captain Savery"—we can tell that he was an early example of a new type of individual that was to become much more common in the coming centuries, the avid (some might say "professional") inventor-promoter. Between 1694 and 1710, he applied for no fewer than seven patents, and he touted devices ranging from a machine for rowing to a ship's log. The only one of these to come to anything, however, was his "fire engine," although even that proved to be have limited significance for future steam engine design.[2]

Savery's device was hardly an engine at all, at least as we would normally expect such a thing to appear—it had no moving parts, except for valves and cocks, and it required hand operation (figure 11.1). As he described it in *The Miner's Friend*, it consisted of a pair of boilers, carefully protected with brickwork due to their high pressures, two "receivers," or metal chambers roughly the same size as the boilers, and two long pipes: one issuing from the receivers down into the water to be raised, one going up from the receivers to the desired outlet. Smaller pipes connected these various parts, and appropriate valves and cocks were put in these. Steam was ushered into the receiver, and then, after the connection with the boiler had been closed, cold water was sprayed on to the walls to cool it down and thus condense the steam. The resulting partial vacuum drew up water from below; when that vacuum was exhausted, the high-pressure steam from the boiler was again sent into the receiver, forcing the water that had been drawn into the chamber up and out and refilling the receiver with steam to begin the cycle again.

In the machine that Savery described in 1702, a number of devices were added to allow the operator to know the water levels in the boilers, to recycle water in portions of the operation, and to simplify operation by linking the various valves to hand levers. In the first years of the eighteenth century, the Savery engine was described and depicted numerous times, as in John Harris's *Lexicon Technicum*, a popular manual of all sorts of practical subjects, whose 1704 edition reprinted the elaborate plate and description of *The Miner's Friend*. It might be noted that Savery was listed among the subscribers to Harris's work, and thus probably made a point of getting his device prominently featured in the article on "Engines." Savery set up a shop for making his engines in London, and he took the costly step of applying to Parliament for an extension of his patent beyond the normal fourteen years, and received rights

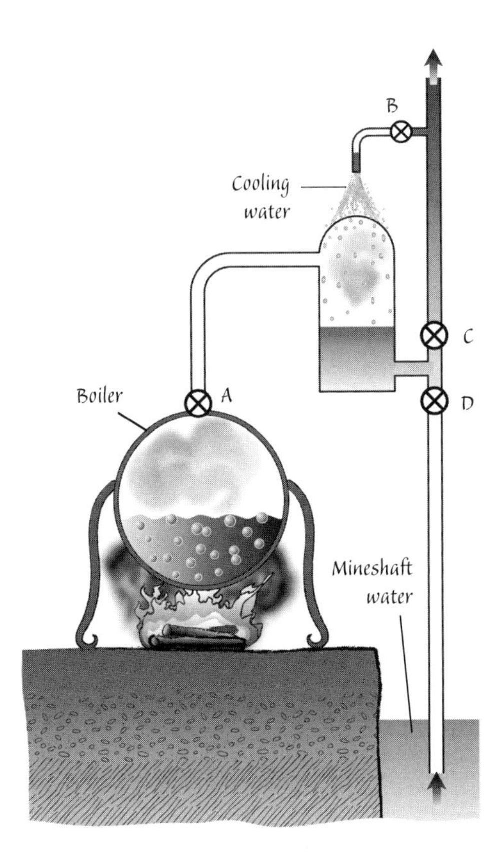

Figure 11.1
In Thomas Savery's "Miner's Friend," high-pressure steam was introduced into a metal chamber (via valve "A") and then condensed when the chamber was cooled down (by water sprayed via valve "B"). The resulting vacuum allowed atmospheric pressure to draw water up from the mineshaft into the chamber. When valve "D" was closed, valves "A" and "C" were opened, and high pressure steam forced water up and out and the cycle was repeated.

extending to 1733. The basis for his application to Parliament was the anticipated difficulty of perfecting his device, a process that, in the words of the Act granting him the extension, "may and probably will require many yeares' time and much greater expence than hitherto." He clearly had hopes for the invention, and while he focused on the well-known problem of mine drainage, from the first he pointed out the wider utility of any reliable water-raising machine, for water supplies, fountains, or supplying watermills. Savery was a self-promoter, of a type beginning to become familiar to some observers of late Stuart England, and this fact alone suggests caution in determining his real technical importance. His economic importance was severely limited by the difficulty of working his engine and by the alternative designs that appeared shortly. There were only a few working installations of Savery engines, either in England or elsewhere. In particular, there is no evidence of it actually being used in mines. As even the early illustrations make clear, Savery was aware that his device was incapable of lifting water up to the engine any more than about twenty feet (since that lifting was only by atmospheric pressure against a partial vacuum). He thus depicted his machine as being installed deep underground, where the high-pressure steam's capacity to force water up beyond the receiving vessels (as much as forty feet or more, depending only on how high a pressure the chambers could withstand) could be useful. The problems of installing and operating fires and boilers deep below the earth, however, were not simple ones, and they were not, as far as we know, ever solved in Savery's time or well afterward. Making high pressure boilers that would be safe for regular operations proved another intractable problem for early-eighteenth-century metalworkers. The real role of Savery's engine was to demonstrate that the promise, touted for almost a century, of devices using fire to raise water or do other work, was not a chimera, but could in fact come to pass.[3]

While Thomas Savery was in fact admitted to the Royal Society in 1705, this by no means meant that he was a philosopher/scientist. From what we can tell of his writings and inventive efforts, he was a practical man of modest background, but with the useful social connections that could make so much possible in the England of his day. His engine, however, did depend for its working on an understanding of some relatively novel perceptions of nature, particularly of the atmosphere. The understanding that the earth was surrounded by a sea of air, extending for miles out from the earth's surface, and bearing down on that surface with a substantial weight was a commonplace by the end of the seventeenth century. A student of Galileo, Evangelista Torricelli, invented a device in 1643 to measure this weight: the first barometer. This also established the fact that even a perfect vacuum could not draw water up more than about thirty feet. In the middle of the century, experiments with air pumps and evacuated spaces were among the most popular scientific endeavors, both for public display and for serious research. In 1672, the German Otto von Guericke, an engineer, avid experimenter, and mayor of the town of Magdeburg, staged a spectacular demonstration of the atmosphere's force, using air pumps to create a near vacuum in-between two brass hemispheres, a foot or so in diameter. Such was the atmosphere's force against the two halves, even teams of horses (up to a total of thirty for one experiment) could not separate them. The notion, therefore, of channeling this great force of the atmosphere's weight into useful work was not a peculiar one by century's end. It was also not a particularly "scientific" one, in that the weight of the atmosphere was by then quite common knowledge, and experiments with it were not the stuff of advanced philosophy.

Experiments with air pumps, along with efforts to improve mine pumps and other devices, yielded improvements in the course of the seventeenth century in a range of designs. The most important ones were those using pistons and cylinders, and by the end of the century piston pumps, small ones operated by hand for experimental apparatus and larger ones operated by waterwheels or other sources of power for mines and water supplies, were far more efficient and reliable than they had been at the beginning. The idea that the atmosphere might be made to do the work of a piston engine was explored by a number of experimenters, but finding a good way to produce a vacuum eluded most of them—some even tried to explode small charges of gunpowder in a cylinder to remove the air, but with no success. The demonstration that the vacuum created by condensing steam could be made to do work, as shown in Savery's "engine," inspired others to experiment with steam in cylinders with pistons.

The first to succeed was an ironmonger and smith from the southwestern English town of Dartmouth, Thomas Newcomen, and the Newcomen steam engine, making its first appearance about 1712, was an invention of great importance. The Newcomen, or "atmospheric," engine was the first reliable device for operating pumps and other similar machines that could be used anywhere that fuel and water could be made available, and which could safely and reliably be scaled up to great size. Even more significantly, the Newcomen engine provided, in the course of the eighteenth century, the foundation for a series of technical modifications and improvements that were collectively to transform the whole notion of mechanical power and its uses. All of the key elements of the Newcomen engine were familiar to any skilled mechanic—especially one working in mines—at the beginning of the eighteenth century. It was a sizable device, typically incorporating a large wooden beam. From one end of the pivoted beam hung the elements of a traditional water pump, with a weight added to it. At the other end was the engine, consisting of a piston and cylinder, connected to a boiler and controlled by a series of valves and linkages. The engine cycle began with the piston raised (figure 11.2). Steam introduced into the cylinder, at atmospheric pressure, displaced air. When the steam valve was closed, a stream of cold water was sprayed into the cylinder, cooling it to condense the steam and produce a partial vacuum. Atmospheric pressure would then force the piston down the cylinder, until the vacuum was spent. When the steam valve was opened again, the counterweight on the pump end of the beam would lever the piston up to the top of the cylinder again, while the cylinder refilled with steam to resume the cycle. The valves could be linked to the piston rods so as to make the pump, unlike the Savery engine, self-acting. Once it was started and in motion, it needed no operator, only someone to tend the boiler. The Newcomen engine was "single-acting," meaning that it exerted force in only one direction of the piston (downward). Since the upward stroke depended on the counterweight of the pump rod, the engine could only pump water; its reciprocal motion could not, for instance, easily turn a crank or do other mechanical work. This limited the Newcomen engine to pumping, but that task it carried out very well and very dependably, provided fuel was plentiful (later modifications adapted the engine to a crankshaft, using a flywheel, but these never worked well).

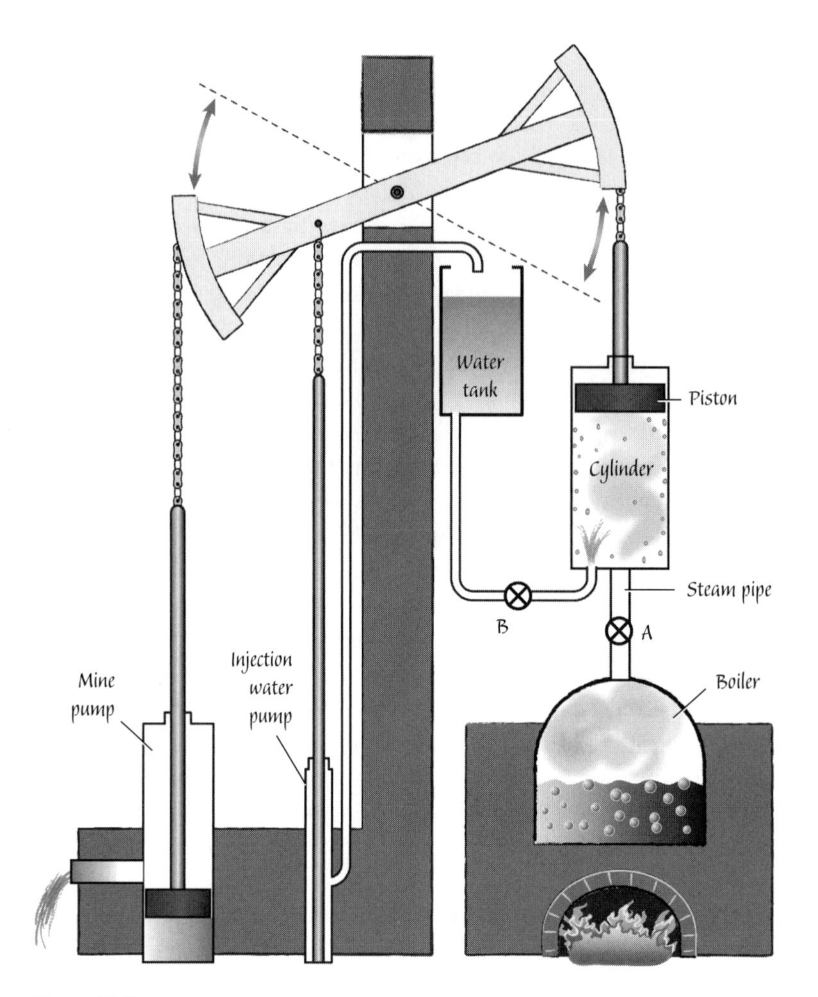

Figure 11.2
In Thomas Newcomen's "Atmospheric Engine," steam from the boiler was introduced into a cylinder (via valve "A"), displacing air in the cylinder. When water was sprayed into the cylinder (via valve "B"), the steam condensed, and the piston was brought down by the atmosphere working against the resulting vacuum. This was the only working stroke, as the piston was brought back to the top of the cylinder only by the counterweight of the heavier mine pump rod.

For all its limitations, the Newcomen engine was quickly recognized by British observers, and then by those from overseas, as an important new technology. Indeed, it is as much the widespread interest in and rapid acceptance of the new engine as much as its actual development that tells us what an important new place technological improvement had in English (and, to a less obvious extent, European) society by the beginning of the eighteenth century. The first recorded engine was erected in one of the coal fields a few miles west of Birmingham, in the English Midlands, and was described by the Swedish engineer, Marten Triewald, as having a 21-inch diameter cylinder, 7 feet 10 inches high. It pumped ten gallons per stroke, running at about twelve strokes per minute, from a mine that was more than 150 feet deep, which was far more than a Savery engine could have managed, but representative of the depth of waterlogged English mines. Triewald wrote that "the rumour of this magnificent art soon spread all over England," and so people came from far and wide. It should be remembered that this was an expensive, somewhat experimental device, and yet within only a few years, according to Triewald (writing in 1734), there were at least three others set in England, and interest elsewhere was considerable. In fact, later scholarship has counted about seventy Newcomen engines erected in England between 1712 and 1725, mostly in coal fields in the north and the Midlands. Since it appears to have cost between £700 and £1,200 to build one engine, this represented an enormous capital outlay for a novel technology. The spread went well beyond England; Triewald himself oversaw the erection of a large Newcomen engine at the Dannemora Mines in Sweden in 1728, and by the 1720s and 1730s there were Newcomen engines in many of the mining areas of Europe and also supplying a few cities and estates with water.[4]

The use of the Newcomen engine spread steadily in the mid-eighteenth century. More than six hundred of them were erected in Britain between 1734 (when all patent protection, under the Savery patent—which was interpreted as covering Newcomen's engine—expired) and 1780 (when it began to be superseded by the new engines of Boulton and Watt). Even after the introduction of the innovations of James Watt, the Newcomen machine remained a popular pumping engine, due in large part to its much simpler construction (and therefore often greater reliability). The basic form of the engine was largely established by the 1720s, but it was the subject of continuing experimentation and improvement. John Theophilus Desaguliers, who made a name for himself as a popular lecturer on a great range of scientific subjects and who became the chief curator of experiments for the Royal Society about 1713, put his hand to improving both the Savery engine as well as Newcomen's. He claimed, in fact, to have made a Savery pump that was twice as efficient as a standard Newcomen engine, primarily by improving the means for cooling the receiver vessels. The most important person to put his hand to the improvement of the early steam engine, however, was John Smeaton. Although trained as a lawyer, Smeaton quickly decided

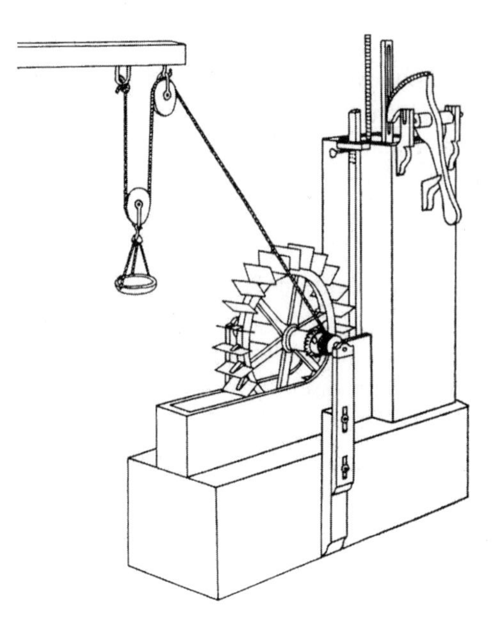

Figure 11.3
John Smeaton built carefully constructed models to measure the efficiency of waterwheel designs. With the model here he was able to show that overshot wheels were almost three times more efficient than undershot ones. From Aubrey F. Burstall, *A History of Mechanical Engineering* (Cambridge, Mass.: MIT Press, 1965).

he preferred working with his hands and so, in 1748, at the age of twenty-four, he opened a shop in London for making scientific instruments. Not long afterward he toured the continent, paying special attention to mills, canals, and harbors, and on his return he quickly established a reputation as one of England's boldest and most successful engineers. He tackled the Newcomen engine with a systematic, experimental approach that was to be held up as a model for engineering research. He gathered extensive statistics on engines scattered all around the kingdom, and then made careful inspections of the construction of many of them, looking for points of inefficiency or potential failure. He adjusted the injection mechanism for cold water into the cylinder, experimented with the ratios between the cylinder bore and its length, and devised several simple devices for managing water levels, the rate of steam admission, and the engine's stroke. Smeaton demonstrated the extent to which the newly complicated mechanical technologies of the eighteenth century would not only lend themselves to much experimentation and improvement, but would in fact demand them, due to their high capital requirements and complexity (figure 11.3).[5]

The work of John Smeaton illustrated well the new directions taken by technological improvement in the eighteenth century, particularly in Britain. In that country, as in no other, the ideas and principles of the seventeenth-century scientific revolution, particularly as exemplified by its English heroes, Francis Bacon and Isaac Newton, had popular currency among a great range of society. Popular lecturers like Desaguliers created entire careers out of making the latest science accessible, and in so doing they also promoted a culture of scientific and analytical thinking that influenced every sphere of life, from politics to music. There was also a growing belief in the ability of new knowledge to solve problems, in just the way that Francis Bacon had promised a century before. The best illustration of this was the approach to the "problem of longitude." Since the days of Columbus and his successors in the early sixteenth century, the advanced art of navigation, combined with improved ships, had made sailing the high seas a much less hazardous and chancy proposition, and entire economies, as well as political fortunes, rose and fell on the command of the seas. Of central importance in navigation, of course, was simply knowing where your ship was at any time, and this posed a problem that taxed the greatest thinkers of the day. The reason was simple. A simple study of the sky, along with charts and tables, is adequate to determine how far one is from the equator (or the poles)—that is, latitude. But determining one's position east or west of a particular point is another matter entirely, simply due to the earth's constant, daily rotation. Observation of heavenly bodies alone will never be enough to determine longitude, since one's position must always be stated relative to another point (or, more precisely, another meridian), and one needs to know the position of heavenly bodies (or, alternatively, the time of day) at that point, at the same moment, in order to make that statement. Without a means of instantaneous communication, this was impossible in the eighteenth century.[6]

The problem of longitude was, of course, well known to all who went to sea or who knew of their business. In the early eighteenth century, however, it took on greater urgency. In part this was simply due to the steadily increasing amount of trade and other business carried on on the high seas, but there were also specific events that made the problem seem more pressing. The most spectacular of these was the gruesome wreck of the British fleet, near the Scilly Isles, off the southwest tip of England, on the night of October 22, 1707. The fleet, led by the Royal Navy's chief admiral, Sir Clowdisley Shovell, ran aground in a fog, largely due to a miscalculation of position. Four ships, including the *Association*, Shovell's flagship, were sunk, with a loss of almost two thousand lives. The event shocked Britain as few such incidents ever had, due in part to the improved communications (such as newspapers) that made this "news" in the modern sense. One of the outcomes was the Longitude Act of 1714, which made solution of the problem a national priority, and offered a reward of £20,000 for the answer. The prize was to be awarded by a committee, which became known as the Board of Longitude, made up of scientists, naval officers, and other members of the government. Over the next several decades, this board was a magnet for proposed solutions, ranging from the outlandish to the merely impossibly difficult. It was early recognized that the simplest solution involved an accurate and reliable timekeeper that would withstand the conditions aboard a rolling ship at sea, along with great fluctuations in temperature and humidity, a timepiece that would remain accurate within minutes even after months at sea. Such a technical challenge seemed impossible to most at the time, and its solution by a skilled and patient clockmaker, John Harrison, was the most striking evidence of the full extent of European mechanical capabilities.

Harrison's marine chronometers were astonishing pieces of workmanship and design. It took him five years of labor, from 1730 to 1735, to produce the first model. This clock was a true marvel of precision, and worked better than any timepiece that had ever been made, passing easily its first test at sea. But Harrison wished to perfect his work, and so instead of seeking the great prize, he simply asked the board's support for another effort. With this in hand, he proceeded to spend another four years or so on another model, only to abandon it before it was fully tested, so that he might go on to yet one more version. Instead of taking the two additional years he anticipated, Harrison's third chronometer was not completed until 1760, nineteen years after it was begun. And it still wasn't satisfactory to its maker. Indeed, by the time it was finished, Harrison had already completed his fourth model, known to historians as H-4, which finally displayed the required accuracy. In its first trial at sea, a voyage to Jamaica, the H-4 lost only 5.1 seconds in about two months to its destination, and this reliability was maintained over many more months of testing. The creation of Harrison's chronometers, and their steady improvement through the decades of the mid-eighteenth century, were dramatic testimony to the capabilities of English mechanics, and of the widespread sense that these capabilities could be applied to the solution of seemingly intractable problems.[7]

It might be easy to dismiss Harrison's accomplishments as representative only of a field of craftsmanship—clock- and watchmaking—that had become a highly specialized area of fine handwork over several centuries of development and that had little to do with larger scale technologies, such as steam engines. Not only would this be too dismissive of the larger implications of improved standards and precision, but there were other fields, dramatically different in scale, that represented similar continued improvement and exacting standards. One of these, not often considered by historians of technology, was musical instrument making. This realm, of course, was a large and complex one, covering many very different sorts of devices, materials, and constructions. Closer to the province of the clockmakers, and a very advanced area indeed by this time, was the making of music boxes and similar machines. On the other end of the scale was the construction of organs. The age of Newton, Savery, and Newcomen was also the age of Johann Sebastian Bach and George Frederick Handel. The Baroque organ was one of the most complicated and advanced technical constructions of the preindustrial age. It displayed skills in shaping materials and designing mechanical, pneumatic, and hydraulic controls that once again suggest that European craftsmen had reached a level of competence—and confidence—by the eighteenth century that opened enormous possibilities of development in a great range of technical areas. It was these mechanical capabilities, as much as the proclivity for experimentation and the comprehension of useful principles of nature, that provided the foundation for the improvement and innovation of machinery in the last third of the eighteenth century.[8]

This combination of a broad awareness of science, both Newtonian principles and Baconian methods, on the one hand, and the high regard accorded practical, mechanical skills, on the other hand, provided the foundation for the rapid development of machine technology in Britain in the eighteenth century. This combination existed not only at the level of communities interacting with one another in a way not seen in any other part of Europe, but also could be seen in many individuals who made a name for themselves in both technology and commerce. The most distinguished representative of this type was unquestionably a Scottish instrument-maker who transformed himself into an inventor, industrialist, and man of the world, James Watt. Watt was born in the small Scottish port town of Greenock, down the river Clyde from the city of Glasgow. His baptismal registry identified his father, James, as a "wright," which is to say a carpenter, and he has been identified variously as a surveyor, merchant, and chandler, and the elder Watt's father, Thomas Watt, was a "mathematician" or "teacher of navigation." In this environment, it cannot be too surprising that the young James should early show a bent both for working with his hands and for what were known at the time as "mathematical instruments." Having no desire to follow directly in his father's footsteps, and probably receiving some encouragement from his grandfather, Watt sought out training as a maker of navigational and scientific instruments, an apprenticeship that was apparently unavailable to him in Scotland. At the age of nineteen he thus traveled to London and acquired a year's training from an instrument maker, enabling him to return to Scotland and set himself up in Glasgow, getting much of his business making and repairing devices for Glasgow University's professors.[9]

The twenty-one-year old Watt thus found himself in an environment that valued science and learning, with the skills of a precision worker in brass, wood, and other materials, and the encouragement around him to experiment and extend his personal capabilities. Not too surprisingly, Glasgow and its university could not really support someone making "mathematical instruments" alone, and so Watt's advertisement of December 1763 reported that he made "all sorts of mathematical and musical instruments, with variety of Toys and other goods." Watt's friends and biographers all reported that he was without any musical abilities, but he apparently taught himself enough theory to be able even to construct some sizable organs. He was thus not afraid to head out into new territory, and that same winter of his advertisement introduced him to the realm that was to occupy him for the remainder of his life—the steam engine. Professor John Anderson asked Watt to repair the college's scale model of a Newcomen engine, which had been sent to London but had come back still malfunctioning. While the young instrument maker was not completely unfamiliar with steam—he had discussed the possibilities of steam carriages with John Robinson, who was to become a professor at the university shortly, and had done some experiments with pressurized devices—the engine was apparently novel to him at the time. The basic repairs, he reported, were quite straightforward, and he was sufficiently confident in his abilities to know that he had fixed it as well as possible. When the model engine still failed to make more than a few strokes before grinding to a halt, he determined to inspect its basic design.[10]

Watt's procedures for investigating the Newcomen engine speak volumes about how the nature of technological improvement had changed in Britain by the mid-eighteenth century. John Smeaton had demonstrated the possibilities of systematic study of mechanical systems, and had effected useful changes in the atmospheric engine's design by applying what he had observed. Watt's study of the machine, however, quickly went more deeply than Smeaton's (whose work he was probably unaware of). It's possible that his natural curiosity might have taken him down new paths, but the nature of the machine he was working with gave him a particular direction. The cylinder of his model was two inches in diameter and had a six-inch stroke, but this tiny cylinder seemed to require an enormous amount of steam to be introduced at every cycle. It became quickly evident to Watt that even though the dimensions of the cylinder were to scale to those of a working Newcomen engine, the scaling down changed the ratio between the volume of the piston and the surface area, increasing the amount of surface area per unit volume (as dimensions are reduced, volume is reduced more rapidly than surface area). This in turn, Watt discovered, meant that a proportionately greater amount of steam had to be put into the cylinder at each stroke simply to heat it up to the point that steam could enter it without immediately condensing on the sides. In a full-sized engine, some steam was lost in reheating the cylinder (which, it will be remembered, was cooled on every cycle to condense the steam and produce the working vacuum), but in the smaller, model engine, proportionately much more steam was lost in this way—the greater surface area meant more steam was required to heat the cylinder back up to working temperature. It was not enough for Watt simply to have this insight; he proceeded in a series of experiments to measure the extent of steam lost to reheating, and was shocked to discover what a significant amount it was, even in a full-sized engine.

Watt resolved then to attempt an improvement of the Newcomen engine that would eliminate, or at least substantially reduce, the loss of steam in the reheating of the cylinder. His first attempt involved changing the material of the cylinder, reasoning that an insulating material, like wood, would hold heat better than the brass of his model's cylinder (or the iron of a large engine). His wooden cylinder failed however, proving insufficiently durable for a working engine. For almost eighteen months he toiled over the problem, continuing experiments on how much heat steam contained and the nature of the processes by which water went from liquid to gaseous states and back. In the course of these he observed that, in his words, "water converted into steam can heat about six times its own weight of well-water to 212°" (boiling point). Puzzled by this, he shared the observation with Glasgow's professor of chemistry, Joseph Black, and learned of Black's recent theory of latent heat, which described the capacity of substances to take up heat during a change in phase (as from water to steam) without increasing in temperature. Results such as these convinced Watt that the only way around the key inefficiency of the Newcomen design was to keep the cylinder as hot as possible at all times. But the actual operation of the engine, of course, required cooling the steam to condense it, and efficiency seemed to call for cooling as rapidly and at as low a temperature as possible. Watt's solution was one of those inventions that seems so simple, but only in retrospect. Years later he spoke of the answer coming to him on a Sunday afternoon stroll across Glasgow Green: if another vessel, with a vacuum, were connected to the cylinder and the steam allowed to enter this chamber, then the condensation would take place separate from the working cylinder. The condensing chamber could be kept cool, and the engine cylinder would remain steam-hot (figure 11.4). The idea of the separate condenser was the product of months of experiment and observation, and its translation into a workable engine would take not months, but years of additional labor, but it proved to be one of the most valuable inventions of the eighteenth century.

Even more striking, arguably, than the success of Watt's invention was the success of the inventor. James Watt was without question one of the most successful inventors, not only of the eighteenth century, but in history. This enormous personal success has interfered with historical appraisal of the invention and of the steam engine's eighteenth-century history, for Watt became the center of a mythology regarding not only the engine's origins but also the true significance of steam power for the economic and social changes that were to sweep over Britain in subsequent decades. The Watt steam engine did not in any way remake British industry, although its availability in the coming decades provided a new flexibility in the provision of power that reshaped thinking about where and how productive work was to be done. The Watt engine did not even completely displace the older model steam engines, for these had virtues of simplicity and reliability that were valued, especially where fuel costs were not a great concern. Nonetheless, the steam engines that Watt designed in the coming years—the separate condenser was but the first of a series of key improvements he introduced—came to be key representations of the ultimate capabilities of British mechanical technology, as well as symbols of a technological future that would measurably surpass all that had gone before.

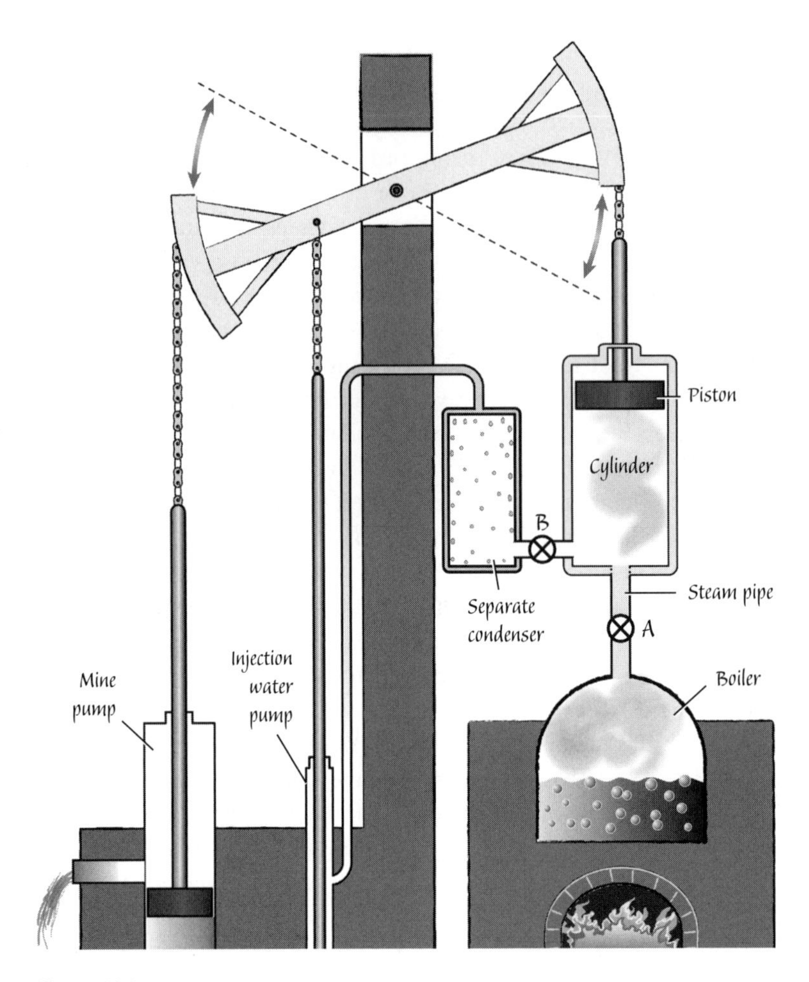

Figure 11.4
James Watt's first important improvement to the steam engine was the introduction of the separate condenser. The engine cylinder was kept hot (typically by a steam jacket), and steam condensed in the separate chamber (through valve "B"), which was itself kept cold by a water jacket. The resulting gains in efficiency were considerable, making the Watt engine the standard for decades. The greater efficiency also made it worthwhile to add further complications to the engine, such as double-action and rotary connections.

Watt's great success, as an inventor, an entrepreneur, and an influential figure in late-eighteenth-century British society, owed a great deal to his relationship with his business partner, Matthew Boulton. The separate condenser was no more than a good idea at first. It quickly became clear to Watt that he would need considerable resources to convert this into a working engine, and these were not readily available to a Glasgow instrument maker. Furthermore, he possessed no personal experience with engines of full size, nor the tools and materials to make one. What he did possess, in the Glasgow community of which he was already a respected part, were connections, social and professional, that he was able over the next few years to parlay into support for his improved engine. Professors at the university were willing to introduce him to promoters—themselves frequently men with a "philosophical bent"—and these in turn provided encouragement and money. Finally, in 1768, Watt visited Birmingham, a "new" industrial town already getting a reputation as a center of experiment and enterprise applied to industry. There he was introduced to Matthew Boulton, who had combined the metalworking skills of his upbringing with the resources of a fortunate marriage to set up in Soho, a small village a couple of miles out of Birmingham, a metalwares factory that eventually employed upward of six hundred. Boulton was quickly attracted to Watt's invention and its promise for making the "fire engine" into a machine of general, economical application (he himself was in need of more dependable power). He sought to join the partnership backing Watt at this point, but was rebuffed by the Scottish backers. This slowed Watt's progress for several years, for the men in Glasgow had no real sense of the resources necessary to make a new engine that was considerably more complicated than the Newcomen design. Nonetheless, with the encouragement of Boulton and others, Watt did manage to take out a patent, issued early in 1769, which turned out to be one of the most famous ever issued.

Watt's 1769 patent was modest enough in its basic claims, which were for "a new method of lessening the consumption of steam and fuel in fire engines." Herein he described the basic principle of his separate condenser as well as a few other modest changes in design. Due to the limited support available from his Scottish backers, it took Watt almost five years to construct a complete model of his new engine, and that one ran very poorly. Matthew Boulton, meanwhile, bided his time, but let Watt know that he had a vision far beyond anything the Scottish partners could support. He wrote to Watt shortly after the patent was issued, expressing disappointment at not yet having a greater hand in things: "I presumed that your engine would require money, very accurate workmanship and extensive correspondence to make it turn out to the best advantage.... My idea was to settle a manufactory near to my own... where I would erect all the convenience necessary for the completion of engines, and from which manufactory we would serve all the world with engines of all sizes." In the coming century, a vision of this scale, centered on a new and promising technology, would become a common part of the European and American economic landscape, but in 1769 it was an extraordinary testimony of faith in improvement and its rewards.

In 1773 Watt's principal Scottish backer declared bankruptcy, and Watt was free to seek more reliable support. The following year Matthew Boulton acquired two-thirds interest in the 1769 patent, and he persuaded Watt to move to Soho and pursue his experiments full-time. He also prevailed upon the Scot to seek an extension of the patent term, which he succeeded in doing, to 1800—a move that was to give rise to much litigation in future years. With the resources of Boulton at his command, Watt abandoned the light and flimsy prototype he had been working on and set out to make a sturdy engine to prove Boulton's faith well placed. This was facilitated by the fact that not far away an iron founder, John Wilkinson, had recently devised a machine for boring cast iron cylinders. Wilkinson's boring mill solved a key problem that had plagued all steam engines up to this point: leakage around the piston in the cylinder. Watt wrote to John Smeaton, whose advice he had begun seeking, that Wilkinson's mill "improved the art of boring cylinders so that I promise upon a seventy-two inch cylinder being not farther distant from absolute truth than the thickness of a thin sixpence at the worst part." Watt owed no small part of his success to recent achievements in machining metals, achievements that presaged a long series of key developments through the next century, on which the further development of mechanical technology was to depend.[11]

Watt's progress toward a fully working demonstration of the superiority of his design was slow, and he needed the ambitious prodding of his partner to move things along. The first working Watt engines were installed in the spring of 1776, one of these being to supply air for Wilkinson's blast furnaces. These were not small devices—one of them had a cylinder fifty inches in diameter—and they were not without problems, but they did demonstrate the dramatic fuel savings of the Watt design. The early Watt engines were calculated to use about one-quarter to one-third of what a comparable Newcomen engine would have consumed for the same work. Such a dramatic improvement quickly caught the attention of possible customers, particularly mine owners in the southwestern county of Cornwall. Here were some of the most important mines in all Britain, but they were for ore and not coal and so fuel was precious. Orders began to come into the Soho manufacturers, and in response, Boulton created a business model that was unique, and itself became a matter of widespread interest and emulation. The firm would design engines for the specific needs of their customers, manufacture key parts of the machines, and oversee their erection on site, while purchasers would provide labor and needed structures. They would also, after purchasing their engines, pay Boulton and Watt an annual royalty, up to the expiration date of the patent, which was calculated as a portion of the savings in fuel costs appreciated by the new engine. As unorthodox as this arrangement was, almost five hundred Boulton and Watt engines were erected in Britain before the patent expired in 1800. Thirty-eight percent of these were pumping engines, but the rest were rotary, mostly for textile mills, and it was in devising the technology to make this practical that Watt made the most important of his further improvements to the steam engine.[12]

By the last quarter of the eighteenth century, the desire for alternatives to waterpower for industrial uses had become acute in some areas of Britain. The limitations of geography, climate, and competition made the waterwheel, still the mainstay power source for milling, grinding, air supply, and the like, simply inadequate for a growing economy. The extent of the need could be measured by the popularity of efforts to use the Newcomen engine, with all its faults and fuel costs, to produce rotary power. It was found that a large flywheel could be connected to a crank at the opposite (pump) end of an atmospheric engine's beam, and if carefully tended this would allow the single-acting Newcomen device to provide rotary motion. It was a dangerous solution, however, for the uneven-running engine could easily stop at the top of a stroke or even end up pulling the flywheel around backward, and so its use was a sign of desperate demand. Boulton impressed the importance of this demand on Watt, and the inventor responded with a series of useful innovations (figure 11.5). These included substituting a "sun-and-planet-gear" for a crank, thus smoothing out the connections with the flywheel and allowing the wheel to turn twice for every engine stroke. Another crucial development was "double-action," which involved leading steam to both sides of the engine piston, alternately, and thus producing power for both the up and down strokes. This used more fuel, but it produced a much steadier engine motion and more power. Watt was particularly proud of a linkage system that he called "parallel motion," which allowed a double-acting engine to move safely and rigidly up and down without, in Watt's words, "chains or perpendicular guides or untowardly frictions . . . or other pieces of clumsiness." Finally, Watt adapted a device that millers had sometimes used, a centrifugal (or "flyball") governor, to keep a rotative engine from running too fast, by using automatic feedback control to cut down steam supply when necessary. The suite of inventions that Watt added to his core innovation of the separate condenser marked him as one of the most brilliant mechanical inventors in history, and made the Boulton and Watt engine the model for advanced mechanism at just the time that Britain could make the most of it.[13]

The nearly five hundred Boulton and Watt engines installed in Britain by 1800 represent probably about one-third of the total number of steam engines erected between 1781 and the end of the century. Of the nearly 1,500 total engines, about one-third were used in mines and another third in textile mills. Others were used for water supplies, in iron works, and for a range of other uses. These were large machines, for the most part—heavy, expensive to build, difficult to operate and maintain—but

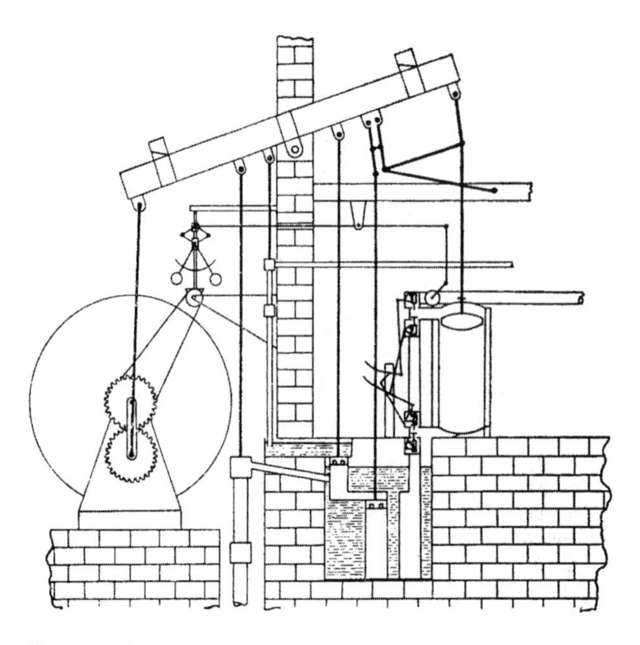

Figure 11.5
By the 1780s, Watt's improvements made the steam engine a complex, reliable, and versatile machine. This one incorporates the sun-and-planet gear (for rotary motion), double-action, a fly-ball governor to control speed automatically, and (in the upper right) parallel motion (rods linking the piston to the beam that assured straight-line movement of the piston rod). From Aubrey F. Burstall, *A History of Mechanical Engineering* (Cambridge, Mass.: MIT Press, 1965).

they represented an important addition to the power available for a great range of tasks. Waterpower continued to be much more important than steam in British—and European—industry, well into the nineteenth century, but to have a power source that transcended the inescapable limitation of water (or wind or animals), one that presented no limit in possible application or extent, was a transcendent event in technological history. In addition, the steam engines of Newcomen and Watt laid the foundations for continuing improvement and extension of engine design.

Even before the extended Watt patent ran out in 1800, radical alternatives were beginning to appear, and they held out their own promise for the future. In particular, the high pressure engine, which immediately held out the promise of overcoming the tyranny of size and weight that the older engines exerted, was an attractive subject for experiment. In the hands of a few bold innovators, such as Richard Trevithick in England and Oliver Evans in the United States, the high pressure steam engine rapidly gained notoriety, if not yet markets, in the first years of the new century. While somewhat lower initial costs and the ease of making smaller, "shop," engines was an attraction presented by the alternative designs, the real promise of these engines lay in portability, for the high pressure engine, unlike those of Watt and his predecessors and rivals, held out the prospect of self-contained propulsion, a dream at least as old as Leonardo, the fulfillment of which would remake the notions of space and time in the nineteenth century in ways hardly dreamed of earlier. For that reason, the story of the new engines will wait for a look at the coming of railroads and steamboats.[14]

12 Fabrics of Change

On the morning of August 1, 1719, four men were put into the pillories that stood in the middle of Spitalfields Market, in one of the eastern neighborhoods of London. This was not an unusual sight in itself, for such punishment was typical of the times, especially for men convicted, as John Humphreys and his associates had been, of "riotous behavior." What was remarkable was the crowd that surrounded the men, for rather than the usual taunting and jeering rabble that normally constituted part of the punishment of the offenders, there stood around them a great group of weavers from the nearby silk looms and shops, mostly, at first, quiet and sullen. But then, as the newspapers reported,

> ... some Women had the Folly to appear in a Hackney Coach dress'd up in Callico, at which the poor Men being too much moved for their small Stocks of Patience to govern, they offered Rudeness enough to their Callico and tore it pretty much, some say stript it off their Backs, and sent them home to dress over again.[1]

The cause of this extraordinary show of rage and frustration, as well as of the "riotous behavior" of John Humphreys, his colleagues in the stocks, and of several hundred Spitalfields weavers through the summer of 1719 was "Callico," or, as we would spell it, calico—simple, light cotton cloth, printed with floral or other colorful patterns. John Humphreys had led a mob in Spitalfields about seven weeks earlier; several hundred weavers roamed the east London streets looking for calico-dressed women, stores that carried the cloth or clothing made from it, and shops in which stocks of plain cloth were printed in calico designs. Throughout that summer similar mobs congregated and rioted in their anger at the cloth, ripping it off the backs of wearers, or splashing them with ink or even acid. During at least one riot, several weavers were killed by gunfire from Guards assembled to keep order. Tracts and pamphlets began, in the style of the time, to flood off presses to join the debate and to insist on government action, one way or another. In the short run, the weavers were put down and order was restored to London's streets. But within two years, the weavers got their wish, and selling, wearing, or using for any other purpose cottons printed or painted, in flowers or checks or any other design, was illegal throughout England.[2]

The "calico riots" of 1719 were actually a late act in a drama that had been playing out for more than two decades in England. The causes were simple, but working out solutions was not. The East India Company, formed at the beginning of the seventeenth century "for the Improvement of Navigation, the Glory of the Kingdom, and the Increase of Trade," had originally aimed to bring to England the great profits of the spice trade, but rivals, primarily the Dutch, frustrated this ambition and the company turned to other Asian commodities. The most important and profitable of these turned out to be cotton cloth, particularly that which had been dyed, printed, and painted in styles and designs hitherto unknown to most Europeans. Since cotton did not grow in the European climate, it had always been an exotic material, little used by any class of society. It was, for the most part, too expensive for the general market, and lost out to silk in the luxury trade. The vigorous efforts of the East India Company changed this in the course of the seventeenth century, as they tapped cheap cotton and labor found in such Indian ports as Madras and took advantage of navigational advances to make the transport between South Asia and England reliable and secure. The prices of Indian cottons, primarily calicoes, declined substantially, and the agents of the Company came to be quite adroit in identifying and promoting styles, colors, and patterns that would appeal to their English customers. At the same time, the silk merchants of Lyon refined their own techniques for promoting fashions in cloth, emphasizing skillful dying and patterning. The luxury trade continued to center on silk, but the Indian calicoes now provided much broader access to styles and fashions for a growing middle class, both in the cities and towns and the countryside.[3]

By the 1660s calicoes were flooding into England in great quantities, and the numbers kept rising in the course of the century. By the last decade of the seventeenth century, the silk weavers of England—their numbers much augmented by the influx of French Huguenot refugees (3,500 into Spitalfields alone)—were protesting loudly and sometimes violently. The first riots were in 1697, and the first legislation to protect the English trade was passed in 1700. This sweeping act attempted to outlaw the importation of all Asian silks and printed cottons, but it failed to stem the tide of fashion. The debate continued throughout Britain, with many arguing that such trade was in fact simply an appropriate meeting of the needs and wants of the people, while others insisted that it struck at the very heart of one of England's core industries (three hundred years later, argument about international trade seem hardly changed at all). Even the makers of woolens, the largest and best established of English textiles, protested and demanded protection (in response, legislation was passed to require that all corpses be buried in English wool). Some spoke with dismay at the encouragement of frivolous consumption in even the middling classes of society, while others railed at the wealth of the kingdom that was being sent into heathen lands to pay for cheap cloth. The key argument always was the need to protect English employment from cheap foreign substitution. In the cacophony of voices, however, there could be heard, just occasionally, a more optimistic note of what the competition of the calicoes might in fact bring. In his defense of the East India trade, Henry Martin, a lawyer, declared that

the *East India* Trade, by putting Persons upon Invention, may be the Cause of doing Things with less Labour; and then, tho' Wages should not, the Price of Manufactures might be abated. Arts, and Mills, and Engines, which save the Labour of Hands, are Ways of doing Things with Labour of less Price, tho' the Wages of Men, employed to do them, should not be abated. And the *East India* Trade procures things with less and cheaper Labour, than would be necessary to make the like in *England*; it is therefore very likely to be the Cause of the Invention of Arts and Engines, to save the Labour of Hands in other Manufactures; these are the Effects of Necessity and Emulation.[4]

Unlike earlier legislation, the Act of 1721 did have considerable effect on the import of cotton cloth into Britain, even if it didn't stop it entirely (and imports for re-export increased considerably in the course of the century). In the beginning the woolen and silk interests celebrated, expecting that their respective trades would henceforth be safe from the hated competition. But in coming decades, the consumption of cotton, either alone or in combination with other fibers such as flax, rose to unprecedented amounts, after a brief setback. The imports now, however, were not imported cloth, but raw cotton, converted to thread and cloth by British industry. By mid-century, the amount of raw cotton used in England each year was roughly 50 percent more than it had been fifty years before. This growth, however, paled next to that of the second half of the eighteenth century, when the annual consumption of cotton reached as much as twenty-five times that of the 1750s. It is in this statistic, more than any other, that one can see the dimensions of an economic and technological change that is the basis for speaking of an "Industrial Revolution" in eighteenth-century England.[5]

Of course, as historians and others have talked about the British Industrial Revolution, there was much more to it than increases in the amount of cotton used, but everywhere else one turns there is ambiguity and debate: about the nature, rate, and timing of economic and social change, about the discontinuity between the world of industrial England and the world that came before, and even about the role of technological change in the transformations of English life and society. Rather than focusing on the debates, however, it makes more sense here to look at the changes that did take place in the trade that had long been the most important industry in England, textiles. Even at the time, Englishmen were aware of change and of the possibility—even likelihood—of more change in the making of thread and cloth, and they had some sense that the scale and reach of the transformations in this industry went beyond anything their society had experienced before, or that any other European country had ever known. On the other hand, it is equally important to understand that to English men and women of the eighteenth century, the technological changes that overtook almost every facet of clothmaking, from the preparation of wool or cotton to the finishing processes like bleaching and felting, did not appear as unexpected, miraculous, or strange. These changes, from inventions that were to become famous to small modifications in techniques or products, were results of a culture in which change—improvements to some, threats, perhaps, to others—had already become an expected part of the world of making and doing things. Change might not be rapid, certainly by the standards of the coming centuries, but change itself was no longer unfamiliar or hard to explain. The pursuit of improvement was accepted, and the fact that it had consequences was understood. Sometimes the perceived consequences were frightening, and they evoked the responses of people worried about their futures, which in this age often meant resorting to violence. But the calico riots help to remind us that these kinds of reactions were not about technology itself, but about livelihoods. The riots also help to make the point that such action—even to the point of breaking up machines and storming factories, such as occurred over the next hundred years—was not always in vain.[6]

Just as the history of steam power has typically been written in terms of great inventions and great inventors, so too has the history of the transformation in textiles been written as a heroic story. But even more than in the case of steam power, where a century of experiment, improvement, and refinement is cast into shadows by the success of men like Thomas Newcomen or James Watt, the history of textile machinery deserves a much broader and more diffuse view of the form and pace of change. Too often the existence of a key patent or the success of a manufacturing enterprise has diverted attention from the long and gradual history of creativity in an industry that supported thousands of men and women throughout Western Europe, and especially in England. Each of the "key" inventions traditionally identified with the industrialization of textiles—the spinning jenny, the water frame, the mule, and the like—were in fact preceded by similar machines that incorporated most if not all of the principles of the famous devices. In addition, many other modifications in textile techniques and tools, often narrowly tailored to the work in a specific fiber, fabric, or region, emerged from the homes, workshops, and factories in which the trade was carried out. The foundations of the "revolution" in textile production were broad and varied, and they characterized a vast variety of crafts and products that were not, in fact, bound by tradition or ignorance in early-eighteenth-century England, but were widely recognized venues for innovation.

One of the reasons that the textile trades constituted such a rich field for innovation lay in the structure of the craft itself or, more particularly, in the relationships among the several distinct crafts that went into the making of cloth. These crafts fell into four reasonably distinct areas: preparation, spinning, weaving or knitting, and finishing. The first and last of these areas actually constituted collections of different activities, the nature of which depended on the fibers and their sources (at the preparation end) and the finished cloth and its uses (at the finishing end). Each of the stages, or even substages, of textile preparation tended to be carried out by different people, using different, specialized tools, and often working in different venues. Some of the crafts involved were simple and had relatively low skills attached to them, while others were associated with a great range of skills, and thus different values attached to different work and products. No other product of the preindustrial world involved so many variations in action, material, knowledge, or use as textiles. These variations gave technological change in textile production a distinctive dynamic; changes in one area, whether it be in one of the stages of manufacture or in one particular fiber or cloth, very often had ramifications in other stages and even for other products far removed from the area of immediate change. That this was widely recognized can be seen in the silk weavers' response to calicoes—the importation of a cloth of a rather different character made in a very different social and technical setting was construed as a direct threat to their well-being. Competition was not, of course, the only kind of force at work in sending out ripples of change; shifts in one stage of manufacture, in supply or demand, could have effects on other stages, although there was rarely, despite some economists' surmise, a simple push-pull dynamic of innovation. That is, changes in technique that affected supplies or demands did not automatically induce technological change in some other area, even though in hindsight that might seem a convenient model. To understand what the pattern of change in the eighteenth century actually looked like, it is necessary to scrutinize in a bit more detail how cloth was made.

For the key European textiles—wool, linen, silk, and cotton—the preparation tasks were distinct and different, deriving from both the original source of the material and from the nature of the fibers themselves. Sheep must be shorn; flax must be cut and beaten; silkworm cocoons must be unwound; and cotton must be picked, seeded, and cleaned. Different fibers then required different additional treatments before they could be further processed, and for the most important European fiber, wool, these treatments themselves varied depending on the breed and quality of the sheep and the eventual use of the wool. For woolens, the most important basic preparation included washing, followed by carding or combing, which involved running the raw wool through cards with short metal pins or longer combs with larger, sturdier metal teeth. This rid the wool of stray matter and, most importantly, aligned the fibers roughly parallel to one another, in a very loose rope, called a roving. It was the roving that was then given to a spinster for conversion into yarn or thread. Preparation of flax, the other traditional European fiber, was quite different, since it required extracting the fine fibers of the flax stalk from the woody portions of the plant before the fibers became available for carding. The extraction process itself involved several steps, including soaking the stalks and then beating them with specialized implements (the technical terms for these steps were retting and scutching). For silk, the processes were radically different still, since the silk fiber came from very long continuous filaments (unlike the relatively short fibers of wool or flax) unwound from the silkworm's cocoon. Cotton required yet another set of preparatory processes, although, as in the case of silk, some of these were done in the exotic locales that were the source of the fiber, rather than on European shores.

All of the short-fiber materials, wool, flax, and cotton, required spinning to make useful thread or yarn. Silk did not requiring spinning, because filaments were quite long, but since they were very thin and delicate, the filaments were typically twisted or "thrown," which was a process of taking two or more silk filaments and winding them around one another to produce a much stronger and sturdier fiber. Spinning for the other fibers was generally quite similar for all three materials, although they differed sufficiently in detail that the tools and the workers were specialized; it would be unusual, in other words, for a flax spinster to shift to working on wool or even for a worker in wool to shift to worsted, which was a longer-fiber wool. The basic action of spinning was twofold: drawing and twisting. Drawing was simply pulling the unspun fibers out from the roving in loose parallel fibers; when these are simultaneously twisted, the thread becomes thinner and stronger. The work of the spinner is aided by the fact that, up to a point, the thread becomes stronger at the same time that it gets tighter and thinner. The thinner the thread the more it resists further twisting; thus, there is a natural tendency, as long as the drawing process is fairly even, for the thread being spun to even out over its length. Nonetheless, more skillful spinsters could produce finer and more even thread, and so this was a craft with an appreciable content of skill. As has been mentioned, in the High Middle Ages, the process of spinning was made easier and quicker by the introduction and improvement of the spinning wheel. The treadle-operated wheel, often complete with flyer-and-bobbin to combine twisting with winding, was a valuable tool, and became an important source of income for many rural women.

Weaving, too, was much improved through medieval inventions, and the further development of looms, particularly for the weaving of patterns, tapes, and other specialized products, was an on-going process. Different clothmaking regions throughout Europe tended to refine their own looms, and the greater mobility of both goods and people in the late Middle Ages and afterward increased awareness of different techniques and tools. It was in weaving, in particular, that different regions, and even towns or parts of cities, were able to differentiate their cloth. The languages of Europe, as applied to textiles, came to reflect this with a great multiplicity of terms for different weaves, blends and weights of cloth, patterning and coloring, and the like. Kerseys, serges, satins, damasks, twills, tweeds, are just some of the terms that survive from dozens, if not hundreds, of distinct cloths that came from variations in weaving techniques. Variations in finishing processes produced corduroys, velvets, velours, flannels, and many other textures, weights, and surfaces of cloth. Different ways of coloring, patterning, felting, fulling, shearing, and the like added still more variety. No discussion of European textiles makes sense without acknowledging this great range of stuff that came from the looms and workshops across Europe. The European textile makers had a long tradition of innovation. The introduction of a wide range of new and generally lighter woolens, for example, gave rise in the late sixteenth and early seventeenth centuries to what came to be known as "the New Draperies," and the creation of new cloths, by new combinations of fibers, weaves, and patterns was an expected part of textile making and selling, giving rise of an enormous, widely recognized variety in textiles. The context of innovation in textile technology was fundamentally shaped by this variety, even as, at the same time, one of the great effects of later innovation was to channel variation into narrower and narrower paths.[7]

Two kinds of innovations characterize most industries, including textiles: product innovations and process innovations. Many of the earlier textile innovations were of the first type, producing novel kinds of fabrics and patterns in an effort to extend markets. In the eighteenth century, however, process innovations became more important, although these too had roots going back for generations, through experiments with different kinds of machines and devices, through the recognition and attempted application of important mechanical principles, or through different combinations of machines and techniques to increase the quantity or the quality of cloth produced. The changes that made such a profound difference in textile production had some basic characteristics, regardless of their precedents. They were generally directed toward (1) increasing the amount of material that could be worked by a single worker, (2) increasing the amount of material worked in a given period of time, or (3) decreasing reliance on worker energy and skill, substituting external power sources and precision machinery. These results were achieved, slowly at first and very unevenly from fiber to fiber, fabric to fabric, but at an accelerating pace from the mid-eighteenth century until late into the nineteenth century, largely through what may be termed "mechanization." Due to the character of the machines developed, and the economic and social context in which these machines were constructed and worked, mechanization of textiles was generally accompanied by a separate, but historically linked creation, the factory system. Because new textile machinery and the power sources that were applied to them were costly, an economy of scale began to affect profoundly the organization of the crafts—now more aptly termed the industry. Some of the new machines could indeed (and did) work in traditional environments, such as small workshops or in homes, but other machines simply were too large, too expensive, and too complex to work in such settings, and they imposed new requirements in terms of buildings, power supplies, worker coordination and organization, and management. Industrialization was the result of the choices made by entrepreneurs, capitalists, and innovators, combined with the social and economic constraints placed on workers, and their responses, in what was for many a rapidly changing world.

Machines could make a difference in textile making even without being large or expensive. Just as the wheel with its flyer and bobbin attachment gave many a farmwife spinster a crucial tool for augmenting family income, so too did just slightly later innovations in weaving and knitting. Knitting was an ancient craft, characterized by the utter simplicity of its tools, typically just two straight needles. Knit goods had a relatively limited set of applications before the seventeenth century, primarily for hosiery (their primary use long afterward as well). In the late sixteenth century, the technique of knitting with four needles was introduced into England, which allowed the making of seamless stockings. Knit hosiery grew in popularity, and knitting schools were established throughout the country to spread the trade. William Lee, a clergyman from a village near Nottingham, introduced the first complex mechanism for knitting, the stocking frame, about 1589. This device, which depended on some very clever mechanical elements to duplicate the work of hand knitting, reduced the skill required for making coarse knit goods to a very low level. Some hand-knitters fought the introduction of the frame, but at first the quality of frame stockings was so inferior that the competition seemed hardly worth fighting. Nonetheless, frames improved steadily over the course of the seventeenth century, and frame knitting shops spread over central England, carving out important segments of the trade. The stocking frame, like the broadloom before it, also complicated the usual gender identifications of the trade, for it was seen as an appropriate appliance for a man to use, unlike the needles wielded by women. Like so many other textile devices later, the stocking frame became the subject of continuous refinement, which altered its technical capabilities, and thus its economic and social significance. In the mid-eighteenth century, a Derbyshire farmer, Jedediah Strutt, made a crucial improvement by devising an attachment to the frame that allowed the knitter to make ribbed hose (or other knitware). So basic was ribbing seen to high quality knit goods that Strutt's attachment was a great commercial success, making frame knitting an industry of some significance, and its inventor one of Britain's first successful textile industrialists.[8]

About the time the knitting frame was introduced, another device that also sped up cloth making on a home or small shop scale made its first appearance. The first record of the ribbon loom can be found in Dutch records at the beginning of the seventeenth century, where protests against the device were lodged. This was a machine that could weave several—up to two dozen or more—separate ribbons on one loom. Previously, it took almost as much work to weave a narrow strip of cloth as it did a broad one, since a shuttle had to be sent back and forth across a warp in either case. The ribbon loom multiplied the number of shuttles that could be used with one motion of the weaver. By a clever set of linkages and the use of small wooden fingers to flick the shuttles across the narrow tapes, the weaving of ribbons could be multiplied and made semiautomatic. By the use of weights and carefully adjusted tensions, the winding up of the finished ribbon while the shuttles were moved back and forth across the warps was accomplished all in one action. While the ribbon loom might seem a rather insignificant and specialized invention, it actually showed the extent to which, by the middle of the seventeenth century, the textile makers of Europe were aware of the possibilities of speeding up or easing work by adding complicated mechanisms to their most basic tools.

The ribbon loom, to some degree, anticipated the somewhat more famous invention of the fly (or "flying") shuttle. John Kay was a reedmaker, which means he fabricated the portions of a loom that beat against the thrown weft, and so he was quite familiar with looms and their mechanisms. His fly shuttle, introduced in 1733, appears quite simple at first glance—an attachment to the loom that allowed the shuttle to be thrown across the "shed" (the opening in the alternating up and down warp threads) by a quick jerk of the weaver's hand, obviating the need to lean over the loom and toss the shuttle across from one hand to the other (a task done in a broadloom by helpers, an apprentice, or children) (figure 12.1). The attachment, however, required considerable effort to make it workable and reliable. In its successful form, it consisted of boxes on each side of the loom. In each box was a paddle or "picker;" a jerk on a cord held by the weaver caused the picker to throw the shuttle out of the box, across the shed, into the opposite box, which then needed to catch it and prevent it from bouncing back out. For this to work, the shuttle actually had to run on a kind of raceway underneath the warp threads. At first Kay provided it with small wheels, but later he removed them so that the shuttle did in fact "fly" across the loom. It actually took a great deal of effort to perfect this invention, and it was met with skepticism and resistance, but properly worked, the fly shuttle could double the rate at which a lone weaver could produce cloth on a broadloom, without the need for assistants. The fly shuttle actually was very limited in its use for several decades, until about 1760, when Kay's son, Robert, introduced an improvement that allowed two or more shuttles, carrying different colored wefts, to be used on the same loom. The more general use of the fly shuttle still depended in large part on the cost of assistants, and in areas where cheap labor remained available, the fly shuttle came into use only in the last years of hand-weaving.[9]

These innovations—the stocking frame, the ribbon loom, and the fly shuttle—served to speed up the processes of making cloth from yarn or thread. Like most of the early textile inventions, they were not seen as "labor-saving" in the modern sense of that term, for reducing the number of hands engaged in productive tasks was not

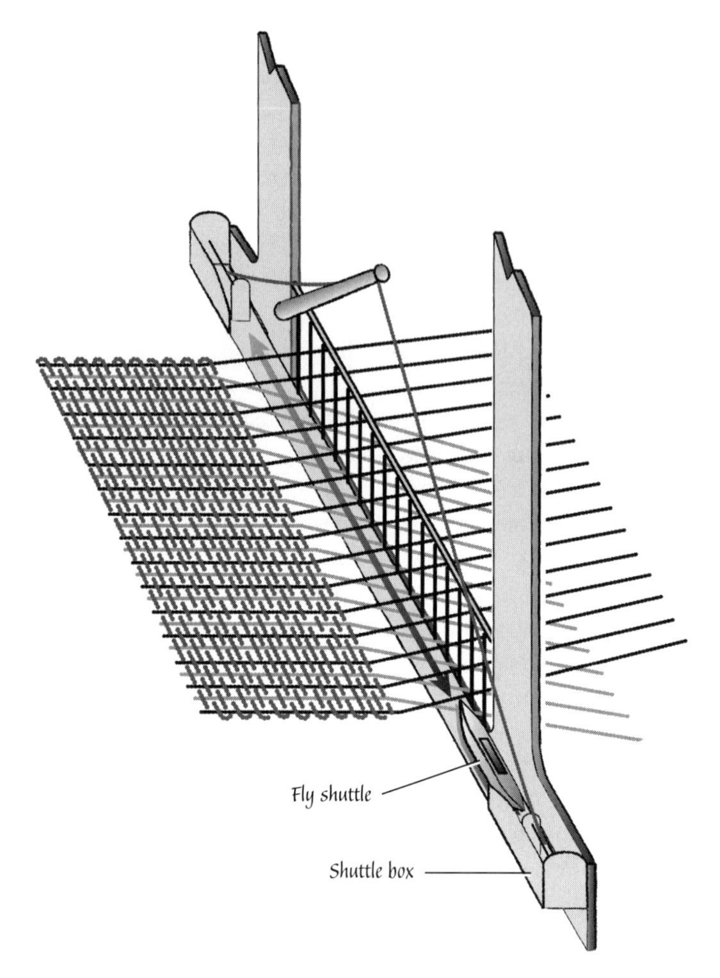

Figure 12.1
The fly shuttle was a simple mechanism that allowed a weaver to throw the shuttle across a broadloom without assistance.

seen as an appropriate goal by most people in seventeenth or eighteenth century Britain. The notion, in fact, ran very much counter to general ideas of political economy, and was, indeed perceived as almost immoral. But to ease labor, to make work more accessible to more people, and to speed up the production of useful products was generally approved and was the accepted goal of many improvements. To modern eyes, the distinction in these goals seems tenuous, at best, but we cannot understand the moral character of technological improvement in this crucial period if we do not acknowledge it. The Calico riots and the passage of the Calico Act of 1721 serve to remind us that the preservation of jobs was not only a popular concern, but was one likely to receive sympathetic political attention. There were cases in which machines were broken—in the early eighteenth century a London mob destroyed about a hundred stocking frames in a riot—but these were not protests against technology, but rather against straitened economic conditions, in which the machines were convenient proxies for the abuse of economic power by merchants and masters. Later in the eighteenth century other inventions received more direct hostility from displaced workers, but these cases too were rarely direct protests against technology or technological improvement itself, but against the perceived immorality of ignoring the larger welfare of the community in the search for short-term profit. The difficulties were to become acute when the technical changes did more than make work faster or easier, but applied large amounts of capital, in the form of waterwheels, for example, to reduce human work to mere machine tending.[10]

The application of power to knitting or weaving turned out to be a complicated business, and was not technically feasible until well into the nineteenth century. Spinning, and some of the preparatory processes leading up to it, turned out to be another matter, and these constituted the core technological changes of the textile industry. It has often been claimed that the introduction of the fly shuttle sped up weaving to such an extent that old spinning technologies could not keep up, thus creating a demand pressure for new machines and techniques. There is simply no evidence for this. The fly shuttle took many decades to spread throughout the weaving trades, and a simple doubling of weaving speed would not have created novel pressures on thread or yarn supply. Spinning had always required more labor than weaving, in terms of the total amount of hours that went into making a piece of cloth. In addition, efforts to improve spinning devices and techniques long predated major changes in any other area. That these efforts should finally produce machines that were effective enough to make the capital investment in factories worthwhile was not a product of demand, but more likely a product a growing competence in designing and making complex machinery (the English clock- and watchmakers of the eighteenth century were widely acknowledged as the best in the world) along with the utility of many years of experimental effort directed toward a wide range of textile processes.

There were two early approaches to increasing the productivity of the spinster. One was to multiply the number of spindles tended at one time; the other was to mechanize the drawing out process of spinning, just as the flyer-and-bobbin had mechanized the twisting action. As early as the sixteenth century, some wheels were equipped with two spindles, and a skillful spinster could manage to increase her productivity by using these "double wheels." Another relevant development of the sixteenth century was that of the silk throwing machine. By the late 1500s, Italian silk makers had devised water-powered machines for twisting and throwing multiple spindles of silk. First at Bologna, and then in the territory of Venice, hydraulic silk mills of considerable complexity became the primary sources of silk thread. In 1607, Vittorio Zonca illustrated the workings of these machines in detail in one of the best known of the late "machine books." A copy of Zonca's book is known to have been in Oxford University's library by 1620, and mills on the Italian model had also been set up in Lyon and other French silk making towns. The idea of the multiple spindle machine was therefore by no means novel in Britain by the late seventeenth century. The first mill using the Italian technology was not set up in Britain until about 1704, when a small one was established in the central English town of Derby. The owner of this mill went bankrupt, however, in 1713, but the mill was leased to John Lombe. He and his brother Thomas built a larger mill on the river Derwent, at Derby, and proceeded to improve the machinery, although probably only with the help of Italian workmen. The issuance of a patent for their machinery to the Lombes in 1718 was due not to their invention, but to their success in importing the Italian technology. The Lombe mill became famous throughout England, primarily because of its size—it has sometimes been dubbed the first textile factory, employing about three hundred women in 1730. At about this time, similar silk mills began appearing throughout England (in spite of the Lombes' patent), and the notion of powered, multiple spindled machines was commonplace.[11]

The spinning or throwing of silk, as has been pointed out, was somewhat different from the spinning of other fibers. Most important, silk is not drawn out while being thrown, since the filaments are quite long. All other natural fibers, however, require the combination of drawing and twisting for successful spinning, and thus making machines in which a spinner could handle more than one or two spindles was a bit more complicated. Nonetheless, as early as the mid-seventeenth century there is some evidence that multispindle machines for spinning "waste" silk (that is, short pieces of silk), linen, and worsted (long-fiber wool) were introduced in some areas of England. The descriptions of these machines (there are no drawings, much less artifacts) are ambiguous, and it would appear that they most likely consisted of devices that used multiple pulleys to allow one person to set as many as a hundred spindles in motion, but which required hand work to draw out each thread. Thus a patent of 1678 described a "spinning machine whereby from six to one hundred spinners and upwards may be employed by the strength of one or two persons to spin linen and worsted thread with such ease and advantage that a child three or four years of age may do as much as a child of seven or eight years old, and others as much in two days as without this invention they can in three days."

These patent claims speak volumes about the attitudes toward children and labor that existed well before the factories of the Industrial Revolution. They also conform to seventeenth century notions that improvements were to expand the opportunities for work, not reduce overall labor.[12]

The first machine that effectively, at least to some degree, multiplied spindles and mechanized the drawing process was a device invented by Lewis Paul, the son of a Birmingham physician, with some assistance from John Wyatt, in the 1730s. Paul was already a moderately successful inventor, having devised a machine for making some of the customary decorations on shrouds, when he took up the spinning problem. His approach to the particular problem of mechanizing drawing was ingenious and important in future machines. The loose roving of fibers was fed through several pairs of rollers, each of them turning at slightly higher speeds than the previous pair. The speed differential, along with the pressure applied by the rollers, caused the roving to be drawn out and thinned. This drawn-out material was then fed to a flyer-and-bobbin arrangement like that commonly used on spinning wheels which then imparted twist to the yarn as well as winding the result on to the bobbin. It took years of trials and redesign to get any kind of satisfactory result from this "roller spinner," and in the end the machines were considered failures. A patent issued in 1738 described several sets of rollers, but no flyer, while a patent of 1758 appears to cover only a single pair of rollers and relies more on the flyer-and-bobbin for its action. Mills were constructed in Birmingham, Northampton, Halifax and elsewhere, and Paul claimed to have realized £20,000 from his invention. Technical information about their working, however, is very scanty, and despite the apparent use of water power in some installations, mills using the machines appear to have had no more than a few dozen workers. Despite these limitations, the combination of rollers and the flyer-and-bobbin was to turn out to be one important approach to mechanizing spinning.

In the 1760s there appeared two different spinning inventions that, unlike their predecessors, achieved a considerable degree of success, the spinning jenny and the water frame. Their success has made them icons of the Industrial Revolution, apparent products of inventive genius solving age-old problems and setting England, first, and then the rest of the world off on the trajectory of industrialization. More often than not, their appearance at this time is accounted for by references to shortages in spun yarn (made acute, it is argued, by the efficiency of the flying shuttle). It has been pointed out, for example, that the Society of Arts in London advertised in 1761 a prize of £50 for "the best invention of a machine which will spin six threads of wool, flax, hemp or cotton at one time, and that will require but one person to work and attend it." The idea that need induced the invention of mechanized spinning, however, is simply too facile. The spinning jenny and the water frame were not, in fact, very different in principle from machines that had been experimented with for more than a hundred years—the Society of Arts prize suggested, in fact, just how obvious the notion of a multiple-spindle machine was. The puzzle, therefore, lies not in the efforts to make these machines but in their actual success, and that is a bit harder to explain. The answer lies in large part in the same accumulation of mechanical skill and experience that lay behind the construction of everything from Huntsman's chronometers to Watt's steam engines. By the 1760s, in other words, British mechanics could take ideas and principles that had been around for some time and, with some entrepreneurial hustle and doggedness, turn them into real machines that could be used and maintained. In addition, the steadily rising importance of cotton over other textile fibers put a material in the hands of the inventors that was, due to its much more uniform fiber length ("staple") and other qualities, more adaptable to machine spinning. It is no coincidence that the jenny, water frame, and other key spinning machines emerged from the Lancashire cotton districts and that they found their success in the transformation of cotton before all else.

The jenny of James Hargreaves and the water frame of Richard Arkwright embodied separately the two already explored means for mechanizing spinning. In the jenny, the work was much like that of an ordinary spinning wheel, without flyer and bobbin, one in which all the work was done by the spinster drawing out fibers from the tip of the spindle by hand, and, after a length was sufficiently spun, returning the hand toward the spindle, allowing the thread to be wound on the body of the spindle (figure 12.2). Hargreaves devised a relatively simple mechanism to hold several sets of roving and spindles—at first eight, and then on up to twelve, sixteen, and

Figure 12.2
The spinning jenny was hand-powered and required skill and preparation, but it could multiply the number of spindles spun at one time.

more—which could be controlled and turned all at once by a skilled operator. The jenny was purely a hand-operated and hand-powered machine, and was thus seen by many as a boon to the home spinner, not a threat. Stories were told of how Hargreaves, a weaver in the town of Blackburn, Lancashire, was inspired by seeing his wife's spinning wheel thrown over on its side and continuing to spin, and the first jenny does in fact resemble a bit a spinning wheel turned over, since its drive wheel was obliquely horizontal. Additional testimony was given that the poor weaver made his prototype using nothing but a pocket knife. The crude machine apparently worked well enough to increase noticeably the output of yarn from the Hargreaves household, and this caught the attention of a neighbor and aspiring cotton merchant and calico printer, Robert Peel. It was Peel who prevailed upon Hargreaves to make his invention public, sometime about 1765, and to continue efforts to improve it. The jenny did not produce a very fine or strong thread, and so its product went only into the weft (the threads carried in a shuttle across a loom) rather than the warp (the long threads stretched lengthwise on a loom). This was nonetheless quickly recognized as a useful speeding up of spinning, and one that could increase the productivity of home spinsters. The implications of this were disturbing to some early observers, and within only a few years of the jenny's public appearance, a mob broke into Hargreaves's workshop and broke up his machines. Soon after this incident, in 1768, Hargreaves moved to Nottingham, where he set up a larger shop for constructing jennies, the product of which was usable by Nottingham's considerable community of frame knitters. He took out a patent for his machine in 1770, although this appears to have done little to protect the device. Indeed, part of the jenny's importance lay in its simplicity and cheapness of construction, and a patent would have been nearly impossible to enforce effectively. Hargreaves died in 1778, in comfortable circumstances but very much in the shadow of his Nottingham neighbor, Richard Arkwright, whose business acumen was to make him the most successful of the textile inventors.[13]

Arkwright's invention, which came to be known as the water frame since it was the first textile spinning machine to successfully make use of waterpower, became, at least briefly, the basis for factory cotton spinning—the core icon of the British Industrial Revolution (figure 12.3). For that reason, and because the man himself became a wealthy and much honored industrialist, Arkwright is usually portrayed as an inventor of the first rank. It also does not hurt his reputation that early examples of his machinery still exist, while other early machines can only be reconstructed from drawings and descriptions. The Arkwright frames that remain are, in fact, well-made and well-designed machines, capable of spinning cotton with a strength and durability of which the jenny was not capable. There was nothing in Richard Arkwright's backgound, however, and very little in his later activities in fact, to suggest that he was capable of building such machines. He was a barber in the Lancashire

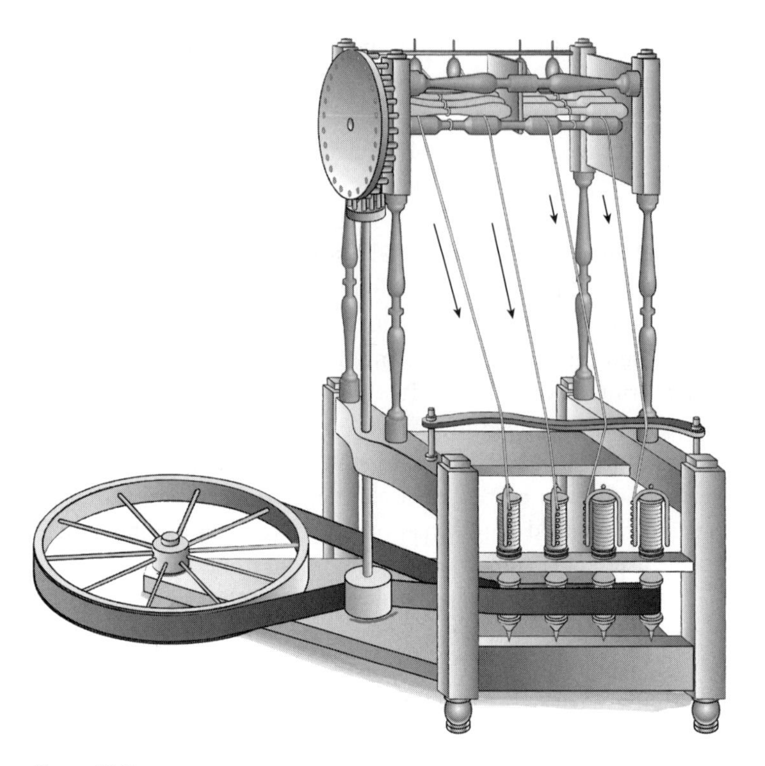

Figure 12.3
The throstle or water-frame used rollers to control the feed of fibers to a flyer-and-bobbin mechanism.

town of Preston, and he first showed his entrepreneurial skills in a wigmaking business. His community was in the middle of the rapidly growing, though still modest, cottonmaking districts, and thus the opportunities attached to improvements in spinning technology were apparent to any man of ambition and drive. Arkwright had the good fortune to encounter a clockmaker, John Kay (of no relation to the flying shuttle's inventor), who was willing to work with him to translate some ideas about spinning improvement into workable machinery. Court testimony came out many years later that Kay had also done work for one Thomas Highs, who, after Arkwright's success, claimed to have been working on similar machine designs of his own. Highs maintained that Kay stole his ideas and passed them on to Arkwright, who was then able to marshal the resources to carry them to fruition. There is simply no evidence to make a judgment on this, but it is clear that the basic principles of Arkwright's frame were, in fact, those embodied in the machines of Paul and Wyatt a generation earlier.

Arkwright always refused to acknowledge a debt to the earlier inventors, but the combination of rollers for drawing and flyer-and-bobbin for twist was very much the same. What are different, and crucially, are the details in how this combination was carried out, from the surfaces of the drawing rollers to the layout of the overall machine. A good example of the significance of this is the spacing between the pairs of rollers. Paul and Wyatt apparently paid little attention to this, and in fact their last design showed only one pair of rollers to control the drawing. Arkwright and Kay clearly appreciated the significance not only of having a series of rollers, each running at slightly higher speeds than the previous pair, but of spacing them precisely to accommodate the particular fiber being spun. If the rollers were too close together, a single fiber would be caught by two pairs at once and be torn apart; if they were too far apart, the fibers would begin to separate and the yarn would weaken. They also understood the need to control carefully the pressure the rollers exerted on the fibers, and so pulleys with weights were attached to each set of upper rollers. It was through the refinement of details such as these, along with the high level of skill from a good clockmaker, such as John Kay appears to have been, that accounted for the difference between technical failure and success.[14]

The differences in the economic fate of the textile inventions lay in the entrepreneurs and capitalists who fostered their emergence. When, in 1769, Arkwright sought to make good use of his new machine, he was already a modestly successful businessman, though at a very low level. This nonetheless eased his approach to those who had more money, and he quickly lined up backers. The capital requirements quickly grew, for the design of the spinning frame, particularly due to the pressure of the many rollers, made it difficult to operate without a sizable source of outside power. While a horse mill was first tried, it was obvious that a water mill would be called for, and this meant a structure of some size and real investment in machinery. In 1771, with the partnership of the knitting machine inventor Jedidiah Strutt and others, a water-powered factory was built in Cromford, in Derbyshire. This proved a technical and commercial success, and was the beginning of an empire that Richard Arkwright was able to build to impressive size over the next decade or so.[15]

The most striking thing about the Cromford Mill and Arkwright's subsequent factories is that there was no hesitation about making the textile business into an industrial-scale enterprise. While Arkwright is rightly called a pioneer in this regard, the speed with which he and his partners moved in this direction, as well as the readiness with which others imitated them, also with much success, are clues that these men did not see what they were doing as particularly radical. At first the product of Arkwright's frame was sold only to the nearby knitters, but it was soon discovered that the thread was strong enough for warps, which had rarely ever been made of cotton at all. Up to this point, English cottons were almost always "fustians," made with linen warps and cotton wefts, or some other blend. There were, in fact, still legal restrictions on pure cotton cloth (from the Calico Act), but soon after Arkwright expanded his Cromford mill, these were lifted (in 1774). The expanded market, along with the effort to make his mill more efficient, no doubt added incentive to Arkwright and his mechanics to continue technical development, directed this time to the crucial business of preparing the cotton for use in the water frame. While the machine produced reasonably strong yarn at a fairly good speed, it was not particularly fine or even yarn. The evenness, in particular, as well as the reliable working of the frame itself, depended on the material going into the spinning machine being uniform and properly prepared. To mechanize as much as possible of this preparation process, Arkwright and company devised an entire series of machines, described in a patent issued in 1775. This patent is sometimes described as for a "carding engine," but in fact attempted to cover ten different machines for processes covering the range of activities required from the unpacking of unclean and densely packed cotton bolls to the final preparation of the roving for the spinning machines. All of the well-known spinning machine inventors, in fact, as well as many unknown others, no doubt, tackled the problem of speeding up and improving these preparation processes, For carding itself, the approach taken was to put the needled cards on cylinders, allowing for a continuous process and one that, it was hoped, could use animal or water power instead of human arms. While there were some technical difficulties, it was actually easier to mechanize carding than it had been spinning, so carding engines became particularly important for the cotton industry, which began growing at an enormous rate in the 1770s. Arkwright's patent for the preparation machinery (devices which, like the frame before them, were generally modifications of earlier designs) was even more resented by other cotton manufacturers than the 1769 patent for the water frame, which had been challenged unsuccessfully. When the 1775 patent was challenged, however, it eventually was voided, in a famous trial of 1785, and cotton machinery became generally available throughout Britain.[16]

By the time of the great Arkwright patent trial, another textile machine had appeared, one that would eventually overtake the water frame, as well as the jenny and all other devices, as the premier machine for the spinning of cotton. Samuel Crompton was a twenty-one-year-old weaver in the Lancashire town of Bolton when, in 1774, he began to experiment with his own spinning machine. The jenny and Arkwright's frame were both very new machines, and it is remarkable testament to the speed with which technical knowledge was spread, at least in the cotton districts, that Crompton would be experimenting with a device generally recognized as a hybrid of these two inventions, hence its name, the mule. It actually seems unlikely that Crompton had himself seen Arkwright's machine, which was still at that time essentially confined to some Derbyshire factories. But, as has been pointed out, the roller principle was well-known through earlier work, and Crompton was able himself to figure out the same subtle tricks that Arkwright and Kay had to make it work in his mule. From the jenny, Crompton took the idea of twisting the yarn by drawing multiple rovings off the tops of spindles, and then dropping the twisted thread down to wind it on the body of the spindles. He avoided the flyer-and-bobbin principle, and by doing so he was able to engineer a machine that made exceptionally fine and strong thread.

The mule, which later was made in massive sizes, with sometimes more than a thousand spindles, began as a modest hand-powered device (with a few dozen spindles), but the principle of its action remained the same (figure 12.4). The spindles were put on a wheeled carriage, placed on wooden rails, which moved away from a set of rollers that controlled the feed of the roving. Unlike the water frame (and like the jenny), the action of the mule was discontinuous: at the beginning of a cycle, the spindle carriage began moving away from the rovings, moving at the same rate as the roving was issuing from the rollers. This drew the roving out particularly effectively, while the spindles imparted twist. When the spindle carriage reached the end of its run, it stopped (the rollers stopped their feed just a moment earlier), a wire came down to move the twisting thread down on to the body of the spindle, which then

Figure 12.4
The mule combined elements of the jenny (back-and-forth motion) with some from the throstle (roller control); it could made in very large sizes, as shown here. From Edward Baines, Jr., *History of the Cotton Manufacture in Great Britain* (London, 1835).

continued to turn to wind the spun thread up. The carriage was then moved back toward the rollers, and the cycle begun again. Crompton was not an experienced woodworker or mechanic, and so his machinery was crude, but the yarn it produced was vastly superior in quality to that made on other machines, even to that made by skilled hand spinnners. This superior product, rather than labor-saving, was the initial virtue of the mule. Crompton first caught the attention of the cloth merchants with cotton thread of a "count" of forty hanks to the pound (a hank is about 2,500 feet of thread), for which he received fourteen shillings per pound (this was the limit of the water frame's fineness, but usually not so uniform as the mule's). Soon, however, he was delivering sixty hanks per pound of cotton, earning twenty-five shillings. Clearly proud of his work, he began offering samples of eighty hanks per pound, for which he could receive forty-two shillings. This, he reported later, gave him incentive to give up weaving and devote all his efforts to spinning sometime in early 1780. He was very secretive about his work. He did not tout his mechanical achievement, but the steady appearance of large quantities of very finely spun yarn from his house on the edge of Bolton prompted persistent inquiries—indeed, harassment—and finally the revelation of a novel spinning machine.[17]

The fate of Samuel Crompton and the exploitation of his mule could hardly contrast more starkly with that Richard Arkwright. Where Arkwright was able to marshal resources to patent his inventions and build large factories for their use, the reclusive Crompton was convinced to give out the knowledge of his machine in return for no more than a promise of a "subscription" from grateful manufacturers (he had some worry that the broad Arkwright patent would prevent him from a patent of his own). The subscription yielded no more than a promise for about £70, and Crompton never even received all of that. He made his living for some years as an accomplished and successful mule spinner, eventually opening up a workshop where he kept several hand-operated mules going. The assistants in the shop, however, kept being lured away by those who valued the intimate knowledge of the mule's working, and so even these modest efforts at expansion frustrated Crompton. To be sure, the man appears to have been almost antisocial, and over the next decade or so he was given several opportunities to work for, and with, some of the ambitious cotton manufacturers in the further development of the technology. He turned down every such chance, always uncomfortable with each cooperative arrangement offered him, but lacking the savvy and drive to take charge of an enterprise himself. Textile machine makers in Lancashire and the Midlands quickly began making mules, and factory entrepreneurs induced them to enlarge the scale of the machines and to devise the means for applying water, and then steam, power to them. Indeed, the mule became the starting point for a rapid and continuous process of technical improvement that lasted more than two generations. One of the earliest observers of British industrial progress wrote in the 1830s, "the man who one year laid out a considerable sum in the purchase of a jenny [*sic* = mule] of the best and most approved make, found himself, in the course of the year following, so much behind hand, that with his utmost industry he could barely turn out a sufficient quantity of yarn to repay him for his present labour, in consequence of alterations which threw the productive power of his machine into the shade."[18]

Later in his life, Samuel Crompton appealed both to the rich cotton manufacturers and then to the British political establishment for acknowledgment and compensation for the enormous contribution his invention and its subsequent development had made, since he himself had realized so little reward. This effort did not in fact yield much return, but it did direct public attention to the great changes that the new machines—Crompton's and many others—had brought about in British society. He surveyed the cotton districts around Manchester and determined in 1811 that there were 4,209,570 mule-driven spindles, in constrast to a little over 300,000 water frame and 150,000 jenny spindles. While these areas of Lancashire did not include some districts that had more water frames, Crompton's central claim that his machine lay at the heart of the cotton industry was well founded. In the 1780s alone, the annual imports of raw cotton into Britain went from just over five million to over thirty million pounds, and the value of exports similarly increased about fivefold. In the next decade, exports of cotton goods increased fivefold again. At the same time, the price of fine spun cotton yarn declined; what had cost 38 shillings a pound in 1786 was less than 9 1/2 shillings in 1800, and the cheapening continued to under three shillings a pound in 1832 (this was for hundred-count yarn). It was with good reason that cotton spinning, above all other industries, became the premier symbol of technological and economic transformation.[19]

The changes in spinning and preparation machinery were followed, in time, by changes in weaving machinery and finishing techniques. It was not, in fact, until the 1820s that weaving by power machinery came to be commonplace, such were both the technical complications and the social resistance to the use of the power loom. While the actions of the weaver might not seem that much more difficult to mechanize than those of the spinner, a number of factors made the successful application of power to weaving complicated. As early as 1785, a clergyman from Kent, Edmund Cartwright, undertook to design a power loom, in spite of, in fact, being largely ignorant of both weaving and machines. Apparently inspired, however, by tales of Arkwright's success, and of the greatly expanded supplies of cotton thread, and by viewing, he said, a famous chess-playing automaton (and thinking that a weaving machine would surely be simpler to make than that), Cartwright put a carpenter and a blacksmith to work in constructing a machine that would automatically weave broadcloth, incorporating mechanisms for feeding and dressing the warp, throwing the shuttle across the shed, beating the weft in, and advancing the finished cloth. He appears to have achieved just enough success to take out a patent (issued in 1785) and get support for continued experiments. In 1786 or 1787 he set up a mill near Doncaster, not far from the Yorkshire town of Sheffield, in which he operated twenty looms, making a variety of fabrics. He continued to experiment, took out a second patent, and set up another small factory, this one powered by a Watt steam engine. That a man with so little knowledge or experience of either the textile trade or of machinery could get the resources necessary for all this is testimony both to Cartwright's very useful family connections and to the atmosphere of rapid, even fervid, technological change that was abroad in Britain in the late eighteenth century.

Cartwright's machines and mills failed, for the technical problems were simply too much, for him and for more experienced mechanical minds. Weaving was difficult to mechanize, at least in part, because much of the craft, while not necessarily requiring great skill or experience, called for constant attention to a myriad little details. The warp, especially for cottons, needed to be kept damp while being woven; the shuttle had to be kept from getting stuck in the shed or bounding off its track; the pressure of the heddle against the weft had to be tight enough to make a uniform weave, but not so tight as to jam or wear the cloth down, and so on. Even when Cartwright's mechanisms worked, each machine had to be supervised carefully, thus saving almost no labor over hand weaving. The fever for improvement and the expanding output of spinning machinery propelled the effort forward, but the actual solutions of the technical problems required another generation—a generation in which handloom weavers actually were among the most prosperous craftsmen in Britain.[20]

The radical changes in cotton were exceptional in their speed and extent. The remainder of the textile industry followed, but a bit more slowly and meeting with more resistance from traditional makers. Nonetheless, the woolen and flax trades did follow the path of mechanization, as did, eventually, the silk makers. The market for cloth of all kinds in Britain, due both to internal consumption as well as, after a hiatus during "the American wars" of the late 1770s and early 1780s, a tremendous upsurge in export trades, grew greatly in the course of the late eighteenth century. One of the reasons that textiles played such a key role in the early history of industrialization is that they enjoyed what the economists call a high "elasticity of demand." As prices fell, and fall they did, not just for cottons but for all cloths, domestic consumption rose. This was, as with production, particularly dramatic with cottons, for cotton clothing was transformed in the second half of the century from an occasional extravagance to a staple, the basis, in fact, for a cleaner, more sanitary, more fashionable citizenry, for whom the fresh shirt, the clean undergarment, a pair of white stockings were no longer the exclusive province of the well-to-do, but which could be found on the factory workers themselves, or the scullery maids, or the fishmongers.

The growth in demand and exports, along with the urgent needs of both army and navy in the Napoleonic period, helped to cushion or hide dislocations in the labor market in the first decade or so of the nineteenth century. But with the downturn in trade during the years of continental blockade and then the reduction of the military in the years after 1815, these became painfully obvious to many. The spread of factories to the older textile areas, such as the woolen regions of West Yorkshire, and, finally, the coming of the power loom in the 1820s, drove thousands of workers out of their traditional livelihoods. This tended to affect women at first, but by 1830 there was no part of the traditional clothmaking labor force that was unaffected. Well before this point, workers had attempted to resist. Indeed, there was no point in the entire history of technical change in textiles at which new machinery did not meet with some objection and, usually, violence. But this reached a new pitch in the 1810s, and a new desperation in the 1820s and 1830s. The riots of the "Luddites" began in Nottinghamshire in late 1811, spread northward into the textile areas of Derbyshire, Lancashire, and west Yorkshire, and petered out by the middle of 1814. These episodes have come to be the classic cases of machine breaking, held up as the most visible resistance to the new technologies of industrialization. Curiously enough, however, the targets of the Luddite rioters were rarely new machines, but instead the factories and workshops of those using familiar machinery, such as stocking frames, to reorganize work and, in many cases, depress wages. The economic conditions of 1811–14 were quite dire, due in some measure to trade blockades and other dislocations associated with the continuing world war being waged against Napoleon and his allies. The later problems, in the 1820s and 1830s, however, were much more directly connected to new machinery, particularly the power looms. But the tragedy of the weavers was not so much the product of technology as it was the result of social and economic displacements with complex and evolving roots. The new machines gave entrepreneurs in some parts of the textile business crucial incentive and means to concentrate production in factories, where they were able to impose a work discipline and a control over wages that could often be oppressive and exploitative. The factory system built around new machines and power sources, in turn, provided a model for other manufacturers, often simply using improved versions of hand-operated machinery, such as jennies, stocking frames, or draw looms, to gain similar control over the work process and the workers. Combined with an acceleration of the long-standing campaign of agricultural reorganization that promoted enclosure (fencing off of open fields and commons) and the massive displacement of agricultural workers throughout Britain, this new system of manufacture, much more than the machines themselves, led to thousands of workers and their families living in crushing poverty, kept on the edge of survival by legal, political, and economic systems that no longer sheltered the worker from change and circumstance.[21]

The textile industry's changes, as dramatic as they were both in size and in the form of manufacture, did not in themselves make a revolution. But they were powerful, highly visible, much remarked-on symbols, and they, more than any other developments of the late eighteenth century, gave evidence of not simply the capacity for technological improvement—improvement for some, at least—but the inevitability of it. The changes in textile production also did more than anything else to spark debate about technology. The "machinery question" captured the attention and both the intellectual and the emotional passions of the age. Ever since the last years of the eighteenth century, serious and important questions have been raised about the social consequences of the Industrial Revolution, which was not so much an event as a process that, in the eloquent language of economic historian David Landes, "began in England in the eighteenth century, spread therefrom in unequal fashion to the countries of Continental Europe and a few areas overseas, and transformed in the span of scarce two lifetimes the life of Western man, the nature of his society, and his relationship to the other peoples of the world." The discussion here has focused entirely on British—really, almost exclusively English—developments, but these were fraught with implications for everyone else, most immediately for other Western Europeans and for their former colonials in North America. These places, too, became caught up in the belief in and pursuit of technological improvement, with fewer and fewer constraints on where it might take them.

As we have seen, at the beginning of the eighteenth century, invention and improvement were morally justified in part by their promised capacity to extend the possibilities of productive employment for those that might not have it. To be sure, even at this point the promises sometime seem hypocritical—it is hard to comprehend from where we stand any compassion behind promising employment to three- or four-year-old children—but at least there was an acknowledged moral content to technical effort. By the beginning of the nineteenth century, these moral justifications no longer seemed necessary. The political economy of Adam Smith and of the manufacturing classes that quoted him found their justification on the basis of production alone, at least once the arguments were said and done. The world that was lost in the transformation that was industrialization was no paradise, and eventually the material well-being of just about everyone living in the pioneering industrial countries transcended anything ever imagined by previous generations. But the transformation itself, it should never be forgotten, was filled with violence done to the bodies and the souls of thousands, even millions, of men, women, and children for whom the promise of improvement rang hollow indeed.[22]

13 Artisans, *Philosophes*, and Entrepreneurs

Technological change in the eighteenth century assumed many forms and displayed many styles, but the awareness of change—of its possibilities and of its unavoidability—was widespread. For some, change was only significantly manifested in the form of invention, and for these patents and novel machines were the fundamental currency of change. For others, however, perhaps a bit more reflective of both the complexity and the depth of change, both in the recent past and in the foreseeable future, change in the arts was at one time cruder and subtler. It was cruder in that it encompassed all of the most ordinary activities of the full range of crafts, not simply the marvelous machines or factories that seemed to reshape entire industries in only a matter of years. But change was subtler for the same reason—that it showed itself in small, often barely distinguishable, departures from older ways of doing things, in whatever manufacture or craft was looked at with just a bit more care than ordinary. No one in the eighteenth century had a greater appreciation of this fact than the son of a cutler from an ancient hill town in the French province of Champagne, Denis Diderot, who counseled,

> We invite artisans for their part to take advice from scientists and not to allow their discoveries to die with them. Let them know that to keep useful knowledge secret is to be guilty of a theft from society, and that it is as bad in these cases to put private interest before public interest as it is in a hundred others where they would not hesitate to speak out. If they are communicative many prejudices will be discarded, above all the idea that their art has reached the highest possible degree of perfection. Their lack of information often leads them to regard as the nature of things a fault which is only in themselves. Obstacles seem insuperable as long as they know no way of overcoming them. Let them make experiments and let everyone share his experiments. Let the artisan contribute his manual skill, the academician his knowledge and advice, and the rich man the cost of materials, time, and trouble; and soon our arts and manufactures will be as superior to other people's as we could want.[1]

Diderot counted himself among the men who called themselves *philosophes*. By this term was meant quite a bit more than simply "philosopher." A *philosophe* was a rationalist, someone who placed the virtues and powers of reason above every other human capacity, including piety or grace. To the traditionalists, he was a "freethinker" in the most dreadful sense of that word, someone who rejected the mores of society, who put his or her own powers of thought above the traditions and wisdom of earlier generations, and who was proud of the fact. Among his great friends, Diderot numbered Jean Jacques Rousseau, a frustrated opera composer who became one of the most eloquent spokesmen for a new philosophy of human thought and conduct. Another early friend was a mathematician, Jean Le Rond d'Alembert, the illegitimate son of minor nobility who lived most of his life with his foster mother, a glazier's wife. From this shaky beginning, however, d'Alembert was able to distinguish himself by his raw talent for rigorous and disciplined thinking and so became at an early age a member of the Royal Academy of Sciences and author of works in physics and mathematics. Diderot and D'Alembert, very different in temperament and focus, became collaborators in an enterprise that was to define for many the core values of the *philosophes*, especially as they related to technology—the great *Encyclopédie*.[2]

In the mid-1740s a French bookseller, André-François Le Breton, decided to publish a translation of the popular two-volume *Cyclopaedia, or Universal Dictionary of the Arts and Sciences*, first published in 1728 and largely the work of a Scot, Ephraim Chambers. Chambers's work had sold quite well, and a French edition promised good profits for Le Breton, but the publisher had trouble finding a suitable translator. Finally, he was able to interest D'Alembert in the project, and a few months later recruited Diderot. In short order the new editors, with the enthusiastic support of their publisher, reconceived the project on a much more ambitious scale, and projected a new, much expanded encyclopedia that would attempt to extend its reach to all knowledge. Of special interest to Diderot was a work that would accord to the practical arts the same attention and respect that previous works had generally reserved for philosophical, historical, or theological subjects. In a "Prospectus" for the series that he distributed in 1750, Diderot wrote, "The entire translation of Chambers has passed under our eyes and we have found a prodigious multitude of things needing improvement in the sciences; in the liberal arts, a word where there ought to be pages; and everything to be supplied in the mechanical arts." To emphasize this last point, he included with the "Prospectus" his article for the first volume on "Art."[3]

The message of this article on "Art," and indeed the entire impression made by the first volume of the *Encyclopédie*, which appeared in 1751, signaled a great shift in the status of technology in European life and culture. While the English and Scottish inventors across the channel were embarking on their own transformation of the culture of improvement, the *philosophes* in France, and the many who read and were influenced by them, were carving out a somewhat different approach to knowledge and change in the technological world. To them, the promise of the future lay in the organization and connection of knowledge, in the integration of technical knowledge into the larger fabric of Western philosophy and the newly enlarged understanding of nature. To a degree, just as the eighteenth-century French physicists showed themselves to be more Newtonian than Newton (in their faith in the capacity of scientific laws and mathematics to describe phenomena in the world), so did the *encyclopédistes* show themselves to be more Baconian than Bacon. They wished, in ways merely suggested by the English philosopher of knowledge almost a century and a half earlier, to organize human thought in a thoroughly comprehensive and systematic way, and they unabashedly placed the technical arts at the heart of this structure. In "Art," Diderot spoke of machines, tools, and techniques; he meant just what we would mean by "Technology":

> In what physical or metaphysical system do we find more intelligence, discernment, and consistency than in the machines for drawing gold or making stockings, and in the frames of the braid-makers, the gauze-makers, the drapers, or the silk-weavers? What mathematical demonstration is more complex than the mechanism of certain clocks or the different operations to which we submit the fiber of hemp or the chrysalis of the silkworm before obtaining a thread with which we can weave?[4]

The cutler's son had no hesitation in making a comparison that the academicians would have scorned, and he went on to quote liberally from Bacon's *Novum Organum*, even repeating the claim that the invention of printing, gunpowder, and the compass were the great turning points of human history—indeed, his language was little more than a paraphrase of Bacon's own. He then went further, with an almost revolutionary fervor:

> The liberal arts have sung their own praise long enough; they should now raise their voice in praise of the mechanical arts. The liberal arts must free the mechanical arts from the degradation in which these have so long been held by prejudice, while royal protection must save them from the indigent state in which they still languish. The artisans have thought they deserved disdain because they were in fact disdained; let us teach them to think better of themselves, only then can we obtain more perfect products from them!

The reference to "more perfect products" (*productions plus parfaites*) is a telling one, a clue not only to the ambitions for improvement that were behind Diderot's words but also to important differences between the French concept of improvement and that displayed by the English. As economic historian John Harris has pointed out, French applications for privileges or subsidies from the government were generally couched in the language of "perfection." One common phrase was to claim that a new process for making something would "give it all that perfection of which it is susceptible." The claim was that an art is being brought closer to some desired ideal form, after which achievement, no further improvement need be sought. In Britain, by the mid-eighteenth century, such an ideal was largely abandoned—technological improvement was an open-ended, never-ceasing process. Patents claimed to be "improvements," but not steps toward some final perfect state. This may all seem to be rhetoric, but it in fact reflected attitudes that were deep-seated in the Continental view of technology and technological change. These attitudes largely gave way in the nineteenth century, but mostly under the assault of British economic muscle and the proliferation of so many novel venues for improvement that perfection seemed largely beside the point. In the eighteenth century, however, the Continental pursuit of improvement was still often characterized by concerns for quality over quantity, for the enhancement of products rather than processes, and for a deeper understanding of underlying principles even if profitability was not immediately increased.[5]

Despite the different paths favored by the French promoters of improvement, their devotion to understanding and describing the "mechanical arts" (that is, all processes of manufacture) contributed to the new European capacities for technological change as much as the more empirical and profit-driven British efforts. Above all, as the *Encyclopédie* so dramatically demonstrated, the *philosophes* and others were successfully promoting a very different attitude toward technical knowledge. In Diderot's discussion of "Art," he proposed "a project for a general treatise on the mechanical arts," and outlined some of the elements of such a work. The excerpt from his article at the opening of this chapter was a plea to craftsmen to abandon their ancient proclivity for secrecy and to join in an effort that would contribute to the common good. To Diderot, "to keep useful knowledge secret is to be guilty of a theft from society." In case his readers did not quite get his point, Diderot spelled it out even more explicitly a few years later, in volume 5's extraordinary article on "Encyclopedia." There he emphasized that a true encyclopedia "should encompass not only the fields already covered by the academies, but each and every branch of human knowledge." Again he pointed out the importance of the active participation of craftsmen in such a project, although he was careful to suggest that even the crafts cannot be truly understood without attention paid to general principles: "It happens inevitably that the man of letters, the savant, and the craftsman sometimes walk in darkness. If they make some small amount of progress it is due to pure chance; they reach their goal like a lost traveler who has followed the right path without knowing it. Thus it is of the highest importance to give a clear explanation of the metaphysical basis of phenomena, or of their first, most general principles."[6]

The balance between "general principles" (by which he meant what we would call "science") and the real knowledge of craftsmen was a delicate one in Diderot's vision of a comprehensive collection of knowledge. He did not trust craftsmen to part with what they knew freely, but felt that only someone widely learned in natural history, mechanics, physics, and chemistry would be able to "smell out secret recipes" that craftsmen would attempt to hide. The successful encyclopedist would be so learned that "workmen will not be able to pull the wool over his eyes, for he will perceive in an instant the absurdity of their lies. He will grasp the whole nature of a process, no motion of the hand will escape him, for he will easily distinguish a meaningless flourish from an essential precaution." But the fullest cooperation of practitioners would be necessary, so he proposed to "sketch out for each workman a rough memorandum whose outlines are to be filled in," and to thereby gather information about materials, tools, and techniques. The encyclopedist would then, by Diderot's hypothetical plan, "confer with them . . . [and] make them supply orally any details they may have omitted and explain whatever they may have left obscure. However bad these memoranda may be," Diderot went on, "when written in good faith they will always be found to contain an infinite number of things which the most intelligent of men would never have perceived unaided, would never even have suspected, and hence could never have asked about." Diderot recognized and acknowledged clearly the importance of what later scholars have called "tacit knowledge," skills and understanding indispensable to successful technical work that simply cannot be written down or described, but are passed on from practitioner to practitioner at the workbench, in the doing.

Diderot articulated more clearly than anyone before the ideal of technical knowledge made accessible to all, and confronted the fundamental difficulty that he saw obstructing this ideal. He held up the open exchange of ideas and theories in the sciences, and the free debate over competing theories and explanations as a model, and then lamented the difference found in applied arts:

> Craftsmen, by contrast, live isolated, obscure, unknown lives; everything they do is done to serve their own interests; they almost never do anything just for the sake of glory. There have been inventions that have stayed for whole centuries in the closely guarded custody of single families; they are handed down from father to son; they undergo improvement or they degenerate without anyone's knowing to whom or in what time their discovery is to be assigned. The imperceptible steps by which an art develops necessarily makes dates meaningless. One man harvests hemp, another thinks of soaking it in water, a third combs it; at first it is a clumsy rope, then a thread, finally a fabric, but a whole age goes by in the interval between each of these steps and the one to follow.

Nowhere will we find more explicit how difficult the advance in technical knowledge and practice had been in past centuries, nor more clearly the new model that, to Diderot, belonged "to a philosophical age." Here without qualification was the modern sense of the importance of capturing technical improvement. "I know," he concluded in the article on "Encyclopedia," "that this desire for an end to secrecy is not shared by everyone. There are narrow minds, ill-formed souls, who are indifferent to the fate of the human race, and who are so completely absorbed in their own little group that they can see nothing beyond the boundaries of its special interests." Those people, according to Diderot, would have the knowledge of the encyclopedia locked away, largely for the good of the nation. But "people who argue thus do not

Figure 13.1
Diderot was interested in rural industry as well as manufactures; this shows a cheesemaker's shed in the Auvergne, in central France. From *Encyclopédie, ou, Dictionnaire raisonné des sciences, des arts et des métiers* (Geneva, Paris, etc., 1754–72).

realize that they occupy only a single point on our globe and that they will endure only an instant. To this point and to this instant they would sacrifice the happiness of future ages and that of the human race." Diderot and his followers would have none of such a narrow view of the world, of knowledge, and of its possibilities. The *Encyclopédie*, to them, was a key element of a larger program that would make technological knowledge part of the enlargement of the province of reason and an element of what some have referred to as a "culture of control."[7]

Almost as eloquent and much more to the point than Diderot's words was the astonishing series of plates accompanying the *Encyclopédie*'s text, which began coming off the presses in 1762 and eventually totaled eleven volumes. To be sure, not all of the nearly three thousand plates depicted the arts and crafts, but a large proportion of them did, and in doing so provided the first comprehensive imagery of the

Figure 13.2
Needlemaking; the *Encyclopédie's* illustrations typically rendered each stage of the production process, suggesting the division of labor even if work was not actually organized in that fashion. From *Encyclopédie, ou, Dictionnaire raisonné des sciences, des arts et des métiers* (Geneva, Paris, etc., 1754–72).

technology of any age (figures 13.1 and 13.2). To create this large-scale picture, Diderot had to bend the rules a bit, "borrowing" (some said plagiarizing) images from numerous older sources, as well as from a long-going effort of the Académie Royale des Sciences, which had been pursuing Colbert's charge to document French industry for more than seventy-five years. The Académie's own efforts had been sluggish and unenthusiastic, however, until Diderot showed up, and so they published their own first set of plates only a year before the *Encyclopédie*'s. Diderot even borrowed from such old authors as Agricola and Vesalius, but the great proportion of illustrations were new and up-to-date. While some areas were not covered very well, particularly in light of their importance—most textile processes, for example—many others were depicted thoroughly and tellingly. The illustrations for pinmaking (*Epinglier*) were typical of the better treatments. There were several plates, showing both work spaces, with the arrangement of different tasks and workers, and the range of tools, both simple and complex, that went into the manufacture. As is the case with many of the *Encyclopédie*'s illustrations of crafts, those for pinmaking show clearly that the principles of the division of labor, so famously outlined for that particular trade by Adam Smith at the beginning of his *Wealth of Nations* (1776), were well known to a range of manufacturers by the mid-eighteenth century. In addition, Diderot's pictures show that a combination of very simple tools and processes along with some rather complicated machines—in this case a device for heading the pins—was not unusual.[8]

The great *Encyclopédie* was a financial success. Its first edition sold about 4,225 copies, about 2,500 above the publisher's original projections—this even though the final cost came to just under a thousand livres, rather than the 280 at which subscribers were first signed up. Five more editions appeared in the years before the outbreak of the French Revolution in 1789, for a total of perhaps as many as 25,000 sets of the work distributed throughout Europe. This success is all the more astonishing in light of the persecution that Diderot and his colleagues brought on themselves through their often clever but just as often too obvious defiance of orthodoxy, in matters of religion, politics, and accepted mores. More than once the great project was brought to a halt by French authorities, only to be rescued by friends of the *philosophes* in high places and the fortitude of both writers and publishers. The influence of the *Encyclopédie* on the entire enterprise of learning, writing, and teaching, throughout everywhere in the world in which European influence was felt, was incalculable. More than any other project since Gutenberg's introduction of the printing press, the seventeen volumes of text and eleven of illustrations that appeared between 1751 and 1772, and the editions that followed, redefined how knowledge—particularly technical knowledge—was to be captured and disseminated. It would be wrong to attribute much real change directly to the *Encyclopédie*'s contents. It was not a work widely available to working men and women, even in its later, cheaper editions, but it is an appropriate symbol of changing attitudes about technology and technological change. In addition, it provided a model that stood for generations of what good and complete technical explanation and illustration was, as well as a model of inclusiveness.[9]

The work of Diderot and his compatriots also represented the clearest statement of the French idea of how technical improvement was related to science and the state. Time and again, when referring to the search for technical perfection, Diderot spoke of the duty of the wise ruler to promote this search. This promotion was not to be, as it was in England, generally in the form of patents and other indirect encouragements, but through the active support of the crown and nobility. This was, in fact, the expected form of promotion for technical improvement on the Continent, and in this period it achieved some important successes. The most notable of these was the creation of the European porcelain industry, which became in the eighteenth century one of the most technologically advanced as well as economically successful enterprises in Europe.

While ceramic goods of considerable quality and beauty had been made in Europe for centuries, these products were thrown in the shade, by common judgment, by the fine porcelains imported from the Far East. At the end of the thirteenth century, Marco Polo had brought back from China a small porcelain bottle, which still can be seen on display in Venice. For the next several centuries, porcelain was a prized import for the European wealthy. Porcelain was distinguished from other ceramics by its pure whiteness and, even more distinctively, its translucency. The material was made by the Chinese through very careful selection and purification of ingredients and the use of very hot furnaces. As early as the fifteenth century, the Venetians attempted to replicate the Chinese product, but with little success. In Florence, toward the end of the sixteenth century, the encouragement of the Grand Duke Francesco de Medici had given rise to a product that resembled the Chinese to some degree, at least in its translucency. This material did not, however, possess the finer qualities of the original, and so quickly disappeared from commerce. This early episode, however, served to suggest how long-standing the active promotion of research on porcelain was among Europe's rulers.[10]

Ceramic manufacture became in the sixteenth and seventeenth centuries an important industry in a number of European cities and regions, and the successful styles ranged from brightly colored majolicas in the Mediterranean areas to blue-and-white Delftware from the Netherlands and gray and brown stoneware from Germany. These manufactures, often in workshops of some size, represented some of the earliest instances of factory production. Just as in textiles, the creation of regional styles and specialties encouraged widespread trading and, eventually, the creation and exploitation of fashions. The tenacious hold of Chinese porcelain on the high end of the market, therefore, was both remarkable and provocative to the promoters of local industries. Before the end of the seventeenth century, French efforts to imitate the Chinese had yielded some good results, the product coming to be known as "soft-paste" porcelain. This could look like the Chinese product, but it could not be shaped as finely or fired as consistently. Nonetheless, the French crown established a number of factories near Paris in the early seventeenth century for its manufacture.

Of greater importance, however, was the work of German experimenters, who were able at about the same time to make a hard white porcelain which became the basis of fine ceramic manufacture in Europe, first at the Saxon city of Meissen, and then throughout central Europe and elsewhere. Two men were primarily responsible for this work: Count Ehrenfried Walther von Tschirnhaus and Johann Friedrich Böttger. Von Tschirnhaus used his independent wealth to pursue the problem of porcelain and developed furnaces that, using large lenses and sunlight, could produce much higher temperatures than known before. Böttger was trained as an apothecary and was an avid alchemical experimenter. His claims to be able to make gold from mercury brought him to the attention of the king of Prussia. When King Frederick I asked to see proof of the goldmaking, however, Böttger fled to nearby Saxony, where the Elector, Augustus, had a long-standing interest in promoting porcelain. The Saxon ruler put Böttger and others to work in a laboratory in Meissen, not far from the capital city of Dresden. After several years of work in the well-supported laboratory, Böttger, with the assistance of von Tschirnhaus, was able to produce a fine white porcelain by early 1708. The key to their success was the mixture of a very pure, white clay—they eventually settled on kaolin—and alabaster, fired at temperatures never before used in European pottery (about 1350°C). It took two more years of perfecting and testing the process before Augustus was willing to announce publicly the successful establishment of porcelain manufacture, but from 1710, Meissen became the center of a remarkably successful industry. While efforts were made to keep the processes secret, it was simply too easy and too lucrative to lure workmen away from Saxony, and so rival factories sprang up in Vienna, Venice, and a host of German cities, and from thence to France and England.

By the time that the new porcelain process reached England in 1745, that country had a pottery industry that had already experienced several decades of dynamic change. In the northern part of the county of Staffordshire, in particular, there arose a vigorous ceramics trade, as potters were attracted to the area by the availability of a wide range of clays, by the proximity of growing coal fields as fuel sources, and by the ease of transport afforded by the River Trent and its tributaries. Improved roads made the transport of the delicate finished product more reliable, although land transport would always limit the market. The relatively late development of the Staffordshire potteries encouraged growth and experimentation free from tradition or guild regulations, and so produced an unusually experimental spirit among the potters. The relatively small scale of the first works also encouraged variation, since entry into the trade was fairly easy, and could even be combined with farming or other occupations. Thomas Wedgwood, of the village of Burslem, for example, was able to continue farming at the same time he became a master potter in the last years of the seventeenth century. The products of the Burslem potteries at this time tended to be stoneware, which could be made brown or reddish, with a glaze made from common salt or lead compounds. Some local clays were nearly white, and could produce a white stoneware, which actually tended to be somewhat yellow in tint. In the first decades of the eighteenth century, with expanding urban markets and the fashions of the cities in mind, Staffordshire potters began making white dipped ware, or slipware, which consisted of stoneware that was then refired with a white "slip" of pipeclay. Later, the quality of the product was improved by the addition of ground flint to the slip, and then to the body of the stoneware. Sixty factories were making Staffordshire pottery by 1750, representing a range of qualities, compositions, and styles. The increase in manufacturers, however, caused a depression in prices, and the industry seemed headed for a decline.[11]

In 1755, a druggist (chemist in British parlance) in the town of Plymouth, in southwestern England, identified a local deposit of clay as kaolin, or china clay, the most crucial ingredient of the new porcelain. Staffordshire potters had already been importing the material from abroad for some uses (although never firing it hot enough for porcelain), but the English supply made it possible to think of new directions for the potteries market. The man who exploited this farthest, and who demonstrated that the English capacity for improvement extended all the way from informed and systematic experimentation to novel and dramatically effective means of marketing was Josiah Wedgwood, one of a number of his family who followed in the footsteps of Thomas. In 1759, the twenty-nine-year-old Wedgwood, who had been working in the potteries in partnership with a number of others, set himself up as a master potter in Burslem. At about the same time, he embarked on a series of experiments, which were to stretch over the next thirty-five years and incorporate almost five thousand tests and trials. He later annotated his Experiment Book with the note that these were undertaken "for the improvement of our manufacture of earthenware, which at that time stood in great need of it, the demand for our goods decreasing daily, and the trade universally complained of as being bad & in a declining condition." Wedgwood was, in fact, successful in this rescue of his industry, to a degree that not even he could have imagined. He pursued extended experiments in every facet of ceramics, attempting to create new colors, new surfaces, new glazes, new furnaces, and, just as significantly as the technical improvements, new markets. He took the work of other Staffordshire experimenters and carried it to greater success, as in the case of a cream-colored stoneware that had been made before, but which in the hands of Wedgwood became a special material indeed. He described this in 1763 as "a species of earthenware for the table quite new in its appearance, covered with a rich and brilliant glaze bearing sudden alterations of heat and cold, manufactured with ease and expedition, and consequently cheap having every requisite for the purpose intended." When, a few years later, Wedgwood received an order from Queen Charlotte for a tea service of somewhat fancier material, he publicized the order widely, gained appointment as "the Queen's potter," and won permission to dub his cream-colored ware "Queen's Ware," by which name it is still known and sold.[12]

Wedgwood continued wide-ranging experiments for many years, developing in their course well-known products such as "Black Basalt" and "Jasper," which were generally used for decorative ware such as figurines and medallions. He extended connections with the informal scientific community of midlands England and gained a considerable reputation as a practical scientist. But Josiah Wedgwood's greatest contribution to the changing technological world of the eighteenth century was his promotion of a new ideal of consumption, and his development of the marketing techniques to promote demand along with the production techniques to fulfill it.

More than any other individual, Wedgwood saw the possibilities of large-scale production of moderately priced goods, combined with good transport systems and sales methods that ranged from endorsements by high-ranking members of society to newspaper advertisements and ever-changing styles and fashions. He took full advantage of the excitement over the recent excavations at the buried Roman cities of Pompeii and Herculaneum. His decorative wares, in Black Basalt, Jasper, and other effects, became the primary vehicle by which many became directly acquainted with the newfound classical heritage, and the neoclassical offshoots it spawned.

What Wedgwood called his "useful wares," however, were of much greater significance, for they redefined for entire classes of Britons (and overseas buyers) the possibilities of consumption. While a broad range of society had hitherto purchased ceramic wares, these had generally been divided into a very small quantity of expensive goods for the wealthy and a great amount of merely adequate and utilitarian wares for everyone else. Wedgwood saw the great profits possible in rationalizing production techniques to manage costs in the production of higher quality products in great quantities, and then offering these to a spectrum of purchasers ranging from the nobility down to wage earners who could be enticed to spend their precious little disposable income on a small quantity of goods that would convey to themselves and their neighbors the possibilities of material improvement. Perhaps only a single teapot could be purchased, or a few plates of simple Queen's Ware, but these would serve to make individuals into aspiring consumers, and Wedgwood recognized the great potential in that for his entire enterprise. Josiah Wedgwood did not get wealthy (he was worth more than a half-million pounds at his death in 1795) by selling cheaply, however. There were always less expensive alternatives to Wedgwood, many of them of comparable quality and in similar designs (rivals copied the popular Wedgwood pieces as quickly as they could), but Wedgwood never tired in the promotions and extensions of his market, making his goods "necessities" for a wide range of European nobility, and then advertising this fact widely back in Britain, to assure the middle-class market that his were the fashions to follow. He kept costs under tight control, and thus profits, by a rigorous organizing of work in his factories and keeping a watchful eye on his sources of material. Many of the techniques later to be associated with the great manufacturing entrepreneurs of the nineteenth century, from rationalized factory management to vertical integration of the supply and distribution networks to never-ceasing promotional campaigns were pioneered by Wedgwood. His work, as is the case with most successful pioneers, was very much part of and from his times. There was very little specifically that he did that was unique, but he brought together the full potential of the techniques emerging around him, and in so doing did more than anyone else to redefine the world of goods for men and women throughout much of the European world.[13]

Substantial as the personal contributions of Wedgwood may have been, they were very much part of larger changes in both the making and selling of goods in the eighteenth-century European world (which included many of the overseas colonies). Ceramics was an ancient class of goods whose technology and economics changed profoundly in this century, but ceramics were not alone in shaping a new culture of production and consumption. A wide range of goods that we now take for granted, from personal vehicles (carriages and coaches in this case) to personal accessories (such as umbrellas or decorative buttons) to a great range of household goods (from coffee pots to sideboards) began to be made more widely available, particularly to city dwellers. New channels of communication (regular postal services, newspapers, periodicals, and the like), new retail and distribution organizations and formats (permanent shops, wholesaling networks, and so forth), better transport (roads, canals, scheduled packet boats), and the growth of a stable urban middle class all combined to foster a culture of consumption, at least in the urbanized areas of Western Europe.

One of the most extraordinary facts of the century was the enlargement of the consuming classes. While this might have been most evident in Britain, it was by no means confined to that country. In prerevolutionary France, despite a still yawning gulf between rich and poor, both the possibilities and the realization of more broadly based consumption of a wide range of goods expanded to more and more people. When inventories of the estates left by people of modest circumstances in Paris are compared, between 1725 and 1785, there is a remarkable expansion of fashionable and trivial consumption, despite a similarity in total wealth. Where only 20 percent of these lower-class estates counted such pieces of furniture as bookcases or tea tables in 1725, almost 80 percent possessed such items in 1785; 5 percent of households in 1725 had gold watches, compared with 55 percent sixty years later; the proportion with fans increased from 5 percent to 34 percent, of umbrellas from 10 percent to 30 percent, and so on. Some of these changes were no doubt due simply to fashions, but the larger pattern is inescapable: in a manner that probably took place in England one or two generations earlier, the French had broadened greatly the material world in which many could aspire to live.[14]

While there was probably some general rise in incomes, particularly in urban areas, in England and France in the eighteenth century, this rise was modest and cannot account for the substantial increases in both the range and the depth of consumption. In textiles, by the end of the century at least, the growing mechanization of production no doubt contributed somewhat to greater consumption in that area, although the larger effect of this was not to be felt until the mid-nineteenth century or so. In other goods, it is not quite so easy to account for the lowering of relative prices for many commodities, but the evidence is strong that technological change or, at least, substantial changes in the organization and forms of production transformed the making of many things besides cloth. The changes, however, were not so much the product of key inventions but were the result of the conscious and wide-ranging development of improvements in many small elements of the production process. Nowhere could this be seen more clearly than in the town that, in 1791, was dubbed by English traveler and observer Arthur Young as "The First Manufacturing Town in the World," Birmingham.

By the middle of the sixteenth century, Birmingham, still a very modest town, had acquired a reputation for metalworking. Iron ore was available nearby, and fuel was not a problem, coal taking the place of wood and charcoal, when possible, from an early date. Tax records of 1683 showed more than two hundred forges in the town, and many shops were devoted to the making of swords, guns, nails, and brass items. With the general expansion of trade and the economy in the eighteenth century, growth accelerated, and the range of metal goods increased further. This growth was, throughout the city's history, facilitated by the fact that Birmingham had no civic charter (until 1838); it was, in other words, not a self-governing city in the legal sense. Among the implications of this were very lax controls over trades; the entry into trades, the products sold, and the techniques used were largely unregulated. The town also was able to accept religious nonconformists at a time when many places made it difficult for non-Anglicans to settle and work. Quakers, for example, became an important part of the town's economic and civic life in the eighteenth century, and as in other areas, their network of communications facilitated the kinds of informal exchanges of knowledge and resources that could make the difference between business success and failure. By the eighteenth century, in other words, Birmingham had developed a remarkable combination of a tradition of skills and marketing in metal goods along with a freedom of enterprise that was unusual in any European locale. The results were already widely evident to visitors by mid-century, particularly with the development of what came to be known as the "Birmingham toy trades."[15]

Birmingham wares were defined in the 1754 *Dictionary of Arts and Sciences* as "all sorts of tools, smaller utensils, toys, buckles, buttons in iron, steel, brass, etc." This included a small amount of what we would today call children's toys, but the much more significant parts of the trade were in making such small metal articles as buckles, buttons, nails, pins, and needles. The heritage of Birmingham was in guns and brass working and making a wide range of iron implements, but the industrial prosperity of the eighteenth century was closely tied to the making of large quantities of goods of similar specification and individually quite cheap. The manufacture of these goods was transformed during the century, not by new machinery (which, in fact, did become quite important in the nineteenth century), but by the organization of production, particularly using the division of labor. In this way, the Birmingham toy trades exemplified the same technical achievements that Wedgwood used in his

Staffordshire potteries to keep costs down and production quantities high, with reliable control over quality. To effect growth using such techniques, the Birmingham metal workers had constantly to be seeking out and applying small improvements to their techniques, tools, organization, and distribution systems. While the technical changes here may not have seemed so dramatic as those that transformed textile production, they were in fact just as profound and were just as novel. More than one observer pointed out that many more patents were issued to Birmingham inventors in the eighteenth century than to those from any other city in the realm, outside London. A rhyming local guide of 1800 had this to say about the city's inventiveness and industry:

Inventions courious, various kinds of toys,
Engage the time of women, men, and boys,
And Royal patents here are found in scores
For articles minute—or pond'rous ores.[16]

Even in the midst of the great variety of technical improvements that were the underpinning for Birmingham's success and growth, it is possible to make some general statements about what kinds of technologies were emerging. The shaping of metal had long, of course, been central to making useful objects. Up to the eighteenth century, however, only a few basic techniques had predominated in metalworking. The customary references to workshops as "forges" simply emphasizes the extent to which most of these techniques involved working heated metal. The most important was, for centuries, simple hammering—for iron, the work of a blacksmith. Since the first appearance of cast iron, molding metal—always important for softer metals such as gold, silver, and copper alloys—became another key technique. After rough shapes were formed by hammering or casting, then grinding was probably the other most basic activity, although this was done to cold rather than hot metal. While all of these techniques remained important, one of the distinctive features of the "toy trades" was the use of thinner pieces of metal, which invited other cold-working techniques, such as cutting, pressing, drawing, and stamping. These were also the techniques that promoted the increased division of labor, the increased use of standard patterns, the employment of children and women, and the growing importance of painting and plating, both for ornament and for protection.

One of the little noted but great material transformations of the early industrial world was growing use of metal sheets, primarily iron, but frequently copper or copper alloy as well. A distinctive material of eighteenth century life was tinplate, from which could be fashioned an astonishing range of practical goods, from coffee pots to candle holders (figure 13.3). This material, first fabricated in Germany in the late Middle Ages, was made from rough iron sheets that were polished and then dipped in molten tin. The tin served both to protect the iron from corrosion and to provide a

Figure 13.3
Tinsmith's shop; other plates illustrated specific tinner's tools. From *Encyclopédie, ou, Dictionnaire raisonné des sciences, des arts et des métiers* (Geneva, Paris, etc., 1754–72).

surface that allowed seams to be joined by simple soldering, rather than the welding required to join two pieces of iron. Since Cornwall, in southwestern England, had been one of Europe's primary sources for tin from ancient Roman times, it is not surprising that German tinplate makers brought their craft to England before the seventeenth century. At first iron sheets for tinning were hammered flat, which was a laborious and rough sort of process. The greatest improvement in the trade came in the 1720s when the first rolling mill was built for making uniform iron sheets, using water power. From this time forth, the British tinplate trade grew rapidly (tinners concentrating in South Wales), and in fact dominated the world market until the late nineteenth century. For Birmingham, the implications were significant, for tinplate gave the metal trades a cheap, easily worked material of enormous versatility. While tinplate resists corrosion, it does discolor and so an ancillary trade grew up for painting tinware, most commonly with a dark, hard varnish resembling, so some said at least, Japanese lacquer, hence the common term, "japanning." Hundreds, if not thousands, of women in Birmingham were employed in japanning, often embellishing their work with colorful flowers and other decorative effects. This work gave rise to another, alternative, material when papier-mâché was introduced in the 1770s. This versatile material could be easily molded, and then made somewhat durable by japanning, and thus was made into everything from coach roofs to chairs and tea trays.

In overall quantity and economic significance, such trades as tinplate and papier-mâché paled next to older materials, like iron or even pottery. But their emergence in the eighteenth century, and their successful integration into the Birmingham metal trades represented well the extraordinary new capacity for innovation and application that made Birmingham the model manufacturing town for the coming century. They also represented well the pivotal place that materials improvements themselves had in the changes of the eighteenth century. Improvements transformed, in both scale and quality, the making and use of materials like iron as much as the smaller-scale substances. The changes in scale were extensions of trends that had been visible since the introduction of the blast furnace in the late middle ages, but new empirical knowledge of how materials behaved in the furnace allowed even more substantial enlargements of iron production. The innovations of the Darbys were only the beginning of a series of improvements in iron making, the cumulative effect of which was to make Britain the dominant iron producing nation for the next hundred years.

Improvements in quality were as important as those in scale. The growth of certain trades, such as the making of clocks and watches, musical instruments, or cutting tools, increased the demand for certain carefully made materials and alloys, the most important of which was steel. It will be remembered that steel is a form of iron that, modern analysis would show, is distinguished by having a carbon content that falls in between that of the product of the blast furnace, pig iron (which tends to have more than 2 percent carbon) and that used in the forge, wrought iron (which generally has less than a half-percent carbon). While the role of carbon was not known to metal workers, it was ancient knowledge that careful and laborious techniques could be used to convert wrought iron, typically a rather soft and ductile material, into a harder substance, particularly valued for taking and keeping a sharp edge—steel. Beyond its ancient application in weapons and cutting tools, steel had by the early eighteenth century come to have considerable value for other uses, as in springs for clocks and watches, and so it should not be too surprising that it was a clockmaker, Benjamin Huntsman, who improved steel making with the introduction of "crucible steel" in the 1740s. Like many materials innovations, this began with better and hotter furnaces. Huntsman discovered that he was able to take old fashioned "blister steel," which was notoriously uneven in quality, and melt small amounts of it in clay crucibles. The resulting molten material would then solidify into a much more uniform and predictable quality steel. The ability to manufacture this material was to become critical in the expansion of tool and instrument making in the late eighteenth and early nineteenth centuries.[17]

For the iron makers at Coalbrookdale and elsewhere, one key challenge that won their continued attention was the conversion of the pig (or cast) iron from their blast furnaces into the much more versatile and useful wrought iron, on which most of the iron trade was based. Due to the same qualities of coal that made it such a difficult material to use for iron smelting, but to an even greater degree, the new fuel simply could not be used in the old refining processes—the coal-based impurities would ruin any iron so made. The answer was to distance the metal being worked from the fuel, but this was not a problem easily solved, and numerous iron makers, at Coalbrookdale and elsewhere, contributed to its solution. While some histories attribute the key development to Henry Cort, who developed his "puddling" process at ironworks near the southern port town of Plymouth, this was really just one of the final stages in a series of innovations that began in the 1760s, some twenty years before Cort patented his process. The key to all the processes was finding effective ways of (1) ridding pig iron (particularly that made with coke) of impurities, such as silicon and (2) separating the pig iron from the coal in firing, so that more impurities (particularly sulfur) were not introduced. The "potting" process was introduced in the 1760s and improved over the next decade or so. This multistage process involved putting pig iron in clay pots, generally with a flux like lime (a calcium compound) to absorb sulfur, and then firing this in the high heat of a reverberatory furnace (one designed to reflect heat back onto the fire). The high heat removed carbon and the pots protected the material from further contamination from the coal fuel. It was found that if coke was used rather than coal, then the fluxes could be left out and the process was simpler and cheaper. Iron makers throughout England began adopting this process with success by the 1780s. Cort's contribution was to simplify the process further by designing the furnace so that larger amounts of iron could be heated, not in pots but in a large pool or "puddle." This had to be stirred to keep exposing the larger mass of iron to the decarburizing heat of the furnace. The refined iron had a higher melting point than the pig, and so congealed on the stirring rod. When this congealed mass was pulled out of the furnace, Cort's process called for it to be put directly into a rolling mill, which could be used to make not only plate or sheet, but also, when grooved, rods and bars. The overall process, as it emerged in the 1790s, was a much faster and more efficient means of making large quantities of iron, in continuous production. It took decades for the new method to transform the British iron industry, but by the early nineteenth century, it was being carried out on a dramatically larger scale than before.[18]

Even before the changes brought about by the new potting and puddling processes were substantial, however, it was clear to many observers that Britain's iron industry was undergoing a significant transformation. The great works of the Darbys and their neighbors at Coalbrookdale, for example, represented to many the same kind of concentration and enlargement that they saw in the textile factories of Arkwright and others. In June of 1776, Arthur Young visited the area, and he wrote the following:

Figure 13.4
The Ironbridge, erected over the River Severn near Coalbrookdale, in western England in 1777, was a celebration of the new structural capabilities of cast iron. Courtesy Institution of Civil Engineers, London.

These iron works are in a very flourishing situation, rising rather than the contrary. Colebrook Dale itself is a very romantic spot, it is a winding glen between two immense hills which break into various forms, and all thickly covered with wood, forming the most beautiful sheets of hanging wood. Indeed too beautiful to be much in unison with the variety of horrors art has spread at the bottom: the noise of the forges, mills, &c. with all their vast machinery, the flames bursting from the furnaces with the burning of the coal and the smoak of the lime kilns, are altogether sublime.

The combination of horrors and sublime was to become a consistent theme through the responses the new industrialism evoked over the coming century. But Coalbrookdale provided yet one more important clue to the future, just barely hinted at by Young, when he remarked on seeing the place "where Mr. Darby has undertaken to build a bridge of one arch of 120 feet, of cast iron." The great Iron Bridge of Coalbrookdale was just being projected at this point, and required almost another five years before completion, but its ambition was already noteworthy. The finished work, to most observers, more than lived up to the expectations—it was a structure just six inches over a hundred feet long, rising forty feet above the level of the Severn, and extending a generous twenty-four feet in width. It was made entirely of iron cast at the works of Abraham Darby III (grandson of the developer of coke smelting), resting on masonry abutments built on both banks of the river (figure 13.4).

While the bridge project had begun years earlier as a simple proposition of economy and convenience for the growing industrial district, in the hands of the younger Darby, it became much more—a symbol of what modern industry was capable of. Even before the bridge opened to traffic on the first day of 1781, it began capturing the imagination and attention of the public, and it quickly became—and to some extent still is to this day—the quintessential symbol of the technological bravado that made British industry the marvel of the world.[19]

14 Airs and Lightning

In the summer of 1783, Benjamin Franklin was living in Paris, attempting to bring to a close one of the last great projects in a life filled with variety and achievement. He was still the ambassador for the new United States to the court of the king of France, having held that post, in fact if not in name, for almost seven years, but more significantly at the moment, he was one of the three commissioners appointed by the Continental Congress to draft a treaty of peace with Great Britain that would bring the American Revolution to a final close and guarantee American independence for all time. Franklin had supported the revolution with great vigor, but he had also been saddened by the conflict with Britain, not least because it came between him and many friends he had made in England. On July 27, he wrote to one of these, Sir Joseph Banks, then president of the Royal Society (of which Franklin was an active and honored member), reflecting on the war recently past: "in my opinion, there never was a good War, or a bad Peace. What vast additions to the Conveniences and Comforts of Living might Mankind have acquired, if the Money spent in Wars had been employed in Works of public utility! What an extension of Agriculture, even to the Tops of our Mountains: what Rivers rendered navigable, or joined by Canals: what Bridges, Aqueducts, new Roads, and other public Works, Edifices, and Improvements, rendering England a compleat Paradise, might have been obtained." For Franklin, as for many of his contemporaries, improvement was largely a matter of effort and resources, not the whim of Providence or fortune. But there were other possible sources of improvement that Benjamin Franklin, more perhaps than any other individual of his time, appreciated. At the end of his letter, in a brief postscript, Franklin indicated to Banks that the bearer of the letter "will acquaint you with the experiment of a vast Globe sent up into the Air, much talked of here, and which, if prosecuted, may furnish means of new knowledge." And so Benjamin Franklin, slowed down physically but still quite alert in his seventy-eighth year, took note of the first reports of the flight of a balloon.[1]

What Franklin had heard were stories about the public spectacle a few weeks earlier in the small village of Annonay, in southeastern France. Two brothers from a local papermaking family, the Montgolfiers, had filled a giant bag of cloth and paper, some thirty-five feet in diameter, with air heated from a fire of shredded wool and straw, and the whole village had watched the bag rise up an estimated three thousand feet into the air (figure 14.1). The bag remained in the air only about ten minutes before it came down about a mile and a half away on a stone wall in nearby vineyards and caught fire. The spectacle was nonetheless sufficiently startling, and to the brothers Etienne and Joseph Montgolfier, encouraging, that Paris, some 340 miles to the north, heard the news swiftly, and it set the scientific literati of the capital buzzing. The matter would in fact have been of particular interest to the man to whom Franklin entrusted his letter to Joseph Banks, Dr. Charles Blagden. Blagden was a physician with a part-time appointment in the British army, but he occupied himself much more with scientific matters than with doctoring, and in fact persuaded his friend Banks to appoint him secretary of the Royal Society. More significantly, however, Blagden was the closest friend and companion to a wealthy gentleman-scientist, Henry Cavendish. It was, arguably, Cavendish who, in the course of the most esoteric experiments more than fifteen years earlier, had set in motion the train of events and ideas that moved the Montgolfiers to make their widely heralded trial.

In 1766 Cavendish reported to the Royal Society the results of experiments on different kinds of "airs" (we would say gases) that he found liberated in several kinds of chemical reactions. One of these, produced by pieces of zinc, iron, or tin dropped into either "spirit of salt" (hydrochloric acid) or "oil of vitriol" (sulfuric acid), was particularly striking for two properties—it burned so readily that Cavendish dubbed it "inflammable air" and it was extremely light—between one-seventh to one-tenth or less the weight of common air. The gas was what we now call hydrogen, and its extreme lightness prompted thoughts in the minds of several people of making something that would float in the atmosphere. The Scottish chemist Joseph Black claimed later to have thought of filling bags with the substance to make them float up, but he could not find bags light enough. The popular scientific lecturer Tiberius Cavallo amused his London audiences by filling soap bubbles with the new gas, which they could then see move rapidly skyward, but he too could find nothing else sufficiently light that would hold the gas, discovering that it moved readily through single layers of paper. Cavallo reported his own results to the Royal Society in the summer of 1782, so it should not be so surprising that Blagden, good friend of Cavendish, Banks, and Franklin, should be writing to a correspondent in France in April 1783 that "theoretical flying has been a topic of conversation among our philosophers as long as I can remember."[2]

Another Parisian who heard the news from Annonay was, like Cavallo, a popular public lecturer on scientific topics. Jacques Alexandre César Charles saw quickly some of the possibilities, for spectacle if not also for science, of the Montgolfiers' achievement. Prodded by fellow members of the Académie des Sciences, a bit irked

Figure 14.1
Contemporary engraving of the first launch of the Montgolfier Brothers' balloon at Annonay, France, 1783. Courtesy Archives Division, National Air and Space Museum, Smithsonian Institution.

perhaps at the provincials' daring, and supported by well-connected members of Parisian society and the court, Charles made plans to duplicate and, in fact, surpass the Montgolfier flying machine. Being well acquainted with the work of Cavendish and other experimenters with new gases, Charles appears to have assumed that the Annonay device used "inflammable air," and so set about designing means to do the same. He contracted with two brothers, Ainé and Cadet Robert, who had devised a new cloth coated with rubber dissolved in turpentine, to make an airtight balloon. The brothers made one twelve feet in diameter that weighed only twenty-five pounds. On August 23, they connected it up to a bulky apparatus for generating hydrogen by dripping sulfuric acid on iron filings, but their gas generating device was very awkward, and in fact allowed large amounts of acidic fumes to escape into the bag, eating small holes in it. It took several days to work out the procedure for making gas and getting it into the balloon, but by August 26, it was full enough to rise up, attached to ropes, about a hundred feet. The sight of this strange apparition above the rooftops in the center of Paris caused a sensation, and so huge crowds gathered the following day to see a balloon rise up from the Champ de Mars (where the Eiffel Tower now stands). The hydrogen-filled balloon rose an estimated three thousand feet, disappearing momentarily into the clouds, and floated through the sky for about forty-five minutes, coming down near the village of Gonesse, where frightened peasants attacked it with pitchforks until the rapidly deflating bag ceased movement.[3]

Benjamin Franklin felt duty-bound to give Sir Joseph a complete account of the entire affair, and so wrote him a lengthy letter three days later. It read more like a scientific paper than a personal letter, and Franklin in fact probably intended it for the Royal Society's *Philosophical Transactions*. After penning his own report detailing the basic facts of the Charles and Robert balloon, he appended several more pages with items reported to him by others, including word of a forthcoming demonstration by the Montgolfiers themselves and the application of a daring young man to be the first man to ascend in a balloon. He communicated the general excitement: "such is the present Enthusiasm for promoting and improving this Discovery, that probably we shall soon make considerable Progress in the Art of constructing and Using the Machines. Among the Pleasantries Conversation produces on this Subject, some suppose Flying to be now invented, and that since Men may be supported in the Air, nothing is wanted but some light handy Instruments to give and direct Motion." He also acquired for Banks a copy of a pamphlet illustrating the ascension over the Champ de Mars. Banks replied to Franklin's enthusiasm: "I consider the present day, which has opened a road into the air, as an epoche from whence a rapid increase of the stock of real knowledge with which the human species is furnish'd must take its date; and the more immediate effect it will have upon the concerns of mankind greater than any thing since the invention of shipping which opened our way upon the face of the water from land to land."[4]

The advance of the new art indeed seemed rapid over the next few months. On September 19, the Montgolfiers sent up a huge balloon from the gardens at Versailles, and sent along as passengers a sheep, a duck, and a cock, all of whom survived the eight-minute flight, although the cock damaged his wing. Less than a month later, on October 15, the daring young man mentioned by Franklin, Francis Pilâtre de Rozier, rose up in a Montgolfier balloon that remained tethered to an eighty-four-foot rope. This was, of course, merely preparatory to his ascent, with a companion, the Marquis d'Arlandes, in the first free flight of an aircraft, which took off from the great park of the Bois de Boulogne on November 21. Franklin was caught up in the general excitement, which was enormous. He immediately wrote a full report to Banks, all the more colorful for the fact that he was able to attend the ascent himself. He confirmed to Banks what he had suspected earlier, which was that the Montgolfier balloon did not rely on some special new "air," but simply on the fact that hot air became lighter than cold, and a sufficient quantity could thus provide the lift needed for a balloon. This also explained why the Montgolfier craft were so much larger than that of Charles. There was widespread anxiety in Paris over the fate of the balloon's occupants, many thinking them quite foolhardy to send themselves up into the air, particularly on such a flimsy craft. But they had a splendid flight, spending about twenty-five minutes in the air and traveling just under three miles from their starting point. The Marquis d'Arlandes, in fact, visited Franklin a few hours after landing, and gave him more details. They had to struggle to keep the great air bag from catching fire from the flames necessary to keep them aloft, as well as work tirelessly to keep fuel on the fire. The limitations as well as the possibilities of the hot air balloon were evident to Franklin and other observers of this first human flight.[5]

In his report of the flight of November 21, Franklin also outlined to Banks what he had heard of the new plans of Charles and his partners for a flight in their craft. The hot air balloon, he explained, was "cheap and expeditious," while filling a balloon, even a much smaller one, with "permanently elastic inflammable air" was "a tedious operation and very expensive." He commented on the competition between the two kinds of balloons, applauding the rapid progress that seemed to be the result: "By the emulation between the two parties running high, the improvement in the construction and management of the balloons has already made a rapid progress, and one cannot say how far it may go. A few months since the idea of witches riding thro' the air upon a broomstick, and that of philosophers upon a bag of smoke, would have appeared equally impossible and ridiculous." He then expressed his dismay at the fact that the English seemed so slow to take up the new art, fearing too much, Franklin suspected, being laughed at: "It does not seem to me a good reason to decline prosecuting a new experiment which apparently increases the power of a man over matter, till we can see to what use that power may be applied. When we have learnt to manage it, we may hope some time or other to find use for it, as men have done for magnetism and electricity, for which the first experiments were mere matters of amusement." Franklin's analogy between ballooning and electricity was a particularly telling one, and no one on earth could have made it more fittingly. It was, after all, Franklin himself who had done more than anyone else to suggest how the parlor tricks of electrical experiment could in fact be made the source of valuable practical knowledge. More than thirty years before the adventures of the balloonists, Franklin had created a sensation in Europe by his reports of experiments to show the electrical nature of lightning, and by the subsequent recommendation that tall buildings be provided with metal rods, connected by wires to the ground, in order to divert lightning strikes from the buildings themselves. The lightning rod was, historian I. Bernard Cohen has argued, "the first major or large-scale example in history of Francis Bacon's prediction that advances in science should lead to practical innovations of benefit to mankind."[6]

The appearance of ballooning in the 1780s was a particularly striking example of how the eighteenth century accommodated a new relationship between theoretical and practical knowledge, presaging the intimate relationship between science and technology that became one of the great engines of change over the next two centuries. The enlarged understanding of nature and the ability to create and use theoretical, mathematical, and experimental techniques for describing and analyzing natural phenomena had influenced technical practice in some areas at least since the early seventeenth century. The steam engine, as we have seen, was in some small part at least the product of experiments and descriptions of the behavior of air and of the atmosphere. Improvements in a range of devices and techniques, from clocks to optical instruments, owed something to the speculations, experiments, and new ideas of the creators of the Scientific Revolution. And philosophers from Bacon to Diderot had preached that the enlargement of the knowledge of nature would necessarily yield great powers over forces and things. But it was not until the developments in electricity and aeronautics of the eighteenth century that the capacity of scientific investigation to turn up phenomena and skills that could be of practical value became widely evident to all. There was no question in the minds of observers that the feats of the Montgolfiers and of Charles and the Roberts were products of philosophical knowledge. To the modern observer there might seem to be a distinction to be made between the two efforts, since Charles was, after all, himself a physicist and his balloons relied directly on the findings of Cavendish and other experiments on gases, while the Montgolfiers had no scientific background and their craft simply used hot air. But at the time the Montgolfiers themselves felt that they were as much experimenters on "airs" as any of the chemists. Heat itself, after all, was generally seen to behave as a particular substance, and the brothers believed that the fires under their balloons produced specific gases that were as distinct as Cavendish's inflammable air, only much less expensive to produce. This, as Franklin himself noted, was the source of considerable debate among the observers in Paris. As was frequently the case in the years ahead, the science that led to practical application did not, in fact, have to be correct science; it merely had to lead experimentation down fruitful paths.[7]

Only ten days after the first human flight, Franklin wrote to Banks yet once more, describing the truly triumphant passage of Charles and the younger Robert in a hydrogen-filled balloon (figure 14.2). Their gaudy craft lifted up from the Tuileries gardens in the view of a crowd that was estimated at half the entire population of Paris. "Never before," Franklin observed, "was a philosophical experiment so magnificently attended." Franklin himself was indisposed that day, but he had his carriage driven down to where he could get a good view. The crowd was indeed enormous—some estimates were as high as 400,000 people—and the elderly Franklin did not want to be waiting in a damp chill in such a crowd. He could see well enough—he called the ascent "a beautiful spectacle," made the easier to appreciate by very gentle winds. He estimated the final height at a bit over 1,800 feet (this was confirmed later by Charles's measurements with a barometer), and he followed it as long as he could with a small telescope, until "when I last saw the balloon, it appeared no bigger than a walnut." He began his report to Banks before he had even heard what had happened to Charles and Robert, but later he learned that "the travelers had perfect command of their carriage." As Franklin noted to another correspondent a few days later, they had in fact gone "thro' the Air to a Place farther distant than Dover is from Calais, and could have gone much farther if there had been more Wind and Daylight." "The Progress," he reflected, "made in the Management of it has been rapid, yet I fear it will hardly become a common Carriage in my time," which was a pity, since he would be grateful to spare himself the roughness of the pavement. Franklin was, of course, correct, although he was to be able to take advantage of one of the balloon's first practical applications when, a bit more than a year after the Charles and Robert ascent, he was handed a letter carried to him from London by Dr. John Jeffries, an American who accompanied Jean Pierre Blanchard on the first air crossing of the English Channel.[8]

If the balloon's promise really lay into the future—it became a source of amusement and occasionally a scientific tool or a battlefield aid, but little more—the eighteenth century did in fact yield up two other more significant products of science for technological application and improvement: electricity and chemistry. The applications of scientific discovery and doctrines in these two areas were modest by the measure of later centuries, but they were nonetheless highly visible and widely appreciated. The applications added significantly to the prestige of science and to the value that both governments and commerce placed on new research, and they pointed to new roles for scientists, roles that in some degree were not that novel, dating at least from the time of Archimedes, but were nonetheless more clearly and widely

Figure 14.2
Jacques Charles and Ainé Robert ascending in their hydrogen-filled balloon, December 1, 1783. Courtesy Archives Division, National Air and Space Museum, Smithsonian Institution.

perceived in a range of institutions. At the same time that rapidly developing sciences were yielding interesting new phenomena and tools for practical application, the growing use of mathematical analysis and analytical techniques was having important effects on the practices in many technological fields, from artillery to agriculture. As was evident to most observers, the making and doing of most things still had little to do with either scientific knowledge or scientific technique, and despite the gradual rise of institutions aimed at bridging this gap, from provincial and national scientific societies and academies to new schools, such as the École Polytechnique that emerged out of the reforms of the French Revolution, this state of affairs seemed likely to change little in the near future. Thus the appearance of novel technologies and devices with unambiguous linkages to philosophical pursuits was the cause of much comment and popular interest. The frenzy over the first balloon experiments, which moved from France through much of the rest of Europe over the next few years, represented an intense, but short-lived example of this. Much more sustained, and much more important for technologies to come, was the growing interest in electricity.

Knowing as we do of the enormous role that electricity was to play in technologies of the nineteenth and twentieth centuries, and of the relatively negligible one that it had before that, it may seem a bit odd that electrical science was already celebrated in the eighteenth century for its contributions to human welfare. The histories of electricity that were being written as early as mid-century, however, were adamant about the practical as well as philosophical value of the science. One reason for this is that the divisions that we make among the subjects of eighteenth-century investigation were much less evident or relevant at the time. It is no coincidence that key experimenters like Cavendish or Joseph Priestley and popular lecturers like Cavallo moved easily between electrical and chemical experiments and demonstrations. Another reason was that the slow but steady accumulation of instruments and theories for dealing with electricity throughout the eighteenth century was seen as growing control over an important natural phenomenon, control that, in the scheme of nature, had to result in more examples of the practical devices represented by Franklin's lightning-rod. In the *History and Present State of Electricity* that Joseph Priestley first published in 1767, his preface captured this belief succinctly:

> It is here that we see the human understanding to its greatest advantage, grasping at the noblest objects, and increasing its own powers, by acquiring to itself the powers of nature, and directing them to the accomplishment of its own views; whereby the security, and happiness of mankind are daily improved.[9]

The word "electricity" was coined by the English physician William Gilbert in his 1600 work *De Magnete* (*On Magnets*) to refer to the property that rubbed pieces of amber (Greek, *elektron*) showed in attracting small lightweight objects. This property had been known to the ancient Greeks and was often compared with that of a lodestone, or magnet, for attracting bits of iron. Gilbert's purpose, in fact, was to highlight the difference between the two kinds of attraction, the magnetic being to him much more important. He pointed out that many other materials besides amber displayed electricity when rubbed, and that the attraction displayed was much more indiscriminate than that of a magnet. In the course of the seventeenth century, "experimental philosophy" took on electricity as a favorite subject, with some effort directed toward making sense of electrical attraction in the context of Newtonian mechanics. Experimenters discovered that electricity was clearly more complex than Gilbert had allowed; not only could a growing list of materials be made attractive by friction, but other effects, such as repulsion, could be observed with patience and technique. In the second half of the seventeenth century, new instruments, "electrical machines," appeared that increased electrical effects and made experimentation both easier and more consistent. By the early eighteenth century, large machines with rotating globes or, later, disks of glass allowed some experimenters to create large sparks (figure 14.3). They also led to the definition of some basic concepts, such as the distinction between conductors and insulators or the observation of two distinct

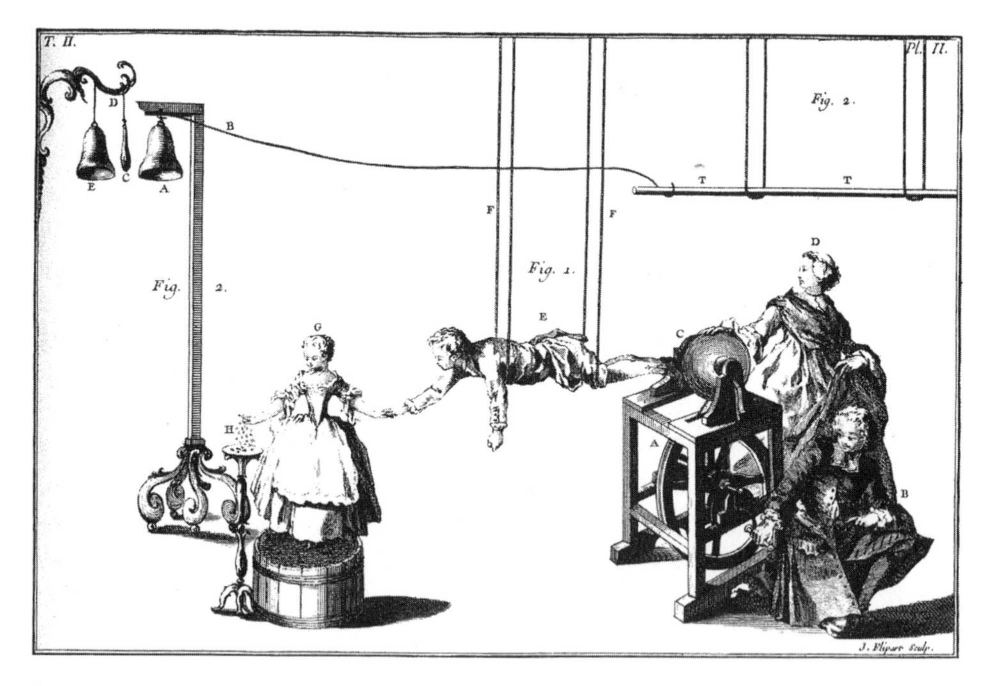

Figure 14.3
Two electrical demonstrations are depicted in this engraving from William Watson's *Expériences et Observations, pour servir a l'Explication de la Nature et des Proprietés de l'Électricité* (Paris, 1748). Burndy Library.

"kinds" electricity, dubbed "vitreous" (that from glass) and "resinous" (that from amber or other resins). In 1745 a couple of experimenters introduced a powerful new tool, dubbed the Leyden jar from the location of the better known of the two, Pieter van Musschenbroek, professor of physics at the university in that Dutch city. Originally simply a glass jar partially filled with water with a metal rod projecting through the neck and held in the hand, the Leyden jar would accumulate electrical charge when held up to one of the frictional generating machines. Better designs soon emerged, using metal foil covering both the inside and outside of glass jars, and very powerful sparks indeed could now be made and discharged at will.[10]

With the Leyden jar, electrical experimentation entered new territory. The large discharges allowed the investigation of long-distance conduction, for example; electricity could be sent several miles and unsuccessful efforts were made to measure its speed. Some also entertained the idea of using electrical discharges for communications, although without usable results. The most important product of the new devices was arguably the impetus given to Franklin's early researches, which were quickly recognized in Europe as major contributions to an understanding of electrical behavior as well as important practical suggestions of how to make buildings safe from lightning. News of the Leyden jar and some of the experimental results from its use spread quickly, and Franklin had a good electrical machine and a jar constructed by Philadelphia craftsmen within months of its announcement. By the middle of 1747, he had experimented and observed enough to begin to formulate his own theories about what electricity was and how it behaved, and had begun to think in terms of "positive" and "negative" electricity rather than vitreous and resinous. Franklin also began experiments with drawing off charges from the Leyden jar with pointed metal rods. He produced the biggest discharges he could, creating a "battery" of alternating layers of glass and metal to store as large a charge as possible. By the end of 1749, Franklin began jotting down notes about the similarities between his large discharges and his observations of lightning, and even before his famous kite experiment of 1752, he had convinced himself of the equivalence and had devised the lightning rod, both to prove his point and to make practical use of it. These ideas were communicated to European correspondents, and while the English were largely cool to these contributions from a colonial, the French reaction was enthusiastic. Franklin's experiments were carried out by French savants, the results supported his conjectures, and his fame grew widely and swiftly.[11]

The lightning rod was arguably the only practical electrical invention of the eighteenth century (not counting the development of experimental and demonstration apparatus, which continued through the decades). Debates continued on into the nineteenth century about the ideal design, but it quickly became an established safety feature for taller structures and ships. Perhaps as significantly, the work of Franklin and others promoted widespread interest in the new phenomena. Electrical demonstrations, using ever larger machines, the Leyden jar batteries, and novel types of conductors, insulators, and glowing glass tubes became popular diversions, and made more and more people aware of some of the new concerns of science. Indeed, it can be argued that the use of electricity as a source of scientific spectacle put it very much into the same category as ballooning: the distinctions between public show and serious research were blurred, those between showmen and scientists were similarly muddied, and the line that might have once seemed simple and clear between philosophical investigation and technical improvement became harder and harder to distinguish, at least in certain realms.

In 1791, Luigi Galvani, professor of anatomy at the University of Bologna, published the results of a decade of experiments on what he called "animal electricity." Galvani had observed the movements of frogs' legs under dissection, and he associated these movements at first with the presence of a nearby static-electricity machine, but later with a host of other electricity-related observations. Those who knew electrical science better than the Italian anatomist were deeply skeptical about this "animal electricity," but the work inspired a fellow Italian, as skeptical as all the others, to try to find out what was really going on in Galvani's experiments. Alessandro Volta was professor of physics at the University of Pavia and had been doing serious electrical research since the 1760s, having devised a charge-holding device called the "electrophorus," which was a simple way of concentrating charges. Volta turned his skepticism of Galvani's concepts to good use, embarking on a long series of experiments, systematically seeking out the source of the electrical action Galvani had located in his dead frogs, but which Volta believed resided in the metallic plates, hooks, and prods in Galvani's setup. He took great care to avoid the kind of precipitous conclusions he associated with his countryman and so did not announce his results until 1800, when, in March, he reported to London's Royal Society that he was able to produce a continuous flow of electricity without the motion of a static generator or the legs of a frog, but simply with a pile of metal plates, alternately zinc and copper, each separated by a disk of absorbent material (such as paper or felt) that had been soaked in brine or a mild acid (figure 14.4). To emphasize the fundamentally chemical nature of his discovery, Volta also devised what he called a "crown of cups," in which a series of cups filled with brine or acid were joined by alternate strips of zinc and copper. When the circuit of cups was completed, it too produced an electric current.[12]

The introduction of the "Voltaic pile" caused a sensation among electrical experimenters and demonstrators. Within a few months of the Royal Society paper, other piles were being used throughout Europe for a wide range of experiments, and interesting results came quickly. There had never before been, after all, a source of steady, continuous electrical current, and the things that could be done with the new phenomenon were marvelous. By May 1800 a couple of English researchers showed

Figure 14.4
The Voltaic pile, as depicted in the *Philosophical Transactions of the Royal Society* (1800). Burndy Library.

that the current could be used to decompose water into its two constituent gases, now called, thanks to the French chemist Antoine Lavoisier, hydrogen and oxygen (this had been done earlier on a smaller scale with Leyden jar discharges). A couple of months later another Englishman showed that salts could also be decomposed by the Voltaic current. This last experiment was carried out by William Cruikshank using a very powerful device, the first true "battery," in which plates of copper and zinc were soldered together and then placed in a wooden trough filled with weak sulfuric acid. The large currents of the battery could do startling things, such as heat iron wires red hot or charcoal to incandescent heat, but these were little more than spectacular effects.[13]

In the hands of a few brilliant experimenters, however, the battery became a wonderful new scientific tool, particularly for chemistry. The most important of these was Humphry Davy, a self-educated chemist barely out of his teens who had already made interesting studies of gases and the nature of heat. These latter studies came to the attention of Benjamin Thompson, an expatriate American who sported the Bavarian title Count Rumford. Rumford summoned the young Davy to London and offered him the position of assistant lecturer in chemistry at a new public forum for science, the Royal Institution. This was envisioned by Rumford as a place for the genteel public to hear lectures, see scientific demonstrations, and otherwise educate itself about new inventions and ideas, and he effectively used his connections in

Figure 14.5
This large electrochemical battery in the basement of London's Royal Institution allowed experimenters like Humphry Davy to investigate new electrical phenomena. Burndy Library.

London social circles to get a substantial list of subscribers to put up money for the new establishment. The first lecture occurred the very same month that Volta announced his discoveries to the Royal Society, and the Institution quickly established itself as a useful link between members of London's growing scientific community and the social circles that brought respectability to their enterprises. The young Davy was a spectacular success in making the lectures popular and informative, and he in turn used the new resources available to him to carry out important research. In particular, he had built a series, over several years, of ever larger and more powerful batteries, and with these he demonstrated both how fundamental electricity was and what it could conceivably do (figure 14.5). In 1807 he isolated for the first time the light metals sodium and potassium, and the following year barium, boron, calcium, and strontium. A larger battery still allowed Davy to separate out iodine, chlorine, and fluorine and to demonstrate the astonishingly bright light of an electric arc—a sustained spark between two pieces of carbon. Such experiments depended on the use of very expensive equipment and materials, and so had no immediate technological consequences, but they gave glimpses of an electric future that would have been unimaginable only a decade before.[14]

That electrical science might yield useful technologies beyond such passive instruments as the lightning-rod was still largely a matter of faith rather than fact well into the nineteenth century. The same cannot be said for chemical science, for the intimate relationship between the rapid reformation of chemistry in the last decade or so of the eighteenth century and the emergence of chemical industry on a new, industrial, scale was demonstrated in numerous ways. It was chemistry, above all, that emerged as the archetype of useful science as the Age of Enlightenment gave way to the Age of Industry. While British chemists such as Joseph Black, Joseph Priestley, and Henry Cavendish made discoveries with practical implications, from soda water to balloons, it was French chemists who led the way in forging the close connections between academic chemistry and industry. The social and institutional settings in which the French worked, particularly the Académie des Sciences, made it easier for the government, in particular, to organize chemistry for both industrial and strategic purposes. In addition, the direct state involvement in certain industries, such as luxury textiles and ceramics, provided additional impetus for the application of new chemical knowledge to techniques. French chemists, too, were in the vanguard in the creation of effective and useful new theories and terminology in the last decades of the eighteenth century, thanks in large part to the work of Antoine-Laurent Lavoisier. Lavoisier worked tirelessly through the 1780s at a general reform of chemistry, an effort that culminated in 1789 with the publication of a broad-ranging textbook, *Traité élémentaire de chimie* (Elementary Treatise on Chemistry), which summarized such key elements of the new chemistry, such as the oxygen theory of combustion and the new nomenclature ("oxygen," "hydrogen," and so on). Lavoisier also provided an important model for how an academic chemist at the leading edge of theory and experiment in his science could also play a central role in making chemistry useful. In 1775 he became a key figure in the French gunpowder inspectorate, and he even moved himself and his family to the Arsenal in Paris to oversee powder manufacture. In this and similar positions, Lavoisier emphasized the close connection between theory and practice that came to be widely associated with chemical science and industry.

By his boldness in overthrowing chemical orthodoxy and in promoting a radical revision of his science, and by the sheer creative genius he displayed in his work, Lavoisier was hardly typical of any kind of scientist in the eighteenth century. More representative, perhaps, was his younger colleague, Claude Louis Berthollet, who became a member of the Académie des Sciences in 1780, at thirty-one. A few years later, in 1784, Berthollet was appointed inspector of the dyeworks and a director at the government Gobelins tapestry factory. Here he became involved in the improvement of a range of textile processes, particularly bleaching. Bleaching was the single most time-consuming process in making cloth; effective bleaching of white linen, for example, required as much as six months, since it consisted of a cycle of vigorous washing, boiling in a lye solution, followed by boiling in an acid solution (traditionally sour milk or buttermilk) and then setting out the cloth in huge bleaching fields for two to three weeks at a time, exposing the cloth to the action of air and sunlight, with a single piece of cloth having to go through this cycle five or six times before it was sufficiently white. Berthollet devised a much quicker and ultimately cheaper alternative by using chlorine. The gas had been isolated by the Swedish chemist Carl Wilhelm Scheele in the 1770s, and Berthollet began investigating it in connection with his efforts to find support for Lavoisier's theory that oxygen was the primary principle in all acids. While in fact the experiments did little to support Lavoisier, and chlorine was not identified as an element until Davy did so about twenty years later, Berthollet observed in 1785 the bleaching action of chlorine solutions, and three years later helped to establish a factory for making chlorine bleaching solutions. Practical difficulties and the protests of traditional bleachers and dyers set this effort back, but work continued since it was easy to demonstrate that a practical "artificial bleach" would have enormous economic advantages. As early as 1791 some experimenters found how to absorb the caustic acid in a mineral, such as slaked lime, making it much easier to handle, although the process was not patented until Charles Tennant in Scotland did so in 1799. The first factory for the manufacture of bleaching powder began production in 1800 in Glasgow, the economics of the process having been much improved by the cheapening of sulfuric acid, a key material in making the chlorine. While making the product useful on an industrial scale required much effort on the part of non-chemists, the initial impetus given to the invention by the discoveries of Scheele and, more significantly, the researches of Berthollet, were held up as prime examples of the economic value of chemical science.[15]

More important than truly novel materials or processes in the industrialization of chemical technology in this period was the scaling up of processes from laboratory or workshop scale to that of the factory, producing, in many cases, tons of product where before there had been only pounds or gallons. The best example of this was the development of a large-scale sulfuric acid industry. Sulfuric acid was long a material of technical importance, being used in everything from glass to soap making, usually under the name (for the English speaking) of "oil of vitriol" or simply "vitriol." Before the eighteenth century, this key acid was generally made from iron sulfate, which occurred in the form of a mineral, green vitriol. Supplies of this mineral were not common, however, and the manufacture of acid from it was a small-scale process. Much more rapid and easier to scale up was the making of sulfuric acid directly from mineral sulfur. By the end of the seventeenth century, vitriol was being made in France and the Netherlands by burning sulfur in glass containers. When this process was introduced to England in the early eighteenth century, it reduced the price to the extent that what once bought an ounce of acid could now buy a pound. Further cheapening followed the introduction of the lead-chamber process in 1746, which substituted large lead vessels for the smaller more fragile glass jars previously used. This apparently simple substitution permitted the scaling up of acid production enormously, which allowed manufacturers to keep up with rapidly growing demand. The growth of the textile trades, the potteries, and of metalworking all depended on expanded supplies of acid.[16]

As important as the enlarged production of sulfuric acid was, the industry that became the key symbol for industrialized chemistry was not acid but alkali manufacture. Alkali was a key ingredient or processing agent in a host of traditional as well as newer industries, from soap to glass to gunpowder manufacture, but supplies rapidly became a problem in the expansion of any of these industries. There were three different kinds of alkali in common use in the eighteenth century: soda (sometimes called mineral alkali), potash (or vegetable alkali), and ammonia (or volatile alkali). For many manufactures, the first two, soda and potash, could be used interchangeably, although one might be much preferred over the other for certain products—soap made with potash rather than soda would not, for example, harden as well, and glass was much easier to work with if made with soda. Soda was sodium carbonate (Na_2CO_3), potash was potassium carbonate (K_2CO_3), and ammonia (NH_3), normally a gas, was used as a solution in water, but of course these chemical formulae and compositions were unknown before the nineteenth century. The French chemist Duhamel de Monceau did discover in 1736 that common salt shared a key ingredient with soda, and this discovery in fact can be see as the beginning of the transformation of an ancient industry. Soda traditionally came from the ashes of certain plants found near the seashore or, in the case of kelp, in the sea itself. Such sources, which required careful treatment and refinement to yield reasonably pure soda, could not be expanded easily and, in the political upheavals of the late eighteenth century, their supply could be quite precarious in fact.

De Monceau's discovery thus inspired both governments and entrepreneurs to seek some way to make common salt into a useful source of soda, and a series of processes emerged in the last half of the eighteenth century. Most of these were of limited value, but just at the time the French Revolution was beginning, a French physician, Nicolas Leblanc, incorporated a number of chemical observations into a process that was to transform not only soda production but the entire nature of chemical manufacture. In its essential form, the Leblanc process was simple: when salt is reacted with sulfuric acid, the products are hydrochloric acid (as a gas) and sodium sulfate ("saltcake"). The saltcake can then be heated with coal and limestone to produce soda and the by-products of calcium sulfide and carbon monoxide. In practice, however, making these reactions work safely, reliably, and economically required much effort. Each stage had to be carefully monitored and controlled, all ingredients had to be carefully prepared and purified, and the by-products had to be managed appropriately. The French Revolution made it at first difficult for Leblanc and his backers (who were, in fact, aristocrats) to make progress. Only when soda supplies were cut off in the course of the Napoleonic wars did the government step in and push artificial soda into large scale production. A similar pattern occurred in Britain, although with different timing for the pressures on traditional supplies, but by the end of the Napoleonic era in 1815, alkali factories could be found scattered around Europe. By the 1820s these factories were appearing on a large scale, and giant alkali works were becoming as much symbols of industrialization as the textile factories.[17]

The alkali factories were heralds of change in other ways as well. They represented the significance and the extent of one of the fundamental shifts of modern industrial technology, that from reliance on organic sources of supply to inorganic, with the consequent liberation of expansion from the bounds set by cultivation and harvest. It was widely recognized that soda made from common salt—found in inexhaustible (relatively) quantities in mines or saltwater—would be preferable, in the long run even if difficult and expensive in the short run, over manufacture that depended on plants of limited supply. Alert individuals in places like France and England understood by the middle of the eighteenth century that production dependent on what could be gathered and grown would always reach limits, and that the expansion of material possibilities would ultimately require breaking from that dependence. The large scale chemical factories, depending of supplies of sulfur, salt, limestone, and the like, from mines and pumps, represented liberation from the organic. The great economic philosophers of the classical school—Adam Smith, Thomas Malthus, David Ricardo—did not grasp the significance of this liberation, and thus underestimated the true potential of technological change for economic growth, but the industrialists of Europe and America did understand the possibilities and they acted on them through the nineteenth century.

Yet another element of the industrial future that the alkali and bleach works gave early and eloquent expression of was the capacity of the new factory system to exact costs, human and environmental, that were in fact not reflected in the market. The workers in the soda factories became among the most pitied in the industrial world. There were, generally speaking, three kinds of workers: process men, yard men, and tradesmen. The last category consisted of those such as plumbers, engine men, wagon drivers, and the like, who carried out particular tasks in a manner not unlike that found away from the factory. The process men were those who worked with the chemical reactions themselves—mixing materials, extracting them from pots and reaction vessels, handling often caustic materials through many hours of the day. These workers, handling fire and acid and alkali for twelve hour stretches day after day were often completely broken in health after a few years. These broken men then became the yard workers, fit only for carrying materials, cleaning up factory areas, and other tasks calling for neither skill nor alertness. The same harsh chemical action that wrecked havoc on the health of workers also caused damage on a larger, if less intense, scale in the neighborhoods and larger regions around factories. The main by-products of the Leblanc process, for example, were the highly corrosive and poisonous hydrochloric acid sent up the great chimneys and thus out into the countryside and the foul smelling and also poisonous calcium sulfide, often dumped into nearby streams. The areas around chemical factories came to be seen as hellish, with foul air and water, all vegetable and animal life destroyed, and ever-present danger presented to inhabitants for many miles around. The alkali factories, in fact, became the precipitant causes for the first serious legislative efforts to control pollution from major industries, but the public agitation, legislation, and remedial efforts were all simply the opening acts of a dramatic struggle that continues to this day.[18]

The giant alkali works, as well as a myriad other smaller chemical factories, were also symbols of a new kind of change. In no other industry was the role of science so openly and readily acknowledged—even boasted about. An evocative effort to describe this new state of affairs came from a British chemist, Thomas Cooper, a friend of Joseph Priestley who emigrated to the United States in 1793 and became the professor of chemistry at tiny Carlisle (later Dickinson) College in Pennsylvania. In a lecture in 1812, Cooper looked back on a half-century of change:

> Let any one examine the state of the arts and manufactures, fifty years ago, and compare it with the situation of the present day, and it will be found, that during these fifty years, more improvements have been made, originally suggested by chemical theories, and pursued under the guidance of chemical knowledge, than in two thousand years preceding. At present there is not a manufacturer of note in England, who is not more or less acquainted with chemistry as a regular branch of education and study.

Thomas Jefferson wrote shortly before this to Cooper, urging on him what he believed to be the most appropriate model for the joining of science and the arts in the new nation, "You know the just esteem which attached itself to Dr. Franklin's science, because he always endeavored to direct it to something useful in private life. The chemists have not been attentive enough to this," Jefferson chided. In the coming years, the example of Franklin—of science as the handmaiden to technology—would be seen as the model of the proper place of science in the American nation.[19]

15 Mobility

When we reflect upon the obstinate opposition that has been made by a great majority to every step toward improvement; from bad roads to turnpikes, from turnpikes to canals, and from canals to railways for horse carriages, it is too much to expect the monstrous leap from bad roads to rail-ways for steam carriages, at once. One step in a generation is all we can hope for. If the present should adopt canals, the next may try rail-ways with horses, and third generation use the steam carriages. . . . I do verily believe that the time will come when carriages propelled by steam will be in general use, as well for the transportation of passengers as goods, travelling at the rate of 15 miles an hour, or 300 miles per day.[1]

Throughout his life, Oliver Evans always saw "obstinate opposition" to improvement, so the frustrated tone of his 1812 prediction of the ultimate triumph of "railways" was very much characteristic of the man who saw himself stymied, usurped, shunned, and exploited at every turn. He was, in fact, often much harsher and damning in his denunciation of perceived opponents and rivals, and the impression he leaves with the modern reader is of a man who simply lacked the measure of social grace and diplomacy that seems so important in smoothing the way for radically new propositions. And Oliver Evans was blessed, or cursed, depending on your point of view, with a head full of radically new propositions, as well as the mechanical talent to make many of them real. He saw and was able to design, at least in rudimentary form, a number of the key inventions that were to fill the life of the first couple of generations of nineteenth-century Americans and Europeans with more fundamental and easily recognized change than any other peoples had ever experienced, or, it could be argued, have experienced since.

While it is customary to think of technological change, and the social change consequent to it, as a steadily accelerating phenomenon from the early nineteenth century onward, at least in relative terms there are good arguments for suggesting that the changes between, say, 1829, when the steam locomotive showed its superiority over alternatives, and 1869, when the transcontinental railroad was completed across the United States, were different in scale and reach from anything seen since. Although the railroad became the most prominent symbol of the transforming power of technology, these were also the years during which the steamboat and steamship transformed water transport, the electric telegraph (including the Atlantic Cable completed in 1866) redefined communications, and a host of other systems and inventions, from mass production with interchangeable parts to gas lighting to photography, gave life in the West a look, feel, and pace profoundly different from that of any other place or time. As we have seen, the expectation and appreciation of technological improvement had been growing steadily among the Western Europeans for centuries, and this experience provided a rich and indispensable foundation for the years of change to come. In the case of the great transforming technologies of steam transportation and electrical communications, there was much continuity from the past, but there also can be no denying—and this was widely perceived at the time—that these particular technologies carried within them the seeds of improvement and elaboration so extensive that they were the first technologies by which people widely sought to define their culture and civilization, and which gave popular credence to the idea that technology was at the root of human progress.

The 1812 prediction of Oliver Evans, as modest as it might seem in hindsight, was actually a bold projection of the direction in which he perceived transport improvement moving. He was particularly well qualified to make such a projection, even if his own contributions to the technology of the railroad and the locomotive were actually quite modest. The insubstantial nature of Evans's own influence on railroad history was certainly not due to a lack of technical acumen or vision, but to limitations imposed by his own social awkwardness and by the economic constraints of early-nineteenth-century America. Evans was, in fact, the finest steam engineer his country produced in its first years—indeed, he could be claimed the finest and most ingenious American mechanic of his day. Born in northern Delaware in 1755 of a leatherworker and farmer of modest means, he was early apprenticed to a wheelwright and wagon maker, but during the American Revolution he turned his attentions first to wire-drawing and then to milling. He spent much of the 1780s designing and promoting modifications to grist mill design that reduced substantially the labor required, incorporating a host of devices for lifting, moving, and collecting the materials in the mill, from the corn kernals to the finished meal or flour. The laborsaving approach gave rise to an "automatic" mill in which processing was continuous, aided by a series of elevators, hoists, screws, rakes, and other devices, which Evans had linked to the waterpower of the mill. The result was a remarkable integrated machine, but one that required a much greater faith and investment in machinery than had ever been asked of any miller before. In a pattern that was to repeat itself often in his life, Evans then spent years cajoling millers for the capital and the opportunity to prove his inventions, as well as seeking out protection through patents and publicity through writings. When the U.S. government passed its first Patent Act in 1790, Evans received the third grant made, and by dint of much promotional effort he was able to sell licenses throughout the middle of the Eastern Seaboard to millers and farmers—George Washington purchased one for his mill at Mt. Vernon and hired Oliver's brother Evan Evans to make the installation. Encouraged by this first taste of success, the inventor then put much effort in broadcasting word of his work through a book, issued in 1795, *The Young Mill-wright and Miller's Guide*.[2]

By the end of the 1790s, Evans was a successful merchant and millwright, and the value of his inventions, and of his book (which went through at least fifteen editions up to 1860), was widely recognized. But there were other visions in his head, and it is at this point that Evans began to resemble a type that was to appear with some frequency in the nineteenth century, particularly in America—the driven, sometimes obsessive, often frustrated, inventor. As early as the 1780s, while he was still developing his millwork inventions, Evans imagined putting steam engines to work in transportation and other tasks. When he applied to several states, before the federal government had a patent act, for protection of his inventions, he included in the requested coverage that for a "steam carriage," although there is no evidence it was more than an idea at the time. As the century came to an end, however, Evans took up experiments with steam locomotion and with designs for a new type of steam engine. There were a few Boulton & Watt engines in America, and other engines of American construction but on traditional principles. Evans directed his attention to building a smaller, lighter engine, perhaps to simplify construction and perhaps realizing that earlier experiments with steam carriages (most notably in France) had foundered in large part due to the heaviness of the engines trying to move themselves. By early 1803 Evans had an engine of his own design running in his Philadelphia store, where it ran a mill he used for making plaster of Paris. The engine probably ran at about five horsepower, using a piston only six inches in diameter in a cylinder of eighteen-inch stroke. Viewers were apparently unimpressed, since the engine was dwarfed by the thirty-two-inch diameter engines just down at the street at the waterworks. These low-pressure engines generated about twelve horsepower with their six-foot stroke. An astute observer would perhaps have noted that Evans had managed to produce almost half the power of the older engines in a machine only a small fraction of the size, but the enormous implications of this would have to be demonstrated much more visibly.[3]

The key to the increased power for size of the Evans engine lay in its use of steam under several atmospheres of pressure. The engines of Newcomen and Boulton and Watt, as well as the many variants that appeared over the years, used steam at or just slightly above atmospheric pressure, and derived their power from the atmosphere working against a vacuum in the cylinder. The Evans and subsequent high pressure engines, by contrast, gained power from the pressure of steam pushing directly against the piston. This, as became evident, sacrificed fuel efficiency, since the creation and maintenance of steam at higher pressures used more fuel. But where power needed to be generated in smaller amounts or in smaller, portable units, the high pressure engine was superior. Later in the century hybrids of these two basic approaches appeared, designed to maximize power, fuel efficiency, or a number of other possible factors, depending on the final applications. Indeed, the design of steam engines, and particularly of the more complex modifications, became the core of a new discipline and a new profession—mechanical engineering—and the problems faced by these engineers also constituted the seeds of a new scientific study, thermodynamics. Oliver Evans, of course, and the other early steam engine inventors, had to work out their designs without benefit of these disciplinary foundations. It should therefore be no surprise that engines in these early years underwent constant modification in every respect, both small and large. The first two decades of the nineteenth century were a period of unending experimentation, and a number of distinct steam engine traditions began to emerge, depending largely on an increasingly diverse range of applications. Evans first applied his engine, as we have seen, to shop work, and high pressure engines turned out to be well suited for such use. Indeed, even through the years ahead in which Evans felt frustrated in a number of his technological ambitions, he at the same time managed to create and operate a successful shop for the manufacture and sale of these smaller steam engines. While this shop disappeared just prior to Evans's death in 1819, and Evans's own engine designs were quickly superseded, it nonetheless demonstrated the American capacity for mechanical self-sufficiency.

The possibilities of using steam engines for transport—particularly lighter and more robust engines than the Boulton & Watt designs—were widely recognized by the beginning of the nineteenth century. The steam carriage ideas that Evans had in the 1780s were to be found elsewhere at the time, and a few experimenters even attempted to construct working devices, but without success. More promising, but still meeting considerable practical difficulties, were efforts to apply steam to water transport. The most important of these were in the United States, where one inventor, John Fitch, got so far as to be able to demonstrate the efficacy of a working steam-powered vessel. In the summer of 1790, after years of frustration, and battles with rivals, Fitch put a steamboat into service on the Delaware River and actually provided scheduled service for several months between Philadelphia, and Burlington and Trenton, in New Jersey. Despite Fitch's technical success—his *Steamboat* ran an estimated two to three thousand miles on its route that season, with few breakdowns—the enterprise was an economic failure and served to discourage similar efforts for years to come. Nonetheless, the popular memory of Fitch's steamboat persisted, and thus observers of Evans's experiments naturally thought in terms of water transport. The first vehicle in which he was able to demonstrate his engine was, in fact, a peculiar hybrid vessel, a steam-powered dredge that he built for the city of Philadelphia in 1805. The *Oruktor Amphibolus* was a very awkward-looking

The Mechanic.

JULY, 1834.

[For the Mechanic.]

STEAM-CARRIAGES.

[Evans' Steam-Engines. See page 1[illegible]6.]

Figure 15.1
Oliver Evans's *Oruktor Amphibolus* was the first steam-powered vehicle in America. From *The Mechanic* (July 1834).

craft, thirty feet in length, on which an Evans engine ran a chain of buckets designed to dredge out the Delaware River near the Philadelphia docks (figure 15.1). To move the vessel from Evans's workshop to the river, it was equipped with wheels that were connected to the steam engine and in mid-July, 1805, Evans sent it out on to the Philadelphia streets, to much publicity, where he demonstrated its mobility for several days before sending it down into the river for work. Despite Evans's passionate espousal of steam locomotion, this was in fact his only experiment in land transport, and it clearly came across to observers as little more than a stunt.[4]

It was at just this point, however, that the steamboat emerged from the realm of stunts and demonstrations into successful application, but the early pioneers were not to be the principal beneficiaries. Through the 1790s and on into the new century, a host of experimenters, in the United States and elsewhere, had projected new steamboat designs and even built working models, but none had gotten even as far as John Fitch in terms of reliable operation. As early as 1789, for example, William Symington, in Scotland, designed a boat that went as fast as seven miles per hour on the Forth and Clyde Canal, but the engine design was badly flawed and the experiment was abandoned. More than a dozen years later, Symington tried again on the canal, this time with somewhat more success. In March 1803, the *Charlotte Dundas* ran for almost twenty miles on the canal's summit level, making the distance in about six hours while tugging two seventy-ton barges. Concern about damage to the canal's walls, however, discouraged further operation there, and Symington could not get other backers to follow through. The early history of steamboat experiments demonstrated forcefully the need not only for mechanical ingenuity but also for capitalist vision and fortitude, and the inventor who was fortunate enough to find this at his disposal was an American painter of modest talent but great personal charm, Robert Fulton.[5]

Robert Fulton's background was somewhat like that of Oliver Evans. He was born and raised in rural southeastern Pennsylvania in the years just before the American Revolution, and received relatively late training, in Fulton's case as a silversmith. But Fulton's talents and personality carried him in very different directions. Just how he came by his initial training as a portrait painter is unclear, but in 1786, at age twenty-one, he set out to make himself into an artist, and he followed a path taken by a number of similar aspirants of his day—he set out for London and the patronage of America's most successful artist, Benjamin West. It was in England that Fulton underwent the transformation from artist to inventor, putting his hand to a range of projects, from canals and marble cutters to submarines. Some of his ideas were good ones, others were not so, but he evidenced a talent for capturing the attention of men of influence and resources, and so had opportunities to demonstrate his ideas that men like Fitch and Evans so often found difficult to come by. Fulton's most important patron was Robert Livingston. Livingston was a prominent member of New York state's landowning aristocracy, and his interest in steamboats began in the late 1780s, originating with his brother-in-law, John Stevens. Stevens was himself a wealthy man who possessed his own mechanical talents and imagination, experimenting on boats and engines about the same time as Fitch and his great rival and antagonist James Rumsey were waging battles for support and public attention. Stevens later did make useful contributions to steamboat design, but not until after Fulton, with his backer Livingston, had shown the commercial promise of the technology.[6]

Fulton possessed one other advantage over his rivals, the virtues of which might not seem self-evident. He understand better than most how technological change really worked. While steamboat development in the last years of the eighteenth century and the first of the nineteenth emerged as a field crowded with men eager for credit, patents, privileges, and recognition as the sole inventor of an important new technology, it was Fulton who recognized the more complex nature of invention. He gave eloquent expression to this in 1796, in a book that he wrote in England to promote his ideas about canal building:

> There is . . . frequently a secret pride which urges many to conceal their speculative inquiries, rather than meet criticism, or not be thought the first in their favorite pursuit; ever anxious to claim the merit of invention, they cannot brook the idea of having their work dissected, and the minute parts attributed to the genius of other men. But in mechanics, I conceive, we should rather consider them improvements than inventions, unless improvement may be called invention, as the component parts of all new machines may be said to be old; but it is that nice discriminating judgment, which discovers that a particular arrangement will produce a new and desired effect, that stamps the merit. . . . Therefore the mechanic should sit down among levers, screws, wedges, wheels, &c. like a poet among the letters of the alphabet, considering them as the exhibition of his thoughts; in which a new arrangement transmits a new idea to the world.

It is indeed ironic that, of all the claimants to the invention of the steamboat (and Fulton himself was no more modest than others), it was the man who comprehended the incremental and derivative nature of technological development who was to receive the mantel of heroic genius from the naïve histories of nineteenth-century progress.[7]

Robert Fulton's first successful steamboat did not sail in America or even in England, but rather in France, where Robert Livingston served several years as the American minister. On August 9, 1803, a vessel of Fulton's design was successfully demonstrated on the Seine before a group of semi-official observers from France's Institut Nationale as well as many of the Paris fashionable. The official newspaper reported a couple of days later that the demonstration had been a "complete and brilliant success," although Fulton himself was disappointed that he was not able to make the boat go upstream more than three or four miles per hour. The boat freely borrowed design elements from a wide range of his predecessors and rivals, but the French setting made this a relatively safe thing to do. Unfortunately for Fulton, at just this time the fragile peace that Napoleon had made with Great Britain and her allies was breaking down, and Napoleon's government, and much of the country along with it, was gearing up for war. Since Fulton made no claims that his vessel could operate on open seas, there was no perceived military value, and, like so many earlier efforts, this too became little more than a show.[8]

Four years later the story was quite different. It was, indeed, four years to the day after his Paris excursion that Fulton put his *Steamboat* (later officially registered as the *North River Steam Boat*) out on to the East River off lower Manhattan for its first trial (figure 15.2). A week later, on August 16, 1807, he brought the boat around

Figure 15.2
There are no contemporary images of Fulton's *Clermont*. This is a conventional late-nineteenth-century image of the craft. Courtesy Smithsonian Institution.

the Battery to a dock on the Hudson (or North River), and the next day he set out for the journey to Albany, 150 miles upriver. The maiden voyage of the *Steamboat* went flawlessly, taking twenty-four hours to go 110 miles to its first landing, at Clermont, Livingston's upriver estate, and a little less than eight hours to go the remaining distance to Albany the next day, averaging about five miles per hour. Fulton lost no time sending out the message that this, finally, was no show. The very next day, August 20, he hung out a sign at the Albany docks announcing a fare of $7 for the trip downriver to New York City. It was probably less the price—about twice that charged by sailing vessels on the river—and more would-be passengers' fear of being blown to bits by an exploding boiler (although Fulton used a low pressure engine) that kept the passenger list down to two bold Frenchmen. But on returning to New York, again averaging about five miles per hour, Fulton clearly made his point—his boat provided reliable, safe, comfortable passage, and when regular service began in early September, business was brisk, and satisfied. The celebrations in the press and elsewhere were actually quite muted, but the age of steam travel had clearly begun.[9]

All of the American steamboat inventors—Fitch, Rumsey, Evans, Stevens, Fulton, and others more obscure—understood that the real significance of working steamboats for the United States lay not on the Eastern Seaboard, where all but a small fraction of the population lived, but in the vast unpeopled interior. The great rivers and lakes of what was then the West always posed the greatest promise and the greatest challenge for those who understood that internal communications would be the key to the country's growth. Fulton himself always acknowledged that the Mississippi and its tributaries were where the great promise of the steamboat lay, and his company moved quickly to transfer the technology thence. Like most of the steamboat ventures, however, the Fulton interests wanted to move only with some kind of monopoly privilege, and so their first vessel did not venture into the inland rivers until the Territory of Orleans (later the state of Louisiana) granted an exclusive privilege in the spring of 1811. That October, the steamboat *New Orleans*, built and guided by Nicholas Roosevelt, the country's most experienced steam engineer, set out from Pittsburgh for an extraordinary journey of more than two thousand miles, reaching the city of New Orleans, after encountering floods, earthquakes, and the sheer immensity of the Mississippi itself, on January 10, 1812. The *New Orleans* proved unsuitable for regular long distance service on the river, however. That called for significant departures in design and construction, particularly to cope with the extremely shallow drafts that characterized the untamed Western rivers throughout their courses. The principal pioneer was Henry M. Shreve, who had spent his life with boats on the great river, and thus had a sense of just what a boat had to accomplish. Shreve captained the steamboat *Enterprise*, the first to make the long journey upriver from New Orleans to Louisville, in May 1815. Some two years later, the *Washington*, a boat of Shreve's design, using a high pressure engine, made the same trip in only twenty-one days. He had to fight the Fulton interests over their monopoly privileges, finally prevailing in 1819. From that point, the steamboat was the queen of the Western rivers.[10]

The steamboat was perhaps the first major technology in which the United States took an early and significant leading role. The European experimenters themselves understood the special circumstances that made the technology seem so compelling on the other side of the Atlantic, although the rapid development of a competent and creative engine-building capability in America was a bit of a surprise. In Europe itself the technology's expansion faced continued obstacles. The first commercial success in Britain was the use of steam vessels in Scotland for pleasure excursions, and it was only after this had exposed a sufficient number of Britons to steam propulsion as a safe and reliable mode of travel that serious transport by steam made headway. In post-Napoleonic France, progress was even slower, with little effort made before the mid-1820s, which was also when the Rhine saw its first regular service. Other places in which great rivers might seem to beckon to the new technology—Russia, India, Brazil, and China, for example—gradually saw the introduction of boats, but they long relied on external sources for both machinery and expertise. Only in America did the steamboat serve the key function of ushering in a period of change in which mechanical technology came to be perceived as the key to widespread and continuing social and economic progress. Elsewhere in the West, and in much of the rest of the world subsequently, this role was taken by the railroad.[11]

As the career of Oliver Evans demonstrated, the promises of the steamboat and of the railroad to some extent ran parallel to one another through the last decades of the eighteenth century and the first of the nineteenth. This, plus the fact that the steam locomotive became the premier symbol of nineteenth-century transportation, tends to obscure the fact that the railroad was not initially linked to steam as a technological concept. The term "rail-road" carried no implication of steam locomotion to its first projectors, and yet the concept of rail travel was a source of considerable excitement and expectation, particularly in Britain. The Canal Age that had begun in the middle of the eighteenth century had largely run its course, at least for profitable enterprises, by the first decade of the nineteenth century, but it left unsatisfied the desire for better transport. If anything, in fact, the canals demonstrated how economically attractive investment in transportation could be while at the same time revealing the limitations of relying on water routes. In the United States, the same lessons came a bit later, but the success of the Erie Canal, which opened in 1825 and linked Albany on the Hudson River with Buffalo at the head of Lake Erie, provided much the same impetus to larger dreams of transport improvement. The canal booms in Britain and elsewhere suggested to many observers several lessons: large amounts of capital could be profitably marshaled in large scale projects; the improvement of cargo (particularly for heavy commodities like coal) movement could pay handsomely; reliability of transport (rather than, say, speed) was the key commercial value; and the reduction of energy costs for transport, along with dedicated, protected ways was the technological key. It was these lessons that drew increased attention to railroads, essentially as kinds of "land canals."

Since their use in Germany in the late Middle Ages, sets of tracks had been laid down by miners to facilitate the movement of heavy cargo in wagons. These were typically strips of wood laid over crudely prepared road beds to reduce the friction of cart wheels and thus allow animals to pull heavier loads at a steadier pace than would otherwise be possible. The first examples in England were constructed in Nottingham coal fields at the beginning of the seventeenth century and their use spread in other English mining areas. By the end of the eighteenth century, the railroad was a well understood means for assisting canal traffic. In the book on canals that Fulton published in 1796, his key suggestions for improvement included the use of railroads for moving narrow canal boats up inclines or across valleys. In the first years of the nineteenth century, railroad projects began to expand. The Surrey Rail-road was a double-track line built from the Thames at Wandsworth, southwest of London, south and southeast for about 18 miles. Using iron rails with flanges on them (what was called a "tram-way"), its first section opened in 1805. It was designed to ease goods traffic running in the southern outskirts of the rapidly growing metropolis, accepting horse-drawn vehicles paying tolls in much the fashion of a turnpike. The model of the turnpike, accepting all traffic with suitable vehicles, was not a successful one, but the Surrey line did demonstrate the technical feasibility of building a railroad for general traffic. One of the first books on railroad technology, Thomas Tredgold's *Practical Treatise on Rail-Roads and Carriages* of 1825, estimated that a horse on a railroad was able to produce "as much effort as eight horses on a turnpike road." He went on to suggest that the advantage was not as much as on a canal, where a single horse could move very large loads indeed, but that when speed was valued, the horse-drawn railway had much more potential. There were probably about three hundred miles of railroad track operating in England by 1810, still heavily concentrated in mining regions.[12]

For most people in the early nineteenth century, greater reliability and the better use of horsepower were the key measures of improvement in transport, and these had been the core incentives for canal building. But another factor, whose real significance was barely perceived at the time, was becoming more prominent: speed. So central is increased speed to our own notions of transportation that it is perhaps difficult to understand what a modest place it had in thinking about commercial transport before the mid-nineteenth century. While military planners had long recognized speed's possible significance, the only available means for several thousand years for improving the mobility of armies were the greater use of horses and the improvement of roads. Road-building had, of course, long been a general's tool, and the late-eighteenth century saw the extension of this practice throughout Western Europe. The French Corps des Ponts et Chaussées systematized road building in France, with military needs paramount. In Britain, John MacAdam and Thomas Telford devised means for improving road design and construction that were used primarily for commercial routes, but which also found favor among English strategic planners, particularly in the improvement of the routes to often troubled Ireland. More important in Britain, however, was the Royal Mail. The improved commercial roads encouraged the British Post Office to seek the means for making faster and more reliable deliveries, and from the 1780s stagecoaches came to displace mounted post riders and the volume, speed, and reliability of mail service rapidly increased, at least within two days ride of London. In 1750 this might mean as far as Bath or Birmingham, in good weather; by the 1780s coaches regularly reached Manchester and Liverpool in the same period of time; and by the mid-1830s a fast coach could reach faraway Edinburgh in less than forty-eight hours. This continued progress was due not only to better roads but also to substantial improvement of the coaches. The elliptical spring, for example, was introduced in 1804, and this allowed coaches to be built both more cheaply and with much more comfort for passengers, allowing for greater sustained speed. All of these developments meant that by the second and third decades of the nineteenth century, new values were being attached to speedy travel.[13]

The eventual significance of the railroad with steam locomotives was so great that it has been easy to overlook the fact that steam was a relatively late addition to railroad technology, one whose utility and practicality seemed questionable for some time after the invention of the locomotive. The first steam locomotive to run on rails was the invention of Richard Trevithick, a machine builder from the mining districts of Cornwall, in the far southwest of England. By the time of Trevithick's youth, in the 1780s, the steam engine had become a familiar feature of the Cornish landscape, and Trevithick became an adept engineer in this environment. After the expiration of the basic Boulton & Watt patent in 1800, he embarked on a series of experiments to improve steam power, primarily for the basic mining applications of pumping and lifting. It was for the purpose of building engines more suitable for hoists that Trevithick designed his first high pressure steam engine, since such devices worked better if they ran at faster speeds than the stately pace of the condensing engine. About 1798 or 1799 he introduced a relatively small and portable hoisting engine, running at a pressure of perhaps thirty pounds per square inch (roughly two atmospheres) with two horizontal cylinders. Having demonstrated both the workability of his basic engine design and its relative portability, Trevithick wasted no time in making models of self-propelled vehicles. In 1801 and 1802 he demonstrated two different steam propelled carriages, which probably worked better than any previous effort but still had difficulty on even a good London street. The wedding of the steam engine to rails was effected in early 1804, when Trevithick used a newly designed engine for operation on a railway that connected the Penydaren Ironworks in south Wales to the local canal, about ten miles away. The Penydaren locomotive was an awkward-looking machine, encumbered with a giant flywheel due to Trevithick's decision to go with a single-cylinder engine, but it managed the expected loads on its railway with ease, running at about four miles per hour, hauling as much as twenty-five tons. This machine and another that Trevithick built for a coal mine near Newcastle were too heavy for the traditional light rails on which they ran, and so never saw sustained service. In 1808 Trevithick built a third locomotive to demonstrate on a circular track in London, but his *Catch-me-who-can* was dismissed as a stunt, and a not very interesting one at that. This ended the Cornishman's involvement with railways, but it planted a seed that grew among English steam engineers.[14]

Just as with the canals, the coal mines, with their heavy loads and large capital, led the way. Through the 1810s, British collieries continued to construct more and more ambitious traditional railroads. In 1812—the year of Evans's prediction—a coal mine near the city of Leeds opened a rail line that used a steam locomotive that moved with a cog and a toothed rack between the tracks. A couple of years later a northern coal mine used a locomotive, the *Puffing Billy*, that ran on smooth rails, and another nearby mine used a locomotive designed by a local engineer, George Stephenson. The success of Stephenson's engine design caught the eye of backers of a more ambitious coal field project, to link a mine at Darlington, in the northeastern country of Durham, with the port on the River Tees at Stockton, a distance of about twenty-seven miles. The Stockton and Darlington was the most ambitious railroad project to date, but it was a substantial success upon its opening in September 1825. Stephenson designed a new engine, the *Locomotion*, for the railway, and the line had the distinction of being the first to carry passengers using steam. But the Stockton and Darlington was something of a half-step, an illustration of just how non-obvious the path to the steam railroad future was even in the 1820s. The *Locomotion* was only used sometimes—at other times and other points on the line horses provided traction and stationary steam engines pulled cars by cables. The line even accepted properly equipped stage coaches, which paid tolls as on the turnpikes or the Surrey Railroad of a generation earlier. Nonetheless, the Stockton and Darlington attracted a great deal of attention and its success promoted the idea of the large-scale investment in tracked ways. It also brought George Stephenson and his locomotive ideas to the attention of even more ambitious promoters.[15]

By the third decade of the nineteenth century, the towns of Manchester and Liverpool had emerged as leaders in Britain's rapidly expanding industrial economy. The port of Liverpool was the key point of entry for the raw materials, particularly cotton, that made Manchester's mills the wonders of the world, and from those mills there flowed back through Liverpool a swelling stream of manufactured goods. Little wonder, then, that the limited transport options covering the thirty-four miles between the two cities became the target of merchants certain that the new industrial order needed something more than old canals and completely inadequate roads. The effort to build a Liverpool and Manchester railroad began in the early 1820s, but the huge capital needs as well as substantial legal requirements slowed the project down until an authorizing act was passed by Parliament in 1826. Things moved quickly thereafter, with George Stephenson as the project's chief engineer, not only for engines but for the extensive construction as well. The line required crossing a large stretch of boggy earth, sixty-three bridges and viaducts, a couple of great inclines, and a tunnel taking it down to the Liverpool docks. Modest as the project may have been by later standards, it gave fair notice of the challenge that expanding railroads posed to civil engineers.

Even with Stephenson overseeing the project, the managers of the Liverpool and Manchester were not completely convinced that steam locomotion was reliable enough for their purposes, and so they staged the world's first railroad trials. The standards required of the entrants are revealing: the locomotive must weigh less than four and a half tons and be able to pull at least three times its own weight for a total of seventy miles averaging at least ten miles per hour. In early October 1829, five engines of strikingly different design were brought to a stretch of the line near the village of Rainhill. Two were eliminated right away as not meeting basic criteria, but three were tested over several days. Timothy Hackworth was the resident engineer of the Stockton & Darlington, and his *Sans Pareil* was a sound machine, but not sufficiently speedy. John Ericsson and John Braithwaite's *Novelty* was speedy, but proved unreliable over the several days of running. The unquestioned champion was the *Rocket*, designed and built by George Stephenson's son Robert, which reached speeds of thirty miles per hour, and was able to average over large stretches of time sixteen miles per hour (figure 15.3). The *Rocket* established the basic form of the steam locomotive, using multiple tubes through its boiler to bring up its steam quickly and twin pistons that were directly linked to the drive wheels. The combina-

Figure 15.3
The success of Robert Stephenson's *Rocket* of 1829 in the Rainhill trials made it the best known of the early locomotives. From Emory R. Johnson, *American Railway Transportation* (New York: D. Appleton & Co., 1903).

tion of speed and dependability also assured that the Liverpool and Manchester, and any later railroad that sought to be "up-to-date," ran exclusively with steam locomotives. When the Liverpool and Manchester opened on September 15, 1830, it was a grand public occasion, with the country's greatest hero, the Duke of Wellington, in attendance, and a crowd of more than fifty thousand spectators to watch eight trains carry more than a thousand passengers out of Liverpool. The celebration was marred by an accident, in which William Huskisson, the MP for Liverpool and a champion of the railroad, fell under the wheels of the *Rocket* and was fatally injured. Nonetheless, the crowds that thronged the route, estimated at a million people, lost none of their enthusiasm for the new technology.[16]

The premium placed on speed was a novelty, another pioneering feature of the new line, largely due to the fact that it was the first time in which it was anticipated from the beginning that passenger traffic would be carried regularly. The Liverpool and Manchester was an instant success. Within a few months of opening, it was carrying a thousand passengers a day, and to the surprise of almost all observers passengers provided most of the revenue for the line. No alternative means of passenger transport could compete with the railroad—a lesson repeated over and over again through the century. The combination of reliability, relative comfort, economy, and speed made the new technology one of the most compelling to appear in history, despite its voracious demand for capital. The Liverpool and Manchester also turned out to be a fine technical model. Not only did the Stephensons' locomotives set the pattern for the future, but other technical decisions made on the road turned out to be sound and enduring. The four-foot-eight-and-a-half inch (1.435 m) gauge of the line's track (chosen because it had been used on the colliery lines where George Stephenson began) became a widely used standard, although certainly not immediately. The wrought iron rails and the rolling stock with smooth iron wheels also became industry norms, although again not without competition over the next few decades. The basic shape of railroad technology was, as we have seen, decades in the making, and its final form depended on many contingencies, but by the 1830s this form was essentially set for all the world to see. And the world took notice.

Even before the completion of the Liverpool and Manchester, other large scale railroad projects were underway, both in Britain and overseas. Perhaps the most audacious of these, and the most important for the future, was in the United States, where the Baltimore and Ohio Railroad commenced construction with great ceremony on July 4, 1828. Oliver Evans and others had been promoting the idea of railways, with steam and without, for decades, but the success, both technical and commercial, of the Erie Canal served as a wake-up call to commercial interests up and down the Eastern Seaboard that investment in transport works would be necessary to keep a city, state, or even entire region competitive, particularly in the contest for the wealth of the West. While canal projects in imitation of the Erie were the most popular responses discussed, the option of railroads was widely known, even if clouded in much more uncertainty. There had been short railroads on the traditional mining model in the United States since the early years of the century, and by the 1820s some fairly ambitious examples could be pointed to, such as Pennsylvania's Mauch Chunk Railroad, which served one of the country's earliest coal mines with a nine-mile line that ran down a mountainside to the Lehigh River. By the late 1820s a number of railroad projects were projected or underway, from Boston down to Charleston, South Carolina. These, like their English counterparts, were typically undertaken without steam necessarily in mind, but with the idea that tracks would, at the least, provide an economical alternative to canals, even if the only traction was horses or stationary steam engines pulling cables.

The commercial interests of Baltimore felt particularly challenged by the status quo, since they had no river westward and faced the impending competition of a canal being built along the Potomac to serve Washington, D.C. A railroad project more ambitious than any undertaken emerged as the only viable option for promoting Baltimore's bid to become an entrepôt for the West, The engineering challenges of a full-scale cross-country railroad—the B&O project called for 375 miles to connect Baltimore with Wheeling, Virginia (now West Virginia) on the Ohio River—were, typically, vastly underestimated, and even when better comprehended, understated so as not to scare off investors. The first operating section of the B&O opened in May 1830, several months before the Liverpool and Manchester was to begin operation, but the Maryland railroad at this point was only thirteen miles long and relied completely on horses. It was two more years before a steam locomotive ran on the railroad, by which time it ran seventy miles west of Baltimore. At this point, the promise of the American railroad was becoming evident, for both passenger and freight traffic.[17]

By the time the Baltimore & Ohio Railroad reached Wheeling in 1852, the United States had over ten thousand miles of track, and other lines had already pushed well to the west. The construction of American railroads was arguably the greatest engineering project of the nineteenth century. As early as 1840 the United States had twice the mileage of all European railroads. By 1860, the total was over thirty thousand miles, and the transcontinental lines were yet to come. Just as important as the sheer scale of the American effort was its distinctive style. In the first thirty years of construction, it was estimated that the cost of building an American rail line was about one-quarter that of the European lines and one-seventh that of those built in Britain. Part of the difference was naturally in the vastly lower costs of land in the sparsely populated United States, but there were other differences as well. Because distances were so great and capital was still relatively scarce, American railroads were built to different standards—steeper grades were allowed on hills, sharper curves around bends, and lighter-weight bridges and trestles carried lines over streams and valleys. Where stone and iron were used in railroad structures elsewhere, the Americans worked wonders in much cheaper wood. These would have to be replaced much more quickly, but this was perfectly acceptable in a land in which getting the line open quickly at a low cost was the primary value. Cutting corners in engineering the way was only the beginning, however. Due to the differences in topography and in the standards for track, the English-style locomotive that became usual through much of the world was quickly displaced in America. Perhaps the most important early change was the swivel or bogie truck. This English invention saw its first application in the United States, when John B. Jervis designed locomotives for the Mohawk and Hudson Railroad (built in the early 1830s to compete directly with the Erie Canal). Instead of the large rigid wheels of Stephenson-designed engines, Jervis put a set of four small wheels on a swivel at the front to give the locomotive the ability to take sharper curves (figure 15.4). His *Experiment* of 1832 could run as fast as sixty miles per hour, quickly demonstrating the American capacity for important innovation in railroad traction. Jervis's designs were taken up by Matthias Baldwin, who became America's greatest locomotive builder, further cementing

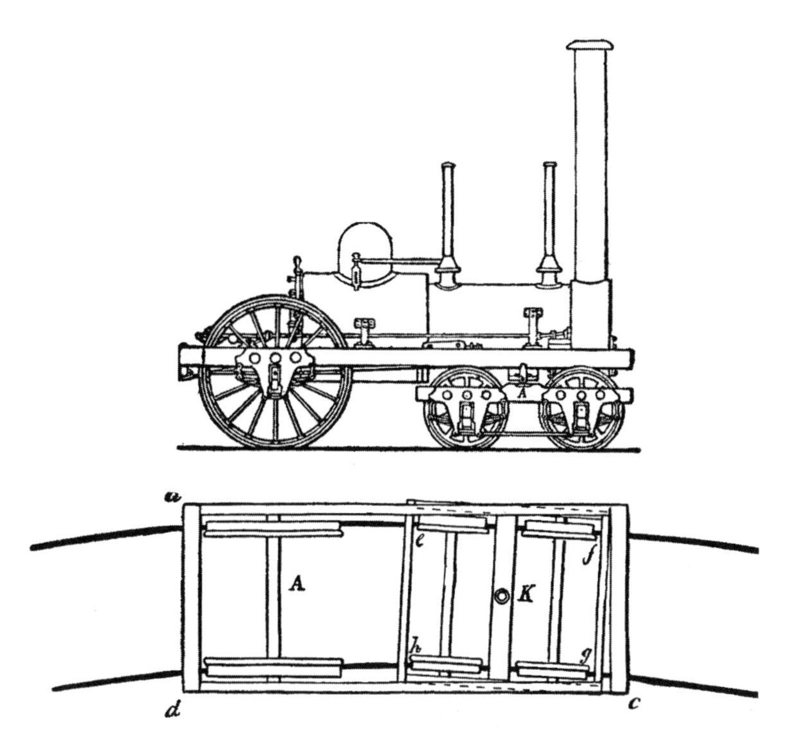

Figure 15.4
John Jervis's locomotive of 1831, with plan of the swivel front truck. From T. C. Clarke et al., *The American Railway* (New York: Chas. Scribner's Sons, 1892).

American independence from the British railroad hegemony that held forth elsewhere in the world.[18]

The rest of the world did follow, but with a bit of a lag. The first Continental railroad following the Stephenson model was a short line built in Belgium to serve coal fields, but by 1832 the government there was making plans for a countrywide system. This, however, was not linked to neighboring German and French lines until the 1850s. In France, coal again provided the initial impetus. In December, 1832 a short line connected the city of Lyon with the coal fields of saint-Etienne, about forty miles away. While this was in some ways as much a marvel of engineering as the Liverpool and Manchester had been, it also displayed the kind of conservatism that slowed Continental railroad technology—literally. Strict speed limits were imposed to avoid accidents, and the last horse-drawn sections of the line were not converted to steam until 1844. Still, French engineer Marc Séguin pioneered use of the multiple-tube boiler, which permitted rapid steaming up and became standard on the world's locomotives. The first line in the German states opened in 1835, in Bavaria, but growth was rapid thereafter, with nearly five hundred miles operating by 1840. Other European countries followed through the 1840s, and elsewhere railroad efforts began as capital, political circumstances, and technical support allowed. Throughout the world, British engineers and manufacturers usually provided the foundations for railroad technology, and even in the United States, British capital was crucial to keeping the railroad building boom going through the middle decades of the century. In its scale of capital and technical demands, in its reach into every sector of the economy, in its demands on a range of resources, from fuel and machinery to land, labor, and skill, the railroad was the most profoundly transforming technology the world had ever seen.[19]

The great influence of the railroad on Victorian culture and technology can blind us to the fact the form and character of railroad technology was by no means self-evident or preordained to those in the midst of its creation. Many choices had to be made, some seemingly trivial, others obviously with great significance, in the course of creating and extending the railroad's domain. While the style of railroad created by the Stephensons on the Liverpool and Manchester remained the most important model for the rest of the world, its nearly universal adoption did not come without contention. The simple matter of the track gauge, for example, remained the subject of both engineering debate and political or strategic calculation for decades. In Britain itself, the virtues of a much wider gauge persuaded the backers of the Great Western Railway—the main line from London to Bristol and points westward—to adopt the scheme of the flamboyant engineer Isambard Kingdom Brunel for a line with tracks seven feet apart, rather than the standard four-feet-eight-and-a-half inches. Elsewhere in the world, other gauges, both wider (for greater capacity and stability on the line) and narrower (for economy and versatility) were adopted, and sometimes the unconventional trackage was maintained for decades in the name of national security (a change of gauge at the frontier meant foreign trains did not run without making adjustments) or simply national pride.

Even the form of propulsion for railroads remained a matter of experiment and contention. In the 1840s, for example, so-called atmospheric railways received considerable attention. These operated by running air-pressure or vacuum tubes (generally the latter) between the tracks. Pistons attached to locomotives or cars were moved by air pressure, which was maintained by stationary steam engines located at intervals along the line. Short lines in Ireland, Scotland, and southwest England demonstrated the feasibility of railroads on this principle, and others were proposed elsewhere. This system was attractive for its quiet operation and smoothness, but the high capital costs and the great difficulty of protecting the tubes from wear and weather made them uncompetitive with steam locomotive technology. At the end of the nineteenth century, electric traction would finally give steam significant competition, especially in heavily traveled urban areas. The uncertainties and constant changes in railroad technology, particularly in its first decades, are important historical reminders that even the most compelling technologies are introduced and developed in a fog of uncertainty and contention, and their final forms are never simply matters of technical or economic rationale.[20]

It is simply not possible to summarize briefly the effect that the railroads had on every aspect of life in the nineteenth century. The entire way in which people defined their relationships with one another, particularly as those relationships were affected (and few knew beforehand how much this was) by where they were and how easily they could move about, was altered. Likewise, the very definitions of commodities and markets were transformed by easy, reliable, and swift transportation. More than anything else, perhaps, what made the railroad so compelling—and such a surprise—was speed. The speed with which people and goods moved had before this seemed firmly bounded by nature and economy. A fast horse represented the basic limit, and that for only short distances and very small weights, at considerable cost. The bold prediction of Oliver Evans that steam-driven carriages on railroads would permit regular travel at fifteen miles an hour represented the outer limit of plausibility to those living before the *Rocket*'s thirty miles per hour or the *Experiment*'s breathtaking sixty. The world of 1860, in which a man or woman getting on a train in New York City was comfortably assured of reaching Chicago in two days, was necessarily very different from that of, say, 1840. And of course, the passage of people was paralleled by that of goods, in quantities and variety that appeared to have no limit to the astute mid-century observer. The passage of people and goods meant, in turn, the movement of armies and materiel, as the Americans learned from 1861 to 1865, and as the Europeans were to demonstrate repeatedly through the next decades, both at home and in their empires. And the swift passage of people and things also meant, in ways no one could have foreseen, a movement of ideas, styles, and expectations such as the world had never seen, for speed, comfort, and economy meant that movement was constant, omni-directional, and without traditional constraints of class, occupation, age, sex, or physical condition. In the United States, for example, people who went westward could now just as easily go back east, people who left their parents behind could visit them again, people seeking out ideas, excitements, and escapes, could do so knowing they would be able to bear back the fruits of their adventures. It is said that the railroads shrank distances, but it would be better to say that they expanded horizons.

For the further development of technology, the coming of the railroad and the decades of intense, sometimes frantic, railroad construction that occupied the Europeans and the Americans in the middle decades of the nineteenth century were the most important elements in developing a "infrastructure" for future growth and change. The scale of civil engineering, of machine design and building, of iron and, later, steel production (for rails, particularly), of fuel consumption and distribution, and of system planning and organization dwarfed anything that had been seen before. The railroad and its capacity to transform the rules of movement and availability commanded vast amounts of capital, both material and human. This capital, in turn, became available in a great range of ways, for innovations in everything from bridges to bureaucracies. The sheer magnitude of capital investments compelled innovations in management, since expensive railroad equipment could not be allowed to be idle or poorly employed, and so many of the modern structures and techniques for organizing and managing large enterprises emerged first from the great railroad companies. Likewise, the enormous growth in demand for iron and steel for producing rails, locomotives, and other elements of the system transformed the making and working of these materials into modern giant scale enterprises, thus making available iron and steel for a host of new applications, such as buildings and ships. The railroad rapidly became the technology that made the difference between being "modern" and not, both because of what it did and made possible and because of what it required. More than the great textile mills and other accomplishments, it was Britain's railroad technology (and the iron industry that backed it up) that made that country the technological center that it remained until the last couple of decades of the nineteenth century. At the same time, however, the Americans used the railroad to declare and establish their technological independence and, eventually, leadership.

16 Messages

Joseph Farington was an English landscape painter at the turn of the nineteenth century, a man of some accomplishment in the generation before the flowering of the genre in England in the hands of such figures as John Constable and J. M. W. Turner. Besides capturing the world around him in paint, Farington was something of a diarist, capturing other things he encountered in words. In September, 1812, during a period of great anxiety for the inhabitants of southern England, concerned as they were about the progress of the war with Napoleon's empire, Farington set down in his diary his impressions of a visit to a peculiar, isolated house on the Norfolk coast, some hundred miles or so northeast of London. Here he found a telegraph station, run by a retired naval officer and his wife. The most conspicuous feature of the station was the tower above it, decked with an array of huge shutters—the very latest in communications technology. With the shutter signals, the station master claimed, he could start a message on its way to London, and the officers in the Admiralty would have it in no more than six or seven minutes. It was, in fact, only to the Admiralty that the station was designed to communicate, for the entire "telegraph" system of England, such as it was, was designed solely for alerting officials of the possibilities of invasion and of news from ships. When the Napoleonic threat disappeared a few years later, the system was shut down, and the telegraph became simply a war relic.[1]

The telegraph—the electromagnetic telegraph, rather than the visual system Farington observed—was to be one of the wonders of the nineteenth century, one of those technologies by which Europeans and Americans were to measure their distance both from a more limited and slower-paced past and from the "less advanced" peoples of the rest of the world. The marvel and utility of rapid communications—of "telecommunications," as we know it—insinuated itself deeply into the Victorian view of what was right and necessary about the world. The telegraph joined the railroad at the apex of transforming achievements, made perhaps just a bit more special by its close linkage with science at its most advanced and esoteric. Our own sense of the origins of the telecommunications marvels that define twenty-first-century notions of high technology is inextricably connected to telegraphy, so much so that one author could publish a history of the telegraph's first decades under the rubric of *The Victorian Internet*. It is therefore easy to lose sight of the fact that, even more than the railroad, the telegraph was a largely unsought-after invention, the uses for which had to be ferreted out and promoted before its place in the social and technical order was secure. After all, the railroad was a marked speeding up and regularizing of movement, of both people and goods, but in forms and patterns not so different from familiar ones. Telegraph messages, however, were a true departure from any previous form of messages; their instantaneous nature was matched by their brevity and by their unconventional handling, such that even after the mid-twentieth century, when the Victorian technology had disappeared, the *idea* of the telegram—of a "wire"—persisted as a distinct kind of message, with distinct purposes and implications.

The history of the pre-electrical telegraph, such as the system that was a marvel to Farington, is useful for showing just how much the telegraph system that emerged in the nineteenth century was the product of the particular circumstances in which it arose. The earlier systems were costly, limited in scope and utility, served very few users, and, for the most part, short-lived. While signaling at a distance was ancient and widespread, from the flares used by sailors in the ancient Mediterranean to the smoke signals of Indians in the North American plains or drums sounding along rivers from the Mohawk to the Congo, the creation of systems meant for regular use over extended distances was the product of the late eighteenth century. The credit for the first workable system that saw extended use belongs to Claude Chappe, a churchman and avid amateur physicist working in western France in the 1780s. Proposals for systems, using a variety of signaling techniques and coding methods, had been made and published for more than a century by the time of the Chappe experiments, but the spirit of innovation and system that emerged from the French Enlightenment appears to have given both extra impetus to Chappe and more opportunities for a receptive audience. The fact that Chappe was able to muster serious support for the development, construction and maintenance of a system of signal towers, which is what the Chappe telegraph was, is all the more remarkable in light of the fact that this was done through the upheaval of the French Revolution. In 1790–91, Chappe, assisted by his brothers, devised and demonstrated a system based on sound and synchronized clocks, but this was soon jettisoned in favor of an optical system. Several alternative designs were experimented with before Chappe settled on the semaphore telegraph that became a model for others, including France's enemies across the English Channel.[2]

The Chappe system continued to evolve, but its essential features were quite simple (figure 16.1). A "telegraph" was a tower topped with a post, on which was mounted a long horizontal beam. At each end of this was attached a shorter beam, pivoted so

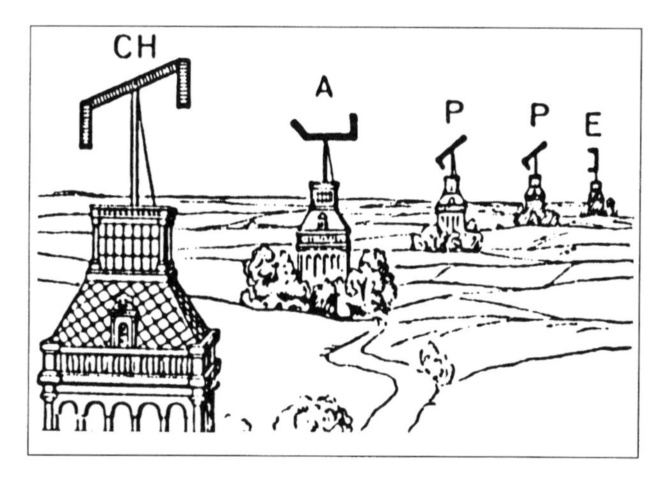

Figure 16.1
Representation of the Chappe semaphore system, showing the inventor's name spelled out in the semaphore code. From Duane Koenig, "Telegraphs and Telegrams in Revolutionary France," *The Scientific Monthly 59*, 6 (1944), 435.

that one end was markedly longer than the other (the short end was often pointed rather than blunt, or was dropped altogether). Signals were set by a code which consisted of different patterns of the main beam (the "regulator"), which could be tilted from its horizontal home position either at 45° either direction or set vertical, combined with tilting of either or both of the side arms (the "indicators"), again at 45° intervals. In its basic form, 256 different positions were possible for the Chappe code. Numerals and letters of the alphabet were assigned basic positions of the arms, and common words and phrases or instructions to the operators were assigned to others. Communications consisted of a message being sent by a sequence of arm positions at one tower and read, usually with the help of a telescope, by an observer at the next tower down the line, to be resent for further observation in the direction of sending. The Chappe semaphore system received its first public test in mid-July 1793—just as the French Revolution was entering a new phase, generally known as the Reign of Terror. Remarkably, the political upheaval seemed to have little effect on the government's interest in promoting the new communications systems, so the Convention—the chief power of the French state—authorized in August the construction of a line of telegraph stations to connect Paris with the city of Lille, near the northeastern frontier, about 120 miles away. All through one of the most violent years in European political history, the Paris-Lille line was built; the first messages were transmitted in the spring of 1794, and on July 16 (almost precisely five years after the storming of the Bastille, and less than two weeks before the Terror reached a bloody climax in the guillotining of Robespierre and other leaders) the first telegraph line was declared officially open. Within weeks, the system had proved its value by rapidly reporting the word of French victories on the northeastern frontier.[3]

The Chappe telegraphs were machines; their use depended on systems of ropes and pulleys for ready manipulation of the semaphore vanes. Their construction also invited continuous efforts at improving not only the mechanics, but the coding systems, and the overall designs were the subject of experiments and alterations. The mathematician Gaspard Monge was consulted, and he recommended increasing the number of arms from two to seven to enlarge the number of possible codes and thus speed the transmission of messages. Chappe rejected this recommendation, as he did many others, preferring instead to improve his two-arm code and to situate his towers more carefully. The system depended on clear lines of sight between towers, and these were disrupted easily by weather and nightfall. But, for all the system's limitations, the French government relied on it more and more, and new lines were built in the late 1790s, connecting Paris to Brest, Strasbourg, Lyon, and Toulon. Some of these lines remained in active use for a half-century, such was the value that the state placed on them. The optical system was, of course, visible to all within sight, but the security of messages was maintained by encrypting them. The relay operators at stations along the line frequently did not know what messages actually meant—they simply passed on the coded signals down the line. Government and military uses were always the primary motives and application of the Chappe system and its imitators. One other use soon emerged, particularly for shorter lines that sprang up near ports, both around Europe and elsewhere—for relaying news of ship arrivals.

These commercial applications motivated entrepreneurs in the United States, and short-lived semaphore systems appeared on Long Island, New York, Sandy Hook, New Jersey, and on the southern coast of New England. Later in the nineteenth century, proposals were floated for a system stretching the length of the eastern seaboard, going as far west as New Orleans. As we have seen, military concerns were the primary motivation in Britain, where a number of lines (using a shutter system, an alternative to Chappe's semaphores), like the one connecting the observation post high on the Norfolk cliffs at Trimingham with London, were operated through much of the Napoleonic emergency. Once the emergency was over, however, the English system was allowed to disappear. Elsewhere in Europe a similar pattern was followed, although in some places, such as Russia, one or two major optical telegraph lines were maintained for decades. In Sweden, a system designed by Abraham Edelkrantz was built at about the same time that Chappe was developing his. Again, military concerns and a sense of vulnerability from the sea motivated government support, and a Royal Telegraph Institution was formed to operate lines radiating from Stockholm. The Swedish system, in one form or another, operated as late as the 1880s, as the optical system proved useful in areas like the Stockholm archipelago, where reliable electrical lines and cables were difficult to maintain. Only in

France, however, did the optical telegraph develop into a full-fledged system of communications, one that reached across borders eventually—as far east as Venice and as far north as Amsterdam. At its fullest extent, in 1852, the French system had 556 stations, stretched over about three thousand miles.[4]

The optical telegraph was, arguably, a minor episode in the history of communications, but its appearance at the turn of the nineteenth century suggests the extent to which the Europeans were now seeking technological solutions for a wide range of concerns. There was nothing in the system of Chappe or others that could not have been put into place nearly two centuries earlier—the telescopes with which the semaphore observers were equipped were the only "modern" devices in the system, and they were no more advanced than the ones Galileo had almost two hundred years earlier. Likewise, while the circumstances of the French revolutionary and Napoleonic wars might seem to have been particularly compelling for rapid military communications, they were arguably no more so than that faced by earlier combatants, particularly after the emergence of national armies in the seventeenth century. The optical telegraph, in other words, was not the product of special capabilities or needs, but the result of a technical mindset, particularly one that was oriented toward the development of systems—even what we might today call networks. Earlier the building of roads and canals, the development of systems of military communications and supply, and even the construction of such urban works as water supply and sewer systems showed the efficacy of coordinated systems for increasing the utility of everything from armies to cities. But the optical telegraph systems were arguably the first "real-time" networks, designed for careful coordination of many elements, both technical and human, over considerable distances, with an immediacy that no other technology had before required or accomplished. The novelty of this concept is confirmed by their wide abandonment. When, in the year after Waterloo, the English electrical experimenter Francis Ronalds demonstrated a telegraph system using wires and static discharges, he was told by the Admiralty, "telegraphs of any kind are now wholly unnecessary."[5]

Ronalds's experiments were within a tradition that was already several generations old. Electrical experimenters in the mid-eighteenth century demonstrated that electrical discharges passed down a wire so rapidly as to appear instantaneous. As early as the 1750s proposals were made for using this phenomenon in a signaling device, and the later decades of the century saw experiments and proposals from a number of quarters, some of them explicitly appropriating the idea of a telegraph, as popularized by the Chappe system. The announcement in 1800 of Volta's discovery of a source of continuous electrical current spurred more intense experimentation, although the lack of suitable detection devices still posed a problem. One solution was the use of the capacity of the Voltaic current to electrolyze solutions. A German physician, Samuel Thomas von Sömmering, designed an instrument that used separate wires for twenty-five letters and ten numerals, and led each wire to a solution from which hydrogen bubbles would emerge when the current was applied. The difficulties of a thirty-five-wire design were obvious, however, and alternatives using as few as two wires and a code were quickly proposed, but none of the chemical systems were adopted.[6]

In the first decades of the nineteenth century, electricity was still a matter of scientific, rather than practical, interest (although the promoters of electrical "cures" might have disagreed). Even the marvels demonstrated by Humphry Davy's giant Royal Institution battery were solely of "philosophical" interest. The Voltaic cell did, however, make electrical experimentation and demonstration more popular than ever, and it was in the course of the wider work that ensued that new electrical phenomena, of both theoretical and practical interest, emerged. In the most famous example of serendipity during this period, the Danish physicist Hans Christian Oersted discovered in 1820, in the course of a lecture on the Voltaic current, the intimate connection between electricity and magnetism. An electric current in a wire deflected a compass needle on Oersted's demonstration table, and observation soon confirmed that an electric current acted like a magnet. While a chance observation, it was one that Oersted was philosophically prepared for, eager as he was to link the varied phenomena in nature to one another. Electromagnetism was quickly recognized by experimenters throughout Europe as a discovery of the first importance, and no time was wasted both in elaborating its theoretical elements and in proposing applications, of which detecting telegraphed signals was one of the first. Through the decade of the 1820s, however, the difficulties of electric telegraphy were more prominent than practical advances. Indeed, there were some who doubted that Voltaic electricity could in fact be transmitted a sufficient distance to make an electrical system worthwhile; others pointed out the ease with which electrical connections could be broken, and thus the insecure nature of an electrical line; still others pointed out the same difficulty that the Sömmering system had faced—the difficulty of having so many wires for transmitting the entire alphabet. Slowly, however, useful experience with the behavior of electric currents and with the electromagnetic effect accumulated to produce a spate of serious telegraph proposals and devices in the 1830s.[7]

Just as with the railroad, the question, "Who invented the telegraph?" does not make much sense. Experimentation and experience drew ever closer to workable devices and systems, drawing on the work of amateur scientists and well-known professionals alike, from Russia to the United States (both, it will be remembered, very much on the periphery of Western science at the time). In Albany, New York, the science instructor at the Albany Academy, Joseph Henry, demonstrated to his students in 1831 and 1832 how he could send an electric current through as much as a mile of wire to transmit signals, and a few years later, now at Princeton, he showed how a relay could be put into the line to cover even longer distances. At the University of Göttingen, in Germany, physicist Carl F. Gauss and his assistant, Wilhelm Weber, built a line more than a mile long that was used to transmit messages as early as the spring of 1833. Their system may have been the first to adopt a binary code for sending the alphabet. Some credit has been given to Pavel L. Schilling, a Russian diplomat in Germany, for constructing the first workable binary code device, but the dating of Schilling's work is uncertain.[8]

The German experiments caught the attention of other observers, one of whom was an Englishman, William F. Cooke. As an anatomy student at the university in Heidelberg, Cooke saw in 1836 a demonstration of a Schilling device and immediately became enthusiastic over the possibilities of long-distance communications. Upon returning to England, the well-connected Cooke sought technical assistance in building an improved device, finally finding it in Charles Wheatstone, who was teaching physics at London's King's College. Wheatstone had been making his own efforts to build a workable telegraph, but had run into problems sending signals over a long line. In 1837, Joseph Henry, in the course of a tour of European laboratories, visited Wheatstone, and he reported later that he spoke then of how he had solved the problem of transmitting over a distance with a properly wound electromagnet and a battery set up for higher voltage. Within weeks of Henry's visit, Cooke and Wheatstone had a workable device, using six wires—five for separate needles that could be used in pairs for a twenty-letter code and the sixth wire as a common return. They patented this device in 1837 and tested it on a short segment of London railway. Basic technical problems plagued the system, particularly since insulation was a real challenge, which encouraged further simplification. By mid-1838, however, Cooke and Wheatstone had a two-needle design in place on a thirteen-mile stretch of the Great Western Railway, and this proved more successful. Nonetheless, even the British railroads, not to mention the government and the public, were skeptical about the larger implications of the invention, and so it was not until 1842 that the Great Western's line was extended on to Slough, the first important junction west of London, only five miles farther down the line. In the next few years, however, the British railroads began adopting the telegraph more readily, so that Dover was connected to London by 1846, and Edinburgh in 1848.[9]

The popularity of experiments with electric telegraphs in the 1830s and 1840s was the product of wide publicity about electrical phenomena, especially electromagnetism, combined with the technological enthusiasm unleashed in large part by the spread of railroads. While electricity was seen as the province of physicists, and knowledge of its behavior was gradually becoming more sophisticated and mathematical, this period was the last generation in which it was still seen as approachable by casual amateurs as well as better trained individuals. A remarkable number of the early telegraph experiments was carried out by these amateurs—William Cooke had started out this way, as had a number of the continental pioneers. The most famous of the amateurs, however, was an American, Samuel F. B. Morse, a moderately successful portrait painter who was able to combine his enthusiasm with social connections and skills and a remarkable persistence of vision into the most successful of the early telegraph technologies. While some claimed later that Morse was, thanks to his Yale education, well trained in electricity, most evidence suggests that his technical knowledge was in fact quite modest. His ideas about using electrical circuits for instantaneous communications, first broached on a sea voyage in 1832, were not particularly original. He brought to the challenge, however, considerable personal enthusiasm, a capacity for getting and applying good advice, and some important notions about the *form* of messages. In particular, from his first attacks on the subject, Morse was convinced that it was not adequate simply to be able to send a signal over a distance, but it was necessary to devise appropriate means for composing and recording messages.

Morse's background as a painter probably influenced his approach to the telegraph. In particular, Morse appears to have been the only early inventor who focused so intensively on the mechanism for preparing and receiving messages, for he was convinced that a complete, workable system required machinery that would put the message to be sent into a real, physical form and that would take the electrical signals and turn them into something that could be "read." His core idea of 1832 would appear to have been to create a device that would allow one to "set" a message, just as a typesetter prepares a line of type for a press, and then provide a means to convert this into electrical signals, to be received by another device equipped to make a permanent record. It is perhaps ironic that the Morse system, in its most popular form, eventually rejected both of these elements, and relied instead on skilled operators sending messages by hand and receiving them by ear. But as Morse tackled possible approaches to the problem as he understood it in the early 1830s, he pursued directions that set his efforts apart from those of others. At first he lacked the skill or understanding of electrical phenomena to translate his ideas into anything original. But he was able to recruit the assistance of, first, a chemist, Leonard Gale, and then a mechanically talented young assistant, Alfred Vail (who also came with his family's well-supplied shop), so that by September 1837 he was able to claim that he could make a device for general messaging. even though all he had accomplished was confined to a single workshop. The first Morse systems sent numbers by a series of pulses, recorded on paper tape. A dictionary marked with coded numbers gave corresponding words for the numbers received. In fact, at this point, Morse and his associates had done little beyond what the European experimenters had already accomplished, and their experimental devices bore little resemblance to the Morse system that later become dominant, but Morse forged ahead in applying for patents.[10]

That an electrical wire could send a signal over a distance had been understood by almost a century before the 1830s. Volta, Henry, and others, designing batteries, relays, and the like, had discovered how to produce enough electricity to make a signal carry reliably. The basic laws of circuits were understood, thanks to the work of men like Ampère, Michael Faraday, Georg Frederick Ohm, and others. And suitable detecting instruments in the form of galvanometers, electromagnets, and the like were well known to all experimenters. All of the mechanical and electrical elements of a telegraph system were, in other words, well known by the late 1830s. What was missing was a practical form in which to put information to be sent. The Cooke and Wheatstone device, with its five needles, and then two needles, was workable, but still a bit slow and cumbersome. The first Morse instruments were also slow and awkward. Messages had to be "composed" beforehand and sent by a mechanism that could take advantage of the speed of the electrical transmission. In its earliest form, this required that the numerical code be put together before being sent, in the physical form analogous to the way in which a typesetter set a line of type, in a holder called a "portrule." This portrule was to be set with metal or wooden pieces to control the movement of a lever that would send signals to control the recording stylus, In the effort to make the recorder a bit less awkward, it was reconfigured to make short and long horizontal marks, rather than the original zigzag marks. These, sometime in 1837 or 1838, probably more due to work by Vail rather than by Morse, became dots and dashes, and the numerical code with its dictionary key was abandoned for the simple binary code which evolved into "Morse Code" through successive efforts at improvement. Just a bit later, Vail designed an elegant and simple "key" for sending dots and dashes, eliminating the awkward portrule transmitter. It was still thought that paper recording was required, although later telegraph operators demonstrated that a trained ear was the most efficient instrument for taking a Morse code message (figure 16.2).[11]

While Morse and his cohorts took out patents to protect their invention, and Morse traveled to Europe in search of support for his system, he and his colleagues operated for some years in the belief that the U.S. government was the most appropriate source of help for demonstrating their system. After years of cajoling, they succeeded in 1843 in getting a $30,000 grant for a demonstration line between Washington and Baltimore, a distance of about forty miles. Stringing the wire along the line of the Baltimore and Ohio Railroad, after promising the railroad company that they would not embarrass them, the demonstration line was completed in the spring of 1844, and on May 24, the Biblical passage, "What hath God wrought" was sent by Morse down the line from the U.S. Capitol to Vail at the receiving station in Baltimore, at a rate equivalent to thirty letters per minute. Even before this official inauguration, public attention was aroused when the telegraph (with the assistance of a train for the yet incomplete stretch) reported the results of the Whig party's presidential nominating convention, meeting in Baltimore, to the assembled politicians in Washington. Despite the technical success of the Morse system, however, it remained

Figure 16.2
Register (receiver) for Morse's 1844 telegraph system. Smithsonian Institution photograph.

unclear to most observers just what use the new form of communications would be put to. There was, after all, no social structure, no economic system, no military or political necessities that lay waiting for the telegraph. Its uses would be as much the product of invention as the device itself.

At this point the telegraph began to trace out an experience that was to be repeated over and over again during the next century and a half, as new communications technologies confronted the challenges of high initial capital costs, the limited uses of small networks, and the uncertainty of the uses of information in new forms. The first challenge lay in the limitations of the network itself. A single line between Washington and Baltimore, for example, would have little effect on people's notions of messaging, so extension was required. But this in turn demanded money, and the U.S. government quickly let it be known that it had no intention of going beyond the demonstration that it had funded. Private capital, on the other hand, was understandably skeptical. What good was the ability to send a short and expensive message over a distance in a matter of minutes? The entrepreneurs, for all their enthusiasm, did not, in fact, have a very good answer to this question. In England, the results of popular horse races were among the first things to be sent regularly (a use that, incidentally, displayed quickly the potential for fraud). Much more important for the technology's reputation, however, was the fortuitous assistance given in capturing criminals. In the most celebrated case, in January 1845, John Tawell murdered his mistress in Slough (the end of the Cooke and Wheatstone line along the Great Western Railway) and escaped by train to London, wearing a drab brown greatcoat. The message was telegraphed to Paddington station in London, alerting authorities to look for a man "DRESSED LIKE A KWAKER" (there being no "Q" in the English code). The telegraph wires became known as "the cords that hung John Tawell," and the publicity was enormous. The real fate of the telegraph at this point was in the hands of entrepreneurs and promoters. In the United States, the Morse interests turned to Amos Kendall, who used political and journalistic connections to help raise money for the "Magnetic Telegraph Company," which by mid-1845 was preparing lines from New York to Philadelphia, Boston, and Buffalo. At about the same time, construction finally began in England of a network of wires radiating out from London, supported at first by an Admiralty contract for communications with the naval base at Portsmouth, on the south coast. In both countries, telegraph lines were constructed at the rate of hundreds of miles per month, and by 1850 the new technology was clearly on its way.[12]

Even more than the railroad, the telegraph demonstrated how new technologies, especially when they were recognized as socially important and the potential sources of great profit, could be the object of enormous experimentation and variation. Now, however, unlike the past, much of this experimentation was within the context of a structure of legal and economic institutions that was to redefine the entire process of invention in the nineteenth century. Indeed, the development of telegraph technology helped to shape these institutions—particularly the patent system and the monopoly corporation—into their modern forms. While patents had been the object of much attention from both the legal and the technical communities since the late eighteenth century, the telegraph, more than any other technology, demonstrated what a tricky instrument the patent system was both for fostering innovation and for protecting intellectual property. Telegraphy was by no means the most popular area for patenting in mid-nineteenth century America (agricultural and railroad-related inventions were more common), but it was here that patent disputes, agreements, pools, and manipulation became key features of commerce. The basic technology of batteries, wires, senders, and receivers was sufficiently simple, and yet the economic pressures for improvement so intense (given large capital costs), that a wide range of variations and styles emerged from the workbenches of enthusiastic amateurs as well as the well-equipped laboratories of the first national and international communications companies. While the modern business corporation, operating on a continental scale, was the result of many historical circumstances and certainly not the product of technologies alone, the telegraph and the opportunities it provided for both expansion and consolidation provided a particularly fertile ground for the growth of new corporate powers and structures. In particular, the telegraph—like other communications modes before and after—fostered monopolization in the nineteenth-century business environment, and thus helped to shape generations of attitudes about the relationship between technological innovation and monopoly.

In a country the size of the United States, the task of extending the telegraph network to cover a substantial portion of the population was initially beyond the capabilities of a single group of capitalists, and so the Morse interests began by licensing other groups to build and operate lines in different regions (figure 16.3). The emergence of new legal instruments in the states that encouraged incorporation at first aided the organization of these regional companies. Initially this system seemed well suited to extending the telegraph's reach while keeping everything under the general control of Morse and his backers, but when one of the organizers decided to strike out on his own and form a company using different instruments and patents, the complexity of the dynamic between corporate capitalism and technological innovation became evident. Henry O'Rielly had initially been a Morse licensee, but when he couldn't get terms that suited him, he broke with the Morse group and acquired interests in other telegraph inventions, including a printing telegraph invented by Royal E. House. The House device was complicated and always had problems, but it worked well enough to allow lines to be built on the eastern seaboard to rival the Morse lines.

Another alternative device was the product of immigrant watchmaker from Scotland, Alexander Bain. This was another complicated system, designed to allow "automatic" sending of telegraph messages using clockwork devices and a recording system using chemically treated paper. These new systems allowed some successful competition with the Morse companies, but their patents were vigorously attacked by the Morse interests, with mixed results. The lines going out to the west that used the House system managed to survive the legal challenges, and they were consolidated in 1855 under the rubric of the Western Union Telegraph Company. Gradually this company acquired some of the valuable licenses to the Morse system, and through adroit corporate management was able to acquire lines, using a variety of technologies, covering most of the United States. This was the source of the Western Union monopoly which, while never complete, was able to control American telegraphy in the decades after the Civil War.[13]

The relationship between monopoly capitalism and innovation became a subject of much debate in the twentieth century, and it continues to ignite controversy. The patent system, which rewards invention by grants of temporary monopoly for specific

Figure 16.3
The telegraph in the United States, 1853. Less than a decade after Morse's first line, the telegraph linked every state east of the Mississippi River except Florida. Library of Congress.

inventions, became an instrument for the perpetuation of monopoly control over a technology or an industry well beyond the period of an initial patent. Dominant firms, from the late nineteenth century forward, were sometimes able to combine their initial economic power with continued, incremental, innovation that could be used to block potential competitors. The monopoly control that resulted was not "natural," as some claimed—that is, the fullest use of the technology did not require that control be centralized in one firm. But complex technologies, subject to many variations and improvements, both large and small, were sometimes prey to a combination of patent management and aggressive business practices that kept their control in the hands of very few, perhaps only one, firm for extended periods. Western Union's control over American telegraphy and over much of the innovation in the industry revealed a pattern that was to repeat itself in a range of technology-oriented industries over the next century or so. The size and complexity of a firm like Western Union, and the large amount of capital required by the technology, generated new forms of innovation and management, presaging the corporate research and development environment of the twentieth century.[14]

The telegraph was also a good example of a technology that seemed to shun "perfection"—there were always elements of the system that seemed amenable to improvement. In the 1840s and 1850s, for example, the problem of insulation on the lines plagued all the telegraph pioneers. The best material and form for wires also posed problems. While copper was recognized as the best conductor, it was expensive, and even more important, it was too weak to hang on poles unreinforced. Iron wire was a much less efficient conductor, but it was cheap and sturdy. Galvanized wire was developed to retard corrosion, since insulating coverings continued to be unsatisfactory for most lines. The insulators used on the poles of the Morse demonstration line between Baltimore and Washington were glass drawer knobs, and glass insulators continued to be favored in America, while Europeans devised ceramic ones. Another area that invited constant experimentation was the design of sending and, especially, receiving instruments. The technically simplest system emerged fairly quickly—that was the simple Morse key and sounder. These were reliable devices, capable, in the hands of skilled operators, of rapid sending and receiving with little chance of breakdown. But an enormous variety of devices emerged from the shops of inventors and would-be inventors to reduce the dependence on the skilled telegrapher. Complicated instruments with keyboards—resembling those of pianos or using circular layouts and other variations—were favored by many inventors, and some of these were put into service by systems that were meant to be used by untrained operators. Even more complex, typically, were the receivers that were designed to produce readable messages directly from telegraph signals. These devices were often intended for users like businesses that wanted direct telegraph communications with offices or sources of news and financial information, without having to rely on inter-

Figure 16.4
A stock ticker, one of Thomas Edison's first successful inventions. Smithsonian Institution photograph.

mediaries to code and decode messages. By the 1870s, devices of this kind had become sufficiently reliable to be common in some settings. The best known style was the stock ticker, which itself was the subject of constant tinkering. The improvement of this machine was one of the key early contributions of the greatest of the telegraph inventors, Thomas Edison (figure 16.4).

Edison was a great inventor not because of his contributions to telegraphy, although these were considerable, but because he exemplified more than anyone else the potential of telegraphic technology, and electrical technology in general, for wide-ranging improvement and elaboration. Electrical invention, like a number of areas, such as photographic and chemical technologies, remained through most of the nineteenth century an inviting field for both scientifically trained individuals and others with little or no schooling. Edison, born in 1847, fit into the latter category, early finding the classroom an uncomfortable and unrewarding place. He was instead lured by the exciting new technology of his youth, and so he proceeded to learn the skills of a telegraph operator in small towns in the eastern Michigan where he grew up. The expansion of the telegraph and its importance in transmitting news during the American Civil War opened up great opportunities for an ambitious young operator, and Edison took full advantage of these, becoming a successful itinerant operator in the years just after the war. Not only did this life allow him to explore the country in a footloose manner suitable to his undisciplined streak, but it also exposed him to the full range of telegraph practices, devices, and problems. Edison was able to take full advantage of the commercial and technical environment in which he immersed himself. Finding a talent for solving the electrical and mechanical problems of improving telegraphy in the variety of forms that it had found in the 1860s and 1870s, he made his way as an inventor not only by solving the technical challenges faced by both upstart companies and the Western Union monopoly alike, but he also was able to play companies and businessmen off one another to suit his own ends, which were primarily not to make money but to make more inventions.[15]

In the early 1870s, Edison worked both as an independent inventor and as a contractor for firms such as Western Union, and he contributed a number of useful inventions. Perhaps the most valuable were those that were used to increase the number of messages that could be sent on the same telegraph line at the same time, such as his "quadruplex," which permitted two messages to be sent along a line while two more messages were being received on the same circuit. Payment for these inventions permitted him to establish a well-equipped and well-staffed laboratory, and in mid-1876 he constructed in the rural New Jersey town of Menlo Park an "invention factory," in which he claimed he would be able to produce "a minor invention every ten days and a big thing every six months or so." While up to this time Edison had dabbled a bit in nontelegraphic devices, and even tried to market an "electric pen" (which was a fairly effective duplicating device, a predecessor to the mimeograph used for much of the twentieth century), it was the new laboratory that allowed free-ranging efforts at invention in many directions. The first important outcome of this, and one of Edison's most original creations, was the phonograph, which he introduced in 1877. This began as a product of yet one more variation on the efforts to make effective telegraph receivers that recorded messages. But the appearance in 1876 of Alexander Graham Bell's first telephone had directed Edison's attentions toward acoustical phenomena, and so when one of his telegraph experiments began to make modulating sounds, Edison pursued this direction, with spectacular results. The phonograph astounded the public in both the United States and Europe—a machine that talked was simply not something most observers had been prepared for. This was the invention that earned Edison the sobriquet, "Wizard of Menlo Park."

While Edison told one reporter—he was the first inventor to become a celebrity, and this at the age of thirty—that he expected the phonograph to be the invention that would "support me in my old age," it in fact turned out to be a difficult one to perfect and market. While it was a true marvel, it was a crude and homely marvel. It reproduced voice and music sufficiently to be recognized, but not sufficiently to be pleasant, and so for the first decade or so of its existence it was little more than an experimental toy, only to be resurrected by a combination of technical improvement and marketing savvy in the 1890s. In the meantime, Edison found the field of invention that would, indeed, solidify his reputation and his financial future—electric light and power.

It was in the creation of a system of electric lighting that Edison demonstrated just what a laboratory like Menlo Park, guided by a brilliant and intuitive experimenter, could accomplish, given the support of capitalists and a world eager for technological marvels (figure 16.5). In the summer of 1878, Edison took a break from work on telegraph, telephone, and phonograph inventions and from public controversy over credit and applications. In the course of this brief hiatus, he was able to discuss and

Figure 16.5
Edison's main laboratory, Menlo Park, 1880. Note the light bulbs in the former gas burners. Edison is seated immediately in front of the pipe organ. Smithsonian Institution photograph.

evaluate proposals for electric lighting systems and concluded both that this was a field still open to significant pioneering and that his expertise in electrical, mechanical, and chemical combinations would serve him well in the new area. He launched into a campaign to create an electric lighting system that would displace the gas system that dominated urban domestic lighting, and he announced this with great confidence and fanfare. The confidence turned out to be ill-founded (Edison's original concepts were fundamentally unworkable) but the resources of his laboratory and the confidence of years of success in solving the technical problems of telegraphy kept him at the problem until, in late 1879, he demonstrated his electric light bulb and a system of generators, conductors, switches, and other apparatus to make it work.[16]

That an electric current could be used to heat up materials hot enough to produce light had been known since the late eighteenth century, but a range of considerations, from the high cost of producing sufficient current to the fact that most materials will burn up as they are heated to incandescence, made electric light impractical. Michael Faraday discovered in the 1830s that a changing magnetic field would produce electric current in a conductor, but it was not until the 1860s that the first large electric generators were devised. Some of these were put to use in lighthouses, where the electric arc, a blindingly bright light produced by carrying large amounts of current over a gap between carbon electrodes, could be put to use. But the arc light had no use in domestic settings. There a smaller, quieter lamp, producing light on the order of a gas or kerosene flame, and easily lit or quenched in any room or part of a room desired, was the only kind of light that would do. It was this that Edison, naively at first, believed he could create. The task of "subdividing the light" took Edison down numerous blind alleys, but he was an enormously imaginative technician, always learning from his mistakes. He discovered, for example, that the first materials he tried out as incandescent elements in his "burners" lasted longer if he put them in a vacuum, and so he had assistants make the best vacuum pumps known. He calculated that for lamps to work well on parallel circuits, circuits that would allow individual lamps to be turned on and off without affecting other lamps, each lamp had to have a relatively high resistance. Some initially promising materials did not have the required resistance, so he set them aside. Finally, in the fall of 1879, he tried out a very simple material he had used before in his lab when he needed a high resistance: carbon. While carbon materials would ordinarily burn up quickly if heated to glowing, in the very high vacuum he was now able to produce, they lasted. When he finally found a form of carbon that was physically robust—simple baked cardboard—Edison had the lamp he wanted. He wasted no time in announcing his success, and the public demonstrations he held at Menlo Park at the end of 1879 were sensations—the newspapers announced the arrival of a new era as the decade of the 1880s dawned.

The technology of electric light and power required much more invention and development in the coming decades, by both Edison and a league of other inventors and entrepreneurs. For several years Edison focused most of his energies into the design and construction of a demonstration central station system in lower Manhattan, where he would be able to show the technology in the heart of American capitalism. His Pearl Street station began operations in September 1882. In the course of building it, he addressed the basic problems of the new system: designing large generators, determining how to lay reliable and safe conductors under city streets, devising reliable means of metering electricity usage, and turning the electric light itself from a delicate laboratory device into a robust, affordable article of commerce. Edison made a practice of covering all of his inventions in detail with patents, but the extent and complexity of the new electrical technology was such that it quickly became a field crowded with other inventors in both America and Europe. Some of these attempted to find alternatives to the technical choices Edison made, while others sought to extend the reach and applications of electric power. Even before the Pearl Street station opened, applications besides lighting appeared. Electric motors could be made in a range of sizes for a host of tasks, from running household fans to operating elevators and hoists. By the mid-1880s, motors were appearing in transit vehicles, and electric streetcar systems began to transform urban travel patterns throughout the world. Readily available electric current produced not only light and motion, but also heat, and so applications ranging from tabletop toasters to gigantic electric furnaces for chemical production made headway. Just as in telegraphy, the new electric light and power technologies proved capable of never-ending elaboration and improvement, and once again corporate structures, such as well-funded research and development laboratories, proved indispensable for sustaining and controlling the impetus of technological change.

The institutionalization of invention in the electrical industries, from telegraphy to power, was accompanied by another change in the nature of invention—the shifting source of innovative prowess from talented individuals, often amateurs, who learned their art by doing, to educated professionals, schooled in science and its techniques and esoteric knowledge as well as in codified practices and procedures. In electricity, in particular, there had always been room for scientists at the inventor's table, but most often the practical application of the phenomena uncovered by a Volta, Faraday, or Henry was very much the work of a Morse or Edison—"amateurs" whose school knowledge of the subject, when it existed at all, was of little relevance to their capacities for technical creativity. Even in the course of the century, however, there were certain technical problems that arose that seemed to call on levels of knowledge that went beyond that easily acquired through practical experience or casual study. The best example is probably that of the Atlantic Cable. More than any other single achievement in telegraphy, the hard-won success of the undersea cable that linked

Europe and America was viewed as a scientific triumph and a milepost in the march of civilization. The very first successful land lines in the 1840s sparked speculation about an oceanic link, but it took almost fanatical entrepreneurial energy, most closely identified with a young New England native, Cyrus W. Field, to push the project into being. The practical obstacles of taking the new technology, at a time when simple river crossings gave constant trouble to cable operators, to the point where crossing thousands of miles of uncharted ocean bottom was economically and technically feasible were enormous. It probably helped that Field actually knew little about the subject. He managed, however, in the course of the 1850s to raise sufficient capital for an attempt. After three failed efforts a link, using more than two thousand miles of undersea cable, was completed between Newfoundland and Ireland in August 1858. The public celebrations were extraordinary, as a new age of international relationships was heralded on both sides of the Atlantic. But the cable was thoroughly unreliable and slow; it took more than a week after the link was completed for a message to be successfully sent across it, and hours were required to put together any kind of coherent communication. Things only got worse, and by the beginning of September, less than four weeks after the link was completed, the cable went dead.

The Atlantic Cable was transformed from the most wonderful technical marvel to the most spectacular technical failure of the age. Out of this failure, however, there emerged not only a deeper understanding of the technology but a growing appreciation of the increasing connections between technology on the one hand and both the content and practice of science. The key figure in the public inquiry into the cable's failure was the professor of physics at Glasgow University, William Thomson. Thomson had helped out during the cable's laying and had analyzed its design thoroughly. He was able to explain a range of inadequacies and poor technical choices. He also designed an instrument for receiving the cable's faint signals—the mirror galvanometer—and he was able to explain the key causes of the line's failure, as well as the most likely routes to success. Thomson's approaches were proven in a cable laid in the Persian Gulf in 1864, successfully linking Britain with its outposts in India. This gave added impetus to Field's efforts to raise funds for a new Atlantic attempt, and in June 1865, the gigantic ship, *Great Eastern*, set out from Ireland with the entire cable in its hold, laying it out towards Newfoundland. About two-thirds of the way across, however, the cable slipped out of reach and sank two miles down to the ocean bottom, eluding efforts to retrieve it. The next year another attempt was made with the *Great Eastern*, and this effort successfully landed the cable in Newfoundland in July 1866. The new, stronger, better-designed cable was an instant commercial success, and it was joined within a month by the 1865 cable, finally retrieved from the ocean bottom and completed. The Atlantic Cable was once again the toast of Europe and America, but just as importantly it showed the key role that science would play in ambitious technological endeavors. Thomson went on to pursue a distinguished career in physics, earning a knighthood and then a peerage, as Lord Kelvin.

Sometimes the role of science in invention was not quite as clear-cut as it was in the case of the Atlantic Cable. The example of the telephone is a case in point. It also illustrated the extent to which successful invention required a flexibility of vision that was sometimes difficult to maintain as technical problems became more and more complex. A deeper understanding of the physics of sound was one of the chief achievements of mid-nineteenth-century physics. While this was at first primarily of theoretical interest, there were a number of ways in which acoustical knowledge could be applied. One of the first of these was in the teaching of speech to the deaf. A pioneer in this application was a Scottish teacher, Melville Bell. His son, Alexander, took up his father's profession, moved to America to make his own way in the world, and decided in addition to seek his fortune as a telegraph inventor. Like others before him, the younger Bell saw possibilities in the new understanding of sound and harmonics for the improvement of the telegraph, particularly in making instruments that could send several messages along the same circuit at once. This "harmonic telegraph" was an idea pursued by a variety of inventors, as an alternative and improvement on the multiplex telegraphy that was being devised by Edison and others, and which was so attractive to rapidly expanding telegraph networks. Bell was very much an amateur in things electrical, but he was a well-read amateur, with a scientist's openness to observing the behavior of his instruments and a good, mechanically skilled assistant in Thomas Watson. He took up residence in Boston, which cultivated one of the most active experimental communities in America, and he readily took advantage of the schools, libraries, and workshops that distinguished that city. By the mid-1870s, Bell had accumulated substantial knowledge of the latest acoustical science as well as of how sound could be made to alter electrical currents.[17]

The transformation of the object of Bell's inventive energies from multiplex telegraphy to telephony was gradual and hesitant, but no one could have been better equipped in knowledge and outlook to make this change. Sound, after all, had been the center of his intellectual and, indeed, personal life. The origins of the telephone lay in a series of experiments and observations, stretching from 1873, when Bell began using steel reeds to produce electrical signals of specific frequencies (to be picked up selectively by similarly tuned reeds at the receiving end of the telegraph line), to mid-1875, when, in the course of experiments with the reeds, Bell observed that his receiving reeds were not simply picking up the expected electric signal but were precisely reproducing the sound, not only in pitch but also in volume and timbre. When Bell substituted a vibrating diaphragm for the reeds, he was still able to reproduce sounds, although weakly. Some months of experiments were required, months of additional observation of how variations on his instruments behaved. The results were still muddy, but it became clear to Bell that the properly designed combination of electromagnets, diaphragms, and resistances should allow him to transmit any combination of frequencies he wished to generate, including, he dared hope, speech. Bell only gradually came to understand the technical and physical principles behind his apparatus's operation, but in February 1876 he applied for a patent covering the transmission of vibrating currents, including sound. Other inventors had suggested similar things, but not as clearly as Bell was able to do. Elisha Gray, another telegraph inventor, actually informed the patent office of his own, similar, ideas on the day that Bell filed for his patent, and working out the real priority over the telephone's invention (there were foreign claimants as well) has been a historical wrangle. But the U.S. Patent Office was satisfied with Bell's claims, and his patent number 174,465, for "Improvement in Telegraphy," was issued on March 7, 1876. It has been called the single most valuable patent in history (figure 16.6).

This value, however, and the value of Bell's telephone generally were by no means obvious to its early observers. Within months of the patent's issuance, Bell had a chance to put the device on a very public platform—the great International Exhibi-

Figure 16.6
One of Bell's experimental telephones of 1876. Smithsonian Institution photograph.

tion held in Philadelphia to celebrate the centennial of American independence. On Sunday, June 25, 1876, Bell demonstrated a modified version of his telephone to a panel of observers at the World's Fair, including Dom Pedro, the emperor of Brazil, and William Thomson, who had traveled from Britain to see what marvels the Americans might have to display. To Thomson, Bell's telephone was the greatest marvel of them all, and the Brazilian emperor was almost beside himself at hearing Bell sing and recite Shakespeare over the instrument. Subsequent publicity was actually quite modest, and for most of the public the Bell telephone was not a memorable part of the Philadelphia fair. But scientists like Thomson recognized the achievement, and Bell's own backers grew anxious to exploit the commercial possibilities. But what were these? As marvelous as the "speaking telegraph" might be, its real uses were not at all clear. Telegraphy was, by this time, a solidly based, ever-improving technology, whose uses were well defined and understood. These depended upon a specific notion of telecommunications, and this was built around the idea of the "message." A message was a specific body of information, formulated at one place, put into transmittable form, sent over wires, received and written down in another place, and delivered to its intended recipient. The telephone did not handle "messages" in this way, and so its contribution to communications was uncertain. Indeed, when the Bell interests, momentarily frustrated with their own commercial prospects, offered to sell their rights to the Western Union Company, they were turned down—in part because the telephone seemed to be perhaps no more than a toy, and in part because the company was confident that if it did turn out to have uses, their own inventors, like Thomas Edison, would be able to do better than Bell. They did not reckon with the fundamental nature of Bell's patent claims, or with the possibility of a radical redefinition of telecommunications.

This redefinition did not come easily or quickly, but at one level the attractions of the telephone were relatively obvious. It "required no skill," in the words of early promotional material, and also provided the most direct communications possible between sender and receiver, with no telegraph operator or messenger boy in between. On this basis, the telephone slowly made its way into American commerce, providing direct communications between offices and factories, between stores and warehouses, between shippers and receivers. In the first years, this all worked well, typically over private, direct lines, as long as they were no more than about twenty miles long. Soon, however, improved equipment allowed longer lines and wider networks, and the technology proved remarkably adaptable to local conditions. In farm areas, for example, some farmers began using wire fences as telephone lines—the quality of transmission was imperfect, but it was still a marvel to most users. In cities, central switchboards and exchanges began to appear as early as 1878, and by the 1880s urban telephone systems, while still too expensive for ordinary folk, were beginning to spread in all cities of any size, both in America and overseas. Within a decade of its first appearance, the telephone had redefined the possibilities of personal communications, even though it took some time yet before it became the social necessity that defined its central place in twentieth-century society.[18]

The transformations in transportation and communications technologies in the middle decades of the nineteenth century, largely through steam-powered travel and the telegraph, reshaped daily life and social relations more profoundly and more rapidly than any changes in history. The functioning and structure of communities and nations were altered as the possibilities of the rapid movement of goods and people and the almost instantaneous exchange of information and ideas reshaped the perception of the spatial and the temporal limitations and possibilities of human action and interaction. As the century drew to a close, these changes began to operate on a different level, with new technologies that functioned as personal tools rather than cooperative instruments were made more widely available. The telephone put long-distance communications (at first defined in dozens of miles rather than hundreds) within easy reach of unskilled middle-class citizens, who explored all the personal as well as economic possibilities that this opened up. A bit later, the automobile (with a slight foreshadowing from the bicycle) was to allow a similar transformation in personal mobility. As important as these changes were, however, they were matched by the changes that the new technologies—joined by such companion technologies as electric light and power—made in the European and American perceptions of the nature and significance of technological change itself. As the culture of improvement took shape in earlier centuries, it was identified with changes that usually (with a few important exceptions) operated largely within the workshop, factory, and other venues of production. But the nineteenth-century transportation and communications revolutions spread the idea that improvement was not limited to tools, materials, and other instruments of production, but that the very quality and pace of daily life could be altered by it, that the individual choices that women and men made to determine how they were to live, where they were to go, with whom they associated, and the myriad other elements that gave life its shape and texture were subject to change—to deep and lasting change—through technology. This perception was itself largely a novelty of the nineteenth century, and arguably was one of that century's most important legacies.

17 Engineering Emerges

May 1, 1851, began with a bright, cool morning in London. After weeks of dreary weather, the sun was finally shining, and people came out in great numbers to enjoy the recently declared holiday. While May Day was an ancient festival occasion, it had not been a formal holiday, particularly in the city, for some time. This particular May 1 was special. On the grounds of London's largest park, a gigantic building, almost 1,851 feet long (popular legend quickly adopted that figure) and a bit over 400 feet at its widest, was opening for its function of housing the "International Exhibition of the Industry of all Nations," known by contemporaries and historians alike as simply the Great Exhibition of 1851. The size of the building, however, was not its most striking feature. The entire structure, designed to be erected quickly and to be disassembled with equal ease, was made of iron and glass—all in all, some 25 acres of glass, almost a quarter-million panes, held up on 3,300 iron columns, 2,150 iron girders, and 372 roof trusses (figure 17.1). It would come as no surprise, therefore, to learn that the building—aptly named the Crystal Palace—was designed by a greenhouse builder, Joseph Paxton, who was actually the head gardener for the Duke of Devonshire. On May Day, the crowds gathered early for the royal ceremony, for what had started as a largely private affair, an industrial fair along the model that had been followed for decades by organizations devoted to promoting trade and manufactures, had quickly turned into a matter of state. This was largely the work of the prince consort, Albert, who took the enterprise under his wing and became its primary champion. It was therefore not simply a matter of formal procedure when, at precisely noon, he led his wife, Queen Victoria, up to the dais in the center of the great building to mark the exhibition's formal opening. It was also Albert's own presentation of the project to the Queen and the nation. Victoria, just a few weeks shy of her thirty-second birthday, had been on the throne for nearly fourteen years already, and she would reign for close to another fifty, but to many observers this day marked the high point of her reign. It also marked the symbolic height of Britain's technological supremacy.[1]

Figure 17.1
Joseph Paxton's Crystal Palace, built for the 1851 World's Fair in London, was a marvel of glass and iron. It also broke new ground in the industrialized prefabrication of buildings. From the World's Fair Collection, Architecture Library, University of Maryland.

For anyone looking at technology in the nineteenth century, the Crystal Palace, both building and exhibition, is both a potent symbol and a powerful object lesson. As a symbol, the Crystal Palace came to represent technological confidence, even swagger, that could only have been possessed by the world's first industrial nation. While Henry Cole, a designer and publicist who became the project's chief philosopher, promoted the Great Exhibition as a means of teaching English industrialists the principles of good design through the example of the world's best, the Crystal Palace was less a set of lessons for the British (although it did present some important ones) and much more a flamboyant display of Britain's own achievements for all the world to see. Although Cole and others worried about the degradation of design through mass manufacture, most were impressed by the achievements of manufacture itself, of the ability of industry to turn out goods in variety and quantity that earlier generations could never have dreamed of. Paxton's building was itself a miracle of manufacture; not the product of anything like construction in the traditional sense, dependent on masons, carpenters, or bricklayers, the Crystal Palace was instead a factory-made product, assembled on its site from thousands of pre-fabricated pieces. It was also itself a celebration of machinery. The building's erection had required machinery at every turn; the iron pieces—columns, girders, arches, and even channels in the glass fittings that led water down through specially designed gutters—all were the products of relatively new types of tools that made the accurate forming of a multitude of identical parts not the product of skilled handiwork (although this still had its place) but the result of very carefully designed and crafted machinery, what we ordinarily call "machine tools." In addition, Paxton made his building itself into a machine. The ventilation units, the screening mechanisms, the drainage elements were all combined in the iron and glass structure to work in a coordinated system—the very definition of "machine" itself.

Later exhibitions, in Paris, Philadelphia, Chicago, and London itself, would celebrate the machine much more centrally than the Crystal Palace, whose centerpiece was actually a great fountain of glass. Nevertheless, that first great World's Fair would still convey to the millions who visited it and to the millions more who read everything they could about it the message that the machine—and machines of every variety and character—had remade the nineteenth-century world. The power of machinery to transform manufacture had, of course, been amply demonstrated by the oft-improved devices that had transformed every stage of textile production since the eighteenth century. The new textile machines had precipitated the concentration of machinery and power in large factories, and buildings that were filled with hundreds of machine-tenders, men, women, and children, were the most characteristic features of the industrializing world. Extending the revolution in production beyond textiles and a few other commodities, however, was a complicated affair, requiring not only new devices, but new approaches to the very idea of what making things involved, and how making things could be improved.

These changes were not to be seen in their final form at the Great 1851 Exhibition, but they were in fact glimpsed and noted by a number of observers. Perhaps the most famous example consisted of the display in the American section of the Crystal Palace of the goods manufactured by Samuel Colt, a Hartford, Connecticut, maker of small arms, and by the Windsor, Vermont, rifle makers Robbins and Lawrence. Colt displayed examples of his patented pistols, along with the claim that they were "made by machinery," while the Vermont gun makers showed off a number of their rifles, accompanied by the claim that they were made from "interchangeable parts." These displays did not, in fact, cause a great stir in 1851, but within a few years the British had appointed committees to investigate what was distinctive about the American methods of making firearms. Over the years, these methods came to be referred to as "the American System of Manufactures," although there was never a precisely defined "system," nor were the methods distinctively American. There were two key elements to the new thinking about making things: mechanization and uniformity. These were historically separate ideas, each important and useful, but it was their combination that was to transform manufacture profoundly. The combination, however, was neither simple nor quick, and in fact it can be argued that the real joining of mechanization and uniformity into a sustainable, economical system of production was not achieved until the twentieth century, when Henry Ford demonstrated the full capacities of assembly line mass production in a complicated but widely desired commodity, the automobile. The roots of this achievement lay almost a hundred and fifty years before, in prerevolutionary France.[2]

The Seven Years War ended in 1763 with a humbling defeat for France and her allies. She lost huge colonial claims in Asia and North America, and was hemmed in on all sides in Europe. It is not surprising, therefore, that the political ramifications were substantial and that calls for reform, particularly of the defeated military, gained much impetus. Among the beneficiaries was one Jean-Baptiste Vaquette de Gribeauval, an artillery officer who led the charge for a complete rethinking of the role of firepower in warfare, and in particular the organization and equipment of the artillery. Gribeauval was chosen to lead the rebuilding of the French artillery service, and in the course of pursuing his reforms, he advanced ideas that were to have profound influence well beyond the military and his own era. At the heart of his new ideas was the use of much larger numbers of smaller cannon, an artillery that could be moved quickly and flexibly to be brought to bear where it could do the most damage to the enemy (figure 17.2). For this program to work, however, the French arsenals had to be able to produce lighter-weight guns in much greater quantities. In addition, for such guns to be of use in the field, they had to be reliable and easily repaired. It was in the course of reshaping the French arms making industry to meet

Figure 17.2
Gribeauval's cannon were simpler and smaller than the traditional guns of the French artillery, and their effectiveness thus depended on larger numbers of reliable, standardized guns. Manufacturing these in sufficient numbers led to important reforms in manufacturing.

these new needs that Gribeauval formulated the principle of uniformity for the parts of his guns and their carriages.[3]

The "uniformity principle," as it came to be known, is so familiar today that it is a bit hard to recognize it for the radical invention that it was. Most simply, it meant that in the manufacture of devices with several or many parts, each part should be made sufficiently uniformly so that it would fit into any of the finished devices without further shaping. This is the principle that we all take for granted in visiting a hardware store or in repairing an automobile or appliance: if one part fails, it can be replaced with an identical piece, made so that it will fit readily into the same spot as the failed part. For guns that would have to be repaired in the field, under the heat of battle, this idea had great attraction to Gribeauval, but its realization encountered tremendous difficulties, both technical and social.

The technical problems with realizing uniformity largely were a matter of fit. In ordinary manufacture, parts are made to fit together just well enough, and craftsmen make one part to fit in just the place that it will finally rest, not in some nonexistent other place where it *might* be used. To make something fit uniformly, as a general rule, is more difficult, and therefore more expensive. If a very fine fit is called for, this is done by shaping—molding, carving, filing—each piece until it mates properly in the final product. This was the traditional, time-honored way in which craftsmen made parts and assembled them. This suggests the social difficulties that the champions of uniformity ran into, time and again: shaping and fitting were traditionally the tasks and the prerogative of craftsmen. The uniformity system, to make it work, required that choices about fit be taken from the hands of individual craft workers and be given to others with the authority to tell craftsmen not only what was to be made, but how it was to be done. This transfer of authority was met with much resistance in most instances, sometimes with sufficient resistance to defeat the effort of those promoting what they believed to be more "rational" manufacture. The shift in control over work also meant a shift in the locus of improvement. As workers lost the capacity to exercise judgment or determine fit, they also lost the opportunity to make their own contributions to change; the worker's ability to define or claim improvement diminished. At the same time, a new agent of change—the engineer or designer—came on the scene, and as he (or, very rarely now, she) gained power over both product and process, he became the chief source of innovation. The implications of this would become partially visible in the nineteenth century, but quite widely evident in the twentieth, as engineering and design became professionalized fields, with institutional values of their own.

To a remarkable extent, Gribeauval and his followers were able to solve the key technical problems of uniformity, but the social and political difficulties would remain largely intractable. The key individual in the technical solutions was a skilled armorer and gunsmith, Honoré Blanc. From the beginning of Gribeauval's efforts, Blanc had been one of the men he relied on to devise the means for uniformity and to instruct gun makers in its practice. In 1777, Blanc devised a complete set of gauges to dictate uniform standards for the parts of a new musket. Gauges were pieces of carefully shaped metal that could be fitted to various parts to determine if they were close enough to the required standard to be used. Neither gunmakers nor any one else were accustomed to the use of gauges at this point, so much of Blanc's work involved instructing, persuading, and cajoling. Still, the 1777 gauges were just one step on a difficult road; Blanc himself acknowledged that their use alone did not yield sufficient uniformity to make parts interchangeable. For this Blanc required many more years of work. Through the various political upheavals of the next two decades, Blanc was able to find support for his work, and was encouraged to set up workshops in several locations to continue refinement of his uniformity methods. His efforts became quite famous, in fact. None other than Thomas Jefferson, when he was ambassador of the new United States in Paris in the mid-1780s, sought out Blanc's workshop. Jefferson absorbed the lessons of what he saw there with great enthusiasm, and wrote back to America that here were methods perfect for the new country's needs. When he returned to the United States in 1789, he arranged to have half a dozen of Blanc's gunlocks shipped on to New York, where members of Congress could inspect them. Blanc himself toiled on, even through the revolutionary storm that soon broke over him. By 1795 he in fact was able to supply hundreds of interchangeable gunlocks to the government, and he continued his work in the manufacture until his death in 1802, after which his factory continued for five more years, achieving an output of nearly a thousand gunlocks a month. Blanc's techniques, however, were soon abandoned, for his product was always expensive and his methods were quite labor intensive. His uniformity was the product of very closely supervised hand labor, not mechanization (although certain forging and stamping methods were adopted to facilitate uniform shaping). It thus exacted costs few were willing to pay, even, in the long run, the military.

At the same time that Blanc was toiling so hard to make his interchangeable gunlock parts, another key improvement in production methods was emerging across the English channel, as France's mortal enemy also attempted to cope with the extraordinary military emergency. In the late-eighteenth century, the largest, most complex and most expensive engine of war was a warship, the successful operation of which depended on its guns and its sails. To make these sails function properly, and to move other parts of the ship into position, ropes and pulleys were indispensable, and a large ship would use as many as a thousand pulley blocks for controlling its lines. These ranged in size from small, simple ones a few inches in length to multipart giants several feet long, used for hawsers and other ropes keeping the ship in place. For centuries pulley blocks were made entirely by hand. They were not hard to make, but they did require some care in shaping and fitting, and making large quantities in a short period of time could pose problems. The expansion of the British Navy in the second half of the eighteenth century encouraged the exploration of ways to speed up block production, and the Royal Navy was always looking for ways to control costs as well. Under the pressures of the Napoleonic emergency, incentives were even greater for improvement in this area, and a small group of British engineers and entrepreneurs sought the means to do so by maximizing the use of machines.

In 1801, Sir Samuel Bentham, inspector general of naval works, sought to set up in Portsmouth, England's primary Channel port, a factory for turning out large quantities of pulley blocks. Bentham had considerable experience with such manufacture, and had in fact invented machinery for it, but when he was approached by a French émigré, Marc Isambard Brunel, with designs for a whole series of specialized machines for block making, he quickly saw their superior virtues. He approached a remarkable machine maker, Henry Maudslay, with Brunel's plans, and ordered from him the full set of twenty-two different kinds of machines. Between 1802 and 1807 a series of machines, eventually totaling about forty-five, were delivered to the

Portsmouth factory, where they were put to work producing pulley blocks of three different sizes. These machines, made of iron, wood, and steel, consisted of a series of saws, gouges, chisels, and drills that could quickly and tirelessly shape wood to the required forms and dimensions. This was the first successful effort to mechanize fully the production of a complex product, and with the Portsmouth system, fewer and lower-skilled workers could produce substantial quantities of the wooden and metal parts needed for a range of blocks. The result was a widely appreciated technological marvel, although it was too expensive and difficult to perfect to give much encouragement to private entrepreneurs making consumer goods, or even to become a model for most military or naval production in the coming decades. But the demonstration of the potential of specialized and carefully designed and made machinery to transform the production process was widely observed and remarked upon. The lessons would be returned to time and again throughout the nineteenth century, most especially in the United States.[4]

Thomas Jefferson's enthusiastic reports of Honoré Blanc's efforts at interchangeability were widely circulated, so that by the 1790s some of the ideas of both the French and the British pioneers were evoking responses from enterprising Americans. The most famous of these was Eli Whitney, a Connecticut mechanic and promoter. In the 1790s, while living in South Carolina, Whitney had devised a machine to speed up the cleaning of seeds out of the short staple cotton that grew in the upland South. Considerable question has been raised about the originality of Whitney's "cotton gin," but it did serve to gain him a reputation as an inventor. He parlayed this into support for efforts, beginning in 1798, to get U.S. government support for the establishment of an arms factory, in which he proposed to make large numbers of muskets with interchangeable parts, on a system like that of Blanc. Despite receiving an advance from the War Department of $10,000, Whitney never, in fact, was able to make good on his contract. He appears to have grasped the two basic ideas that were to be the key to mass production—interchangeability and mechanization—but his own inventive and managerial talents were simply not up to the task of bringing these concepts to fruition. His self-promotional abilities, however, were considerable, and so he was able to link his name to the ideal despite a long series of technical and financial failures.[5]

The real origins of a system of production in the United States lay in the government armories, which, like the French armories a generation earlier, responded to the deficiencies made manifest by a war (the War of 1812, in this case) and attempted to reorganize production to rationalize the provision of arms to the military. The experiences of Blanc and other French efforts, as well as the travails of Whitney, informed the armory efforts, but considerable innovation and risk-taking was still required to create a truly workable system of production. Just as in the French case, progress was made chiefly because some key individuals became convinced that interchangeability was worth the great initial effort and expense. Persistent experiments and trials were necessary over the course of several decades before the Americans were able to display the products that so impressed visitors at the 1851 Crystal Palace. In 1815 Roswell Lee became superintendent of the U.S. Armory at Springfield, Massachusetts, and he spent the next eighteen years championing "the great object of uniformity." His superiors in Washington were generally supportive, which was fortunate since many obstacles, both technical and social, remained. The greatest difficulties were encountered at the government's other armory, at Harpers Ferry, Virginia (now West Virginia), where arms were made by craftsmen who resented any attempts by supervisors to impose standards or other limits on their work. At Springfield, on the other hand, Lee and his workers were able, bit by bit, to show how the goal of interchangeability could be pursued. This progress depended on the contributions of men working outside the armories, such as John H. Hall, who came to the attention of the army's ordnance department by his invention of a breechloading rifle in 1811 (figure 17.3). From 1819, Hall manufactured his rifle under contract at Harpers Ferry (conflicting often with the local gunsmiths), and made considerable progress toward interchangeability through the use of gauges. The significance of this began to be evident when another contract arms maker, Simeon North, working in Connecticut, was able to produce rifles in his shop whose parts could be readily exchanged with those from Hall's Virginia factory. Here, through the use of gauges and very careful supervision, was the beginning of a real payoff for uniformity. By the mid-1840s, when the armories began producing new rifles and muskets designed

Figure 17.3
Breech and lock of John Hall's rifle as made at the Harpers Ferry Armory. Smithsonian Institution photograph.

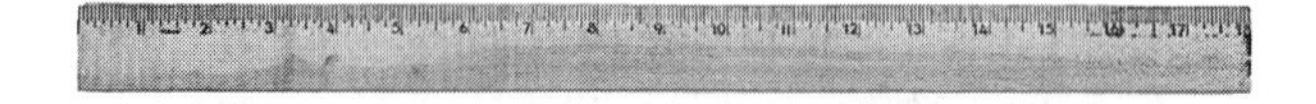

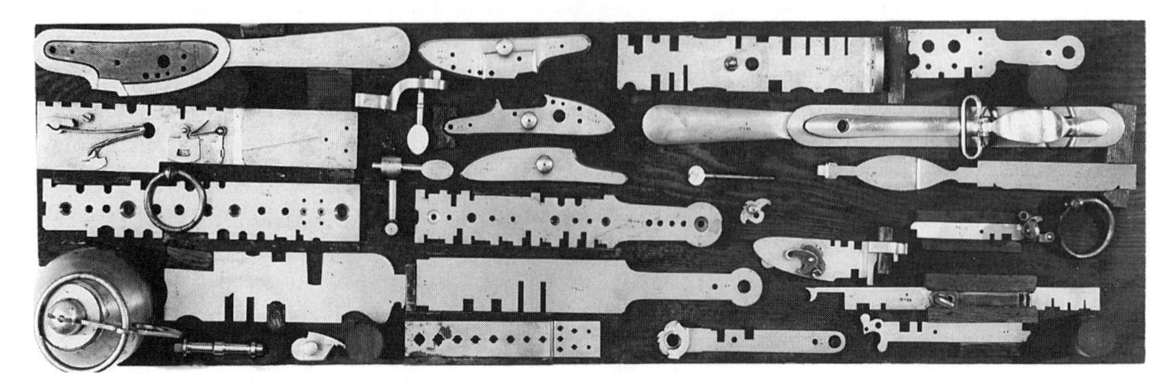

Figure 17.4
Inspection gauges used at the U.S. Army's Springfield Armory for the Model 1841 rifle. Smithsonian Institution photograph.

around uniform gauges, interchangeability on a large scale became standard practice for U.S. government arms makers (figure 17.4). Their success paved the way to the more highly publicized triumphs of the private manufacturers, like Colt and Robbins & Lawrence, whose Crystal Palace displays drew such attention.[6]

Interchangeability was not the only area in which American manufacturers were able to make early contributions. The virtues of mechanization were touted quite early among American industrialists, first in the textile industry, where they followed rapidly in the footsteps of the British pioneers, and then in an ever-widening circle of other products. One special inducement to increased use of machines in American production was the relatively high cost of skilled labor. In the United States, the ready availability of land on the western frontier made factory labor, at least among American-born male workers, more costly, as a general rule, than in Europe, where occupational opportunities were more limited. There was, in fact, a complex interplay of many factors determining the real cost of labor in the United States, and their relative importance has been the subject of considerable debate among economic historians, but the fact remains that mechanical ingenuity very quickly came to be seen as a special American value and capability. One of the best examples was the work of Thomas Blanchard. Blanchard was a mechanic in Worcester, Massachusetts, when he was approached in 1818 by a local armorer for help in machining the somewhat oval shape of the barrel of a breech-loading musket. His successful solution depended on the use of cams to guide the cutting tool around the workpiece. When Blanchard turned his attention to making the even more irregularly shaped wooden stocks of the guns, he was inspired by what he knew of the Portsmouth block-making machines, but he took their action one step further. Whereas the Portsmouth machines were each designed for very specific forms and cuts, the so-called Blanchard lathe was designed to make a wide variety of irregular shapes in wood. The key to Blanchard's approach was to make a metal model of the desired wooden shape (gun stocks were soon followed by hat blocks, shoe lasts, and other shapes) and to use this model to guide the movement of cutting tools against the unformed wooden blank. The machine was not a lathe in the usual sense, for cutting was not done by turning the wooden workpiece against a stationary cutter, but the cutter was a set of rotating blades that was moved laterally against the workpiece, following a curve determined by the pattern piece. After the cutters had moved the length of the piece, the blank was turned just a fraction (and the pattern turned the same fraction), and the cutters once again set to work along the length. Often this produced a relatively roughly shaped form, but it was trivial to then smooth it down and finish it. This approach required much less skill and performed the work of irregular shaping much more quickly than traditional methods. Working as a contractor for the Springfield Armory, Blanchard, again inspired by the Portsmouth block-making factory, produced a series of fourteen machines to mechanize the entire process of gun stock making, thus completely eliminating skilled labor from the process.[7]

An important part of the achievement of large-scale interchangeability came about through the gradual but substantial improvement of the machines used for making parts. As the Portsmouth block-making experience demonstrated, one key to mechanizing production was specialization—the design and application of machines that were dedicated to specific tasks in the course of the shaping of wood or metal. Another key, equally important but more elusive in realization, was precision. Clockmakers such as John Harrison had already demonstrated the possible benefits of the very precise shaping and fitting of parts, and scientific instrument makers and fine craftsmen had long made precise fitting a core goal of their workmanship, on aesthetic grounds as much as practical ones. But making precision routine was an entirely different matter, and it was the achievement of a number of ingenious and dedicated mechanics working in England from the late eighteenth century into the middle of the nineteenth. Their achievements, known collectively as machine tools, do not betray their originality and significance easily, for at first glance they would seem to be modest improvements on well-known, sometimes ancient, devices, such as lathes, drills, and planers. But by making precision a prime value and by determining means for achieving ever greater precision and doing so in ways that made it achievable more simply, quickly, and with less skill, the new machine tools became the indispensable means for the mass mechanization of the manufacture of almost everything made. One of the most important achievements of nineteenth-century technology was the application of machines to making everything they could be put to, and at the heart of this self-conscious mechanization was the creation of dependable easily worked devices for repetitively cutting, grinding, drilling, planing, stamping, forming, molding, and otherwise giving shape to a wide range of materials and doing so with a uniformity of shape and fit that such close students of technology as Diderot and the *encyclopédists*, for example, could never have imagined.

Joseph Bramah was a farmer's son who was saved from the agricultural life by an accident in the fields. He learned the trade of a cabinet maker and opened up his own shop in London. Quickly, however, he was lured by the excitement and promise of invention, and he took out his first patent in 1778 for improvements in the design of the water closet (or flush toilet), which at the time was quickly becoming a necessary appliance in houses of the well-to-do. Turning to more intriguing mechanical problems, Bramah devised an ingenious lock in the early 1780s, While a bit simple by the later standards of lockmaking, Bramah's lock nonetheless represented a substantial advance over traditional designs, but its manufacture, requiring close fitting, posed problems that bedeviled his early efforts. In 1790, however, there came to Bramah's shop a young apprentice, Henry Maudslay, whose skill at metalworking, even at the age of eighteen, gained him a substantial reputation. Bramah was delighted to take advantage of Maudslay's skill in devising machines that would turn out parts of the lock with the needed precision and consistency. Over the next few years, Maudslay continued work in Bramah's shop, combining extraordinary standards for precise working with clever mechanisms for reducing variations or guiding the worker's hand or tool to the right work point. When, in 1795, Bramah devised the first workable design for a hydraulic press, it was Maudslay's precise fitting that made the machine, with its demanding tolerances under high pressure, possible to produce. Soon thereafter, Maudslay set up shop on his own and proceeded to produce a series of machines that are generally regarded as the true beginning of modern machine tools.[8]

Maudslay's contribution to the technology of production, and in particular to the design of machinery, can be encapsulated in two core principles, principles embodied in material form rather than words. The first was the accurate screw. The movement of the parts of a machine tool depend fundamentally on the accuracy of the screws used to bring together or separate the portions of the tool, or the workpiece and the tool edge. If the guiding screws are not cut with great precision and care, all following actions are suspect or, at least, much more difficult. The second materialized principle was the flat plane surface. The metal surface on which the workpiece and/or cutting or grinding edge of the tool are moved must be truly flat and smooth, or else all other calibrations will be inaccurate. Maudslay thus took infinite pains to see that his screw threads were cut with extreme precision and that his plane surfaces were ground and polished to the utmost levelness, and he preached this gospel to everyone who came to see or work in his shop. It was, in fact, as a teacher, master, and apostle of precision that Maudslay made his greatest mark on the technology of the nineteenth century. In the first years of the century he made the famous series of machines required by Brunel's designs for the Portsmouth block-making factory, and the fame of these spread Maudslay's reputation even wider.

Through Maudslay's shop came a series of apprentices, pupils, and simply observers who then took the gospel of precision away with them, often making their own contributions to a rapidly expanding mechanical technology or perhaps simply demanding of their suppliers or workmen a higher standard of precision than anyone had cared to seek before. Joseph Clement was a weaver's son who started out in Bramah's shop but shortly moved to work under Maudslay and then, in 1817, set up a shop of his own, where he produced particularly finely made lathes and planers, some of strikingly original design. A shoemaker's son, Richard Roberts, worked for Maudslay for two years before setting up his own business in Manchester, where he designed lathes, planes, gear-cutting machines, and an astounding variety of drilling machines. He began also to focus on uniformity, using gauges and templates to standardize products. Roberts also advanced the design of automatic machines that could carry out multiple tasks at one time, following a kind of "program" embodied in a model placed in the machine. James Nasmyth was a Scotsman who began working for Maudslay shortly before the master's death in 1831. A few years later he set up business near Manchester, where he became a key supplier of machines, tools, and locomotives for the new railroad industry. The much larger scale of machinery needed by the railroads, not only for engines but for the shaping of thousands of miles of rails and the manufacture of millions of spikes and sleepers and other parts, pushed Nasmyth and others after him to enlarge their machines and the size of their works. He became most famous for his steam hammer—a giant device that used the power of steam to drive huge weights down on the large plates, rails, bars, and rods that went into the ever larger machines and engines of the railroads, the steamships, and the shops that served them.[9]

The most famous of Maudslay's disciples was Joseph Whitworth, who not only proved to be a superb machinist and a clever improver of tools, but also became the most prominent champion of the principles of precision and mechanization for the new generation that would see Britain through its decades of domination of the world's machine-building industry. Perhaps Whitworth's most notable specific contribution was the development and promotion of standard screw threads. Before him, each machinist or machine shop used the screws that suited their own work, without regard for the work of others. Whitworth wanted to carry the principles of rational machine design beyond the individual shop, however, and produce standards that would have national, if not international, scope. The particular choices that Whitworth made in developing his screw standard—taking the national average of the all of the angles for thread sides that he found as the basis for his proposed standard, for example—were not widely accepted outside Britain, but his efforts still influenced practice throughout the world. In Whitworth we can see clearly an important transitional figure: the machinist-craftsman of the eighteenth and early nineteenth centuries becoming the mechanical engineer that would redefine the way in which machines were made and used in the industrial age. It was the railroad, more than anything else, that increased the responsibilities and influence of machine builders to the level that they wished to call themselves engineers. But it was the dogged pursuit of rational, standardized practice, as exemplified by Whitworth and others of Maudslay's followers, that truly formed a discipline and a profession.[10]

Even before the mechanical engineers began to emerge as an identifiable group, however, the technical and economic changes of the early nineteenth century had put the spotlight on another group of engineers, the civil engineers. The building of railroads, in particular, provided the impetus for extending the capacities for building bridges, tunnels, viaducts, and other transport structures well beyond their traditional limits. Even before steam railroads, the bridge builders of France and Britain, in particular, demonstrated that modern knowledge of materials and structures made possible constructions that had been unthinkable before the late eighteenth century. The best early representative of this new kind of engineering was Thomas Telford, born in 1757 the son of a shepherd in northern England's Lake Country. After a small amount of schooling, Telford became a stonemason, but his ambition soon took him out of the rugged but poor country he was born in, first to Edinburgh, and then to London. Considerable native talent along with some fortunate introductions allowed Telford to quickly gain experience and commissions for building throughout England. In the 1790s, he began to go beyond the construction of mansions and churches and to work on engineering structures for canals and roads. Some of the canals required aqueducts, and Telford began experimenting with the use of cast iron for carrying water channels up over large stretches. The most spectacular result of this work was the astonishing work built on the Ellsmere Canal, on the western edge of England, to carry it over the River Dee. This Pont Cysyllte Aqueduct rises almost 130 feet over the river valley, carrying an iron trough more than a thousand feet long. It was the largest aqueduct of Britain's canal age, and it was spectacular testimony to Telford's mastery of arch bridges, as well as of cast iron. Work such as this cemented Telford's reputation as the finest British bridge builder of his age, and this gained him the commission for designing his country's most ambitious engineering project before the coming of the railroad, namely, the road between London and the port of Holyhead, on the northeast coast of Wales and a key embarkation point for Ireland. The strategic importance of this route encouraged the government to give much direct assistance to the project, which in turn led Telford to attempt a series of daring constructions to make the road swift and safe. The most spectacular of these was the iron suspension bridge built over the narrow straits that separate Holyhead, on the island of Anglesey, from the British mainland. These Menai Straits required an often hazardous ferry crossing, which Telford avoided by the most ambitious bridge ever constructed, using wrought iron chains to suspend a road deck more than 550 feet long between two huge stone piers, 153 feet high. The resulting bridge took more than eight years to construct, and when it was completed in early 1826, it was easily the longest clear span bridge in the world, and long remained a symbol of the superiority of British engineering. Thomas Telford himself became a symbol of the style and daring of the self-taught British engineer, the archetype of the "heroic engineer" of the nineteenth century.[11]

Railroad engineers, such as George Stephenson and his son Robert or Isambard Kingdom Brunel, son of the Marc Brunel who helped designed the Portsmouth block making machinery, carried the legacy of the earlier canal and road engineers much further. The enormously larger loads represented by railroads, as compared with older coach and cart weights, and the much greater value placed on speed and directness by the railroad promoters posed considerable challenges that were overcome through a combination of much more capital, much more systematic study of structures and materials, and the use of tools, machines, and new materials that continued to be developed and improved over the course of the century. The bridges and tunnels of the railroad age were larger, heavier, longer, and bolder than anything ever seen before. Telford's use of iron, like Darby's iron bridge several generations earlier, demonstrated what new materials could make possible, but the ambitions of the new engineers required that the materials and the particular structures using them be studied and understood in ways never before attempted.

The best early example of the new ambitions and what they required was the bridge that Robert Stephenson designed to carry the railroad over the same body of water that Telford's great bridge had spanned, the Menai Straits. A suspension bridge of the earlier sort did not seem practical to Stephenson in 1845, when he began thinking about the problem in earnest, for the loads of a railroad were much greater than any Telford had to accommodate. Stephenson's alternatives were constrained by the British Admiralty, which considered the Menai Straits a strategic waterway and thus insisted that any new bridge not add obstacles to ships passing through. The one location for a bridge that thus presented itself was about a mile down from Telford's, where the so-called Britannia Rock stood in midway between the banks. A large central pier could thus be constructed on the Rock, but the Admiralty even voted against allowing arches, even temporarily, for those large enough to accommodate the railroad bridge would interfere with the passage of ships. This left Stephenson with the option of a flat girder or cantilever bridge, but no one had ever attempted such large loads over such a long unsupported span. After looking at the dwindling set of options, the engineer settled on a box girder of wrought iron. But such a structure would be quite novel, and Stephenson thus took steps to learn more about the likely behavior of such a structure before actually investing in it. In a series of investigations and experiments, largely carried out by William Fairbairn, a mechanical engineer with great experience in shipbuilding, and Eaton Hodgkinson, who had spent decades studying the properties of iron, the engineers of the Britannia Bridge showed not only that Stephenson's design was sound, but also how systematic engineering research should be done, through the building of models, the testing of materials and forms, and the careful calculation of likely forces and stresses. The design that emerged from the laboratory was radical indeed; it was a box girder of such large dimensions that the railroad ran, not over, but through the girder, actually a tube. This design also allowed Stephenson to construct the bridge entirely before it was actually put into place. Four giant tubes were assembled on the shore of the straits, and then in a spectacular construction feat, they were floated out to the strait, and raised to be attached to the giant stone piers that had been built on either shore and on the Rock (figure 17.5). A smaller bridge of the same design was built to cross the River Conway, some miles up the train line from the straits. This gave the engi-

Figure 17.5
The Britannia tubular bridge over the Menai Straits, Wales; a prefabricated tube is being lifted into place. Courtesy Institution of Civil Engineers, London.

neers an opportunity to experiment with the actual construction and raising techniques. Finally, the Britannia Bridge itself, twin tubes fifteen feet wide, 1,511 feet long, more than a hundred feet above the water, opened to railroad traffic on March 18, 1850. When Paxton's Crystal Palace opened in London just over a year later, Stephenson's achievement was still the benchmark by which British engineers measured their ability to overcome any obstacle.[12]

As bold as Stephenson may have been, he did not long hold the title as Britain's most daring engineer. That would soon pass to his lifelong rival (and admirer) Isambard Kingdom Brunel. It had been the young Brunel, in the 1830s, who had defied the standard gauge of the Stephensons in the design of the seven-foot-wide track of the Great Western Railway, and he never lost his capacity for defying the conventional. While many of the Victorian engineers were exceptionally versatile, tackling everything from locomotive design to tunnels, Brunel was superlative in scope as well as in reach, for he as readily plunged into the building of great ships as he did of great bridges. His most important bridge was probably the one he designed to cross the River Tamar, at Saltash in southwest England. Here Brunel encountered the same sorts of Admiralty restrictions on obstructing navigation that Stephenson had faced, but with the additional difficulty that there was no convenient rock placed in the center of the Tamar channel. He thus had to design novel means of building a central stone pier on bedrock that was almost eighty feet below water and mud. After meeting that challenge, Brunel too had to fabricate his bridge on shore, but he chose not the flat box girder of Stephenson but a sweeping pair of shallow arches, made of wrought iron tubes, oval in section and almost seventeen feet at the widest. The two 465-foot spans were then hung from the iron tubular arches, about a hundred feet above the water. This work took almost seven years to build, not being completed until the year of Brunel's death in 1859. By that time, however, the engineer had become deeply engaged in an even more daring—some said foolhardy—project of a very different kind: the largest ship ever built.

From the time of his youthful work on the Great Western Railway, Brunel had been fascinated with the use of steam to drive large ships, and he saw his first ship, the *Great Western*, as a logical extension, "from Bristol to New York," as he put it, of his railway. That ship, then the world's largest, was followed by a much larger one, the *Great Britain*, which was the first large iron steamer and the first to rely on screw propulsion. This successfully made the run from Liverpool to New York in 1845, but was soon put on the much longer Australian route. This inspired Brunel even further, desiring to design a ship so large it could carry all the fuel it needed for voyage to Australia and back. His *Great Eastern* was arguably the greatest single structure built in the nineteenth century—it was the largest ship built until the twentieth century's great ocean liners, dwarfing anything ever before attempted, or built for many decades after (figure 17.6). At just under 19,000 tons weight and 32,000

Figure 17.6
I. K. Brunel's *Great Eastern* under construction. When it was completed in 1859, its tonnage of 19,000 was almost six times the size of any ship yet built, and it was decades before this was surpassed. Courtesy Institution of Civil Engineers, London.

tons displacement, almost 700 feet long and over 110 feet at its widest, the *Great Eastern* was six times larger than any previous ship. It had a double iron hull and used sail, paddlewheels, and screw propellers to move. Its engines were so large they had the capacity equal to all the cotton mills of Manchester. Among the novel technologies incorporated in the *Great Eastern* were steering gear that was powered by steam and used hydraulic servo-mechanisms that became the standard of 20th-century vessels (this gear was installed in 1867, after the original equipment failed).

Like the challenges of the Britannia Bridge, those of the *Great Eastern* took engineers down important new experimental paths, such as the first modern experimental tanks for testing stability and resistance of structures moving in water. Simply the launching of a ship of this size turned out to be an engineering challenge of enormous proportions, requiring many months and almost driving the promoters into bankruptcy. The great ship was finally floated in January 1858, but there was no money left to outfit it, and so it did not sail until mid-1859. The vicissitudes of the *Great Eastern* did not end there, but the life of its engineer did, for Brunel, exhausted from his efforts and its frustrations and suffering from Bright's Disease, died at the age of fifty-three in September 1859. His ship was a colossal failure as a passenger ship, particularly since within a few years the Suez Canal (completed in 1869) made it superfluous for the Asian and Australian routes, but it played its own crucial role as the greatest cable-laying ship of its time.[13]

As the nineteenth century wore on, the heroic engineer continued to be a figure of importance, not only in Britain but elsewhere. In the United States, it was again the railroad that posed the greatest of early engineering challenges, and it was both in the design and construction of locomotives and in overcoming the obstacles of carrying track across thousands of miles of barely charted territory that the American engineering profession was roused into being. Idiosyncratic inventors like Oliver Evans and Robert Fulton gave way to men whose reputations rested not on flashes of inspiration and novel ideas but on the dogged pursuit of technical competence and excellence. Perhaps the first American to be widely recognized as an "engineer" was John B. Jervis, a farmer's son from upstate New York. As a youth, he worked clearing land for the construction of the Erie Canal, and quickly learned the rudiments of surveying, canal construction, and general engineering. By his mid-thirties his talents and expertise were well known, and he designed the country's first operating railroad and the first American-built locomotive, the Stourbridge Lion. He adapted English locomotive designs for American conditions and economies, devising the "bogie truck" that allowed the front wheels of a locomotive to swivel, thus accommodating the sharper curves of American tracks. His experience, first in canals and then in railroads, set a pattern followed by many in the first generation of American civil engineers. In the first half of the century, American engineering was more notable for its reach than for its daring—the sheer size of the country made the job of crossing it with canals and railroads formidable indeed, and engineers had to focus on economical rather than spectacular works. Large railroad crossings, for example, were much more likely to be achieved by long wooden trestles than by bold new structures. Indeed, perhaps the greatest achievement of the early American engineers was simply getting where they wanted to go with the relatively scarce capital made available to them, so that 30,600 miles of track crossed the eastern part of the continent by 1860. When, in 1869, the first transcontinental railroad was completed with the driving of a spike at Promontory Point, Utah, the feat was a stupendous one, but it was clearly the product of dogged, often conservative, engineering that simply found practical ways around all obstacles rather than seeking to draw attention to itself by novelty and daring.

The next generation of American engineers, however, were to show their own sense of spectacle and radical innovation. James B. Eads, for example, first made

Figure 17.7
James B. Eads's railway and road bridge over the Mississippi River at St. Louis was one of the greatest American structural achievements of the nineteenth century. Courtesy Institution of Civil Engineers, London.

his fortune devising means to salvage sunken steamboats on western rivers. During the Civil War, Abraham Lincoln asked for his help in utilizing the rivers for the war effort, and he responded by quickly building a series of armor-plated gunships. Shortly after the war's end, he took up the challenge that had been avoided by more established civil engineers to build a railroad bridge across the Mississippi at St. Louis. The result was a radical design featuring three arches, each over five hundred feet in span and largely made of steel, a novel and largely untried material for construction (figure 17.7). The Eads Bridge was the first truly significant American contribution to civil engineering design, but not the last.

Figure 17.8
Roebling's Brooklyn Bridge towered over every other structure in New York when it opened in 1883. Library of Congress photograph.

Even before Eads's bridge had been completed, a German immigrant engineer, John A. Roebling, was busy promoting what was to become the most famous of all American bridges, the Brooklyn Bridge (figure 17.8). Unlike Eads, Roebling actually had considerable experience in bridge construction at this point in his life (he was already in his sixties when he began the project). This reputation rested on his use and extension of the suspension bridge form, particularly through the use of iron and later steel cable, rather than the chains that earlier engineers, such as Telford, had used. Suspension bridges at Pittsburgh, Niagara Falls, and Cincinnati demonstrated Roebling's capabilities, beginning in the 1840s. But the challenge of linking Manhattan with the Borough of Brooklyn and Long Island beyond required a structure of much greater daring. When they were completed in 1875, for example, the towers of the Brooklyn Bridge were the largest edifices ever built in North America—276 feet above high water; they loomed over the buildings of America's largest city in a way hardly comprehensible today. The central span of the bridge was 1,595 feet long, and the two side spans were each over 900 feet. It was by any measure a stupendous structure (much longer than any earlier bridge), and more than anything else it marked America's own heroic age of engineering.[14]

18 Stuff, Reality, and Dreams

There exists in our images of the nineteenth century a great divide. On one side, the world and its people are as they had always been in history, seen and remembered through the eyes of artists, poets, and chroniclers, through impressionistic representations of the arrangement, texture, and shape of life. On the other side of this division, the world and memories of it take on a very different character; it is somehow harder, more concrete and real, more precisely rendered in the mind's eye. The source of the divide is no mystery; it is a technology that profoundly reshaped the ways in which people viewed not only the world around them, but also their sense of how that world related both to the past and to the future. That photography transformed our ways of looking at things—and people—is a commonplace, but it is perhaps less widely appreciated that this novelty, from the time it appeared publicly in 1839, was also instrumental in the profound alteration of the European and American comprehension of technology and technological improvement. Nothing did more than photography, for example, to blur the line between science and art, and to identify technology and invention as mediating instruments between these realms. Few other inventions of the early and mid-nineteenth century were so closely identified in the public mind, and in the community of practitioners, with science and the growing sense that nature's secrets were truly at hand for the creation of new devices and powers. Only electrical inventions, themselves just beginning to emerge at photography's birth, wore a similar mantle of philosophical origin. The artistic implications of photography were also immediately apparent to its first viewers: The French historical painter Paul Delaroche was quoted as saying, upon seeing the daguerreotypists' handiwork, "From today, painting is dead!"

The scientific basis of photography was, at one level, quite simple—the discovery of certain materials that were transformed by exposure to light. As early as 1728 the German chemist Johann Heinrich Schulze, in an effort to produce a phosphorescent substance, produced a mixture of chalk, nitric acid, and a small amount of silver salts. Upon exposure to light, Schulze observed that his materials darkened, and he experimented enough to determined that it was a silver compound that was reacting to the light. It was about a half-century later before this phenomenon was investigated further with any seriousness. In 1777, the Swedish chemist Carl Wilhelm Scheele embarked on a series of experiments on silver chloride with the intention of showing that light contained phlogiston, his hypothesized agent of combustion. The darkening of the silver salts, which Scheele proposed was caused by light reducing the salts to metallic silver, seemed to make the point that the Swede aimed at. Over the next couple of decades, experimenters found other photosensitive materials, including several silver compounds, as well as salts of mercury, gold, and potassium. But the idea of using the phenomenon to capture an image permanently did not readily emerge from this work. The other key technical element, besides photosensitive substances, for a true photographic process, was some means to stop the darkening action on the material. Making an image would require control over the light-sensitive surfaces, so that once an image was acquired it could be preserved, and this proved an elusive discovery.[1]

While photochemistry caught the attention of a number of experimenters in the first decades of the nineteenth century, its application to imaging was slow to emerge. Perhaps the most important impetus was provided by the appearance of another imaging technology, lithography. This was the creation of a Munich playwright, Alois Senefelder, at the end of the eighteenth century. Senefelder, apparently disgusted with the unreliability of the printers of his plays, sought an alternative that would allow rapid and simple reproduction of both text and images. A man of considerable talent and imagination, Senefelder discovered how to apply a printable image to the surface of a finely grained stone (*lithos* is Greek for stone) and then to use this for printing. This was the most important new printing technique to appear since Gutenberg, and it quickly caught the imagination of European artists and printers alike. Lithography distinguished itself from other printing techniques by relying not on raised or etched surfaces, but on chemical action. Lithographic stones (and the polished metal plates that replaced them) were flat surfaces, with ink adhering to some parts and not to others, depending on the action of either repelling or absorbing substances applied to the stone or plate to make an image. Senefelder himself referred to this as "chemical printing." It should thus not be too surprising then that early experimenters with the new process should think of other ways of using chemical action to make or transfer an image.

Among these experimenters was an enthusiastic amateur mechanic and inventor, Joseph Nicéphore Niépce, who, with his brother Claude, sought to extend the possibilities of lithography by finding a photosensitive material that would use light in some way to transfer an image to the lithographic stone. This search for what would later be called photolithography was unsuccessful, but it sent the two brothers down a path that led to the first crude form of photography. One of the accepted lithographer's materials was a tarry substance known as bitumen of Judea, and this substance turned out to be photosensitive—it would harden when exposed to light. It also became susceptible to the action of certain solvents, and it was this property that led Niépce to use it to make an image using a camera. The camera obscura was an old device, dating from the sixteenth century or so, which simply used a pinhole in a box to cast an inverted image on a flat surface. It took hours of exposure of the bitumen in bright sunlight to create an image, but by the mid-1820s Niépce had succeeded, and the oldest known photograph is a view, showing rooftops, walls, and trees, from the window of his house near Chalon, in central France, made about 1826. By this time, Niépce's experiments had become more widely known in France, and he had been approached by an clever entrepreneur and showman, Louis Jacques Mandé Daguerre.[2]

Actually, Daguerre would probably have preferred to have been thought of as an artist, and this is how he had begun his career. But his creative talents found other directions, first as a stage designer, responsible for sometimes spectacular stage scenery and effects, and then as the creator and promoter of the "Diorama," a popular entertainment based on large realistic paintings of dramatic outdoor scenes, presented in a special Paris theater in which lighting, smoke, and other effects could be used to provide viewers with a sense of verisimilitude that was startling and thrilling to early-nineteenth-century audiences. It was efforts such as these that put Daguerre on the path of attempts to depict nature with a realism that had hitherto eluded painters. The camera obscura was a familiar tool of painters trying to capture realistic scenes, and Daguerre outfitted his with fine lenses, attempting to sharpen the image from which he was working. At the same time he was attracted to the idea of somehow finding a material that would render the camera obscura's precise depiction permanent. Daguerre's lens supplier informed him that Niépce appeared to have succeeded in doing just this. For several years Daguerre attempted to wheedle the process out of the country experimenter, but, failing this, he then persuaded him to form a partnership, in which Niépce agreed to share his secrets with Daguerre, and in return the showman agreed to back further development and elaboration of the still difficult and crude process.[3]

For several years the two partners exchanged information about their experiments, making slow but unspectacular progress in capturing images through long exposures—typically seven or eight hours. In the course of these experiments, Daguerre began to focus his attention on metal plates coated with silver. When the plates were treated with iodine, they became light-sensitive, and ghostly images could be made out on them, but the images were still difficult to produce or to make sharper. Niépce died in 1832, but Daguerre forged on, and in 1835 he discovered that when his exposed plates were treated with mercury vapor, the images became much clearer. Many more months of experiments were required before Daguerre discovered how to fix his images permanently by washing off the photosensitive chemicals with a salt solution. The result was the daguerreotype, which the inventor announced publicly toward the end of 1838. The images were astonishing to their early viewers—sharp, clear, beautifully toned in black, white, and shades of gray, shining out from the polished silver plates on which they were made (figure 18.1). Well aware of how skeptical most people would be about the notion of such pictures being made, not by artists, but by chemicals and light, Daguerre quickly sought the blessing of France's most prominent scientists, and so it was not Daguerre himself, but François Arago, the director of the Paris Observatory and a major figure in the Académie des Sciences, who made the announcement of Daguerre's discovery to the assembled scientists and the press on January 7, 1839.[4]

Despite the public announcement of the daguerreotype and the sensation that it caused in France and beyond, Daguerre was very secretive about his process for many months, as he negotiated payment from the French government for his invention. His wariness was probably increased by the rapid appearance of other claimants to the invention of photography. The most important of these was an Englishman, William Henry Fox Talbot, who had been experimenting on different processes for almost as long as Daguerre. Fox Talbot was a wealthy amateur scientist and scholar whose interest in photography came from his frustration over a lack of artistic ability. Thinking that perhaps nature, aided by chemistry, could do what his unskilled hands could not, he began in 1833 to experiment with photosensitive solutions on paper. He was the first to comprehend the possibilities of using negative images to create positive photographic copies, but his work was hampered by the lack of an effective means of making his images permanent. Nonetheless, when he read of Arago's announcement of Daguerre's process, he promptly wrote to his friends in England as well as to the French scientists of his own work, which he had been in no hurry to perfect or announce. One of the English scientists who responded to both Daguerre's and Fox Talbot's announcements was the astronomer John Herschel, whose own efforts to duplicate or better the new invention led him within weeks to the discovery of the use of sodium thiosulfate solution (to) to fix photographic images. This is the chemical that photographers still use, typically referring to it as "hypo" (from Herschel's term, "hyposulfite soda"). Herschel also proposed the word "photography" for the new art, as well as the terms "negative" and "positive." Once Fox Talbot had a proper fixative for his images, he was able to develop the negative-positive process to turn out many copies of the splendid landscape and still-life photographs that he made.[5]

After 1839, photography came to be an active and sometimes controversial arena of technical improvement. While initially the world saw, for the most part, two competing processes in the metal plates of Daguerre and the wax paper negatives and salt paper prints of Fox Talbot, soon these processes were joined by numerous others. In addition, photographic apparatus became an active field of experiment and change,

Figure 18.1
Daguerreotype of Michael Faraday. Library of Congress photograph.

from developing solutions to camera lenses. In 1851, an English sculptor, Frederick Scott Archer, introduced a new method of photography in which glass plates were coated with a photosensitive solution in a new substance called collodion. This "wet-plate process" transformed photography from an expensive, slow, and complicated medium into one that was versatile enough to be taken anywhere and used almost anytime. Its most spectacular applications were on the world's battlefields; it saw its first use to record war in the Crimea just a few years after its introduction, and in the 1860s it gave American photographers the means to capture the national agony of the Civil War (figure 18.2). The chemical key to wet plate photography was itself a substance that had come from a chemist's laboratory in the mid-1840s. Collodion was a solution of nitrated cellulose in alcohol and ether. While French chemists had experimented with reacting cotton with nitric and sulfuric acids beginning in the 1830s, it was the Swiss chemist Christian Friedrich Schönbein who announced the discovery of the properties of heavily nitrated cellulose in 1846. The most spectacular property was the violence with which the new material burned. In fact, it could be made into an explosive and became the basis for guncotton. But equally intriguing to chemists was the fact that nitrocellulose would readily dissolve in common organic solvents, yielding a clear viscous liquid that, when dried, produced a transparent pliable film. This film, on glass, became the basis for wet-plate photography. But collodion, as the dissolved nitrocellulose was called, had another future as well, and in this it demonstrated other ways in which chemistry would transform nineteenth-century life.[6]

The rise of mass manufacture in the nineteenth century—the extension of the techniques of both production and marketing that emerged from the pioneering work of Josiah Wedgwood and of the Birmingham small wares makers of the late eighteenth century—encouraged a wide range of efforts to extend and improve the kinds of materials available for making relatively inexpensive commodities. Some of these efforts focused on effecting useful economies in traditional sorts of materials, such as metal alloys, so useful compounds of tin, zinc, copper, and nickel emerged from metallurgical experiments, and enlarged in important, if typically uncelebrated ways, the range of goods affordable to middle-class and even working-class buyers. Others directed their attention to making inexpensive materials more suitable for a wider range of uses or tastes. Machinery for manufacturing wood veneers, for example, gave furniture makers the opportunity to fashion their goods from relatively cheap or imperfect woods and then give their products a finish and look that would appeal to increasingly discriminating and fashion-conscious purchasers. Even more spectacularly, the discovery of electroplating created an entirely new industry for turning base metals into precious-looking ones, and thus helped reshape consumer aspirations towards appearances that mimicked the tastes of the very highest reaches of society. If a cheap copper alloy could, for just a few cents, be made into something

Figure 18.2
Collodion print, Union battery, Yorktown, Virginia. Collodion emulsions were used in "wet-plate" photography, requiring shorter exposures than daguerreotypes and producing negatives from which unlimited numbers of prints could be made. Library of Congress photograph.

that looked like the finest silver, then the look and feel of middle-class life need not be so very different at all from that of the rich.

Alexander Parkes was one of the pioneers of electroplating, clearly reveling in both the commercial and the artistic possibilities of what was not only a new industrial process, but a new medium of expression. Employed as a metallurgist and chemist by the Birmingham firm of Elkington and Mason, Parkes found plenty of opportunities to explore not only the possibilities of electroplating (he once "gilded a lily" just to show that it could be done), but also other means of expanding the material universe of the makers of what was often called "fancy goods"—toiletry articles, costume jewelry, household ornaments, and the like. Beginning sometime in the mid-1850s, Parkes experimented with collodion, intrigued by the "plastic" nature of the filmy layer and inspired to see if he could not produce something more solid and substantial from it. The result was a material he put on display at the London Exhibition of 1862, immodestly dubbed "Parkesine." He made a range of artistic products, such as medallions with the effigy of the late Prince Albert, and explored a number of interesting effects, making his material in imitation of ivory, shell, and other semiprecious materials. In 1865, Parkes published a description of his material, although he was careful to omit key technical details of its composition. Over the next several years Parkes attempted to commercialize his invention, but without success. Parkesine was simply not a very useful material. Neither Parkes nor his partners could figure out what it was really good for, nor did they have a sound idea of how to improve its qualities.[7]

Even though a failure, both technically and commercially, Parkesine deserves historical notice because it was the first recognizable effort to create what we would call a "plastic." Plasticity is a quality shared to one degree or another by many materials, from iron to clay, but the later nineteenth century saw the appearance of a class of substances that distinguished themselves from others by making plasticity the most prominent and prized value. While plasticity literally refers to the ability to be shaped, generally but not necessarily by molding, the so-called plastics extended the notion to include the capacity not only to assume a variety of forms but also to display a range of visual effects, most important in imitation of other, more familiar, substances. The first material that was a commercial success at exploiting these properties was celluloid, an improvement of Parkesine invented by an American mechanic, John Wesley Hyatt. It is not in fact clear that Hyatt knew about Parkesine when he set out in the mid-1860s to make a material that could substitute for ivory in billiard balls, but it is evident that Hyatt traversed much the same path that Parkes had, but went just one or two crucial steps further. In particular, Hyatt learned the usefulness of adding camphor to his nitrocellulose mixture, which added a dimensional stability to the product that resulted when the volatile solvents evaporated.

Hyatt's celluloid was not a successful billiard ball material, but in a pattern that repeated itself often in the development of novel materials, its properties were intriguing and attractive enough to invite experimentation, both technical and commercial, to develop other applications. This development stage for novel materials is often convoluted and uncertain, for the precise mix of properties, price, and novelty that the market will accept is a notoriously difficult thing to predict. Typically, failures in both applications and markets must be sustained before success emerges. In the case of celluloid, failures went from billiard balls to dental plates before Hyatt and his companions began making collars, cuffs, combs, toys, and other small goods that slowly but steadily found markets, in both North America and Europe, through the 1870s. The most important property of celluloid was its capacity to be made into faithful facsimiles of a great range of valued substances, including not only ivory and tortoiseshell, but also coral, mother-of-pearl, and even, in collars and cuffs, starched linen. By the last decades of the nineteenth century, celluloid, and a host of imitators and competitors, in both North America and Europe, had insinuated themselves into bourgeois life, recognized means for giving domestic existence a fashionableness that would otherwise be out of reach.[8]

In a pattern discernable at least since Wedgwood's day, the expansion of production, the lowering of prices, and the subsequent extension of markets into new segments of society was often accompanied by a growing concern for fashion. This affected a great range of commodities and a wide spectrum of markets, from fine china and furniture to celluloid collars and factory-made clothing. Some remarkable technical achievements accompanied this extended reach of fashionableness, none more remarkable than the new dyes that transformed the look of Victorian material culture. Before the middle of the nineteenth century, cloth was dyed using animal or vegetable products, and the processes for coloring cloth were the product of centuries of empirical experience and experimentation. As the textile industry exploded in size and complexity through industrial mechanization, the varieties of cloth and the range of markets similarly expanded, placing new demands on manufacturers for novelty and distinction. Ever more sophisticated power looms made it possible to make novel weaves and textures cheaply, and the mechanization of calico printing made patterned cloth commonplace. Experiments with coloring, however, yielded relatively few useful results, however, until academic chemists became directly engaged in the search for kinds and sources of dyes.

The emergence of the artificial dye industry from the late 1850s was the most dramatic and important example of the new capacity of chemical science to give shape, direction, and momentum to technological improvement. Where the appearance of photography and the first plastics at about the same time made important use of new chemical knowledge, these efforts were largely carried through by non-chemists, by educated amateurs and well-informed craftsmen and mechanics. But the transformation of the dye industry, which was followed by the development of a host of new and artificial substances, from aspirin to vanillin, was the first notable instance of entire crafts, centuries old, being displaced by not only the products but also the people from scientific laboratories. This is thus a particularly important episode in the history of improvement, as it convinced large segments of society that the advancement of science necessarily translated into the advancement of technical products and processes. This conviction arose not only from the highly visible nature of the chemists' new products, but also from the speed with which novelty seemed to cascade from the laboratory, once a few key substances and techniques were developed, and the speed with which hundreds of farmers and other traditional dye producers found themselves technically redundant.

William Perkin was a very bright young East Londoner, showing talents both artistic and scientific at an early age. In the early 1850s, barely a teenager, he began dabbling in the new technology of photography, and this stoked a growing interest in chemistry. He thus gained admission in 1853, at the age of fifteen, to the new Royal College of Chemistry in London, where the German-born and trained August Wilhelm Hofmann had been recruited to bring the latest advances in continental scientific education to England. Hofmann was a student of Justus Liebig, whose laboratory at the German university of Giessen established a reputation both as a center of advanced chemical research and as a source of useful knowledge, particularly in agricultural chemistry. Among the subjects explored in Liebig's laboratory was the chemistry of traditional dye-stuffs. Hofmann himself discovered early in his career that an oily substance that could be distilled from coal tar—the smelly waste product of coal gas production—was identical to a substance that could be derived from the key blue dye, indigo. This oil became known as aniline ("anil" being an old term for indigo), and it joined a series of chemical products derived from coal tar, such as benzene and toluene, that organic chemists, with justification, classified as "aromatic" compounds. That these aromatic compounds were found both in coal tar residue and in valuable substances such as dyes and drugs (like quinine) intrigued many chemists, and gave Hofmann, in his new London laboratory, a mission to demonstrate the nature of, and, if possible, make use of, the link. The young Perkin thus found himself in an environment in which the practical possibilities of advanced chemical science were very much an article of faith. With youthful enthusiasm, Perkin set out to make a name for himself, equipping a small laboratory in his parents' home, and tackling a problem that both Liebig and Hofmann had themselves found intractable—the synthesis of an "artificial quinine." Natural quinine, derived from Andean plants, was the only effective antimalarial drug known, and was thus of critical value to a British imperial effort that found itself spread through much of the world's tropics.

Beginning in late 1855, Perkin set about his experiments in an informed and rational way, but no quinine—a white, soluble alkaloid—appeared in his test tubes. Instead, when he reacted a salt of aniline with oxidizing agents (in an attempt to combine two aniline molecules to produce quinine, in what chemists call a "condensation reaction"), he got a black sludge precipitating in the bottom of his reaction flask. When this precipitate was dissolved in alcohol, it yielded an intense purple color. Well acquainted as Perkin and his fellow students were with the widespread interest in dye-stuffs, it is not surprising that he sought to see what happened when he placed a piece of silk in the solution. The result was "a beautiful purple which resisted the light for a long time." This discovery took place in the spring of 1856, and Perkin was enthusiastic but cautious about exploring the possibilities of what he came to call "Tyrian purple," after one of the most famous of ancient colorants. Samples were sent to a Scottish dye firm, which reported back that the material, if it were not too expensive, could be "one of the most valuable that has come out for a very long time," particularly since purple was a much sought-after color, one extremely difficult to produce in silk. Months of work were required to determine if, in fact, the dye could be made economically in larger quantities, and more work still was directed to improving the fastness of the dye, first in silk, then in wool, and finally, in the greatest prize of all, cotton. As it happened, the mid-1850s saw an upsurge in the fashionability of purple goods, using relatively new but somewhat imperfect dyes, known in Britain as "Roman purple" (murexide) and "French purple." The new aniline purple of Perkin & Sons (William Henry's father stepped in to start a factory, and his brother also threw himself into the business) reached the market in late 1857, at just a moment when the rage for what the French dubbed "mauve" (after the purple flower of the mallow) was reaching a fever pitch. The new dye did not create the fashion, but its great economy and technical superiority allowed it to ride the fashion to new heights, so that aniline purple became completely identified with "mauve," and the foundations of a new industry were secured with stunning speed.[9]

The real significance of mauve lay less in its effects on fashion or the textile industry but more in the impetus that it gave to chemists to extend the magic of the artificial and novel as far as they could manage. The aniline-based reactions that Perkin had started with were amenable to all sorts of variations, and a wonderful range of these produced marvelous new colors of their own. A little more than a year after the commercial debut of Perkin's mauve, the manager of a dye works near the great French silk-making center of Lyon, François Emmanuel Verguin, discovered how to produce a brilliant, novel red from aniline. The actual details of Verguin's work remained obscure, but by May 1859, the brothers Renard began manufacturing the red they called "fuchsine" and selling it to the eager Lyon silk makers. French, British, and then German chemists all joined the hunt for new colors and new processes.

Fuchsine was soon joined by "magenta" (named after a battle in June 1859, in which the French had defeated the Austrians) and by an aniline blue that emerged from a Paris laboratory. Much of this early experimentation was largely hit-and-miss, as standard reagents were mixed with aniline and related compounds simply to see what would result. The person who brought a bit of order to this chemical chaos was none other than Perkins's teacher, A. W. Hofmann, who made a systematic study of the reactions involved, and was able, in mid-1863, to describe the compositions of the new dyes with some precision, adding in the process one of his own, a color that came to be known as "Hofmann's violet." Hofmann used his achievement to make explicit calls for the support of pure science, arguing that the new industry, which had appeared and established itself in the key textile centers of Europe in less than five years, was fundamentally dependent on laboratory chemistry informed by the highest levels of theory. His argument was heard widely, but no more persuasively than in his native Germany, which in 1865 called him to the chair of chemistry at the University of Berlin.[10]

While the new dye industry was born in British laboratories and factories and given further impetus by French textile makers, its larger significance lay in the transformation of German industry. Hofmann was, after all, a representative of a tradition in German academic chemistry that already had several decades of distinction behind it, thanks to the contributions of men like Friedrich Wöhler, Robert Bunsen, and Justus Liebig, who lost no opportunities themselves to demonstrate the practical uses of new chemical knowledge. The artificial dyes opened up an entirely new avenue for this demonstration, and German chemists were quick to exploit it. In the same year that Hofmann went to Berlin, the chemistry professor at the small University of Bonn, Friedrich August Kekulé, proposed a structure for the simplest of the aromatic compounds, benzene. His famous ring structure—picturing benzene, known to have the empirical formula C_6H_6, as a ring of six carbon atoms, each with a hydrogen attached outside the ring—was one of the great triumphs of nineteenth-century chemical theory. More complicated aromatic materials could then be pictured, often through substituting other substances, including chains of carbon atoms and even other rings, for the hydrogen atoms of the benzene ring.

The essential reactions of substituting and joining aromatic compounds were mastered best in the German laboratories, nowhere more thoroughly than in that supervised by Adolf Baeyer in Berlin's Gewerbe Institut (Trades Institute). The first triumph of Baeyer's laboratory was the synthesis of a key natural dye, alizarin, in 1868. This material was the essential ingredient of madder, the most important red dye, and its successful production in the laboratory was the first case of a known colorant being synthesized from mineral sources. Considerable work was required to make artificial alizarin cheaply, but once the processes were developed, the traditional agricultural sources—acres and acres of farmland dedicated to madder production—were quickly made obsolete. None other than William Henry Perkin, mauve's discoverer, developed his own approach to making artificial alizarin, so that the British were able to keep up with the Germans for a few years longer, but by the end of the 1870s, German dominance of the synthetic dye industry was widely apparent. The laboratories of Baeyer and Hofmann in Berlin, and those of German firms such as the Badische Anilin und Soda Fabrick (BASF), exploited chemical discoveries and possibilities with much greater efficiency than anyone else. The successful synthesis in 1880 of the key blue dye, indigo, in Baeyer's laboratory and then elsewhere, was perhaps the most spectacular achievement of the new industry. Even more potent for the future, however, was the simultaneous emergence of an even more novel application for the new chemistry—pharmaceuticals. An aniline derivative, salicylic acid, was determined in 1873 to have value as a pain killer, and this discovery spurred other research into possible therapeutic materials. The most spectacular success came at the end of the century, when a variation of salicylic acid was brought to market as "aspirin."[11]

Pain relief was only the beginning of the new chemistry's contributions to medicine. Louis Pasteur was a brilliant French chemist working in the northern city of Lille when he turned his attention to the mechanism of fermentation, the biochemical process indispensable to the manufacture of beer, wine, bread, and cheese, among other commodities. Interest in the microorganisms that he found to be the agents of the process turned Pasteur to the study of disease. His success in 1871 in halting a damaging disease of silkworms that threatened Lyon's ancient silkmaking industry led to other efforts, and in 1881 he was able to halt an outbreak of anthrax in sheep, and four years later he performed his most famous feat, treating a young boy that had been bitten by a rabid dog. That specific chemical substances could be found for attacking bacterial diseases was even more emphatically demonstrated by Robert Koch, who first showed how aniline dyes could be used to identify specific bacteria. Working in Berlin in 1882, he isolated the bacterium that caused tuberculosis. While his own efforts to cure the disease proved largely ineffectual, he steered the way for his onetime assistant, Paul Ehrlich, to identify artificial dyes that were effectively bactericidal, and beginning in 1891 Ehrlich found chemical therapies for malaria, sleeping sickness, and syphilis. The arsenic-benzene compound he discovered for use against this last disease was introduced commercially in 1909 as "Salvarsan," and it became the model for wonder drugs that would transform medicine in the new century.[12]

Chemistry changed in the nineteenth century from a relatively straightforward investigation of the constitution and behavior of the material world into the core of an active campaign to transform and improve upon the materials of nature. The models presented by the makers of plastics and of dyes was followed by others. The perfumers found in the laboratory copies of the essential fragrances of their craft (coumarin in 1875; citronellal in 1889; and varieties of musk oil beginning in 1888). Food and drink purveyors did likewise with flavorings like vanillin (1879) and sweeteners like saccharin (1879). The latter became an important industrial commodity—one of the first American discoveries in this area and the initial product of the company that became Monsanto in 1902. Artificial fibers joined the ancient list of natural ones when nitrocellulose fibers were introduced in the 1880s. These were only novelties, however, due to both their flammability and their cost. Only after the turn of the century did several forms of "artificial silk," most particularly viscose and rayon, make their way to market, slowly altering tastes in cloth and clothing. At about the same time, a new class of plastics emerged that owed more to the laboratory tradition of the synthetic chemical industry than to the mechanical tradition that had spawned celluloid. The most prominent example of the class was Bakelite, invented by Belgian-born chemist Leo Baekeland in 1907 and introduced in the next decade as an artificial resin, a substitute for shellac and similar insulating compounds. The material was a product of reacting phenol, one of the basic materials of the coal tar chemists, with formaldehyde, a very simple hydrocarbon, under pressure. Bakelite turned out to be a remarkably stable and versatile material, although lacking some of the imitative possibilities for which celluloid had been noted. It thus became better known as a utilitarian material, soon indispensable where good heat or electrical insulation was needed. It also could be used as a liquid resin or binder, and thus helped usher in an important class of laminates and composites, eventually including such well-known twentieth-century staples as formica and plywood.[13]

By the early twentieth century, the development and use of new materials had come to be widely perceived as a mark of modernity. While the chemists were seen as the heroic figures in this regard, their contributions were as yet modest in the bulkier, more obvious elements of material life, in technology's big things. And yet here too there had been great changes and much novelty, but these were still largely the product of empirically derived improvement, of the experiments of craftsmen and enthusiastic amateurs rather than the laboratories of the chemists. The spreading importance of large machines and engines and the coming of the railroad, in particular, made iron and steel vastly more prevalent in the Victorian world than before. And newer metals, either useful new alloys or other novel combinations of well-known metals or completely new substances, some of them unknown before the nineteenth century, appeared and spread through the course of the century, giving the worlds both of industry and of everyday life a very different feel, texture, and weight.

No single material became most closely identified with material progress in the nineteenth century than iron. Even before the century's beginning, the quantities and uses of iron expanded significantly, thanks to the innovations of the coke smelters and others, primarily in Britain. In addition to iron bridges, aqueducts, and a host of smaller goods, the steam engine itself played a remarkable role both as producer and user of iron. While the engines of Newcomen and Watt had used iron only where they had to, in their boilers, cyinders, and pistons, for example, the expansion of steam engine technology in the early nineteenth century and, particularly, the coming of high pressure engines, reduced wood and other nonferrous materials to negligible amounts in the new engines. At the same time, the growing availability and power of steam engines in factories enhanced the ability of iron founders and the workers of the metal to shape it in greater quantities, speed, and economy. Giant rolling mills, for example, became possible with the installation of heavy-duty steam engines, and they were used for the production of prodigious quantities of sheet, plate, and, above all, rails.

The railroad, more than anything else, made the nineteenth century an age of iron in both Europe and America. The "Iron Horse"—the locomotive—became larger and larger as the years went on, until by century's end, the largest of them were over ninety tons (the largest locomotives, built later in the twentieth century, weighed some six hundred tons). An even greater consumer of iron, however, was the thousands of miles of track built to carry railroads across the world's continents in the course of the nineteenth century. To supply the insatiable appetite of the railroad promoters (except, of course, during the periodic busts that punctuated the continuing railroad booms) for iron rail, ironmakers devoted themselves to a new scale of manufacture and fabrication. In the spate of construction that followed the end of the American Civil War, iron and steel rail consumption in the United States exceeded at one point 1.5 million tons per year; in the next boom, which peaked in the early 1880s, annual consumption reached close to 2.2 million tons of rail. The industrial structure required to supply such a demand dwarfed anything that had been seen in any industry earlier. Entire cities, such as Pittsburgh and Lehigh, Pennsylvania, grew large and prosperous (and utterly dependent on one industry) during the boom years. An industry of such scale and importance naturally drew much attention to efforts at improvement. This was largely incremental, as ironmakers struggled with the very specific and immediate challenges of increasing output, reliability, and economy. The railroad itself became a key instrument in increasing scale. Large capital investment in ironmaking and in large machinery and furnaces could be supported by a transportation system that could reliably move raw materials (iron ore and, even in the wood-rich United States by midcentury, coal) to furnaces and workable metal to factories as well as finished products to final consumers. The simple increase in the scale of iron production in the first two-thirds of the century was prodigious and already made the era an age of iron.[14]

The most important single event in the great transformation of the nineteenth-century material world was the introduction and spread of cheap steel. The name most closely associated with this is that of Henry Bessemer, an ambitious English artisan who, before he was twenty-five, began an inventive career that extended over a wide range of fields and interests, from electrometallurgy and metal casting to textile machinery, paints, pumps, and projectiles. This latter interest was spurred further by the outbreak of the Crimean War in 1854. The lack of preparation of the British forces was widely reported, and Bessemer turned his inventive hand to the effort, proposing new kinds of ordnance for the troops. When he was told that the untrustworthy iron of the guns made his new shells dangerous to fire, Bessemer turned his attention to the qualities and manufacture of ordnance iron. At this time, the cast iron product of a blast furnace was typically converted to the desired wrought iron through the laborious and fuel-hungry process of puddling. In the course of his experiments, Bessemer observed that adding air to a batch of cast iron in a furnace appeared to decarburize part of the batch, without the labor or extra heat of puddling. Following up this observation with experiments, he devised a crucible with molten iron in which he was able to blow air directly into the melt. This did indeed convert the pig iron into wrought, but still consumed a great deal of fuel. Devising yet another container for the molten iron, this with openings that allowed air to be blown in at the bottom, Bessemer discovered the air itself was capable, without additional fuel, of producing a violent heat in the molten iron (figure 18.3). After this spontaneous fire had burned itself out, a relatively pure wrought, decarburized, iron was left in a semi-molten state in the bottom of the container. Controlling this violent reaction and figuring out processes and mechanisms to control the materials and extract the products required many more months of work, but by the middle of 1856, Bessemer was ready to announce his discoveries publicly, which he did with a widely-reported presentation to the British Association for the Advancement of Science, in August 1856, on "The Manufacture of Malleable Iron and Steel Without Fuel." Within weeks, British ironmakers were experimenting with Bessemer's process, aiming to cut costs drastically with his methods. In the United States, at about the same time, a Kentucky ironmaker, William Kelly, pursued experiments in secret on a very similar process. His work had little real influence on later developments, except to cloud Bessemer's patent rights in the United States.[15]

The Bessemer process, as it came to be called (even in the United States), turned out to be much more difficult to exploit than its discover first promised. There were a number of problems, the most important of which was the fact that pig iron containing phosphorous was useless in the process, yielding a brittle, unworkable metal. Since most British iron ores are phosphoric, this was a near-fatal flaw for native producers, and users of the process, including Bessemer himself, had to resort to importing pig iron from Sweden and other places with ores of a very different character. This obviously undercut the promised economies of the innovation. The difficulties, however, spurred on a host of systematic attempts, in both Europe and America, to devise alternatives, processes that would still save on fuel but would allow the use of a much wider range of ores. Just as significantly, the desired processes would be

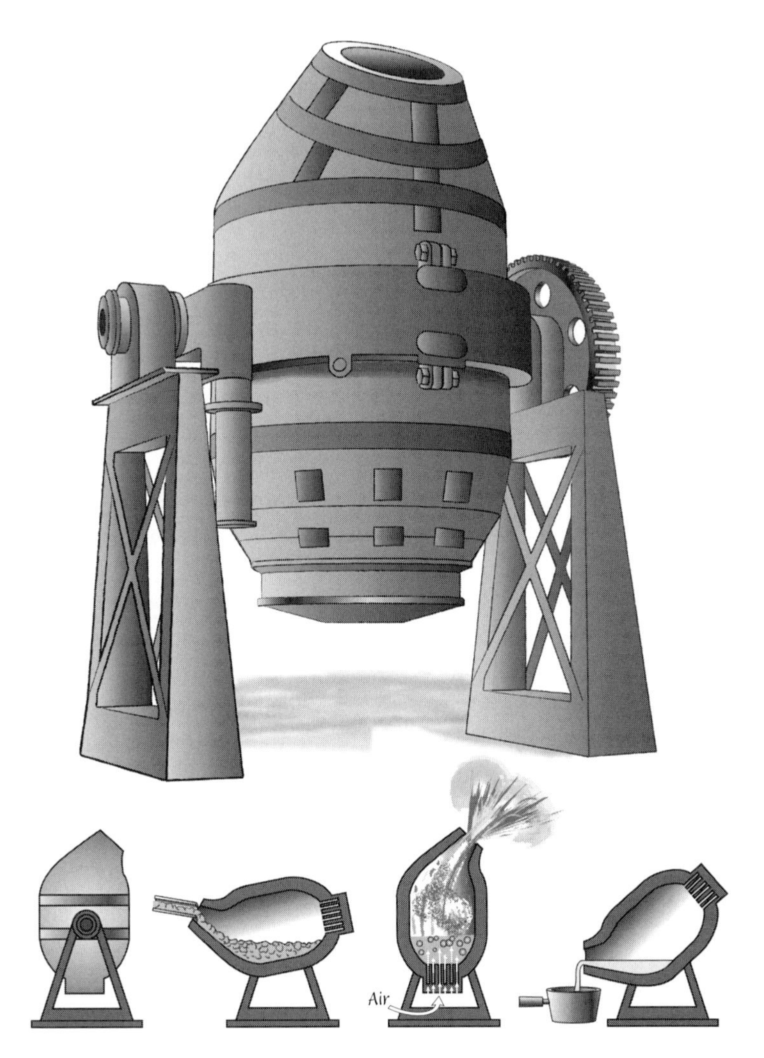

Figure 18.3
The Bessemer converter permitted the production of large quantities of relatively cheap steel. Air was blown though the bottom and the vessel was then tilted to pour out the liquid metal.

applicable to the production of both wrought (or "malleable") iron and steel. This latter was something of an afterthought for Bessemer, but he early discovered that it was possible to manage the conversion process sufficiently that the decarburization of the pig iron stopped short of total elimination of carbon, and left just enough in the finished product to give it the qualities of steel. Steel differs from other forms of iron by containing a moderate amount of carbon—too little carbon and the material is wrought iron, too much and it is cast or pig iron. Up to this time, steel had been a much valued metal, but one notoriously difficult and expensive to make. It was the product of labor- and fuel-intensive processes, the chemistry of which was little understood. Small batches were made by skilled craftsmen, and the product was highly valued in uses that required a combination of hardness, toughness, and resilience, as in blades or machine dies. The Bessemer process provided the first means for making steel in large quantities, and slowly the value of this became evident.

The last four decades of the nineteenth century witnessed the transformation of the ancient craft of making iron—characterized most fundamentally by the ascendancy of steel as the dominant metal of Western life. After its introduction, many improvements were made to the Bessemer process, involving the addition of other materials to the iron or to the converter, thus improving the qualities of the product and the range of materials that could be used. An alternative means of making steel on a large scale was soon developed, the so-called open hearth. This depended on the development of new furnaces, which were able to use hot gases in the combustion chambers to create higher heats than iron or steel makers had ever had available. This "regenerative furnace" was introduced by William and Frederick Siemens, but it was another pair of brothers, Pierre and Emil Martin, in France, who used it, in 1863, to make steel by simply using the high heats to melt and combine pig iron (with its high carbon content) with wrought iron (with its negligible amount of carbon) to make a steel with whatever carbon content was desired for a particular use. While the open hearth used more fuel than the Bessemer process, it could use cheap, low grade coal, and, more importantly, it could easily be fed cheap scrap iron. Its relative slowness was actually an advantage, since it allowed much greater control over the furnace's product. For two decades or so, the Bessemer and Siemens-Martin processes competed in the world's iron and steel districts. They were joined in the late 1870s by the "basic process" devised by Sidney Thomas and his cousin, Percy Gilchrist. By adding limestone to the lining of the converter (and to the charge of iron itself), they solved the problem of using phosphoric and other low-grade ores. Many districts by-passed by the earlier technical revolutions in iron making were at once given their own successful process, and the steel revolution spread around the world.

From the vantage point of more than a century afterward, it is difficult fully to appreciate what a difference steel meant to the world's economies. The first place this became apparent was in railroads, where the greater durability of steel rails over iron ones made railroad lines that would have been uneconomical due to maintenance requirements suddenly feasible. The coming of steel led to the greatest of the railroad building booms; almost thirteen thousand miles of track were laid in the United States in the year 1887 alone. The extension of rail lines fed supply as well as demand in the steel industry. Birmingham, Alabama, for example, which became the iron and steel manufacturing center of the southern United States, grew from a small valley town into a "magic city" in only a couple of decades after the Alabama & Chattanooga and Louisville & Nashville railroads linked it to supplies of fuel and ore as well as to markets in the early 1880s, and this phenomenon was repeated on various scales from Germany's Ruhr district to Tietsin, China. The railroads, in fact, overextended themselves in many cases, but this too was the product of cheap steel as much as anything else. Just as dramatic, if less widespread, was the entry of steel into all kinds of construction, from dramatic bridges, such as the great arch bridge over the Mississippi at St. Louis designed by James B. Eads and completed in 1874, to the skyscrapers that began to give a new form to American cityscapes by century's end. Steel truss bridges made railroad and road construction more rapid, economical, and safe. From the 1870s, iron and steel reinforced concrete joined the options of engineers and architects, and dramatic new building forms began slowly to emerge. While experiments with iron frames began in mid-century, as in James Bogardus's New York City factory building, the real possibilities of large iron-frame structures began to emerge clearly only with the reconstruction of Chicago after the great fire of 1871. The ten-story Home Insurance Building building designed by William LeBaron Jenney in 1884–85 is sometimes called the first skyscraper (159 feet high), as it incorporated a number of key features, each of which had been experimented with in other structures. It had, for example, elevators—a real necessity for taller buildings (these were hydraulic lifts; electrically powered ones came later). By 1891 Chicago had six buildings of ten to fourteen stories in height, made with iron or steel skeletons, and the United States had made its most distinctive contribution to architecture, based largely on the use of steel.[16]

In the twentieth century, the economic role of steel expanded even further, and its technical development accompanied and assisted this. In the last years of the previous century, new kinds of steel appeared that allowed machines and machine tools to work with unparalleled speed and efficiency. The high-speed tool steels, made by alloying the metal with novel additives such as chromium and tungsten and then carefully heat treating the result to increase hardness and durability, made possible machining, stamping, drawing, and other operations at rates that manufacturers had never before imagined. In the hands of visionaries like Frederick Winslow Taylor, one of the new steel's developers with novel concepts of a "scientific" management of work and production, the material began to effect slow but profound change in industrial practices. Other alloys, such as vanadium steel, and combinations with chromium and molybdenum, produced metals that turned out to be crucial to the development of such key twentieth-century industries as the automobile and aviation. The new steels also established the metallurgical laboratory as one of the key agents of technological improvement in the new century.[17]

Metallurgical novelty did not end with steel however. Despite the key role that steel, in all its ever-expanding forms, would play throughout the twentieth century, it never possessed the air of innovation and modernity that surrounded even more suprising new metals in the last decades of the nineteenth century. The most important of these was aluminum, which had never even been seen at the century's beginning. For some decades, chemists had endeavored to derive the elemental base of two familiar and related substances, alum or alumina, but without success. In 1808, Humphry Davy, working with the huge new battery he had installed in the basement of the Royal Institution in London, was able to isolate the new light metals, sodium and potassium, through electrolytic action. He thought the same process should work for the sought-after material, which he dubbed, before ever seeing it, aluminum (others, thinking this was not completely consistent with the terms just coined for the other new metals, adopted "aluminium" instead). Davy was never able to isolate pure aluminum, but in the 1820s, chemists found nonelectrical means of getting somewhat impure examples. One of these was Friedrich Wöhler, in Germany, who resumed his experiments with some new techniques almost twenty years later, and in 1845 was able to announce that he had made pure globules of the metal. He was the first person to be able to determine its properties, and he remarked particularly on its astonishingly light weight. Chemists were aware that the sources of the new, elusive metal were actually very common—ordinary clay contains alumina (the metal's oxide) or other aluminum compounds, and some minerals, such as bauxite, could be substantial sources of relately pure alumina. Spurred by the combination of seemingly wonderful properties and the prospect of abundant sources, a few chemists, from the 1850s, devoted great efforts to determining how to make the metal in quantity.

In 1853, Henri St. Claire Deville, in Paris, improved on Wöhler's process and made some sizable nuggets of aluminum. This inspired him to seek out help to commercialize the metal, as he was certain that the novel properties, particularly its light weight and its apparent resistence to corrosion, combined with the availability of ores, promised to make aluminum one of the key metals of commerce, if only economical processes for making and working it could be found. Support for his venture was not difficult to come by, particularly after he had the blessing of Emperor Napoleon III, who was swayed by naive notions of the military possibilities of the metal (such as lighter-weight helmets for his soldiers). Deville was in fact able to convert his laboratory process into one usable in a small factory and to reduce the price of the metal to about twenty-seven dollars per pound—not cheap, to be sure, but still

DECEMBER 20, 1884. HARPER'S WEEKLY. 839

THE LATE GENERAL GERSHOM MOTT.
Photographed by J. E. Smith.

COLONEL THOMAS L. CASEY, GOVERNMENT ENGINEER OF THE WASHINGTON MONUMENT.—Photographed by Bell.

THE LATE REUBEN R. SPRINGER.

B. R. Green, Civil Engineer. Capt. G. W. Davis, Engineer. P. H. McLaughlin, Superintendent. Col. T. L. Casey, Government Engineer. James Hogan, Rigger. Lewis O'Brien (colored).

FEET HIGH—SETTING THE CAP-STONE ON THE WASHINGTON MONUMENT.—From a Sketch on the Spot by S. H. Nealy.

Figure 18.4
Placing the aluminum cap on the apex of the Washington Monument, 1884. Aluminum was still a precious metal at this point. From *Harper's Weekly*, December 20, 1884.

affordable to any who might have special uses. He quickly, however, ran into the problem that aluminum was simply too novel—it was not, in fact, a material that could sell itself by dint of its properties, but needed to be promoted while at the same time made more cheaply. Deville attempted to imagine, describe, and try out every application possible, from the arms for precision scientific balances to opera glasses and mustard spoons. For the next several decades, aluminum remained available for these applications, but its use was typically measured in ounces or grams, rather than tons. It was perceived as a metal for jewelry or fancy goods, but not for everyday life (figure 18.4).[18]

For almost three decades, the French makers of the metal had the field largely to themselves. By the 1880s, however, the challenge of making aluminum into a cheap metal was taken up by others, and in 1886, in a remarkable example of simultanenous discovery, a novel process that took the metal back to its electrical roots transformed the economic, and thus the technical and social, status of the metal. In France, Paul Héroult, a twenty-two-year-old amateur chemist, discovered that when alumina was dissolved in molten cryolite, an aluminum-based mineral that was found in Greenland, pure aluminum metal could be extracted from the mixture by a strong electric current. The same discovery was made in Oberlin, Ohio, by Charles Martin Hall, likewise twenty-two years old and fresh out of college. Working in the family toolshed, Hall doggedly worked on various possible means of using electricity to separate aluminum from its oxide, convinced that the discoverer of such a process would gain wealth and fame. While it was not that difficult for Héroult and for Hall to get backing to exploit their electrolytic process, it was a task of some size to develop markets for the new metal. The world, in fact, did not have many obvious uses for a light metal, and the applications that had long been satisfied by the expensive French processes remained for some time the primary market for aluminum. The metal remained a rather minor part of the scheme of things until very novel sources of demands, the most important of which was aviation, emerged in the twentieth century. But from the outset, aluminum, which promised to be as cheap and versatile as iron (although it never was), was seen as the metal of modernity, a representation of the newfound capacity to make the world over in completely novel ways.

19 The Improvement of Violence

Sometime shortly before dawn in the morning of September 7, 1776, a loud thump sounded against the hull of the sixty-four-gun frigate *Eagle*, anchored in New York harbor, not far from where the Statue of Liberty now stands. If any of the British sailors aboard took particular notice, they raised no alarm. Beneath the ship, under the mild late summer waters of New York Bay, Sergeant Ezra Lee, of Old Lyme, Connecticut, worked feverishly in pitch black at his task, which was, simply, to blow the *Eagle* to smithereens. In this effort he was assisted by perhaps the strangest war vessel ever to take to the seas, the world's first fully operational submarine. Designed by David Bushnell, a Yale graduate who had long been fascinated with the possibilities of underwater weapons, the *Turtle*, as it was usually referred to, was an awkward craft, with just enough room for its pilot, who was kept busy maneuvering it with sets of pedals and hand-operated propellers and only about thirty-minutes of air (figure 19.1). The Americans had waited at New York for almost two months, looking for an opportunity to try out the new weapon, which they hoped would effectively pierce the British blockade of their ports. Unfortunately, Lee's plan, which was to attach a 250-pound gunpowder charge and a timer fuse to the bottom of *Eagle's* hull, was thwarted when the augur that he was using to penetrate the hull hit a metal strip. After repeating his attempts, Lee, running out of air, had to give up. The *Turtle* bobbed to the surface in the breaking dawn, where it was spotted by a British lookout. Lee jettisoned his charge and hastened a retreat to the American-held position at Manhattan's Battery. According to some reports, the charge blew up, damaging nothing except British equanimity, which was enough to cause their commanders to lift anchor and make their way farther out into the bay.[1]

David Bushnell's *Turtle*, while it worked, was a failure. At least a couple more efforts against British frigates were also frustrated, largely by the tides and currents of New York Bay, and it is unknown what actually happened to the vessel after these attempts. Bushnell himself eventually gave up on the *Turtle* and turned his attention to improving the underwater explosives—which he sometimes referred to as "torpedoes"—that had sparked his original interest (although, again, to no

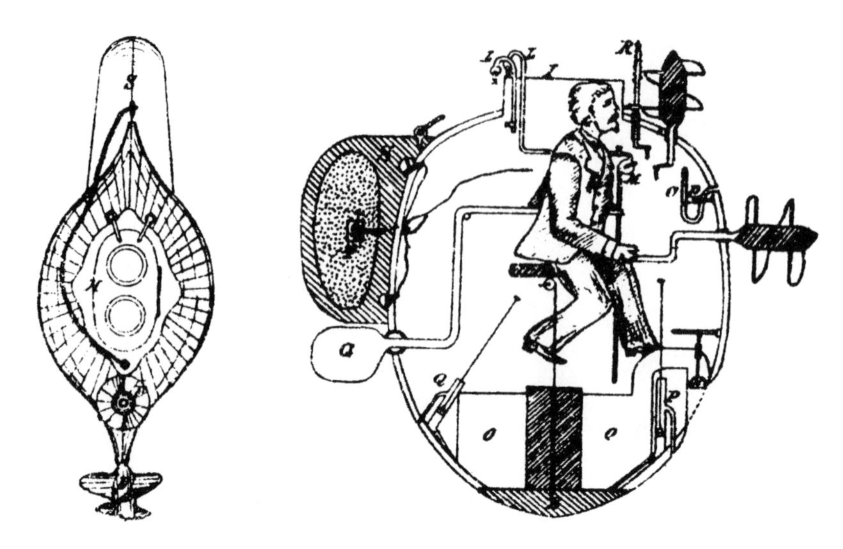

Figure 19.1
This drawing of Bushnell's *Turtle* was prepared by Lt. Francis Barber, USN, in 1875 to illustrate a lecture on submarines. It is a nineteenth-century imagining of an eighteenth-century craft. Courtesy Chief of Naval Operations, Submarine Warfare Division, U.S. Navy.

discernible effect on British ships). His craft was certainly not the first effort at making a submersible vessel; these went back several centuries at least. It was not even, arguably, the first such vessel to demonstrate the practicability of the idea. The honor of constructing that probably belongs to Cornelius Drebbel, a Dutchman working in the service of the English crown in the first decades of the seventeenth century. There are no full descriptions of Drebbel's vessel, however, but only allusions to its purported use. Other devices were similarly designed and described over the next century and a half, but it was Bushnell's *Turtle* that demonstrated that technology, by the late eighteenth century, was finally up to the ambitions of underwater warriors. Part of its significance, therefore, lies simply in being the first realization of what was to become, in the ensuing century, an important development in naval warfare, until the submarine emerged in the First World War as a weapon of considerable importance.

It is not difficult, however, to see more than this in the *Turtle*. While from one perspective, it was simply one in a very long line of attempts by the wagers of war to gain advantage through new machinery, it could be argued that, in the light of the two centuries following that early morning in New York Harbor, Bushnell represented just a hint of something new in the improvement of violence. The great wooden horse used by the Greeks at the walls of Troy, the giant mirrors designed by Archimedes for the defense of his native Syracuse, and the various machines dreamed up and described by the Renaissance engineers were all precedents for this awkward little submarine. But these were, in fact, little more than dreams and myths and onetime wonders. When, just a decade before the American Revolution and Bushnell's experiments, an anonymous English writer penned the imagined account of "The Reign of George VI, 1900–1925," he could not conceive of change on a modern scale. In the description, for example, of the destruction of the Russian fleet in Stockholm harbor in an imagined 1920, the instrument of conquest was the fireship: "On a dark night [the British commander] sent in six fire-ships among their squadron; eleven ships of the line were burnt and seven frigates, four sunk and seven taken." After the wars at the end of the 18th century, after Bushnell's submarine and the balloons of the Montgolfiers and Charles, it would no longer be possible to sustain such an image of unchanging weaponry and military capability. The *Turtle* is a convenient point to recognize that, as the character of technological change itself changed in the last years of the eighteenth century, so too changed the relationship between the waging of war and the pursuit of improvement. Not all at once, by any means, but nonetheless bit by bit the warmakers came to see that their capabilities and strengths, and ultimately their fates, were inextricably linked to the novel weapons and other tools that emerged from factories, laboratories, and inventors' workshops.[2]

The sources of military improvement were myriad. Bushnell represented one kind—the independent, entrepreneurial inventor who saw military needs as a set of opportunities that would open the doors to fame and fortune. While this source is well worth exploring and understanding, and certainly gained notoriety in the nineteenth century, it was not in fact particularly important relative to others. Closely related, but distinct and more significant, were inventors and engineers who were, in fact, not independent agents, but who functioned as technological entrepreneurs within the military establishment itself. Military engineers, gunnery or ordnance officers, armory workers or managers—such men sometimes found themselves in the position where important breaks with past practice seemed urgent and advantageous, and some had the intelligence and the drive to overcome the natural conservative instincts of the military command to effect change. As the nineteenth century went on, many of the agents of change were institutionalized: schools, arsenals, and shipyards undertook not only their traditional functions, but also the roles of innovators, experimenters, and testers. The institutionalization extended to forge linkages between the military and naval establishments and growing corporate and industrial interests. The "military-industrial complex" was not, in fact, an invention of the twentieth century, but was readily recognizable and quite powerful by the century's beginning.

Much change, also, was effected by adaptations of and adjustments to other elements of the technological world that were transformed in the period. The greatly expanded capacity to produce goods in quantity, particularly goods of uniform quality and character, enlarged the scale of warfare beyond anything known before the nineteenth century. This material expansion, it should be pointed out, supported but did not cause the social expansion of war, in which much fighting was done not by a few thousand professional or semiprofessional soldiers carefully moved into battle like pieces on a chessboard, but by hundreds of thousands of largely conscripted soldiers that were arrayed against one another in great masses. This change is most closely associated with the French Revolution. To defend the new republic, the revolutionaries called for a "*levée en masse*," a general conscription of the entire population to national service, including putting all able-bodied single men between eighteen and twenty-five under arms. Another important underpinning for this new kind of warfare was the astonishing change in transport capabilities, particularly in the first half of the nineteenth century. Whereas Napoleon's armies traveled in a fashion that would have been familiar to Julius Caesar almost two thousand years earlier, the troops of Generals Grant or Sherman in the American Civil War were supported by a network of railways that, even when they didn't move troops, completely changed calculations about supply lines. A similar transformation took place at sea: the reliability and speed of steam transport slowly changed long habits of naval battle, but more rapidly overturned assumptions about the movement of men and supplies over the seas.

In the decades following Bushnell's *Turtle*, Europe underwent almost twenty years of protracted warfare. In some ways the character of the French Revolutionary and the Napoleonic wars was traditional—the last great preindustrial warfare in the West, with horse and sail providing the only real assistance to foot or oar, with musketry and artillery that made battle a fairly short-range affair, in which soldiers and sailors could still see clearly the faces of their enemies, until hopelessly obscured by smoke. After 1815, general warfare disappeared from the European continent for almost a hundred years, interrupted only by relatively brief and confined clashes between countries jockeying for some specific advantage in the great power rivalries. For the first half of this century of relative peace, changes in military technology or practice were generally modest, ordinarily confined to improvements in specific capabilities, from the accuracy of firearms to the composition of ammunition. During the second half of the period, however, from about 1865 to 1914, change was rapid and profound, extending through all elements of warfare from rapid fire guns to gigantic battleships. In few areas of society did the power of the culture of improvement make itself more deeply felt than in the enormous efforts and wealth expended on the means of waging war by the Western powers in the decades before the First World War. The tragedy and even irony of this fact was not lost on contemporary observers, as they remarked on the puzzle presented by a civilization that considered itself the most moral and "advanced" in human experience arming itself with powers for killing that had been unimaginable just a couple of generations earlier. While this puzzle has many parts, at least some of it becomes more comprehensible by recognizing the hold that the culture of improvement had on the Western mind by the middle of the nineteenth century.

Remarkably, perhaps the clearest early expression of the importance of this for the experience of the West came from the man who followed closely in Bushnell's footsteps, finding the lure of underwater warfare irresistible and convinced he knew how to effect it, Robert Fulton. Frustrated at mediocre success as a painter, studying in London under the expatriate American Benjamin West in the 1790s, Fulton turned his attention to engineering, initially directing his enthusiasm for schemes improving the design of canals (it was the middle of Britain's canal boom). When he stopped off in France for what was envisioned as a few months in the middle of 1797, Fulton became completely captivated, for reasons never made clear, by the vision of a submarine. He spent the next seven years in France, largely in pursuit of government support for his *Nautilus* (figure 19.2). He got enough to build the device—essentially an enlarged and more sophisticated version of the *Turtle*—and he tried it out several times on blockading British vessels, but with no better results than Bushnell. He struggled mightily to convince the French authorities, which, by the end of his stay largely meant Napoleon himself, to support further development and to pay him for additional efforts. Finally, however, Fulton gave up on the French and, in May 1804 arrived in London with the intention of selling his invention to their mortal enemies. From a technical point of view, Fulton enjoyed no more success in England than he had in France, although he made out much better financially, becoming, in fact, wealthy on his contracts with the British government.[3]

More significant for our story, however, are the arguments that Fulton put forth in his British negotiations, for here we can see particularly clearly the emerging role of technical improvement in the way that warfare came to be viewed in the new century. Early in his stay in England, for example, Fulton drafted a long letter (apparently never sent) to Prime Minister William Pitt. He argued for support of the *Nautilus* as a demonstration of how "Science in its progress towards the improvement of Society has now begun the destruction of military navies." His submarine would be capable of "putting an end to maritime wars with all the dreadful catalogue of crimes which they entrain." Not for the last time in the nineteenth century do politicians hear the claim that a new weapon will make war too horrible to contemplate. A couple of years later, still pressing his arguments to the government, Fulton wrote to the Foreign Secretary, Lord Grenville, with an even more vivid image of the world he saw opening up: "every year exhibits new combinations and effects: steam engines, cotton mills, telegraphs, balloons and submarine navigation and attack have all appeared almost within our memory: and only vulgar minds harbor the thought that a physical impossibility is impracticable because it has not already been done."

In these first years of the nineteenth century, Fulton's sentiments were by no means universal opinions; as he himself was able to observe, conservatism still ruled the day in the counsels of war. But their expression, often repeated in one form or another by Fulton through his years in France and England, and reiterated by him

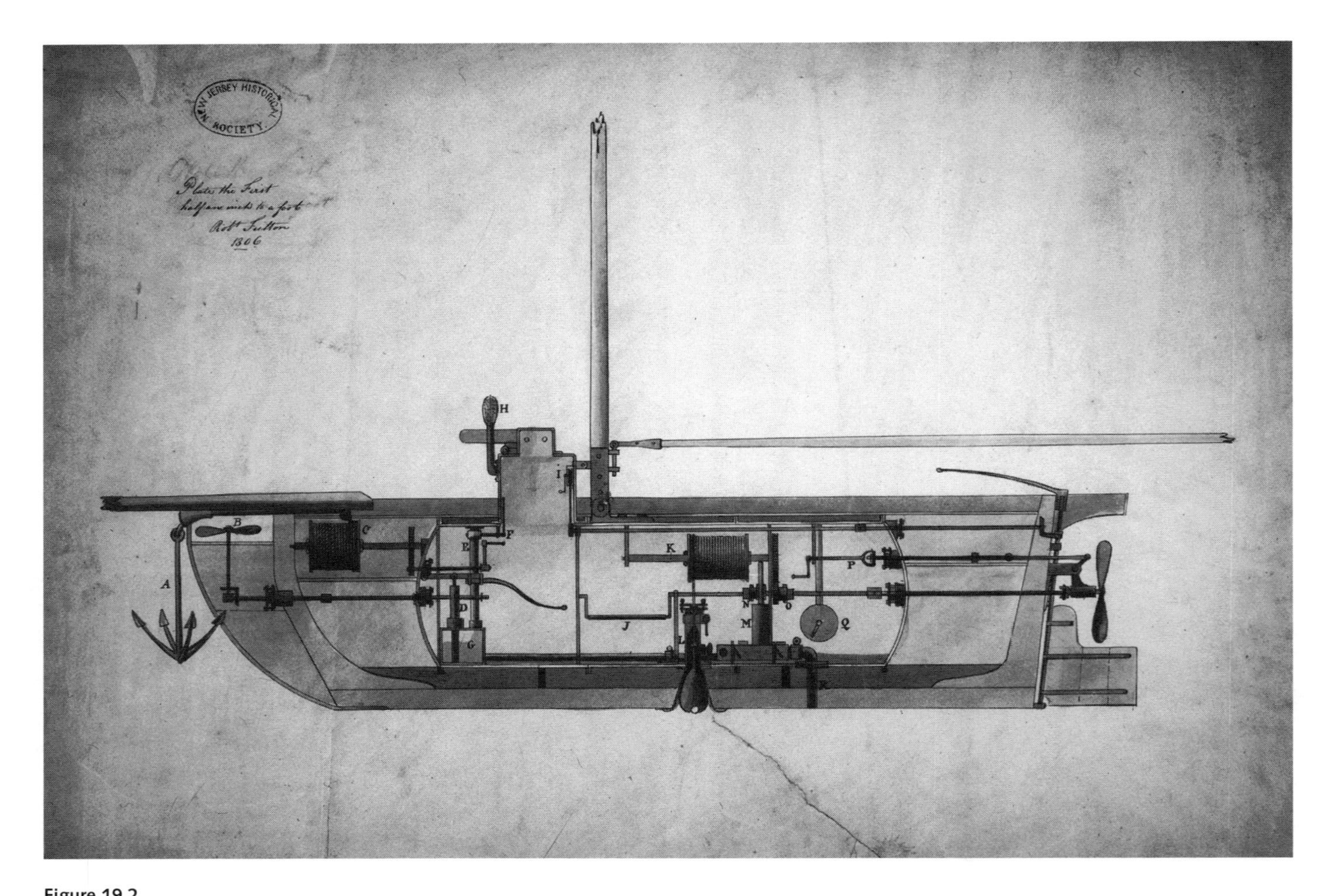

Figure 19.2
Robert Fulton's proposed submarine, *Nautilus*. From Center for Architecture, Design, and Engineering, Library of Congress.

almost a decade later when he designed the world's first steam-powered warship for the U.S. war against Great Britain, were echoed more and more frequently by others in the decades to come.[4]

Fulton may be seen as a convenient representative of another stream of thought that was also barely visible at this point, but which was to become more significant in the coming years—the idea that a single weapon could spell the difference between victory and defeat. This became an especially important manifestation of the kind of thinking that has come to be called "technological determinism," but whereas this usually simply refers to the way people explain why things are as they are, in this particular, military, form, this idea has been, at least since Fulton's day, the basis of policies and decisions that have influenced much in history. Fulton argued that his *Nautilus* would be so powerful that it would make naval warfare obsolete: "[when] warships [are] destroyed by means so new, so secret, and so incalculable, the confidence of the sailors is destroyed, and the fleet rendered worthless in the age of the Jeremiahs of fright." So powerful and different did he see warfare with his vessel that he wrote an essay, "Observations on the Moral Effects of the *Nautilus* Should It Be Employed with Success," in which he acknowledged that its use broke the usual rules of battle, but in so doing it would lead to the eventual abolition of warfare altogether. A hundred years later, another famous American inventor would make much the same argument in favor of applying the most advanced science and technology to warfare. Thomas Edison, who headed a panel of inventors and industrialists to encourage experimental weaponry during the First World War, defended his efforts after the war's end in an interview entitled "How to Make War Impossible." Here he encouraged governments to "produce instruments of death so terrible that presently all men and every nation would well know that war would mean the end of civilization." In the century between Fulton and Edison, numerous "superweapons" were promised, and a few were built, but the spiral of weapons and of violence never reached an end.[5]

Superweapons did not make warfare in the nineteenth century more awful. Instead, it was the application of industrial technology that increased enormously the killing power of armies and navies, along with accelerating, typically incremental, improvement in almost all familiar weapons. In the improvement of weaponry we can see perhaps most clearly the novel character of change in the nineteenth century. Unlike many tools, after all, the attraction of the better sword, bow, or gun had always been evident, and yet change in the past was typically small, uncertain, and temporary. The "better" weapon almost always came with a trade-off in terms of the demands of cost, skill, strength, or reliability. There had, of course, been some very important changes in the past—the most important of them being the coming of gunpowder and weapons to use it in the late Middle Ages. But since that time, while the firearms of the soldier and the cannon of the artillery had certainly undergone important changes, these effected only incremental alteration in either the forms of fighting or the character of battle. The new mechanical capabilities of nineteenth-century technology, combined with the means of mass production that were first effected in arms making and the new materials that also received an extra spur from military needs, overthrew completely this traditional pattern of slow change. The transformation worked at every level, from the arms of the most humble infantryman to the giant weapons brought to bear in fortresses and ships.

To comprehend the revolutionary nature of change in military technology, it is particularly important to remember that weapons are only effective if they can be properly used, and the conditions of their use are not idealized ones, but rather those created by the heat of battle. The creation of mass armies at the end of the eighteenth century meant that the users of weapons were no longer so predictable in their character or skill, but were more likely to be quickly trained temporary soldiers. This compounded problems that had always dogged the commanders of infantry. In the Napoleonic wars, for example, the procedures for firing a gun were largely those that had been followed for centuries, and these defined the form and pace of battle more than anything else. The typical infantry weapon was a muzzle-loading smoothbore musket. To fire it, a long sequence of steps was required: the infantryman would take from his pouch a paper packet containing powder and ball. After biting off one end, he would typically put the ball in his mouth, sprinkle a little of the powder into the firing pan of his gun, raise the gun vertically and rest the butt on the ground while he poured the rest of the powder into the barrel. He then spit the ball down after the powder, and followed this with the wadded up paper. The ramrod was removed from its slot in the gunstock and used to force ball and paper down the length of the barrel. In an ordinary musket this was fairly easy, since the ball fit very loosely (the wadded paper, in fact, was necessary to prevent the ball from coming out when the gun was raised). With luck remembering, in the heat of battle, to remove the ramrod, the soldier then raised the gun, pointed it (true aiming was rare, and pointless since the ball would hardly ever fly straight), and pulled the trigger which then caused a flint to spark, igniting the powder in the pan, and thence the main charge in the barrel. Even when this sequence went without a hitch, the effective range was rarely more than a hundred yards, and accuracy was so poor than often only volleys were truly effective. There were alternative weapons, most significantly rifled guns that were much more accurate (it was well understood that giving the bullet a spin as it left the barrel greatly assisted it in flying true), but these were expensive to make, and, more important militarily, were much harder (and slower) to load, since the bullet had to fit tightly and thus needed to be pounded down the barrel (which also increased the likelihood of misfires). It is small wonder, then, that to Napoleon and his rivals the bayonet charge was still thought of as the most effective of all infantry maneuvers.[6]

Even before Waterloo, the key elements of Western firearms began to change, and a sequence of improvements ensued that was, after decades of experiment and development, to transform completely the nature of battle. Most of the important innovations arose independently from one another, and each of these typically required much work to make effective, and then the process of integration with others took still more experiment and effort. Chemical experiments in the 1790s discovered compounds that would ignite with a hammer blow (most significantly, mercury fulminate), and after a couple of decades of experiments, armies began adopting the percussion cap by 1820. Soon after this, several experimenters discovered how to shape a gun's ammunition so that its base expanded when fired. A bullet that was small enough to load easily could thus fit snugly against the gun barrel when fired, which meant that rifles could now be used without the need to pound down the bullet. Another means of speeding up loading, through the gun's breech rather than down its muzzle, was experimented with for decades, but designing and manufacturing an effective breechloader required much effort, and any number of abortive attempts were promoted before reliable breechloaders appeared about midcentury. The first breechloading rifle to receive attention due to its battlefield performance was that devised by a German gunmaker, Johann Dreyse, in 1836. This detonated a paper cartridge by penetrating it with a needle upon firing. While there were some reliability problems (the needle often broke on impact), it not only allowed six shots to be loaded and fired in the time a muzzle-loader took for one, it also facilitated loading while the soldier was prone or kneeling, thus vastly reducing his vulnerability. Other breechloading rifles appeared from French, British, and American arsenals, so that by the time of the Franco-Prussian war in 1870–71, they were in general use, and they were being elaborated into repeating and rapid-fire designs. By the last quarter of the nineteenth century, the speed and accuracy of battlefield fire had made open battle practically impossible between Western armies. The transformation of the battlefield became complete over the next decades with the perfection of the machine gun. The American Civil War had seen the introduction of the Gatling Gun, and the French had adopted the *mitrailleuse* by 1870, but these versions, relying on multiple barrels and hand power, paled in effectiveness next to the invention of Hiram Maxim of the first fully automatic machine gun, introduced in 1885. One squeeze of the Maxim gun's trigger was enough to discharge 250 rounds, which it could do in less than half a minute. At first the Europeans celebrated the gun's efficiency, demonstrated in the short bloody confrontations that marked the powers' grab for Africa. Later they were to realize that the machine gun had sounded the final end to land warfare as generations had known it.[7]

The development of ordnance—larger guns—in the first half of the nineteenth century followed a similar trajectory. Here, too, the effectiveness of rifling to improve accuracy was well known, but the problem of designing an effective artillery shell that could be loaded both tightly and quickly in a rifled cannon barrel turned out to be difficult. While the first effective rifled cannon were introduced in the 1840s, smoothbore guns were used well into the late nineteenth century. The course of improvement was slow in the first part of the century: the fabled effectiveness of Napoleon's artillery, for example, was based on a superb tactical use of batteries of medium-weight guns (twelve-pounders) in conjunction with massed infantry. The effective range, however, was no more than about a thousand yards (and if accuracy was important, was much shorter). The ammunition was either solid cast-iron balls, which could be used against walls or the sides of ships, or canisters filled with musket balls or other bits of metal. These last were purely antipersonnel weapons, and their improvement received as much attention as any other problem. The most famous example was the spherical case designed by British artillery officer Henry Shrapnel and used by the British in the early Napoleonic campaigns. The round casing would disintegrate in midflight and spread as many as 170 musket balls over the target, thus reducing the need for accurate firing. Breech loading in cannon, just as in firearms, was understood as a desirable means for increasing rates of loading and fire, but this too turned out to be more difficult to make safe and reliable—the consequences in self-destruction, after all, of a leaking or burst breech were even greater for cannon than for rifles or muskets. In the years after Waterloo, there was much experimentation with both guns and their ammunition, but effective changes were relatively minor.[8]

This situation changed dramatically after midcentury, and arguably the changes in great guns were the most dramatic of all military improvements in the nineteenth century. While gun design and fabrication techniques saw important improvements, perhaps the most crucial step was in materials, particularly the manufacture and working of steel. At the Crystal Palace Exhibition in 1851, German gun founder Alfred Krupp displayed a six-pounder cannon cast from steel. Its virtues, given the great strength of steel over traditional iron or bronze, were so obvious that Krupp soon had buyers all over Europe. His own path of improvement had just begun, however: at the Paris Exhibition of 1867, he displayed steel cannon weighing about fifty tons and capable of delivering shells of about a thousand pounds (figure 19.3). When the huge increase in gun size was added to the greatly improved accuracy of rifled barrels, an entirely new age of artillery was clearly in the making. The new guns could fire, with reasonable accuracy, about two miles, with shells of enormous power and deadliness. In addition, the breech-loading mechanisms added to new means for controlling recoil meant that guns could be fired much more rapidly, without laborious reaiming. The primary problem for the military for the second half of the nineteenth century was not spurring improvement in weaponry, but in controlling costs, as newer and ever more expensive armaments were quickly rendered obsolete by yet more improvements. European armies found themselves in sporadic warfare

Figure 19.3
Guns from the Krupp foundry, on display at the 1876 American Centennial Exhibition, Philadelphia. Library of Congress photograph.

throughout this period, beginning with the Crimean War and continuing through the quick wars that variously engaged Prussia, Austria, Italy, France, Russia, and a host of smaller states, and on to the imperialist conflicts overseas, as the European powers grabbed for pieces of Africa and Asia. Through all of this conflict, it is not clear how much difference the new weapons actually made, since any technological advantage on one side was typically quickly matched on the other side (with the obvious exception of African tribes and ill-equipped Asian rulers), but the one thing that is certain is that the sequence of improvement made warfare much, much more expensive.[9]

The great costs of modern warfare came to be evident by the mid-nineteenth century. The war that France, Britain, and Turkey waged in the Crimea gave some hint of how increasing technological capabilities were also raising the costs of warfare. Railways, steamships, and telegraphs were all used to effect by the allies against Russia, as were the best arms then available. But the natural advantages that the Russians had in defending their own territory, along with astonishingly poor generalship, prolonged the effort, and the sheer task of creating, maintaining, and supplying the infrastructure of war made the enterprise immensely costly. The American Civil War totally destroyed the economy of the South, and even drained the great resources of the North to near exhaustion. The arms races between the European powers after the 1860s continually put great strains on their economies, even while industrialization fueled great economic expansion. The developmental costs of new weapons, a matter of much concern and debate in our own day, was likewise a matter of constant concern for politicians and citizens alike in the latter half of the nineteenth century. Only the relative brevity of European conflicts between 1815 and 1914 kept the costs of military and naval efforts within bounds. The extended nature of the American Civil War gave some clue of the devastating economic consequences of modern warfare, but the special and well-recognized damage of civil war masked the full implications for European observers.

One area of improvement in particular drew attention to the key elements of cost, destruction, and an endless technological spiral that accompanied the nineteenth-century improvement of violence: the emergence of the modern warship. While at the opening of the century, the large naval vessel, such as the *Victory* that Admiral Nelson led in the battle of Trafalgar in 1805, was arguably the largest and most complicated technological artifact known, within a single lifetime such ships were completely obsolete and had been displaced by new craft whose complexity, cost, and dependence upon industrial technology far exceeded that of their predecessors. To some degree it is possible to see that the huge and powerful steel ships that closed the century were simply industrialized versions of their wooden predecessors—large, complicated, and heavily armed machines of war. But the kind of technological change represented by the great navies at the end of the nineteenth century was at a level of speed and extent that transcended what was to be seen in most other realms of life and commerce. So urgent did the achievement and maintenance of the technological edge appear to be to the champions of seapower that naval innovation and construction became the model for just how far the culture of improvement could now reach into both the material and the mental lives of the West. It is thus no surprise that it was here that it first became apparent that the Western technological ethos had broken beyond its traditional geography and had been transmitted, at least in part, into other cultures, most notably modernizing Japan. The Japanese destruction of the Russian fleet in the straits of Tsushima in May 1905—a scarce century after Trafalgar—was the most dramatic possible sign that modern technology was no longer a monopoly of the West.[10]

The technological spiral that turned into the naval arms race can be said to have had its origins with none other than Robert Fulton. In the waning months of the War of 1812, Fulton completed construction of the world's first steam warship, which he originally called *Demologos* ("voice of the people"), but which was known on launching as *Fulton I*. It was a formidable vessel, clearly designed from the beginning for fighting. Its paddlewheel, for example, was placed in a well in the center of the ship so that it would not be vulnerable to enemy guns, and it had five-foot thick walls for protection and twenty-six guns of considerable size. *Fulton I* was completed too late to see action, and the U.S. Navy, in a fashion that was to be typical in the nineteenth century, never followed up on the pioneering effort. *Fulton I* pioneered in another way as well, as it cost nearly a quarter-million dollars to build, a hint of the tremendous expense that cutting-edge naval technology would continue to incur. While the British navy experimented in subsequent years with steamboats for towing some of the larger sailing ships, it did not construct its first steam warship until 1830. Early designs were hampered by unreliable engines, high fuel consumption, and the problem of using paddle wheels in fighting ships. Only when the screw propeller was proven a reliable alternative in the 1840s did steam warship design begin to flourish. The French showed the way with their *Napoleon* in 1847, but the British followed quickly, so that by the outbreak of the Crimean War in 1854, the British battle fleet relied on steam. That conflict did not so much prove naval steam power as it demonstrated how far there was yet to go to design ships that would make effective use of the new power. There was little question of the benefits of steam in battle—the difference between ships that had to rely completely on the wind and sails for maneuvering at sea and those that could change direction and speed at will was as great a change in naval capabilities as history was ever to see. But, as with so many great technological changes, there was much to be done to understand how to design a ship to take full advantage of steampower. There was also much to be done to improve the key elements of the vessel, from its power plant and propulsion to an understanding of how guns were to be placed to take full advantage of the new flexibility in movement.[11]

The modern form of the warship emerged through a series of experiments and innovations, at first somewhat hesitant and notable as much for what they retained of the past as for their novelty. But gradually through the century there emerged a very new form, in which the warship became more and more a machine. The history of this development is particularly remarkable for the extent to which it was driven by reactions and responses as much as by planned directions of development. New technological developments in one area called for responses in others, which then yielded counterresponses. Nowhere was this spiral more evident—and more notorious—than in the interplay between guns and armor. Steel in the guns of Krupp and other makers allowed naval guns to reach sizes and ranges that could hardly have even been dreamed of before. Not only did the size of the guns change, but their ordnance as well. In the 1820s, a French gunner, Henri-Joseph Paixhans, devised exploding shells to take the place of older forms of shot and canisters, and by the 1830s and 1840s Paixhans shells were being used by all modern navies. These wreaked particular damage on wooden ships, and thus hastened the demise of both wood and sail. The French warship *La Gloire*, launched in 1859, was the first iron-clad warship, with 4.5-inch thick wrought iron armor on its wooden hull. The British answered quickly with their *Warrior*, completed in 1861. Its iron hull with added armor incorporated watertight compartments and achieved a speed of over fourteen knots, but it, like all larger ships of its day, also carried a full complement of sail (using steam outside of battle conditions was considered wasteful). By the end of the 1860s, iron armor had become standard features for the vessels of modern navies.[12]

Just how iron and guns were to be effectively wedded, however, was still far from obvious. This is clear in a look at perhaps the most innovative vessel of the mid-19th century, the famous *Monitor*, completed for the U.S. Navy in early 1862 (figure 19.4). Here, finally, was an all-steam, iron-hulled war vessel, designed as a fighting machine from the outset. John Ericsson was a Swedish-born professional engineer working in New York City when he approached the U.S. Navy with his proposal for constructing an "Iron Clad Shot-Proof Steam Battery." Ericsson's vessel was not the only ironclad contracted by the U.S. Navy early in the war; news that the new Confederate Navy was constructing (or, in the case of the *Merrimack*, reconstructing) ironclad ships in the Norfolk Navy Yard pushed the very conservative Federal establishment into financing a number of efforts, but none so avowedly experimental as Ericsson's. The *Monitor* was unlike any ship ever built, for Ericsson intentionally made it as innovative as possible; it was said that it contained at least fifty "patentable inventions." In form it really was more submarine than ship, with the only thing protruding above the water being its pilothouse and steam-powered gun turret, with twin eleven-inch Dahlgren guns, the most advanced in the U.S. Navy. The twenty-one-foot diameter turret was round to deflect shot off its eight-inch-thick armor plate. On March 9, 1862, the *Monitor* encountered the Confederate ironclad

Figure 19.4
Contemporary engraving of the battle between the *Monitor* (in the foreground) and the *Virginia*. Currier & Ives print; Library of Congress.

Virginia (the rebuilt and armored *Merrimack*) in the waters of Hampton Roads, off Norfolk, Virginia. The two well-armored ships hammered at each other for several hours with rather little effect. There were no fatalities, and only one casualty—the *Monitor*'s captain was blinded by a shell that happened to explode just outside the slits in the vessel's pilothouse. Both ships withdrew, and both sides later claimed victory in what was essentially a draw. The draw, however, was telling, as it reflected the difficulty of finding a decisive edge, even in using the most innovative technologies.[13]

Neither the *Monitor* nor the *Virginia* was a model for the warships to come. One was a bold experiment that, to be sure, led to further "monitors" for the American navy, but the design was always flawed—it was particularly vulnerable in heavy seas, which is what sank the original on the last day of 1862. The Confederate ship, for its part, was no pattern for the future, but a makeshift use of iron armor on a wooden hull—effective against old wooden vessels, but no match for any modern ship. In the years to come, a new kind of ship would emerge, with designs that evolved from older ship forms but with capabilities that were dramatically new. A number of gun designers followed in the footsteps of Krupp and others, and the range, size, and power of naval guns increased beyond the capabilities of the navies to use them to full effectiveness. In the twenty-five years from 1860 to 1885, for example, the largest British guns grew in size from under five tons to more than 111 tons, and their maximum ordnance grew from 68 pounds to an astounding 1,800 pounds, which was capable of penetrating 34-inch armor at a range of a thousand yards. By the end of the century, navies were using guns with ranges as much as six miles (shore battery guns reached ranges of upwards of twenty miles, with shells exceeding a ton in weight). At the same time, the maximum armor on ships grew from about 4.5 inches to 24 inches at the waterline, while experiments on new kinds of steel continued to improve resistance to shells. As the spiral of gunpower and armoring continued, navies sought advantages in other directions. The most auspicious of these was the emergence of a new kind of power plant, replacing the reciprocating steam engine with the steam turbine, which was initially only of use on smaller vessels but was rapidly developed into larger and larger versions, increasing the speed of the best naval vessels. To the arsenal of weapons was finally added effective self-propelled underwater torpedoes, as well as reliable and destructive mines, realizing the promises made by Fulton almost a century earlier.[14]

By the last decade of the nineteenth century, the constant cycle of improvement in the weapons of war, particularly in the great ships and their armament, had become an accepted part of the modern world. The great European powers, joined to some degree by the United States and Japan, continued their jockeying for empire and for strategic advantage, in the belief that both political and economic power rested on military and naval strength. While it was not universally accepted, the doctrine of sea power enunciated by the American Captain Alfred Thayer Mahan in 1890 was sufficiently influential that politicians, businessmen, and officers alike made their calculations of the respective strength and prosperity of nations based more and more on available naval power. This, in turn, came to be seen as increasingly dependent upon constantly changing technology. When, in the last years of the century, the German Empire began a sustained program of building great warships to challenge British command of the seas, the policy was driven in large part by the perception that the cycle of improvement meant that no nation, including that with the greatest power, could rest on its established strength, but all had to confront the implications of constantly improving technological capabilities. If this was true, so German naval chief Admiral von Tirpitz's logic went, then even Germany, which at this point had no high seas navy to speak of, could compete within a number of years with Britain, provided only that the nation was willing to keep building the best possible ships. Over the fifteen years from 1898 and 1913, the Germans sustained a massive effort in this regard, constantly pushing naval technology ahead. The British, for their part, saw no choice but to respond. This surprised no one, but what was widely unexpected was that the British, too, sought to rely as much on technological change as on numbers or size to sustain their effort. Their most signal achievement, which changed everyone's calculations about the foundatons of naval power, was the construction of a dramatically new class of battleship, represented by the *Dreadnought*, which was launched in 1906.[15]

The battleships and other modern vessels constructed in the decade before the outbreak of the First World War represented the last phase of the great naval arms race. The *Dreadnought* was universally recognized as the symbol of this final turn of the technological spiral. It can be argued, however, that this class of battleships was not in fact the real harbinger of the future but the last effort of a "big ship" mentality that would never again be the true focus of innovation. The *Dreadnought* brought together several key technologies and attempted to weld them into a carefully designed integrated fighting machine. The most important of these was great speed and large, long-range guns. The steam turbine had been developed into a power plant that could effectively move even the largest ships, so the *Dreadnought* was equipped with the largest ever built, which moved her at twenty-one knots, far faster than other large ships. While she did not have, by the standard of other battleships, very many guns—only ten—these were all huge twelve-inch guns mounted on turrets, allowing an eight-gun broadside that was twice as powerful as any other ship's. The *Dreadnought* was well armored, although less than some of its predecessors. The British logic was that its great speed would be more protection than any armor, and they advanced this thinking through the introduction of another class of warship, the battle cruiser, that was even more lightly armored and even faster—sustaining as much as twenty-five knots. The other key to protecting these ships was their great range of fire, which reached over fifteen miles. At this distance, the key challenge to making these ships effective was accurate gunnery, and this turned out to be a problem of enormous complexity, which gave rise to some of the most advanced control technology ever attempted, and, in fact, to some of the first, pre-electronic, "computers." The British enjoyed their advantage only briefly, for the Germans and then others quickly built their own "dreadnoughts," "superdreadnoughts," and other advanced ships. By the beginning of the world war, Britain had twenty-nine large new ships in service (and thirteen more under construction), while the Germans had eighteen on the seas and eight more being built. It was this continued disadvantage, despite an enormously expensive construction program, that drove the German Navy to rely so heavily on submarines once war broke out.[16]

The Germans did not originate the submarine technology that they were to use with such devastating effectiveness. First the French and then the American navies pioneered the realization of the dreams of Bushnell and Fulton. While the Confederate Navy had had its own submersible, and had actually used one to sink a Federal ship in 1864, the real leaders in submarine development were the French, who saw the type as a means to get around overwhelming British naval superiority. The U.S. Navy gained its own submarine in 1900, and this design, by J. P. Holland, became popular with a number of others, including the British and Japanese. German development came rather late, as most resources were being put into the expensive battleship program, but when the German Navy began to develop its U-boat, it rapidly became the best of the class. This was due in part to the use of the relatively new Diesel engine, and in part to their formidable armament—huge torpedoes with gyroscopic directional controls. While at first the submarine was seen largely as a defensive weapon—a means for undermining the battleship threat in home waters—its technical refinement, particularly in terms of speed, range, and firepower, made it a formidable offensive weapon. In the years just before World War I it was recognized by some that such a weapon would have only one use—to sink ships—and that this had dreadful implications for warfare against merchant and passenger shipping. Once again, it was becoming clear that technology would rewrite the rules of war, whether the combatants wished it or not.[17]

The most dramatic lessons about technology taught in the First World War, however, were not learned on the seas, but in the forests, fields, and trenches of the fronts. The improvements in guns, both large and small, that had continued through the nineteenth century and on into the twentieth, along with a series of other, often little noticed innovations, such as barbed wire, completely changed the conditions and consequences of battle. In particular, the new firepower gave advantages to defensive positions, advantages that field commanders were tragically slow to understand. The machine gun, in combination with barbed wire and trenches, made open field attack—the historic form of infantry action in European warfare—impossibly costly. What everyone had anticipated in the summer of 1914 to be a war of movement and maneuver became instead a war of stalemate and position. One machine gun, operated by perhaps a half-dozen men, did the work, it was estimated, of about ninety riflemen. When they were concentrated in companies, such as the Germans learned to do early, machine guns were almost impregnable. Even more deadly, as it turned out, was the new artillery. While remaining hunkered down in trenches could keep a poor infantryman safe from small arms or machine gun fire, it could do little to protect from the massive long-distance shelling that became for many survivors the most profound memory of the battlefield. It has been estimated that as much as 70 percent of all wounds came from shellfire, and the experience of "shell-shock" came to represent both the individual and the collective trauma of the war. Trench mortars, howitzers, and heavy siege guns, typically firing high explosive shells, were massed behind the lines in astonishing numbers, and attempted to make up for the inability of the lines themselves to move forward by battering at the enemy day in and day out. Relatively late in the war, guns became more mobile with the introduction of the battlefield tank, but these were relatively ineffective in this conflict. The tank did, however, betoken one other technical element that supplemented much of the war's support structure—the internal combustion engine. From the Diesel engines of the submarines to the high speed radial piston engines of the aircraft to the heavy duty engines of the trucks and other vehicles that showed up in quantity for the first time, the technology of the "motor" showed itself indispensable to modern warfare.[18]

Of all the weapons introduced into battle during the First World War, there was one that came to represent most powerfully to both soldier and civilian the overwhelming presence of new technologies in modern war and the sense of menace that these carried with them—poison gas (figure 19.5). As early as the first weeks of the war, the French had sent grenades filled with tear gas toward the German lines, and the Germans tried similar tactics some weeks later, but these efforts had meager results. It was one thing to use chemicals to attempt to disrupt enemy action, and it was quite another to use them to kill, as the Germans finally did at the Second Battle of Ypres in April 1915. On the afternoon of April 22, French troops became wary as the German artillery, which had bombarded their lines heavily that morning, resumed fire late in the day. A yellow-green cloud began drifting toward the French lines. Thinking that this was some kind of smoke screen for a possible attack, the French officers instructed their men to stand in their trenches and expect an infantry attack. As the cloud reached the soldiers, however, it turned out to be chlorine, and within seconds the Frenchmen were dropping, their lungs permanently seared by the poison. The attack was, in fact, a great success for the Germans, although they failed to follow it up thoroughly. The news of the use of gas, however, drew widespread condemnation. The war had already shown how porous the "rules of war" were for

Figure 19.5
Bringing in casualties from a World War I gas attack. Library of Congress photograph.

the increasingly desperate combatants—the passenger liner *Lusitania*, sailing from New York, would be sunk by a German U-boat a little more than two weeks later, with the loss of almost 1,200 lives. But the deadly gas still carried a very special horror with it. This did not stop its continued use. The English launched their first chlorine gas attack at Loos in late September, although shifting winds yielded about as many Allied casualties as enemy. Alternative gases were sought. Phosgene was adopted because it did not bring on coughing—soldiers would thus inhale more of it. In 1917, the Germans used mustard gas, an almost odorless substance that caused horrible blisters both internally and externally. Countermeasures were quickly adopted and improved, so that only a small number of gas victims were fatalities. Nonetheless, the weapon created such revulsion that it was outlawed internationally by convention in 1925—a prohibition that, remarkably, stood even through the greater horrors of World War II.[19]

The period from 1815 to 1914 is typically characterized in European history as a century of peace. That it clearly was not, as there was hardly a year in all this period in which one or another of the great (and minor) powers were not engaged, somewhere in the world, in some level of warfare. From the wars of Latin American independence to the great land grab for African colonies, this was in fact a century of great violence, even if that visited upon the Europeans themselves tended to be modest and sporadic. In a century in which the arts of peace were celebrated as never before, the arts of war received more than their fair share of attention, as weapons, warships, and the basic technical underpinnings of warfare underwent constant innovation, development, and refinement. The arts of peace themselves turned out to have profoundly significant uses for the makers of war—railroads and steamships moved troops and materiel, the telegraph and the telephone were battlefield instruments, the factories that could turn out great quantities of cloth or nails could likewise produce bullets and guns. No industrial technology of significance lacked its military application in total war, as was first demonstrated in the American Civil War and brought home with emphasis in the Great War of 1914–18. It was further demonstrated that there was indeed hardly any province of human knowledge that could not be turned to the improvement of violence. The First World War came to be known as "the chemists' war," as laboratories provided not only poison gas but also newer and better explosives; superior alloys for guns and ships; improved fuel for airplanes, ships, and tanks; even substitute fertilizers and ersatz fibers to keep the war machine fed and clothed. The disillusionment with the rosy Victorian promise of moral improvement and uplift through technological, economic, and scientific progress that emerged from the exhaustion of the world war provided its own profound shock to the culture of improvement. At the same time, it confirmed that technology itself indeed appeared to have no limits.

20 Learning

John Anderson, professor of natural philosophy (that is, physical sciences) at the University of Glasgow, was by all accounts a somewhat disputatious fellow. Everyone would acknowledge his intelligence—the son of a minister in rural Scotland, he served in the army before commencing his studies at Glasgow, and before he began teaching natural philosophy in 1760, he was the professor of oriental languages. He dabbled in military matters all his life—writing essays on war and weapons and designing a sophisticated device to control the recoil of cannon which he offered to the French revolutionaries in 1791 after being turned down by English officers. Perhaps it was these revolutionary sympathies that made him so intensely unpopular with his fellow professors or possibly it was just a prickly personality. Whatever it was, John Anderson was able to maintain a grudge and he determined to keep making his point beyond the grave. When he died in early 1796, it was discovered that he had left everything he had, primarily a wonderful collection of scientific instruments, along with a fine library and a modest museum of antiquities and other scholarly specimens, to found an institution to rival the university for which he taught for so many years, but for which he obviously had no affection. This rival institution was, above all, to carry on and extend a tradition that he had established in his own teaching—it was to make learning available to all who might benefit, and particularly to working men and to women. Soon after he had begun his science teaching, Anderson had made a point of carrying on a series of lectures on Tuesday and Thursday evenings, dealing with a great range of scientific and antiquarian topics. His new institution, therefore, was particularly designed to serve an unconventional audience. He made a point, in fact, of spelling out the importance of encouraging female attendance at lectures and demonstrations. Despite the fact that Anderson's estate was woefully inadequate to fully support the ambitions he outlined, the magistrates of Glasgow moved quickly, and within less than six months of his death "Anderson's Institution" had been incorporated, and a few months later its first professor hired.[1]

To a substantial degree, Anderson's Institution was a natural outgrowth of trends and practices that were well established by the late eighteenth century. Popular lectures on scientific topics were to be found in most major cities of Western Europe, and the dissemination of scientific and technical knowledge through lectures, informal courses, demonstrations, and encyclopedias had become common. What made the events and practices in Glasgow exceptional (although certainly not unique) and important for the future was the audience. Whereas the lectures and demonstrations earlier in the century had been directed toward groups of gentlemen (and gentle ladies, to some extent), the new efforts were for the benefit of a much wider audience, and in the coming decades the provision of popular education for working men and women, in technical subjects and in more general ones, became one of the nineteenth century's most important social innovations. One reason for beginning this story with Anderson and Glasgow is the appearance at the Institution just a few years after its founding of George Birkbeck, who took over the duties for lectures in the sciences in 1800. Birkbeck was a physician and a graduate of the University of Edinburgh. He pushed Anderson's Institution even further than the town fathers had originally wished to open its doors to poorly paid workers. After only a few years in Glasgow, Birkbeck moved to London, where, aided by a fortunate marriage, he became a successful and well-to-do physician. The Glasgow institution continued to thrive and develop and gave rise in 1823 to the formation of the Glasgow Mechanics' Institute. The success of this endeavor stirred great interest in London, and Birkbeck plunged back into the cause. At this point, however, he was able to use his own wealth and influence to give the effort more than simply intellectual support. The London Mechanics' Institute was formed in 1824, with Birkbeck paying much of the initial cost. Others soon followed, both in Britain and elsewhere, and the mechanics' institute became the nineteenth century's most distinctive form of popular education.[2]

Mechanics' institutes always reflected the intellectual and social tensions of their times. In most places they were initially established through the efforts and resources of members of the merchant and manufacturing establishment and of the professions, such as medicine, the law, and the clergy. Their audiences were usually quite broad, incorporating many of the same members of polite society who attended the popular scientific lectures of the eighteenth-century salons as well as the "mechanics" and other workers for whom they were ostensibly formed. The topics covered in lectures, courses, and demonstrations ranged widely in many establishments, going beyond the core topics of mathematics, chemistry, and "mechanical philosophy" (physics) to incorporate natural history, agriculture, astronomy, and even languages and literature. Their purpose was the diffusion of "useful knowledge," but what was "useful knowledge," and who should decide this? The purposes that motivated men like Anderson, Birkbeck, and dozens of other "respectable" members of society to promote these institutions for the working class were varied, but they were often a mixture of charitable, political, and intellectual concerns that related in very general terms the larger peace and welfare of society—and relations among the classes—to imparting a shared culture and knowledge of the general order of the world. Birkbeck explained further that he had been frustrated as a lecturer in Glasgow at having to explain to craftsmen and mechanics the purposes and designs of his instruments when he felt they could have simply learned what they needed from his Andersonian lectures. This idea that working men and women would become more useful, more productive elements of the industrial order if they knew more about the workings of nature was another common motivation for the support of the mechanics' institutes. For the mechanics and other working people, the new institutes offered both a means of inexpensive education and a new, "respectable" form of association, one which they often struggled to gain control of for themselves, although with success that was often short-lived owing to the lack of resources.[3]

The spread of mechanics' institutes was rapid. A French nobleman, Baron Charles Dupin, visited the London and Glasgow institutes in 1826 and returned home full of enthusiasm for them. In the manner that was typical for such endeavors in France, he turned to the state for assistance, and by the end of the year he could report that government encouragement had enabled ninety-eight towns to begin organizing "instruction to the working classes." In the United States, news of the British efforts also spread quickly and was widely noticed. Perhaps the most important vehicle for this dissemination was a new genre of periodical that began to appear at this time. The *Mechanics Magazine* was first issued in London in the fall of 1823 (one of its first numbers included the call that resulted in Birkbeck's organization of the London institute). By early 1825 an American imitator had appeared, also called *Mechanics Magazine* at first, but soon adding the appellation "American" to its title. About a year later it was further transformed into the *Franklin Journal and American Mechanics' Magazine*, and it carried widespread notice of the mechanics' institute movement. At about the same time, there appeared the first issues of the *American Journal of Education*, and it too made much of the new British efforts. Its editor, William Russell, was a native of Glasgow, and so was pleased to report on a movement traceable to that city. The most seminal article in the *Journal of Education* appeared in October 1826 from the pen of Josiah Holbrook, a thirty-eight-year-old Yale graduate with a checkered career as a scientific lecturer and geologist. In his anonymous article, Holbrook called for the formation in every town and village of "societies for mutual education." While he placed the sciences first among the subjects to be taught, he drew no boundaries, and he envisioned a broad-based system of associations for popular education. A skilled and persuasive salesman for education, Holbrook then went on the lecture circuit, and in a matter of months had begun a broad movement for the formation what became known as "lyceums." There were an estimated six thousand lyceums and mechanics' institutes founded across the country by mid-century (but only about one hundred of these were true mechanics' institutes—the vast majority were typically short-lived literary societies). These were the key elements in a movement that lasted through the nineteenth century in America, providing the foundations for popular, accessible adult education in small towns and large cities throughout the country.[4]

The mechanics' institute and lyceum movements were the most visible institutional elements of a larger nineteenth-century concern for promoting "self-improvement" and "mutual improvement." In both America and parts of Western Europe, the culture of improvement expanded to incorporate not simply the means of production and daily life and the institutions of government and society but also the lives and capacities of individuals, both as functioning members of society, with important (and often changing) economic and political roles, and as human beings with innate talents and endowments. One or two centuries before, much of the same energy might have been directed towards moral improvement, towards assuring the welfare of the individual's soul and of the community at large through the promotion of piety and religiosity. But the industrializing West readily shifted its attention toward more measurable forms of improvement, both in material terms and in the visible means by which men and women earned their living and enhanced their lives. Education became one of the century's most powerful and pervasive dogmas, and education itself underwent constant redefinition, away from an ancient focus on "higher learning" and towards "useful knowledge," particularly in its materially manifested forms.

The range of institutions nineteenth-century men and women devised to pursue and extend the dogma of education was considerable, from grammar and Sunday schools to universities and polytechnics. The mechanics' institutes found their own place within this range, although they never achieved the long-lived stability of the more formal schools. It is hard to assess just what role they actually played in technological development in the century: some successful engineers and inventors spoke readily of the influence of an institute's lectures or the encouragement and inspiration found in an open library, but in many cases the influence was probably less obvious—the availability of books, the occasional opportunity to listen to others' ideas, the informal network of other ambitious mechanics that could be found in such a place. For the larger mass of working men and women, particularly those aspiring to a skillful craft, the possibility of access to low-cost learning was meaningful, whether taken advantage of or not. In the United States, at least, the mechanics' institutes also served as important way stations or testing grounds for the development of different kinds of technical education. As more formal engineering education developed over the course of the century, the informal institutions provided a kind of indicator of what needed to be included in the emerging colleges and engineering schools and what, ideally at least, should be relegated to more elementary venues. In a few cases, such as Philadelphia's Franklin Institute, a mechanics' institute acquired both missions and resources that went well beyond the norm, and thus was able to carve out broad, long-lived functions for the technological community.[5]

The fortunes of lyceums and mechanics' institutes waxed and waned in the course of the nineteenth century. Such was not the case with the more formal institutions of technical education that emerged, for this was the period in which, after a slow and halting beginning, the collegiate education of engineers, farmers, and other independent technicians became a social and political priority of the first order, particularly in the United States. As was so often the case, the American attempts to organize technical education in schools and colleges were not the earliest ones, but instead initially borrowed quite openly and candidly from a range of European examples. Different European countries devised nationally distinctive types of formal technical education, reflecting economic, political, and social dynamics in institutions and systems that were the subject of continuing experimentation and innovation throughout the century.

The most important model for all champions of formal technical education was that provided by the French, particularly after the Revolutionary reforms of the 1790s. The *ancien regime* organized several schools for training officers in both its military and civilian services. The artillery schools organized in the seventeenth century were transformed into training centers for military engineers by the mid-eighteenth. Bernard Forest de Belidor, for example, was an instructor at the military schools at La Fére and Metz and wrote key textbooks in a range of fields, promoting the use of the most advanced mathematics (such as the relatively new calculus) for solving engineering problems. The École du corps royale du génie (School of the Royal Corps of Engineers) established another mathematics-oriented curriculum, influenced particularly by the work of instructor Gaspard Monge, whose principles for the application of geometry to engineering influenced practice for two centuries. Together with the École des ponts et chaussées, the value of these schools was evident to the reformers of the First Republic. The result was the formation in 1794 of a new, unified institution based in Paris, the École Polytechnique. The school instituted a new curriculum, which was designed to provide technically trained individuals for all the services of the French state. Admission was by competitive examination, and successful candidates (about 150 per year) were supported by a stipend and engaged in a rigorous technical, mathematical, and scientific training, complete with laboratory exercises, shop work, drawing exercises, and the like. No other institution exercised such influence on the educational practices of other countries as this first polytechnic.[6]

The École Polytechnique was not the only significant contribution to technical education to emerge from the French Revolution. At about the same time that the Polytechnique was being formed, a proposal emerged for the creation of a public collection of models, instruments, and tools that would exemplify the methods of all of the useful arts. Some years passed, however, before action could be taken, but in 1799 the new Conservatoire des Arts et Métiers took over the buildings of the old priory of St. Martin des Champs, and began installing what remained of the royal collections of models and instruments, including those of the disbanded Académie des Sciences. The legislation establishing the Conservatoire directed that "the construction and use of tools and machines employed in the arts and trades shall be explained there," and after a couple of decades public lectures and courses were introduced. The Conservatoire was the first public technical museum (and it remains, at its original location, the world's oldest), and thus provided a model for future efforts elsewhere. Despite this example, technical museums grew very slowly and haltingly in the course of the nineteenth century, and when they did emerge in places like London, Vienna, and Washington, they were typically not efforts to emulate the Conservatoire directly, but more frequently were the offshoots of the great international exhibitions of the century. Their contributions to popular technical education were, like those of the mechanics' institutes, difficult to measure and were more likely indirect and inspirational rather than specific and utilitarian.[7]

It is much easier to assess the importance of the French innovations in formal technical education, for they did not stop with the École Polytechnique. From the beginning, this was considered an institution serving primarily the interests and needs of the state. After the fall of Napoleon and the restoration of the monarchy in 1815, this orientation came to be identified (as to some extent it still is) with the preparation of an elite corps of government officials, rather than with providing the country with a source of well-educated practical engineers. The limitations of this were apparent to some, particularly as France attempted in the years after the fall of Napoleon to catch up with British industrialization. To meet the evident demand for industrially oriented engineers, a group of well-connected individuals organized in Paris in 1829 the École Centrale des Arts et Manufactures. While partaking of the École Polytechnique's instructional model (the founders included several *polytechniciens*), the Centrale explicitly sought to orient its curriculum to produce engineers who would be of immediate use to industry. After a year of theory-based training, students entered programs that focused on mechanical engineering, structures and architecture, inorganic chemistry, or mines and metallurgy. The École Centrale was unusual in France for being the product of private efforts rather than state initiative (although it received the required state authorization readily enough). Its subsequent wide influence on other technical schools in both Europe and America was perhaps enhanced by this initial independence from political agendas. Instead, the Centrale was perceived as particularly responsive to the new needs of the rapidly changing industrial economy, able to produce engineers for the new railroads, iron structures, and giant chemical works as industry required.[8]

The country that was initially most influenced by the French experience was not a country at all, but the loose collection of states and cities called Germany. Educational reform was a part of the political and social agenda in Germany beginning in the seventeenth century, spurred in part by the passions of the Reformation and in part by concern for the weakness that seemed to be an unavoidable consequence of German disunity. Technical education did not figure strongly in these reform debates, however, until after the Napoleonic wars. The world's first school of mines was formed in Freiburg, Saxony, in 1765, and the Bauakademie opened in Berlin in 1799 to train civil engineers, but it was during and immediately after the Napoleonic years (during which much of Germany was under French authority) that the push for technical education gathered momentum. Some of the newer universities, such as Berlin (1810) and Munich (1827) made some accommodation for scientific education, but real engineering training became the subject of separate institutions in most of the key German states. By the second half of the nineteenth century, these had begun to be organized as *technische Hochschulen*, or polytechnics. Gradually these acquired status and powers equal to those of the traditional universities, to the extent that the only difference in the training required for admission was that the *technische Hochschulen* did not require Greek. When the Polytechnic School at Charlottenburg, near Berlin, opened in the 1880s, it was widely regarded as the finest school for science and technology in the world. The Germans were exceptional in going further to organize an extensive system of formal noncollegiate technical education—the *Realschulen*, for ages ten to sixteen, which led to two years of further training in technical colleges in engineering, chemistry, construction, and commerce. By the end of the century, it was the German system, so extensive and comprehensive, rather than the French that reformers elsewhere pointed to as the premier model.[9]

So firmly embedded in the modern American psyche is the faith in schooling that it takes some effort to comprehend the extent to which the increasingly complex engineered world of the nineteenth century was largely the product of men (and very few women) who gained their skills in building, designing, and manufacturing in the field and factory and not in classrooms and laboratories. Sometimes these people were described as self-taught, but it would be more accurate to say that they were educated by a combination of experienced mentors and the necessity of solving technical problems on the job. Their tools—intellectual or material—were not necessarily rudimentary or crude, but they were simply not the product of formal schools. Even so rough and ready an inventor as Thomas Edison, who celebrated his lack of formal education, read thoroughly and carefully in the books available in free libraries and elsewhere, and took advantage of whatever practical instruction he could get in the Western Union shops where he first worked. Some of the great early engineering works in America became famous as "schools" of a sort—the Erie Canal, the Baltimore & Ohio Railroad, and the Western Union telegraph network gave rise to the first generations of American engineers, and the "shop culture" that they represented remained important and strong throughout the nineteenth century. Slowly during the course of the century, the shop culture represented by these practically trained engineers gave way to a "school culture," which tended to be more open and democratic, more comfortably accommodating the "new Americans" of the late-nineteenth-century immigration, for example. The source of this school culture was a system of engineering higher education unlike any in the world.[10]

The initial inspiration for the American system came from the French. While the style of state-organized engineering that the French developed from the seventeenth century onward was very different from that adopted by the Americans, there was one area in which the linkage between engineering and state interests was clear and compelling—the military. This was also an area in which American regard for French models was high from an early date, due largely to the French assistance in the American Revolution. It comes as little surprise, then, that when the U.S. Army decided that a permanent corps of engineers was needed and that this corps required some means for uniform, advanced training, that the French experience should inform the styles and approaches adopted. As early as 1776, John Adams wrote to a correspondent, "Engineers are very scarce, rare and dear... we want many and seem to have none. I think it is high time we should have an Academy of this Education." The Corps of Engineers was initially stationed at the fort at West Point, on the Hudson River some miles north of New York City, and a program of training began as early as 1794, under George Washington's orders. It was not until 1802, however, that the United States Military Academy was formally constituted at West Point. The institution had relatively little effect until it was reorganized soon after the end of the War of 1812. Sylvanus P. Thayer was appointed superintendent in 1817, and the graduate of Dartmouth College and the Military Academy quickly undertook far-reaching reforms. These consisted largely of attempting to duplicate the training provided in Paris's École Polytechnique by using textbooks written for the French school and hiring such instructors as Claude Crozet, a graduate of the Polytechnique who became one of America's most influential civil engineers. The distinctive instructional program of the French engineers, with its rigorous curriculum, laboratory exercises, and strong emphasis on mathematics, geometry, and drawing, left an imprint on the style of American technical education for years to come.[11]

Many West Point graduates went on to provide guidance and expertise for the civil works that a growing country needed. During the first half of the nineteenth century, the academy remained America's primary source of formal engineering training, particularly in construction and related areas. The creation of a private, nonmilitary system of engineering education was more halting and difficult. The most notable efforts were individual initiatives, and these often ran into difficulty with funding, staffing, and enrollment. Alden Partridge, for example, was an early graduate of West Point and taught there for several years (it is said that he was the first person in the United States to have the title of "professor of engineering"). He became acting superintendent of the academy in 1815, but left the army upon Thayer's appointment in 1817. Two years later he established in Norwich, Vermont the American Literary, Scientific, and Military Academy. Despite the name, Partridge clearly had in mind nothing less than a private version of West Point, but the effort struggled for years with meager results, only granting its first degrees in engineering in 1837, by then under the name of Norwich University. In 1834, Stephen van Rensselaer, a Harvard graduate and product of one of New York state's old Dutch establishment, commissioned Amos Eaton, a broadly trained professor from Williams College, to organize a school at Troy, New York, to promote "the application of experimental chemistry, philosophy, and natural history, to agriculture, domestic economy, the arts, and manufactures." At first, Eaton's school operated at a fairly elementary and general level. Over the next fifteen years or so, however, it developed along lines that emulated the European model, until it was by midcentury a true polytechnic in fact and name.[12]

Slowly in the decades before the American Civil War, scientific and technical education insinuated itself in many of the college-level institutions scattered across the United States. Many of the older schools in New England established scientific courses as early as the late eighteenth century, and these became common features in American universities. Courses in chemistry, natural philosophy, natural history (biology and geology, for the most part), and medicine were widely available. Engineering and technical courses appeared more slowly, but by the 1830s these too were appearing in traditional schools. It is no coincidence that this coincided with the first great railway construction boom in America. The University of Virginia established a school of civil engineering in 1835, the College of William and Mary did the same the next year, and the University of Alabama followed suit in 1837. Two successful military schools appeared in the South at this time—Virginia Military Institute and the Citadel, in South Carolina, and both of these placed some emphasis on engineering, following the École Polytechnique model closely. Just as the railway boom spurred these developments, so too did the bust of the late 1830s deflate them, and many of the new programs failed to survive their first decade or so.[13]

With the economic recovery of the 1840s, and the resumption of investment not only in railroads but in the new telegraph, shipbuilding, and the like, interest in promoting scientific and technical education resumed, and the results were far-reaching. In some of the older universities, well-funded efforts established technical training on a more ambitious basis than previously. Abbott Lawrence, a successful Boston area manufacturer and investor funded Harvard University's effort to establish a scientific school for "instruction in theoretical and practical science," and the first courses included civil engineering. Yale and Dartmouth followed only a few years later, although Harvard's own commitment to engineering waned quickly. The University of Michigan, at this time the best supported state institution west of the Appalachians, began a course for the degree of "Civil Engineer" in 1855. Even more significant was the establishment of a host of new institutions for which technical education was part of the core mission. William Barton Rogers was a geologist who quit his post at the University of Virginia to move to Boston in 1853. There, failing to exert any influence on Harvard's plan for its scientific school, he spearheaded an effort to establish a separate Massachusetts Institute of Technology, which was chartered in 1861. The new MIT was very much thought of as a "school of applied science," and it drew support from the Boston manufacturing community. The Civil War interfered with the school's early years, but by the late 1860s it was carving out a mission for itself. By that time, however, the movement for technological education was beginning to swell. The most important influence was the passage in 1862 of an act sponsored by Senator Justin Morrill of Vermont which granted to each state the proceeds from the sale of thirty thousand acres of Federal land for the purpose of endowing colleges to support the teaching "of agriculture and the mechanic arts." This was arguably one of the most far-reaching pieces of legislation passed in the nineteenth century, and it gave voice and substantial resources to an ideal of practical education that became central to the American creed.[14]

Giving a clear character to this new liberal creed of practical education was not a simple matter. There remained much tension and conflict, as defenders of classical education fought what they believed was a reduction of standards and a misdirection of educational energies. There was always a political as well as an intellectual dimension to the battles, as the competition for power and influence in a rapidly growing and industrializing America washed over the debates over the purposes and methods of education. Ezra Cornell was a successful builder of telegraph lines who decided that traditional educational establishments were instruments of an elite that did not fully contribute to the common welfare. "I would found an institution," he famously declared, "where any person can find instruction in any study." He recruited to assist him Andrew Dickson White, a former history professor who was serving in the New York State Senate alongside Cornell when the issue of how to apply the state's Morrill Act grant came up for debate. Armed with a generous endowment from Cornell and with the land grant moneys, White established an ambitious new university in upstate New York with the intention of providing an influential model of secular, liberal, accessible education, one in which science and other forms of knowledge would be at the service of practical ends. Cornell University was open to all, without regard to gender or race and with as much access to the poor as possible. For White, "scientific and industrial studies" were at the very core of the university's mission. The effort was indeed enormously influential, providing a model of nonsectarian, accessible education that shaped the paths of many of the state universities that followed as well as some of the distinguished private ones that emerged in the later decades of the century, such as Stanford, Johns Hopkins, and Vanderbilt Universities.[15]

The character and substance of engineering education was molded by changing technological needs. When West Point was formed at the beginning of the century, "engineering" meant what we would call civil engineering—the construction of a range of works on the land or the water—roads, canals, fortifications, dams, bridges, harbor works, and the like. By midcentury the meaning had expanded significantly, and with it the scope of engineering education. The best representation of this was the founding of the U.S. Naval Academy at Annapolis, Maryland, in 1845. The coming of steam for ship propulsion was recognized as one of the most profound changes in naval technology in centuries, but the service found itself ill prepared to operate vessels dependent on steam engines. While the formation of the Academy fulfilled a range of purposes, a key one was to provide future officers with the rudiments of steam engineering. The teaching of mechanical engineering was a substantial novelty, and the Naval Academy experience was profoundly influential. Immediately after the Civil War, the academy hired Robert H. Thurston to teach science and engineering. Within a few years, Thurston devised a comprehensive curriculum in mechanical engineering that he publicized widely. His work came to the notice of Henry Morton, then editor of the *Journal of the Franklin Institute* and, in 1871, the founding head of the Stevens Institute of Technology, in Hoboken, New Jersey. This was the first private institution founded with an orientation around the teaching of mechanical engineering, and Thurston was a natural choice to develop the program. Over the next several decades, Thurston wrote many articles and textbooks directed toward establishing the field as an academic discipline as well as a respected profession.[16]

Thurston is a good example of the kind of men who strove to make engineering in America into a respected profession as well as a field of learning. Engineering always faced considerable ambiguity in its struggle for respectability and authority—it depended on specialized knowledge while at the same time relying on satisfying clients and employers. Its association with outdoor work, with mechanics and builders, made its drive for professional status all the more difficult. As college-based training came to be seen as an appropriate, and then as an indispensable, foundation for an engineering career, the basis for engineering's claims to status shifted. The professionalization of engineering—the organization of the work and craft around standards of action and knowledge and the development of means for self-government and policing as well as controlling access to practice—began with the organizing of largely self-taught and apprenticed civil engineers in the first half of the nineteenth century. In 1818, the Institution of Civil Engineers was formed in Britain. An earlier Society of Civil Engineers dated back to the late eighteenth century, but this did not claim to have any larger role in governing or speaking for the profession as a whole. The Americans had greater difficulty organizing themselves, and the American Society of Civil Engineers was not organized on a permanent basis until 1867. At first, the organized civil engineers claimed to represent all of engineering, but fragmentation set in. In 1847, a group of railway engineers organized themselves in Britain as the Institution of Mechanical Engineers. American attempts at a similar organization soon followed (although the American Society of Mechanical Engineers was not successfully formed until 1880). Aptly enough, the next national organization (local ones quickly proliferated) to appear was a group of mining engineers, for the exploitation of America's mineral resources was one of the mid-century's great technical missions. The American Institute of Mining Engineers was organized in 1871, kicking off a decade that saw national associations formed for disciplines as varied as history, chemistry, and economics. By the final decades of the century, the professional organization had taken root throughout the West as more and more specialists sought to define their authority and their domain.[17]

Professional associations played a particularly important role in defining specific technical disciplines and specialties. As industries increased in size, and new technologies, from engines to electricity to chemical production, achieved greater importance in the last decades of the nineteenth century, the demands and opportunities for specialized technical expertise fostered the emergence of a variety of specialists, who then had to struggle to identify their particular sources of value and authority. In general, new fields of engineering emerged from two sources. Some came from older engineering areas, particularly civil and mechanical engineering. As engineers were called upon to tackle more complex tasks, often adopting new technologies, tools, and materials, new specialties arose and subgroups of practitioners sought to define exclusive domains for themselves. In this manner, for example, sanitary engineers came out of the civil engineers who were engaged to design and build the water supply and sewer systems of rapidly growing cities and towns. An equally important source of new fields comprised groups of technicians who were initially associated with a lower-status craft tradition, but whose technical range and importance increased to the point that they demanded a different status. Electrical engineers, for example, were designers of electrical devices and systems who wished to differentiate themselves from electricians, telegraph operators, and the like. Chemical engineers, as another example, organized themselves in the first decade of the twentieth century to distinguish themselves from the chemists that industry hired by the thousands to perform relatively routine analytical work. For all of these engineering groups, an emphasis on design as a key work product and on specialized, increasingly science-based knowledge and schooling served as the most identifiable elements of professional identity and status.[18]

The importance of the organized engineers for the pursuit of technological improvement was considerable, but not always straightforward. The engineering institutions, both the schools and the professional societies, came to acquire great authority over how improvement was to be defined. In an increasingly complex technological world, this was arguably a salutary development, although it should not be looked upon as an inevitable one. One example of this authority was the emergence of engineering standards—codes defining everything from the forms of screws to the means for measuring electric current. Such codes were widely seen as necessary in many technical areas as the extent and complexity of machinery, structures, and techniques grew rapidly. When, for example, in the 1820s the problem of boilers exploding on steamboats began to be seen as acute—in the United States, forty-two explosions killed 273 people between 1825 and 1830—the technical community felt the need to respond. The Franklin Institute began a series of systematic studies on the causes and means of preventing disastrous explosions, and it presented results to Congress in 1836 with recommendations for legislation. While doubts about federal authority and federal power made many legislators reluctant, continued disasters moved the legislation forward and the doctrine that public safety demanded technological constraints was firmly established (although it was many years still before steamboat inspection was effective). In subsequent years numerous important, if less spectacular, concerns required safety and inspection measures and the imposition of standards defining how things were to be designed, constructed, and manufactured. Sometimes government agencies—often within or prompted by the military—acted directly. The United States formed its National Bureau of Standards in 1901. Frequently engineering societies, manufacturers or builders associations, insurance groups, and other "non-governmental" agencies and associations stepped in to frame the key questions and formulate technical specifications of every stripe. To this day, the technical standards that are indispensable for the large-scale pursuit of every technology are created and maintained by a peculiar mix of public and private agencies and associations.[19]

A particularly important instrument in propagating new standards as well as in promoting a wide variety of other technical concerns, from the efficient and fair operation of the patent system to the organizing of technical fairs and expositions was the expanding technical press. While the *American Mechanic's Magazine* appeared as early as the 1820s, and the *Journal of the Franklin Institute* became another early vehicle for technical information, the single most important new medium in the United States for spreading information about new technologies as well as for teaching a wide audience about basic principles of new machinery and techniques was a new kind of journal, *Scientific American*. For most of the nineteenth century, *Scientific American* was operated as an adjunct to a patent attorneys' firm, and it devoted much effort to encouraging inventors, advocating strengthening of the patent system, and disseminating news about the latest areas of inventive effort. Beyond this, however, the magazine sought to present clear and accessible explanations of a wide range of technologies, and it thus served not only as a means of furthering the education of mechanics and other would-be inventors but also as a channel of information about technological change to investors, other businessmen, government workers, and educators. It served as a digest of information gleaned out of a wide range of other sources, from foreign newspapers to scientific journals, and thus became a kind of periodical of record for all the popular areas of technological effort in America. Toward the end of the nineteenth century and into the twentieth century, more specialized technical journals, typically sponsored or promoted by the engineering societies (both national and local) or by particular industries (such as railroads or construction), proliferated and took on much of the burden of informing practitioners of the latest work and novelties in ever-more narrowly defined fields. The technical press, particularly in the United States, served as key promoters, arbiters, and recorders of the modern culture of improvement. The decline of the mechanic's institutes and the popular lecture circuit owed much to the ready availability of up-to-date technical periodicals.

The spread of a popular and semipopular technical press also fostered another development that presaged some of the ways in which technological improvement would change its character in the twentieth century. With the spread of industrialization and the rapid growth of a comfortable middle class, which now included many factory workers, there appeared a new class of technical actors, perhaps best referred to as "amateurs." While there had always been individuals, typically well-to-do with substantial leisure, who had dabbled in various crafts or in inventive efforts, the new social and economic circumstances of the late Victorian period, in both Europe and America, allowed vastly greater numbers of amateurs to engage a great range of technologies at varying levels of technical expertise. The new transportation technologies that appeared at century's end were the most prominent venues for this; first the bicycle and later the automobile were personal mechanisms that encouraged a certain amount of self-education and mutual education for would-be repairers and improvers. In the first decades of the twentieth century the emerging technology of wireless communications (radio) provided yet another important realm for the avid amateur. Hobbyists supplied readily identifiable, enthusiastic audiences for magazines varying greatly in quality and substance, and these outlets, in turn, further promoted amateur technical efforts. The rise of the middle-class technical amateur further encouraged the strongly masculine identification of most of these new technologies, as strong social pressures continued to make such avocations much more acceptable for young men than for young women.

In the course of the twentieth century, the amateur innovator continued to be an important source of innovation and change. The communications mechanisms built around them—popular magazines, clubs, and even national associations—became themselves distinctive and significant elements of the culture of improvement. The amateurs also became important outlets for individuals with technical expertise that might have been redundant otherwise—men who, for example, were trained in radio, automobiles, or aviation in the course of the world wars but whose skills had little or no commercial value at war's end. Later in the twentieth century, in electronics generally and later in computing, the line between amateur and expert skill was blurred considerably, and it became accepted that certain types of technical products, from shortwave radio receivers to computer software, might easily come from the host of hobbyists who spent endless unpaid hours "tinkering." In a world in which much technical knowledge appeared to be increasingly esoteric and out of reach of those without special education, these alternative sources of improvement became useful and important outlets, often crucial sources of fresh input and unorthodox ideas in an increasingly corporate technical culture.

21 Dynamics

Our knowledge of the properties and laws of physical objects shows no signs of approaching its ultimate boundaries.... This increasing physical knowledge is now, too, more rapidly than at any former period, converted by practical ingenuity, into physical power.... From this union of conditions, it is impossible not to look forward to a vast multiplication and long succession of contrivances for economizing labour and increasing its produce: and to ever wider diffusion of the use and benefits of those contrivances.

John Stuart Mill's *Principles of Political Economy* first appeared in 1848, and it was, arguably, one of the most lasting products of that year of upheaval. In it, the prodigiously overeducated Mill, finally getting out from under the overbearing influence of a famous father, James Mill (who had been determined that his son should be an example for his theories of education and psychology) sought to bring together the best of liberal economic thought as he saw it. In terms of economic ideas, there is not that much startlingly original in the *Principles*, but the work long remained one of the most popular expressions of the core economic and social philosophy of liberal England. Mill's observations on the new capacities of science for effecting improvement were also not original with him, but they represented a particularly clear statement of the popular perception of the new sources of improvement in the mid-nineteenth-century world. At the time that Mill wrote, the relationships between "physical knowledge" and "physical power" were being dramatically transformed. There were many contributors to this transformation, but perhaps the most important instigator was another famous son of a famous father, writing a generation before Mill.[1]

Sadi Carnot was the obviously brilliant son of one of France's most accomplished engineers, Lazare Carnot. The father had been swept up in the republican fervor of the French Revolution, and when his eldest son was born in 1796, Lazare was a member of the Directory that governed France until the coup d'etat that eventually led to Napoleon's dictatorship. Sadi entered the École Polytechnique at the unusually young age of sixteen, and had no difficulty in finishing the school's courses in two years, whereupon he entered the École du Genie, carving out a career for himself, largely unaffected by the shifting political fortunes around him, as Napoleon was set aside and the Bourbon Restoration returned monarchy to France. What the younger Carnot did respond to, however, was the shifting technological world he was able to observe, especially when he left active service to spend some time in Paris. There he sought to extend his education, and in so doing he wandered into the Conservatoire des Arts et Métiers and there discovered the steam engine. At the Conservatoire, Nicolas Clément was studying the efficiency of steam engines, attempting to find better ways of measuring the heat put into the engines and the work gotten out. Carnot found the studies fascinating and embarked on his own series of investigations, which resulted in 1824 in a 118-page book that was one of the most important ever written by an engineer.

Machines which are not driven by heat, those which are driven by the power of men or of animals, by a fall of water, by the movement of air, etc., can be analyzed down to their last details by mechanical theory. Every event is predictable, all possible movements are in accordance with established general principles which are applicable in all circumstances. This, of course, is characteristic of a complete theory. A similar theory is obviously required for heat engines.[2]

These lines announce the key aim of the *Réflexions sur la puissance motrice du feu et sur les machines propres à développer cette puissance* (Reflections on the motive power of fire and on machines designed to develop that power). The better informed of the book's French readers would have recognized the special aptness of Carnot's goal, for his father, Lazare, had been one of the key contributors to the theory of machines "not driven by heat." Sadi now set out to do for steam engines what his father and other French Newtonians had done for machines that could be described using classical mechanics. The analogy went well beyond goals; Carnot's entire approach to understanding how one gets work from heat was based on using the same tools of analysis that earlier investigators had applied to water power. A key passage from the *Réflexions* made this quite explicit:

In accordance with the principles we have now established, we can reasonably compare the motive power of heat with that of a head of water: for both of them there is a maximum which cannot be exceeded, whatever the type of hydraulic machine and whatever the type of heat engine employed. The motive power of a head of water depends upon its height and the quantity of water; the motive power of heat depends also on the quantity of caloric and on what may be called—on what we shall call—the height of its fall, that is on the temperature difference of the bodies between which the caloric flows.[3]

It did not matter that Carnot spoke in the older language of "caloric," applying the notion that heat was some kind of substance. The analysis that emerged depended not on what heat was, but on how it behaved, and most specifically on how it moved. In framing his questions in this way—just in the way his father had asked about how machines in general could be studied—Sadi Carnot was providing an example that would be followed in the coming decades by a host of improvers, not simply of heat engines but of everything from water wheels to electric motors and even entire factories. The key was to measure, analyze, and control *flow*, whether the subject be heat, water, air, electricity, or even human beings themselves. The word adopted to describe the study of such flows was "dynamics," and the new character of technical improvement in the nineteenth and twentieth centuries was best exemplified by the emergence and application of new kinds of dynamics—thermo-, hydro-, aero-, electro-, and human. New and better ways of doing things were to be discovered by concentrating on things that could be measured; the immeasurable and unquantifiable were to be expunged as much as possible from the vocabulary of improvement.

Carnot himself was not able to realize the improvements in steam engines that were ahead; he died only a few years after the *Réflexions* appeared, at the age of thirty-six. His book was not widely understood, so it was not until it was popularized by a fellow Polytechnique graduate, Emile Clapeyron, that the implications of Carnot's approach came to be appreciated. Clapeyron devised a somewhat more useful mathematics to apply Carnot's analysis, and he also introduced to a wider audience a graphical means—the so-called indicator diagram—for depicting the working cycles of a steam engine (figure 21.1). This had been used by a few earlier engineers, including James Watt himself, but they guarded the tool, while Clapeyron clearly received great pleasure in publicizing its utility. Clapeyron's diagram helped to emphasize one of Carnot's great conceptual contributions, that the working of an engine—of any engine—is ideally thought of as a *cycle*, and that this cycle was, again speaking in ideal terms, completely reversible. In this ideal cycle, heat is added to water, turning it to steam; the steam does work by continuing to expand in a cylinder; the steam then loses heat as it is condensed back to water, and the cycle begins again. The perfect heat engine was one in which the first part of the cycle was done with all the heat going to increase the volume and then the pressure of the steam, and the latter part of the cycle was carried out with all the heat being removed from steam and then condensed without any heat being lost outside of the engine. All the work in such a device comes from the heat "falling," as it were, from a high point in the furnace to a low one in the condenser. This so-called Carnot cycle was to be the model of what "efficiency" meant in an idealized world, a kind of limit and yet also goal for the improvers of the steam engine and the other prime movers to come.[4]

In the coming decades, a number of European scientists—both amateur and professional—developed the implications of Carnot's work into the new science of thermodynamics. In the 1840s, the German physician, Julius Robert Mayer, and the English brewer, James Prescott Joule, each published experiments in which they announced measurements of the "mechanical equivalent of heat." That such a value existed was implied by Carnot's work—a given amount of work (measured in, say, foot-pounds) should equal a specific amount of heat (the amount, for example,

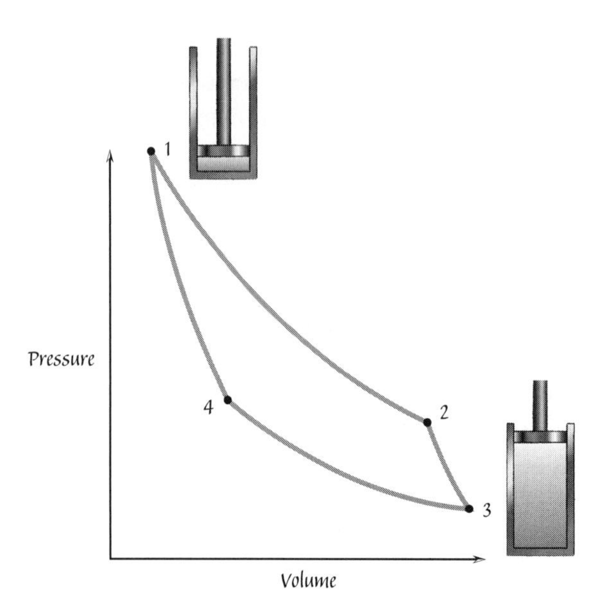

Figure 21.1
The indicator diagram is a means of showing the relationship between the pressure and the volume of gas in an engine cylinder (as a percentage of the stroke). At 1 the pressure is highest as the engine begins its working stroke, expanding the volume of gas as the piston rises; at 2 the working stroke largely ceases, the pressure drops sharply (to 3), and the pressure gradually begins to rise again as the piston falls. At 4, the piston has dropped most of the way but steam continues to enter the cylinder, increasing the pressure back to point 1. This version shows an idealized steam engine cycle.

required to raise the temperature of one pound of water by one degree Fahrenheit). A further implication, suggested by Joule, was that heat or work or (as the result of other experiments carried out by Joule) electricity were all different forms of the same thing, which Joule called "force." In 1847, the remarkable German scientist—physiologist, physicist, philosopher—Hermann von Helmholtz published a paper that made this idea explicit and mathematical, calling the principle the "conservation of force." Back in Britain, the new professor of natural philosophy at the University of Glasgow, William Thomson, was wrestling with the contradictions that he perceived in Carnot's notion that heat must be conserved in heat engines and Joule's demonstration that heat was lost when it was converted into work. Almost in passing, in one of his first papers on the problem, Thomson remarked, "Nothing can be lost in the operations of nature—no energy can be destroyed." This was the first use of the term "energy" in its modern physical sense, as a fundamental and indestructible component of nature. A few years later, in 1850, the German physicist Rudolf Clausius brought together the implications of the earlier work by enunciating two fundamental principles, later put into the form of two "laws" of thermodynamics.

The first of these was that energy could be neither created nor destroyed; the second stated simply that heat always tends to flow from warmer bodies to colder ones (and that work must be done to make it do otherwise). Within a few years, the new science of thermodynamics (Thomson's coinage) was a recognized means for examining the action and conversion of a host of different energy forms.[5]

Over the next several decades a number of observers noted the larger implications of the laws of thermodynamics for the Western worldview. In particular, much was made of the implications of Clausius's second law, which came to be stated in a number of forms, but which was popularized in the notion that the universe was "running down." Since heat could only naturally flow in one direction, from hotter to colder, there would come a point at which everything was the same temperature, heat no longer flowed, and no work at all could be done. This so-called heat death of the universe appeared to put a finite limit—at least in theory—on history, life, and time itself, and philosophers of all stripes had their take on just what this meant for their views of the human future, morality, and progress.

Almost as striking as the notoriety given to thermodynamics by the philosophers was its general irrelevance to the improvement of the steam engine itself. It was said many years later that "science owed more to the steam engine than the steam engine owed to science." This was, in fact, largely true, for through most of the nineteenth century the power, efficiency, and reliability of steam engines improved steadily and markedly, but this was due to incremental, largely empirical improvements rather than significant theory-based redesign. As the steam engine became more essential for every aspect of productive life, from factories to water supply systems to the ubiquitous railroads and steamboats, the construction of engines in an enormous range of sizes and designs became a competitive and fast-changing industry itself. In this environment, the design and manufacture of engines was the subject of constant experimentation and refinement. The experiments included new and more complicated designs, such as double- and then triple-expansion engines, in which successive cylinders were used to eke every bit of work out of a given quantity of steam. Just as important, however, were smaller scale and less obvious sources of improvement, such as better quality iron (and later steel), better materials for packing pistons or for lubricating surfaces, and more precise machining and closer fits of critical parts. The typical steam engine at century's end was easily twenty times more efficient than an early Boulton & Watt product, and far surpassed the earlier machine in reliability and power. While thermodynamics could often help explain *why* one design worked better than another, the new science was hardly ever the actual source of innovation. The engineers gradually absorbed the language and the mathematics of the physicists, adapting it for their own practical purposes, but they did not give ground to the scientists in the generation of novelty and improvement—this was still, for the steam engine, at least, their domain.[6]

The contributions of thermodynamics to technology were less to the improvement of steam engines and more to the encouragement of alternative paths for designing prime movers. In particular, the focus on heat and the ideal ways in which heat might be made to do work inspired some to seek ways of making the action of heat more direct than converting water to steam. Whereas a steam engine relied on "external combustion"—the fire in a boiler—thermodynamics encouraged finding means of harnessing "internal combustion," and the resulting inventions launched a new industry with profound implications, particularly for transportation. Possibly as long ago as Leonardo da Vinci, experimenters had speculated about the possibility of building an engine that would use explosive power, typically gunpowder, as a source of useful work. By the end of the seventeenth century, a number of designs had emerged, and such well-known experimenters as Christian Huygens and Denis Papin had described how they might build such gunpowder engines. Practical problems, with both fuel and materials, combined with the success of the steam engine in the eighteenth century to divert attention elsewhere, but the lessons of Carnot and the thermodynamicists gave renewed life to experiments in this direction.[7]

Even before Carnot's ideas were widely known, another factor had encouraged a few experimenters to revisit internal combustion—the discovery of gaseous fuels. In 1766 Henry Cavendish, the reclusive heir to one of the greatest fortunes in England, but also one of the most skilled scientific experimenters of his generation, announced the discovery of a gas that he dubbed "inflammable air." It will be remembered that the light weight of this material, which we know as hydrogen, was one of the primary stimuli to the French balloon experiments that took place a few years later. The inflammability of the substance, too, was the subject of technical experiments. It was too expensive to produce to be used as a general-purpose fuel or illuminant, but it did provoke some ideas about engines that could use a gaseous or liquid fuel. Much more important in this connection was the discovery that the gaseous by-product of producing coke from raw coal was both cheap and a good lighting source (it is primarily hydrogen and methane). While this had been observed in the mid-eighteenth century, its technical significance was not exploited until the French engineer and chemist Philippe Lebon demonstrated a small lighting system in Paris and an English mechanic, William Murdock, independently, devised his own gas-lighting system in the Birmingham factory of Boulton & Watt. Lebon, who was murdered on a Paris street before realizing much return from his efforts, was well aware that his illuminating gas opened up the possibilities for an internal combustion engine. After the mid-1810s, gas lighting spread rapidly in the cities of Europe and North America, and experiments on gas engines followed. Various versions of these incorporated, individually, numerous useful features, such as valves, compression mechanisms, and electric spark ignition. Nothing, however, could compete with the ever-improving steam engine for power, economy, or reliability.[8]

The possible attractions of an internal combustion engine kept experimenters returning to the concept, and the lessons of the thermodynamicists provided additional encouragement, for the ideal cycle of Carnot fostered efforts to keep the working elements in an engine as close as possible to the source of heat. More significant to the emergence of practical internal combustion engines in the 1860s, however, was the increasing demand for efficient, flexible power sources. As the efficacy of machines grew, with new machine tools and new standards of manufacture making powered machines attractive features of every kind of production, from the giant factory to the small craft shop, machine users became more sophisticated in the demands made of their power sources. By midcentury, waterpower still moved much of industry in Britain and America (not to mention elsewhere), but the simple capacity of steam engines to be used almost anywhere made the attractions of flexibility manifest. Since the days of James Watt, however, it had been recognized that reducing the size of steam engines reduced their fuel efficiency, and the need to keep boilers going whether machines were running or not would always be a drain on economy. Hence, when the Belgian-born J. J. Étienne Lenoir introduced for sale a gas engine of one-half horsepower in 1860, he achieved a remarkable measure of commercial success, particularly considering the high cost ($500 for the half-a-horsepower engine in the United States) and the rough running, high fuel costs, and unreliability that characterized these first engines. Lenoir actually did not claim great originality for his product, but rather an effective combination of the inventions of a half-century of experimentation with internal combustion. Production of his engine lasted less than a decade, but Lenoir's extensive efforts to popularize and market the machine gave the idea of a compact, easily operated, flexible engine, working on readily available fuel, great notoriety. The students of Lenoir's engine created a new technology that made the look and feel of both industry and transport in the coming twentieth century far different from that of the nineteenth.[9]

Lenoir displayed his internal combustion engine at the London International Exhibition in 1862, and received considerable attention, as well as the contacts that soon translated into English, American, and other foreign licenses. Within a few years other manufacturers appeared, primarily in France, attempting to improve on Lenoir's design, whose operation had plenty of obvious faults, from noise to fuel consumption. Experimentation on internal combustion was widespread, for the sales of the Lenoir engine and the favorable press attention it received signaled a latent demand for a small, versatile prime mover. Among those trying their hand at improving the machine was a German traveling salesman (largely of tea, sugar, and kitchenware), Nicolaus August Otto. Within months of reading descriptions of Lenoir's invention, Otto devised modifications and had an instrument maker in Cologne make a small model engine. Over the next few years Otto continued his experiments, and in early 1864 he attracted the attention of a Cologne sugar refiner, Eugen

Langen. Langen caught the internal combustion fever and put his money and business acumen on Otto's experiments. By 1867, Otto and Langen had an engine, which they displayed at the Paris International Exposition. Their machine was awkward and noisy, but when the World's Fair judges discovered that it consumed less than half of the gas per unit power of competing machines they awarded it a gold medal, and industry orders from around Europe quickly followed. Meeting the orders was difficult and within a few years additional capital was imperative. Thus in early 1872 was organized the firm of Gasmotoren-Fabrik Deutz (from the Cologne suburb where it was located). A few months later the firm hired Gottlieb Daimler, a polytechnic-trained engineer, to oversee production. Daimler, in turn, brought with him a young designer, Wilhelm Maybach; together the two of them were to parlay their experience with Otto's firm into the creation of the automobile industry.[10]

Before the internal combustion engine was to usefully propel motor cars, or anything else, for that matter, its basic design and principles would still have to be changed. The Deutz firm manufactured the Otto and Langen engine, with modifications by Daimler and others, for several years with great success. But whatever improvements it may have represented over Lenoir and other early models, the design still had drawbacks that were particularly obvious to an old school friend of Langen, now an engineering professor in Berlin, Franz Reuleaux. The machines being made at Deutz were limited in power to a maximum of about three horsepower; more powerful ones simply experienced too much shock in working to be reliable or smooth running. It was in an effort to transcend this limitation, above all, that Otto sought to change the principles of the machine's operation. After consulting with Reuleaux, Otto produced a revolutionary new design, introduced in 1876 as the "Silent Otto" engine (figure 21.2). The key innovation was the so-called four-stroke cycle, in which each explosion of fuel and air in the cylinder was accompanied by four strokes of the piston: (1) intake, (2) compression, (3) ignition, and (4) exhaust. Many technical problems had to be solved before this engine was ready, but when it appeared on the market in early 1877, it was immediately a commercial success. With only minor modifications over the next couple of decades, thousands of silent Ottos appeared, not only in Germany but widely abroad, where the Deutz firm was able to sell lucrative licenses to machine manufacturers anxious to provide a prime mover that would allow small shops and other users to have reliable, versatile, and economical power available whenever they needed it.[11]

The internal combustion engine became one of the most popular targets of improvers over the next few decades. Dugald Clerk, in Scotland, introduced a practical two-stroke engine in 1880 which provided a simpler, if less economical, alternative to the Otto engines. Daimler showed how to make versatile and powerful engines using liquid, rather than gaseous, fuels. Even more important, however, the drive to make engines more efficient—to achieve more closely the thermodynamic

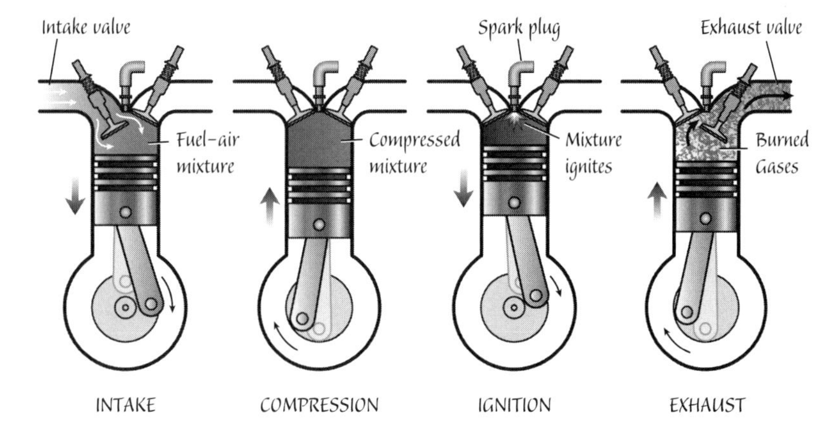

Figure 21.2
The four basic strokes of the Otto internal combustion engine. The adoption of the four-stroke engine made internal combustion power plants smoother running and more fuel efficient.

ideal—inspired a young German mechanical engineer named Rudolf Diesel to strike out in an important new direction. When he was a student at the Technische Hochschule in Munich, he heard lectures about the thermodynamic inefficiencies of the steam engine, and he said later that this provided the initial spur for attacking the problem of designing what he called in an 1893 treatise, "a rational heat engine." Diesel's Munich teacher, Carl Linde, achieved his own success at applying thermodynamics to practical problems by designing the first practical refrigerating machines. Using academic and social connections, Linde was able to get backing in the 1870s and 1880s, particularly from the Bavarian brewing industry, for the manufacture and sale of his refrigerators. Diesel found it a bit more difficult to get support for his own radical modification of the internal combustion engine, but in 1893 he was able to persuade the Augsburg Engine Works (primary manufacturers of Linde's machines) to construct a working prototype of his engine, which differed from other internal combustion engines by working at very high compressions—so high that ignition in the engine required no spark or other heat source but came from the heat of compression itself.

The technical problems of achieving this result were formidable, and it took more than four years of experiments before Diesel and his backers had a motor that could conceivably be marketed. The fuel savings per unit of power over the Otto-type engine were substantial—the Diesel engine, when working properly, used only about half the fuel of even the best of the traditional motors. But making an engine that worked in the real world, that operated readily and reliably, required still more years of experimentation. The development effort, in fact, turned out to be more than Diesel himself could manage, and it was only through the dogged efforts of other

German engineers that the engine slowly, in the first years of the twentieth century, found markets and users, particularly where large sizes of reliable and economical internal combustion engines were wanted. By this time, however, Diesel's own opportunities to reap the rewards of his vision had slipped away, and, apparently in a fit of depression, in 1913 he threw himself overboard from a ship in the English Channel.[12]

Internal combustion was not the only direction for improvement of heat engines, although it was the direction most influenced by the new science of heat. Another kind of steam engine appeared in the late nineteenth century, inspired by improvements that were being made in water power, particularly by the emergence of the water turbine as a practical machine. The impetus for the development for the steam turbine came largely from the desire to produce an engine that could sustain high rotary speeds. In Sweden, for example, Gustav de Laval invented a reaction turbine in 1882 to run a butter separator that he had introduced. The principle of using the motion imparted by jets of steam exiting from nozzles had been known as far back as Hero of Alexandria in the second century BC, but from Hero's time forward there had been no particular utility to a machine that revolved at very high speeds. De Laval's butter churn was hardly much incentive for further inventive effort in this direction, but the generation of electricity emphatically was. This was recognized by a young British engineer, Charles Algernon Parsons, who had something of the fever to become a famous inventor. In 1881, Parsons set out to design the best machine for powering the electrical generators that were just then beginning to be installed to power electric lamps, either the bright arc lights that had begun to illuminate public spaces or the brand-new incandescent lamps that were emerging from the laboratories of Thomas Edison and his rivals. Electrical generators were machines that gained greatly by running at high speeds, but such running put unusual stresses on traditional steam engines, particularly in small sizes. In a visit to America in 1883, Parsons observed some of the advanced water turbine technology that was being used in the more modern water-power establishments, and he took what he saw back to his workshop to see how he might translate what he had seen into a steam device. In 1884 Parsons constructed a turbine that operated at 18,000 revolutions per minute and generated 7.5 kW of electricity. Over the next few years the steam turbine gave the engineers yet another key field for improvement. De Laval himself designed a machine operating at 30,000 rpm before the decade was out, and in 1900 Parsons built a turbine generating 1,250 kW while achieving a fuel efficiency seven times better than his original machine. Some of the wider implications of the invention became widely public when, in 1897, Parsons ran his 100 ton vessel, *Turbinia*, at a hitherto unheard-of speed of 34.5 knots. In the twentieth century, both steam turbine and diesel engines completely rewrote the rules of marine propulsion, with implications for both civilian and military uses.[13]

The work that had inspired Parsons came from another area of technology in which scientific investigation was remaking the basic principles of design and operation, water power. Here, too, we can see the complex interaction between steadily advancing practice and emerging theoretical fields, in this case, hydrodynamics. Waterwheels continued to be the most important prime mover for industry in both Britain and America well into the nineteenth century, despite the growing importance of steampower. Their design was therefore a constant target for improvers. Early in the eighteenth century some observers began attempting to apply some of the principles and methods of Newtonian physics to waterwheel design, but both conceptual and mathematical tools were simply too crude to be useful. In midcentury, however, systematic study of waterwheel design and performance, using models and ingenious methods for measuring power, began to make a difference. In England, a young London instrument maker, John Smeaton, carried out a series of experiments on models in the 1750s, checking and re-checking his results with great care before publicizing them in 1759 (see figure 11.3). He established that overshot wheels were by their nature more efficient than undershot wheels, and were to be preferred whenever practical. French authors (among them, Sadi Carnot's father, Lazare) took up the same work and published a number of treatises, attempting to outline a theory of water-powered machines. In the first decades of the nineteenth century these efforts began to bear fruit in practical designs, such as the wheel of J. V. Poncelet, described in the 1820s, which used careful control of water impact and of the shape of the vanes of the wheel to reduce shock and to maximize the usable force on the wheel. French engineers and their sponsors promoted the idea of the "rationally designed" waterwheel, and this encouraged further, more ambitious experiments. The result was the first appearance of the modern water turbine.[14]

While the development of steam turbines carried a distinctly British stamp, and that of internal combustion engines a largely German one, the emergence of the water turbine as an important prime mover—today still important in hydroelectric generation—was largely a French and American affair. The successful improvement of the conventional waterwheel by Poncelet and other systematic experimenters encouraged more French engineers to seek further advances. Benoît Fourneyron was a young mining and metallurgical engineer who was in charge of waterpower installations. After a lengthy series of experiments in the mid-1820s, Fourneyron built and tested a small (6 hp) outward-flow radial turbine, which demonstrated the usefulness of the turbine concept. Over the next several decades Fourneyron continued to design and build turbines of generally the same form, which was soon superseded by alternatives. In the 1840s a number of French engineers introduced axial, inward-flow designs, and the field of turbine invention was flung wide open.

The most enthusiastic entrants came from a surprising quarter, the United States. The textile and other mills of American industry ran overwhelmingly by waterpower, and by midcentury the improvement of power technology was of widespread interest, nowhere more so than in the mill town of Lowell, Massachusetts. There, Uriah Boyden, an engineer for one of the larger mills, designed and installed a Fourneyron-type turbine in 1844, and he proceeded to make the water turbine a respectable power source in New England mills. Boyden charged customers on the basis of how much water he saved them, so he made very careful measurement of the performance of his turbines. He made the data available to the chief engineer for the Lowell Manufacturing Companies, James B. Francis, who proceeded to make an extraordinary series of tests of the various wheel and turbine designs available to him. Francis analyzed his data with great care, deriving his own design for an ideal turbine, which combined aspects of both radial and axial designs ("mixed-flow" it was called). The Francis turbine was arguably the first American contribution to technology that owed as much to systematic testing and measurement as it did to mechanical ingenuity. Francis published the results of his experimental series in *The Lowell Hydraulic Experiments* in 1855, and thus lay the foundations for continued American achievement in hydraulic engineering.[15]

The rather sudden appearance of American expertise in waterpower innovation and in the systematic investigation of turbine design was remarkable, and the Francis experiments became well known both at home and abroad. But it must not be forgotten that improvement in water power was still largely an empirical affair. The other great American contribution to design, for example, was largely the work of Lester Pelton, a California manufacturer of waterwheels who had been drawn west in the 1840s Gold Rush. The so-called Pelton wheel was a modification of a homegrown design of the gold fields, known as a "hurdy-gurdy," which used the high pressures of the fast-flowing streams of the Sierras, directing water through a nozzle directly on to triangular buckets. The turbulence of water in these simple buckets was an obvious source of inefficiency, so experimenters produced many alternatives, with curved designs, particularly useful in iron, as opposed to wooden, wheels that became popular in the 1860s. Pelton's contribution consisted largely in studying the shape of the buckets with great care, varying them in a range of experiments, until he devised a distinctive curved split bucket that became the hallmark of his products. Patenting this design in 1880, Pelton was able to turn this into a remarkable commercial success, supplying much of the world with Pelton wheels within a decade of its introduction. The Francis and Pelton turbines owed little to the rapidly advancing field of hydrodynamics, but, just as in thermodynamics, the new science allowed investigators to explain with confidence just why the American innovations worked so well.[16]

Aerodynamics was yet another field of science that grew out of practical concerns and which, under the influence of Newtonian physics, was developed into an increasingly sophisticated mathematical specialty over the course of the eighteenth and nineteenth centuries. And like the other "-dynamics," this field had relatively little practical influence for most of this period. But in the closing years of the nineteenth century and the opening ones of the twentieth, aerodynamics gave birth to a dramatic new technology that became one of the core object lessons for those seeking examples of how modern invention was becoming more and more reliant on advanced science.

As long ago as Leonardo da Vinci, it had been recognized that any aspirations that humans might have about flying would need to be better informed about the realities of bird flight and of movement in the air. Leonardo was captivated by the idea of flight, and he made extensive studies of birds and their forms and movements. He was acutely aware of what was lacking in human understanding of flight. In his *Codex on the Flight of Birds*, he wrote, "A bird is an instrument working according to mathematical law, an instrument that is within the capacity of man to reproduce with all its movements." But Leonardo, of course, lacked the very mathematical tools that would make such a reproduction feasible. These tools were supplied, in large part, over the course of the seventeenth and eighteenth centuries by the like of Isaac Newton, Daniel Bernoulli, and Leonhard Euler. The same physical principles and mathematical tools that emerged from the study of fluids could, in large part, be generalized to incorporate gases and air.[17]

In the course of the eighteenth century, it became evident to some that there was more practical use to be made of the systematic study of movement in air than simply trying to emulate the birds. Benjamin Robins, for example, was a brilliant English military engineer who grasped that a full understanding of a projectile in flight—such as ball shot from a cannon—required the study of the interaction of the moving body with the air around it. He devised experiments to measure the different forces of air as well as the friction or drag experienced by bodies in motion. For Robins the importance of his work lay principally in the improvement of gunnery, but for others it contributed to an emerging science of air and motion. Not long after the death of Robins in India at the age of forty-four, John Smeaton extended the pioneering work he did on improving watermills to windmills, adopting Robins's experimental apparatus to allow him to measure the forces at work on a windmill's blades. Smeaton was also apparently the first person to appreciate the effect of curving the surfaces over which air flowed, that what we would call "lift" increases when air flows over a curved surface rather than a flat one. For all of the claims for practical uses for aerodynamics, it continued to be the vision of flight that spurred on the most important work in the nineteenth century. Yet another Englishman, George Cayley, captured this vision most clearly at the century's outset, applying his understanding of the findings of Robins, Smeaton, and other experimenters to the design, in 1804, of the first practical glider (figure 21.3). Cayley's work illustrated one of the recurring themes of nineteenth-century technological improvement—while

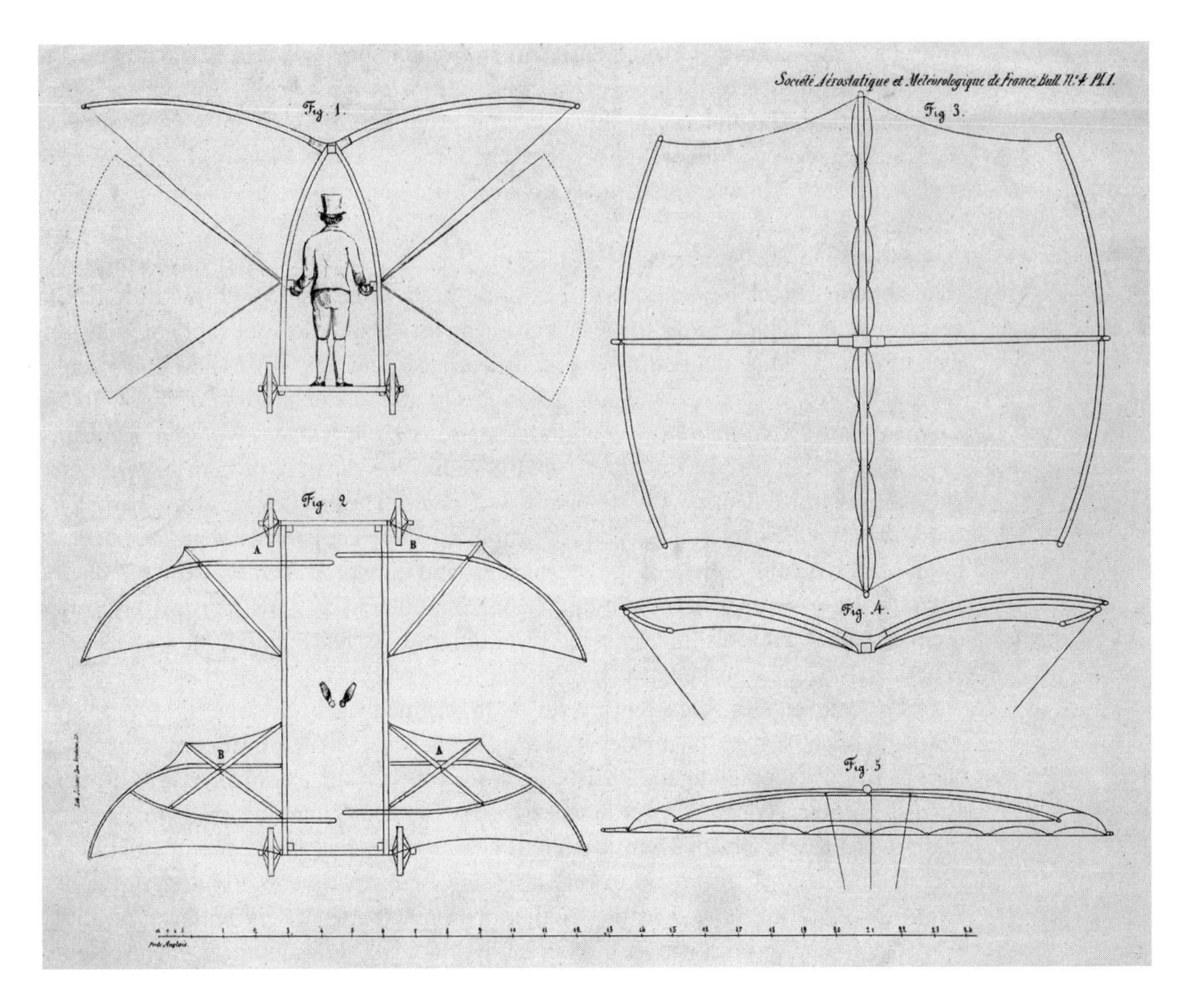

Figure 21.3
George Cayley's drawing of a human-powered flying machine, 1853. Library of Congress.

he knew of the mathematical tools that had been created by the Newtonians and others, he found them of little or no use in the actual design of his glider. Experiment and experience wedded to ambition provided the real impetus for improvement.[18]

Cayley himself did not win immediate fame for his pioneering work, but the principles and the goals of flying machines nonetheless slowly gained popularity in the course of the nineteenth century. In the latter half of the century, organizations sprang up in a number of countries with the goals of making aeronautics respectable and making aerodynamics truly useful. The Aeronautical Society of Great Britain, for example, began in 1866, following an earlier but more modest French effort,

Figure 21.4
Otto Lilienthal glider, ca. 1895. Library of Congress photograph.

and rapidly drew attention to the outstanding problems of understanding flow in the air. One of the society's first papers was a work by Francis Wenham that sought to set out some basic principles for determining a wing's size and shape, with remarkable insight on the fundamentals that were to turn out to be the most useful. At this point, however, both the experimental and the theoretical tools were still too primitive to make clear what was truly useful and what was not. Thanks to Wenham and some of his colleagues, however, the experimental tools began to become substantially more powerful. The first wind tunnel was supported by the Aeronautical Society in the 1870s, and although much of the early data was flawed, the new tool gave encouragement to many seeking to understand what challenges had to be overcome.

The most significant user of such data was a German mechanical engineer, Otto Lilienthal, who became the first man to fly with wings (figure 21.4). The application of aerodynamics to flight took an enormous step in the late spring of 1891 when Lilienthal constructed a glider large enough to carry him and made a lengthy series of airborne hops off of a small hill near his home in Berlin. These glider flights were taken only after two decades of careful measurements and tests; this is what differentiated them most clearly from the long history of poorly informed, and often ill-fated, flying efforts that had made aeronautics largely a joke among a skeptical public. By the mid-1890s Lilienthal's fame was widespread, and heavier-than-air flight had achieved a measure of respectability. This was further increased when, in May, 1896, Samuel Pierpont Langley, head of the Smithsonian Institution, announced the success of his "aerodrome," a steam-powered craft catapulted from a houseboat on the Potomac River down from Washington, D.C. The unmanned propeller-driven machine was able to climb almost a hundred feet into the air and sustained a flight of almost a minute-and-a-half before it ran out of power. The promise of aerodynamics applied to true powered flight appeared to some observers to be only a matter of time.[19]

Despite the progress that had been made in aeronautics, particularly during the 1890s, the accomplishment of the Wright Brothers was nonetheless a substantial one. Their achievement—manned, powered, sustained, controlled flight—was the product of several years of extraordinarily intense and intelligent research. As singular as their accomplishment may have been, they were clearly well aware at the time of the degree to which they were improving on earlier work, particularly that of Lilienthal and Langley. Their success has been analyzed and explained at great length by a number of scholars, but it can be simplified into three basic achievements. First, they understood applied aerodynamics better than anyone else, thanks to careful and perceptive experiment and measurement. Second, they took advantage of new achievements in engine technology to equip themselves with the very best power plant available for the purpose. Third, they adopted an approach to control of their aircraft that was distinctly different from that of other pioneers, an approach that, it can be argued, was a technological dead end but that still, in 1903, was just what was needed to make the available technologies work in a flying machine.[20]

Starting as late as 1899, Wilbur and Orville Wright, owners and operators of a bicycle shop in Dayton, Ohio, began their research by gathering together all they could find (in English, since they knew nothing else) on flying and flying machines. The people at Langley's Smithsonian, used to the kind of inquiries they received from the brothers, were particularly helpful, and in short order they felt equipped to build themselves a glider. The shortcomings in the unmanned glider's performance, followed by difficulties in a model that was large enough to carry Wilbur aloft in a very strong wind, suggested to the brothers problems in the aerodynamics of their design. This had been based on the best data they could find, particularly in tables constructed by Lilienthal. After almost two years of glider experiments, they concluded that the test measurements and tables used by Lilienthal and everyone else were so flawed as to be essentially useless, and they proceeded to construct a wind tunnel and make measurements of their own to determine the proper shape of a wing and the proper angle of attack for flight. This was an enormous task—and a bold rejection of widely accepted scientific data—but it was a key to the brothers' eventual success. By mid-1902 they had enough information to design an entirely new craft, and when they began testing it at Kitty Hawk, North Carolina, they confirmed the lessons they had newly learned about wing shape and size. By the end of that year they felt confident enough to begin shopping for a suitable engine, something that would, as they specified it, generate up to nine horsepower but weigh no more than 180 pounds. They eventually had to construct their own engine, one that was marginal in performance but adequate. At the same time they applied their new aerodynamic knowledge to the design of a much more efficient propeller, and by late 1903 they were ready to test the whole.

The craft that the Wright brothers and their helpers constructed at Kill Devil Hills in November, 1903 was quite large—a biplane with a total wing area of 510 square feet and a span of 40 ft., 4 in. It differed from the designs of Langley and others in many important ways, but the most significant was perhaps that the Wrights had made a point of making their machine intimately controllable by the pilot. A whole series of linkages were built into the plane so that the pilot could adjust the machine's attitude in every dimension. Most novel were the cables connected to the ends of the wings that allowed the pilot to twist or "warp" the wings in response to the plane's motion in the air. Their plane, in other words, was built to be flown somewhat like a cyclist rides a bicycle, with the capacity to respond instantly to changing motions of the air or the craft. Whereas earlier experimenters, such as Langley, had focused on making the flying machine as *stable* as possible, the Wrights were determined to make it as *responsive* as could be. For several weeks they were plagued with engine or propeller problems, but on December 17 they were ready. With Orville at the controls, their "Flyer" lifted off the ground, Wilber running alongside to steady the wings, and flew for twelve seconds, covering a distance of 120 feet (figure 21.5). This brief fling was enough to tell them that their calculations had been correct and their contraption was a true flying machine. Three further flights that day confirmed the fact; the last one lasted almost a full minute and covered 852 feet. Manned, powered, and controlled heavier-than-air flight—aeronautics in its true modern sense—began on that day.

Aeronautics advanced rapidly in the next decade or so, although from the beginning, patent restrictions, national rivalries, and widespread uncertainty about the real usefulness of human flight interfered with some efforts. The Wrights themselves were greeted with incredulity in their first attempts to promote their invention, and it was really not until an enthusiastic reception in Europe—fueled, no doubt, by greater sensitivity to possible military implications—made them famous that Americans began to hail their achievement more generally. By World War I, however, less than eleven years after the flights at Kitty Hawk, the airplane had emerged as a technology

Figure 21.5
Perhaps the most famous photograph in the history of technology shows the Wright Brothers' Flyer lifting off on its first flight at Kitty Hawk, North Carolina, with Wilbur at the controls and Orville running alongside. Library of Congress photograph.

worthy of great attention and improvement. The war itself saw only modest departures from the basic model provided by the Wright Flyer (more powerful engines probably made the biggest difference at this point), but the years following the war's end witnessed dramatic changes in every aspect of the airplane's design. Advances in materials and structural engineering contributed significantly, but the role of an ever more sophisticated understanding of aerodynamics was equally key.

Even before the war, mathematicians and physicists made progress in better understanding the basis for the Wright Brothers' success and began constructing the theoretical tools that would inform aircraft design for years to come. In the German university city of Göttingen, Ludwig Prandtl, for example, began elaborating his concept of the boundary layer (the layer of air dragged along with a moving body) as early as 1904, and his laboratory churned out data from wind tunnel and other experiments that gave German airplane builders some advantages in their designs. Only after the war, however, did the real utility of Prandtl's concepts become more widely evident. Similarly, while some steps were taken earlier, only in the immediate aftermath of the war did the institutional foundations for both theoretical and applied research emerge (outside of Germany). In the United States, for example, the National Advisory Committee for Aeronautics (NACA) was created as an interagency coordinating group even before the country entered the war. The need to go further, however, became quickly evident, so the funds for a laboratory were appropriated, and a facility opened in Hampton, Virginia, in 1920. Aeronautics was for the rest of the twentieth century to become perhaps the premier example of how the strategic implications of new technologies would push governments to take greater control of technological improvement.[21]

Aeronautics illustrated an important fact about how governments responded to technological novelty: when a technology makes boundaries obsolete, the state gets involved. At the same time that aeronautics made the move from theory to practice, another technology emerged that followed a parallel trajectory in a remarkable number of ways, despite the fact that it occupied a very different realm. Just as aerodynamics provided a foundation for the creation of an aviation technology, largely at the hands of practical men, so too did electrodynamics—the study of the forces created and propagated by electrical circuits—give rise to an important new technology, once again largely due to efforts of practical men to find utility in the hitherto unsuspected properties of nature. The electrical technologies that emerged in the last decades of the nineteenth century—telephony, lighting, and power—were largely the creation of individuals whose experience with electricity was direct and utilitarian, not academic or theoretical. Most of these gained their initial exposure through the technology of telegraphy, which itself relied on remarkably simple and straightforward electrical phenomena: electromagnetism and currents conducted through wires. Through the middle decades of the century, the technical capabilities of the telegraph grew enormously. Practical inventors such as Thomas Edison contributed ingenuity, imagination, and hard work to this growth. At the same time, scientific investigators, exemplified best perhaps by William Thomson (Lord Kelvin), did not hesitate to tackle some of the more difficult technical challenges, such as designing instruments sensitive enough to be used for undersea telegraphy. The usefulness of electrical theory, of the systematic description of electrical phenomena by means of rigorous mathematical laws and formulae, was accepted by the entire range of electrical inventors, and this became even more true as electrical technology entered other areas, such as light and power.

The pursuit of theories and tools to describe electrical phenomena with mathematical precision began in earnest at the moment in 1820 that the Danish professor Hans

Christian Oersted announced his discovery that a electric current produced magnetic attraction around a wire. Within weeks of the distribution of Oersted's paper, his experiments were being repeated and elaborated upon. The most fruitful work was that of a French mathematician, André-Marie Ampère. Already forty-five years old in 1820, and normally indifferent to the work of physicists, Ampère was fascinated by Oersted's experiment—particularly with the fact that the magnetic action could readily be seen to act in a curve around the current-bearing wire. Forces that worked in curves were simply unknown in the physics of the Newtonian world as it was understood. Forces operated in straight lines, and when curves appeared, as in, say, the orbits of planets, they were readily understood as the result of combinations of straight-line forces (typically gravity and some tangential impetus). But Oersted's electromagnetic forces clearly operated in curves (he described them as spirals), and Ampère sought to devise precise mathematical descriptions. In only a few months he managed to make experimental apparatus that allowed him to make measurements (figure 21.6) and was able to provide at least preliminary versions of the basic laws of electrodynamics.[22]

The greatest builder of the experimental tradition in electrodynamics was Michael Faraday. Like others, he was stimulated by Oersted's discoveries to extend his understanding of the kinds of motions created by electric currents. Faraday, however, was no mathematician, so his approaches were very different from Ampère's. Faraday was one of the greatest experimental designers of the nineteenth century, and his follow-up on Oersted illustrated this well. He devised an ingenious way to show the circular motions around a wire: he suspended a wire dipped into a cup of mercury; in the middle of the cup he placed a permanent magnet; when an electric circuit was completed running through the suspended wire and the mercury, the wire rotated continuously around the magnet. This was, in a sense, the first electric motor, although it was incapable of doing any actual work. Over the next decades, Faraday contributed a host of devices and observations that helped lay the experimental foundation for electrodynamics. His descriptions were always highly visual, rather than mathematical: indeed, he even boasted of his nonmathematical approach, writing to a correspondent, "experiment needs not quail before mathematics but is quite competent to rival it in discovery."[23]

Over the next decades, through the middle years of the nineteenth century, a clearer picture, both in images and in mathematical relations, emerged of the behavior of electric currents and the forces associated with them. More like thermodynamics than hydrodynamics or aerodynamics, electrodynamics advanced even as the question remained quite open as to just what electricity really *was*. What was important was its behavior and its relationships to a growing list of other natural phenomena. Just as Oersted demonstrated a relationship with magnetism (and Faraday eleven years later showed that this relationship worked in both directions, with his

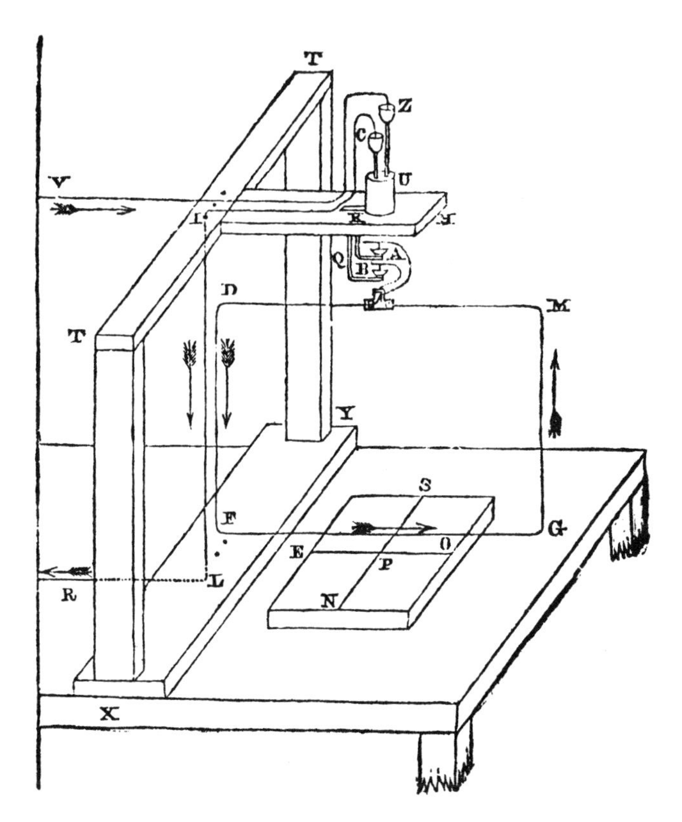

Figure 21.6
Ampère constructed this apparatus to measure the magnetic forces represented by electrical current in a wire. Burndy Library.

demonstration of electromagnetic induction), so did subsequent experimenters reveal electrical connections with heat, material composition, and light. In the efforts to make sense of these connections, a number of models of electrical behavior were formulated. The most influential of these was Faraday's conception of lines of force, a means of characterizing in a roughly quantitative way the magnetic activity—or "field"—produced by an electric current. In the 1840s and 1850s, Faraday designed experiments, published studies, and wrote up speculations in ways that provoked responses in a number of more mathematically inclined researchers. One of the first to attempt to do something with Faraday's lines was William Thomson, whose own way of modeling the world seemed particularly sympathetic to Faraday's models and speculations. But it was another Scottish-born mathematician—James Clerk Maxwell—who was to take these and turn them into the foundations of not only the most powerful model of electromagnetic behavior but also the stimulus for the creation of some of the most important new technologies of the next century.

Like Faraday, and like almost all the electrical inventors of the nineteenth century, Maxwell sought to devise visual, mechanical images of electrical phenomena, models that would help to translate the unseen electromagnetic world into a more familiar visible realm that worked by the familiar dynamics of Newton. Perhaps this was why he was able to grasp the implications of Faraday's lines-of-force image of electromagnetism better than most of his colleagues. His mathematical bent, however, demanded that a mathematical rigor be imposed on this image, and if there were elements of Faraday's model that did not lend themselves to mathematical harmony, Maxwell was not hesitant to supply additional pieces that would make the numbers work. The result of this modeling was summarized in a set of equations, put forth in a remarkable 1865 paper, "A Dynamical Theory of the Electromagnetic Field," in which Maxwell sought to explain, above all, how an electric current gave rise to a magnetic field and how changing magnetic fields induced electric currents. The resulting model indicated that the effects of these changing fields were propagated through space at a finite speed, and this speed, it was easily demonstrated, turned out to be exactly that of light. The conclusion was inescapable, both to Maxwell and to those physicists willing to go along with his model (which, at first, excluded most European scientists): light was itself an electromagnetic phenomenon. More specifically, light traveled through space as an electromagnetic wave and, it also followed, other electromagnetic waves also existed, or could be created, given the proper conditions of changing electrical or magnetic fields.[24]

Maxwell's mathematical description of the behavior of changing electrical and magnetic fields was remarkably simple and elegant for those who understood it, but for those who lacked the skills to comprehend or use his equations—and this included almost all of those who actually worked with electricity—the work was simply too abstract. Even those who understood the mathematics debated heatedly about the "reality" of Maxwell's model; there was much of it that seemed to classical physicists to be poorly supported by experimental evidence or that seemed to defy any real observational test. No one for at least twenty years imagined that Maxwell's electrodynamics held any significant implications for the technology of communications. Even when, in 1886–87, the young German physicist Heinrich Hertz devised the means to demonstrate the reality of Maxwell's electromagnetic waves, it at first seemed to be little more than a contribution to esoteric physics. Hertz's success required that he devise a means of generating the waves predicted by Maxwell, of detecting them, and of demonstrating that they were indeed of the same nature as light. He used a spark discharge to create a high frequency oscillation that would produce the predicted phenomena, and a wire loop antenna that could, faintly but visibly, respond with a slight spark when oriented correctly in the region of standing electromagnetic waves (figure 21.7). The experiment was carefully designed to eliminate other possible electrical effects as the cause of the antenna's response, and Hertz

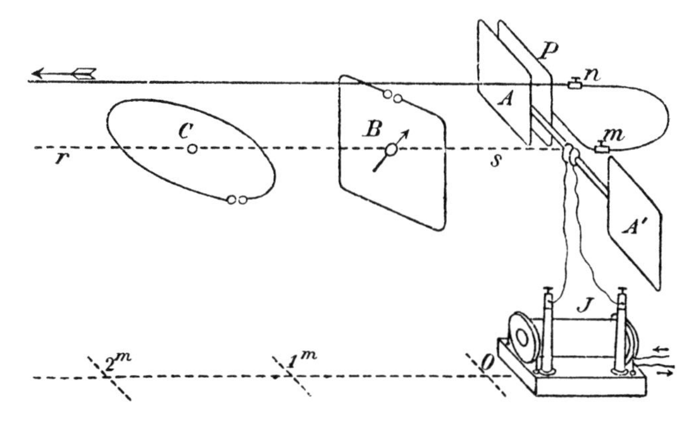

Figure 21.7
Heinrich Hertz provided this illustration to show the apparatus he used to generate and detect electromagnetic waves. Burndy Library.

was able to demonstrate a range of optical-like effects, such as reflection from plane surfaces and focusing with parabolic "mirrors." The effects were convincing to the physicists, but depended on very precisely constructed apparatus and close observation, all in the space of a single large room. That these same effects might give rise to important new technologies was by no means obvious.[25]

The physicists of Europe and America took instant note of Hertz's experiments, and a number of them, like Oliver Lodge, professor of physics in Liverpool, improved the apparatus and the results in producing what came to be called "Hertzian waves." But several years passed before even a glimpse of technological possibilities appeared. Early in 1892, William Crookes, one of Britain's best-known physicists, published an article in the popular *Fortnightly Review* on "Some Possibilities of Electricity." To him, the Hertzian waves were filled with possibilities:

> Rays of light will not pierce through a wall, nor, as we know only too well, through a London fog. But the electrical vibrations of a yard or more in wave-length of which I have spoken will easily pierce such mediums, which to them will be transparent. Here, then, is revealed the bewildering possibility of telegraphy without wires, posts, cables, or any of our present costly appliances.

Crookes went on to describe how such telegraphy might actually work, how, for instance, two people wanting to communicate would have to agree on the frequency at which their apparatus would be tuned. The idea of "wireless" telegraphy was not itself new—both famous inventors, like Edison, as well as obscure ones had attempted to use electrical effects that operated between open space, like induction, to communicate, but with no success. By the time of Crookes's article there no doubt had been a few experiments in using the Hertzian waves, but these had been of no importance.

The announcement and description of the possibilities by one of Britain's most eminent scientists, however, gave a new credibility to such efforts, and directed even some scientists, like Lodge, down technical paths.[26]

It was not to be a scientist, however, that would make a name for himself as the "inventor of wireless." It was, instead, an enterprising young Italian student who fixed on the goal of using the Hertzian waves over useful distances, absorbed readily what he could find about equipment, circuits, and antennas, and then used every resource—personal, social, and financial—available to him to push his vision forward. Guglielmo Marconi was a wealthy amateur experimenter, not yet twenty years old when, in 1894, he read the obituary of Heinrich Hertz written by his former physics professor. The professor, Augusto Righi of the University of Bologna, described Hertz's experiments and discoveries in some detail, and these descriptions inspired Marconi to tackle the problem of sending and receiving the waves as far as possible. His own knowledge of electricity was limited and his technical abilities were not exceptional, but the young man did have the knack of adapting the rudimentary instruments that were available to make them more practical and robust, to allow them to survive and function outside of the laboratory. His teacher, Righi, had already made his own improvements in Hertz's spark generator, and Marconi was able to simplify the best detecting device then available, a glass tube with iron filings between two metal electrodes, known as a Branly coherer (after its French inventor). Even a weak electromagnetic signal would make the iron filings clump together, allowing them to conduct a current. After a signal was detected, the filings could be knocked loose again and the device was ready to detect another signal. When the coherer was connected to a bell or some other sounding device, it could act as an audible telegraphic receiver. Marconi's style always was to take the best device already known and improve on it in ways that made practical, rather than laboratory, application possible.[27]

Besides a certain technical cleverness, Marconi brought to his efforts two other signal assets: social connections and a focused ambition. At first he concentrated on increasing the distances at which he could detect the signals from his spark transmitter. This led him to make larger and higher antennas. The design of these antennas, his decision to ground the antennas to the earth, and his adoption of lower frequency (longer wavelength) signals were the products of empirical experiments, all directed towards increasing the distance of his signals. Between 1894 and 1896 Marconi increased this distance from the space within a room to nearly two miles. At this point his social connections—specifically, those of his well-born Irish mother to the British establishment—came into play and Marconi took his apparatus and his experimental knowledge to Britain for further development. The young Italian received a serious hearing from officials of the British Post Office (which was in charge of the country's telegraphs) and the Admiralty (which was concerned with sea-borne communications). In 1896 and 1897, Marconi demonstrated his system in a series of well-publicized tests. While there were obviously many problems yet to be solved, these went well enough to associate Marconi's name firmly with the new technology and to encourage investors to form the "Wireless Telegraph and Signal Company Limited." Marconi was able to get key patents on the technology, particularly on means of tuning, which turned out to be especially critical as more people began transmitting signals. There were, in fact, numerous others whose contributions to key technical problems were, if anything, more substantial than Marconi's. Adolf Slaby, a German professor, promoted experiments in his country, and when he was joined by another professor, Ferdinand Braun, the Germans had in short order a system technically the equal to Marconi's. In Russia, Alexander Popov demonstrated apparatus using Hertzian waves as early as 1895, although he failed to develop it into a system. In the United States, a range of experimenters, from the brilliant but eccentric Nicola Tesla to far lower-key John Stone Stone, demonstrated some of the possibilities of wireless telegraphy. But only Marconi, at this early stage, was able to marshal resources and reputation on a sustained basis for the creation of a communications system.[28]

In the fifteen years between 1897 and 1912, wireless communication became a viable and then important technology. At first technical limitations were severe: signals were weak and frequencies hard to control, receivers were very slow and imperfect, but wireless quickly became a very popular target for improvement. Marconi kept pushing the distances of transmission, and in 1901 claimed to have achieved what most thought impossible—wireless transmission and reception across the Atlantic (figure 21.8).

While these early claims were hard to verify, within a decade large stations were being built on all oceans for radio communications. Weather was always a factor in clear communications, as was the presence of competing signals. Tuning technology steadily improved, receivers became ever more sophisticated, and a range of new approaches, including solid-state (crystal) apparatus and the first electronic vacuum tubes, emerged and found application on a rapidly expanding technological front. The early limitations were such that Marconi (and his competitors) had to seek out with care the applications that would make sense. Early on, communications with ships at sea gave the new technology a unique niche, and the Marconi company exploited this to the hilt. It even exacerbated international tensions by at first refusing to receive signals sent by non-Marconi operators, but the first International Wireless Conference, called in 1906, forced systems to open (although the British and Italians initially resisted the terms of the agreement). In the next few years the importance of maritime radio communications became manifest, highlighted spectacularly by the role that wireless played in bringing rescue to the iceberg-struck RMS *Titanic* in April 1912. When it became widely known that more lives might have been saved

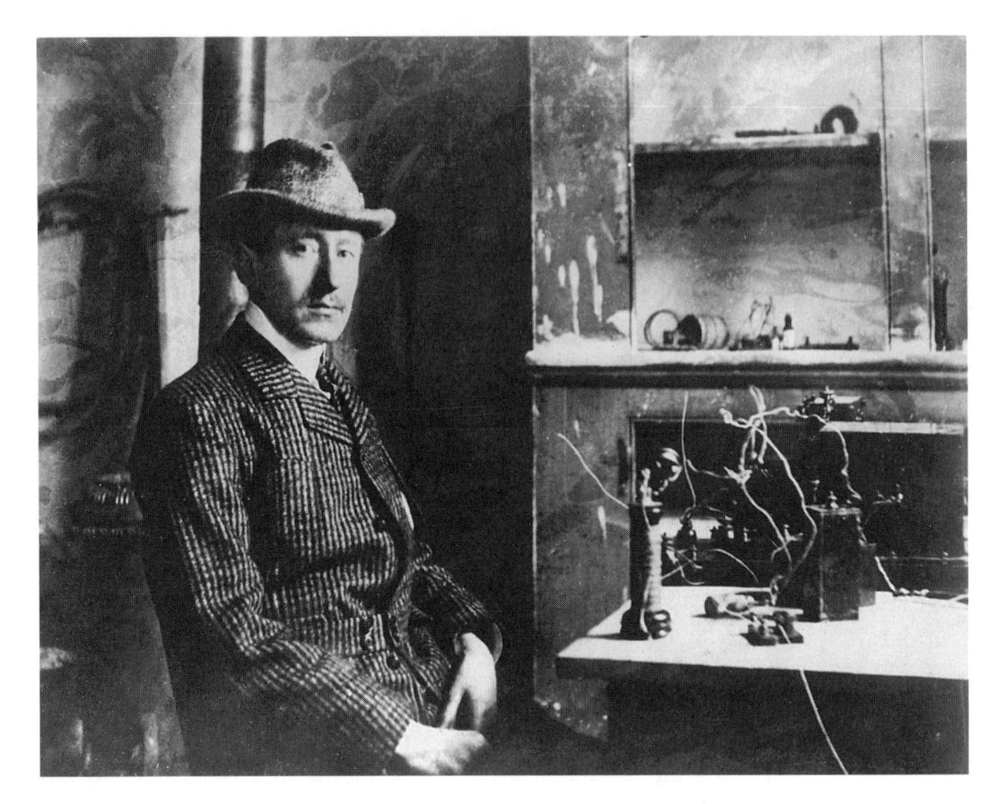

Figure 21.8
Guglielmo Marconi with receiving apparatus, St. Johns, Newfoundland, 1901. Library of American Broadcasting, University of Maryland.

had nearby ships been monitoring radio transmissions that night, regulations were put in force to require round-the-clock radio service at sea.

By 1912 the improvement of radio technology was pursued on many fronts. As early as 1906, Reginald Fessenden demonstrated the possibilities of radio telephony—voice communication by wireless—in a "broadcast" at Cape Cod, Massachusetts, and by the middle of 1907 he had sent voice 180 miles across water. The spark transmitters of the first pioneers gave way to alternators, giant adaptations of electric power generators that allowed truly reliable transoceanic wireless. New circuitry, such as the heterodyne—another invention of Fessenden—allowed much clearer reception of signals, reducing static and increasing sensitivity at the same time. Another American maverick, Lee DeForest, graduated from Yale in 1899 with a thesis on wireless telegraphy, and then threw himself into the task of making his own technical contribution. The result, after about six years of effort, was the three-element vacuum tube, or triode. This was the first electronic amplifier, and found use not only as a key element in sensitive receivers but as the most reliable generator of radio frequencies as well. When used in increasingly more sophisticated circuits, such as the regenerative, or feedback, circuit invented independently by De Forest and yet another American individualist, Howard Armstrong, around 1913, the triode and its successor vacuum tubes became the foundation for much of the subsequent improvement of radio. By the outbreak of World War I in 1914, radio was one of the key technologies of the modern age, even though its future in broadcasting, was yet to come.[29]

The term "dynamics" comes from the Greek word *dyne*, meaning ability, power, or force. By the nineteenth century, dynamics had come to refer to a branch of Newtonian mechanics that sought to describe with mathematical precision the motions and related forces of bodies. To a large degree, when applied to water, air, heat, and electricity, dynamics involved understanding, describing, and predicting flow and the effect that changes in force would have on flow. Mathematicians, beginning in the seventeenth century, constructed ever more complex and powerful tools for dynamics. In many cases it was understood that the equations of the mathematicians, the differential equations of motion and force, were often not directly soluble, but they could lead to approximate solutions that were useful in describing why certain forces resulted in the actions observed. More than anything else, from the technological point of view, dynamics in its various manifestations made it possible to describe the real and the ideal—to see how the various elements in a system of motions or forces worked together to make a machine "imperfect," as some forces acted against others in any given design. The nineteenth-century rise and spread of dynamical ways of looking at the world and of technical systems was very much part of what the German sociologist Max Weber so evocatively called "the disenchantment of the world." A core element of this disenchantment, according to Weber, was the ascendancy of "rationalism," which he defined as "an increasing theoretical mastery of reality by means of increasingly precise, abstract concepts." Such rational mastery of the technological world was the core goal of the applications of dynamics. In this way dynamics defined and redefined "improvement" in every field it touched.[30]

The ultimate influence of the dynamical worldview lay in its extension to the realms of humanity and society. The American sociologist Lester Frank Ward published his *Dynamic Sociology* in 1883, and while his "dynamic" terminology was not widely adopted and his influence was minor compared with some of his contemporaries, such as Emile Durkheim or Ferdinand Tönnies, his use of dynamics to refer to the means by which social science might be applied to useful ends would have been readily understood by his readers in both Europe and America. It made perfectly good sense to extend the dynamical dream—the idea that measurement and analysis would yield an understanding of the difference between ideal and actual systems—to human and social relations as well. Indeed, Ward actually expressed the goal of his "applied sociology" in terms of reconciling "achievement," by which he meant the accomplishments that individuals pursued for personal goals, with "improvement," which to Ward referred to defining and reaching appropriate goals for society as a whole. Such a notion was common among the social scientists of the mid- and late nineteenth century, whether sociologists, economists, or psychologists.

In industry such dynamical thinking reached its clearest expression at the end of the century in the work and writings of Frederick Winslow Taylor. Taylor's scientific management was nothing less than dynamics applied to human work itself. Each of the steps of the work process was to be defined and then measured to the utmost detail. Once this was done the scientific manager's goal was to determine the "one best way"—the ideal that represented every dynamicist's vision, from the perfect heat engine of Carnot to the dragless lifting wing of the Wright Brothers' wind tunnel. Taylor was famous for using a stopwatch to measure every little element of even the simplest task, such as shoveling coal or loading bricks. Data on time and motion could be compiled and studied and the elements of each movement and combination of movements in a task could be analyzed to determine what was "wasteful" and what "efficient." Taylor's method relied on identifying the most skillful or productive workers and paying the closest attention to their methods, and at the same time trying to identify sources of waste even among the best. The result would be a rationalized, ideal method of doing any particular task. Incentives could then be used (both positive and negative) to get workers to adopt this ideal and conform to it as closely as possible. In the words of one of Taylor's champions, Louis D. Brandeis, "calculation is substituted for guess; demonstration for opinion." In subsequent years industrial psychologists and other students of industry would cast considerable doubt on the efficacy of Taylor's approach, and from the beginning the representatives of workers, in their unions and political organizations, objected heatedly to the depersonalization of work. But at the beginning of the twentieth century, scientific management was the most logical extension of the dynamical worldview.[31]

22 Land and Life

In vain may the *tongue* and the *pen* be employed to satisfy the practical man of the errors, which a life of labor has confirmed, and the experience of ages has consecrated. He is either deaf, and blind, and dumb to your appeals, or answers you in the language of distrust, and with the reproach that they are theories of idle speculation only. But make for him the experiment, explain to him the method, exhibit to his natural senses the successful result—he will hesitatingly yield credit to ocular demonstration, and tardily follow in the footsteps of improvement.[1]

The conservatism of farmers has been a notorious stereotype for centuries. The rural peasant was always used as the butt of jokes about reluctance to change old and proven ways. The self-styled "agricultural improvers" of the seventeenth and eighteenth centuries were often not even farmers, or, to be more precise, they tended to be gentlemen of wealth and substance, whose actual contact with the soil was frequently at one or more steps removed. It is all the more remarkable, therefore, that in the course of the nineteenth century, farming, especially in the United States, came to be not only the locus for radical and thoroughgoing technological change, but also where the greatest faith was placed in the improving powers of science. Indeed, well before the century was out, the devotion to science-based agricultural improvement was so deep and widespread in America that it had laid the foundation for some of the key institutional changes in the support of improvement, particularly through the active intervention of government.

Agriculture's widespread and enthusiastic embrace of science and invention in America was one of the most telling episodes in the emergence of the culture of improvement. Every aspect of farming was affected and came to be seen as appropriate targets for inventors and experimenters of all stripes. While many tasks, such as restoring soil fertility or the large-scale mechanization of plowing or harvesting, resisted quick and simple fixes, there was sufficient success on many fronts from the first decades of the nineteenth century to entice farmers, singly and in the associations they formed by the hundreds, to expand steadily the faith that their labor could be made easier and more fruitful over time. The essentials of that faith were particularly significant for the form and character that improvement was to take in the future, for the farmers embraced with an astonishing enthusiasm the notion that scientific knowledge and techniques were the basis for long-term progress. The fact that farmers usually acted outside of any corporate or collective context meant, in addition, that they were particularly ready to invest their faith in government institutions for the development and dissemination of new knowledge and techniques. The agricultural sector, therefore, led the way in forging a new style of improvement that was to increase in importance through the twentieth century, with the active and visible hand of government seen everywhere. The consequences of change in agriculture were enormous, for all other transformations of the economy rested on astonishing increases in the labor productivity of farming. While in 1790 about 90 percent of Americans were employed in agriculture, by the 1870s farmers constituted a minority of the working population, and in the twentieth century the percentage continued to fall until at century's end it was in the single digits. Despite this, the amount of food and other products of the land available to the typical resident of the Western world (and, increasingly, of the rest of the globe) grew far beyond anything earlier centuries could have imagined. While agricultural change lacked the "sex appeal" of other technologies, its role in making the modern world was second to none.

As with most other technologies in the industrial period, the greatest change in agriculture was the sense of the possibilities of change itself. From the late seventeenth century, agriculture was a favorite target of improvers, but real change, by and large, took place only at the margins. New tools, new crops, and new breeds all emerged from the discussions of the gentlemen farmers of Norfolk and elsewhere, and they received notice in such quarters as the Royal Society and in the provincial scientific and philosophical societies that were so popular in polite quarters from the mid-eighteenth century onward. The new circumstances in which Europeans found themselves in the New World and in other colonial settings evoked much experimentation with unfamiliar plants and new methods, but this did not translate into the sense that cultivation and husbandry themselves were subject to significant advancement. The primary need was seen less for experimentation and systematic study of agricultural systems and more simply for better means of spreading information about alternative practices. When, for example, George Washington, who always saw himself foremost as a farmer, recommended the formation of a national "board of agriculture" in his last message to Congress as president, he suggested that such an agency "contributes doubly to the increase of improvement by stimulating to enterprise and experiment and by drawing to a common center the results everywhere of individual skill and observation and spreading them thence over the whole nation." The means at hand were generally seen as limited to prizes and premiums for good ideas, both because the Constitution appeared to limit quite strictly federal involvement in such things and because these were the only means for encouraging improvement recognized by most of eighteenth-century society. Advances in both knowledge and technique were the products of individual effort and occasional genius, and institutional means of fostering improvement were largely limited to recognizing and rewarding individual achievement.[2]

Even more than most technologies, agriculture can best be viewed as part of a complex extended system. This system begins with the soil and the plants and animals raised on it, but production from that soil requires tools, energy, and knowledge. In addition, the farmer's products enter a transportation, storage, and distribution system that was itself radically altered in the course of the nineteenth century. Changes in either the input or the output elements of this system can have profound effects, and in the nineteenth century farmers found themselves in the midst of changes from every direction, some spurred on by their own needs and demands and others generated by change in everything from iron manufacture to chemistry to railroads. The American circumstances of the nineteenth century, in which vast stretches of territory were being opened up and subjected to cultivation for the first time, were particularly ripe for the ready reception of novelty, even among tradition-bound farmers. In addition, particular problems in agriculture, such as declining soil fertility, gave new impetus to campaigns for improvement, as gradually scientific knowledge came to be seen as indispensable to fulfilling the promise of prosperity.

The two key concerns of American farmers at the end of the eighteenth century were decreasing soil fertility (and thus crop yields) in more and more areas of the country and the challenges of spreading agriculture to "virgin" lands across the Appalachians. The thin, rocky soils of New England (outside of a few river valleys) began to show signs of exhaustion only a few generations after their first clearance, encouraging farmers to begin the American pattern of pushing ever westward in search of fertility. The richer soils of the South—the Carolinas, especially—rewarded farming much more than the New England ground, but the widespread choice of tobacco as a cash crop wrecked havoc on this region as well, as intensive cultivation resulted in the destructive removal of soil nutrients after only a few harvests. Because tobacco farmers believed animal manure imparted undesirable tastes to their crop, there was no way to maintain fertility and so the pattern was set of abandoning fields after less than a decade. It could take as much as twenty years of letting a field lie fallow before it became briefly profitable to farm it again. Other crops, such as rice and indigo, flourished in the lower south, and they, more than tobacco, propelled the spread of slave cultivation through the region. When, after the American Revolution, these crops became less profitable through the loss of British markets, the investment in slaves and commitment to the plantation system encouraged the shift to cotton cultivation that became the mainstay of the southern economy in the first half of the nineteenth century.[3]

The relationship between cotton and technology is one of the most familiar ones in American history. The idea that Eli Whitney's cotton gin, introduced in 1793, was responsible for the rise of the cotton South is one of the classic examples of a technological-determinist argument. Like almost all such arguments, there is a grain of truth in it, surrounded by the chaff of oversimplification and hyperbole. The gin idea—that mechanisms with sets of rollers could be used to separate cotton fibers from the seeds to which they were invariably attached—was an old one by the late eighteenth century. Cotton growers in south Asia had used such devices for centuries. They were most useful in handling black-seed cotton, of which so-called Sea Island cotton was the best-known American example. This crop grew well in the Atlantic coastal lowlands of the South, but not beyond. In the Piedmont and highlands of the South and farther westward, only green-seed (short staple) cotton thrived, and the separation of fiber from seed in this variety was much more difficult. Whitney, a young, mechanically inclined, but not very experienced Connecticut Yankee, improved the older form of gin for the purposes of treating green-seed cotton by increasing the violence with which the gin's rollers attacked the inserted bolls. Wire teeth in one cylinder caught the cotton and pulled it through a comblike screen, leaving the seeds behind as another roller then brushed off the fibers. One virtue of this simple device was that it could be scaled up, depending on how much energy could be applied to one machine—one person could crank a single small gin, but animals and even steam engines could be harnessed to larger devices, which became parts of the southern agricultural landscape in the first half of the nineteenth century.

The kind of agricultural improvement that Whitney's gin represented was at the time exceptional. Self-styled "improvers" of the eighteenth century generally focused on new crops and breeds or new methods for restoring soil fertility. In this sense, Whitney's contribution was seen in the light of the new variety of cotton that it fostered, but couching improvement in terms of a mechanism was itself a novelty, although one that became increasingly common in the nineteenth century. At first there were occasional suggestions for new forms of plows and the like (most famously by Thomas Jefferson). After 1815 or so improvers moved more and more into the realm of tools and machinery. Jethro Wood of Scipio, New York, for example, sought patents for a couple of cast-iron plow designs in 1814 and 1819, incorporating interchangeable parts as well as the strength of an iron frame. Wood's real improvements were actually quite modest (there had been other cast-iron plows, and other plows with changeable parts); what set his efforts apart was the success that he had in selling them. Plow manufacturers purchased the rights to make Wood's plows, and by 1819 thousands of them were being sold. Wood's commercial success sparked what one author has called "a flood of innovations" in the 1820s, directed toward every conceivable variation in plow parts. Over the next several decades, up to the beginning of the Civil War, the design and manufacture of American

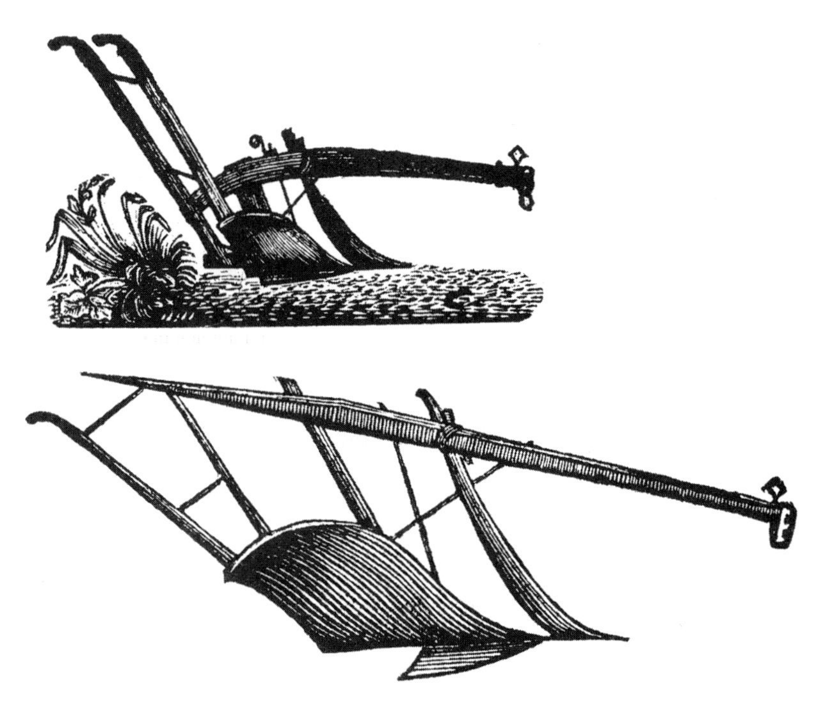

Figure 22.1
American patent plows, 1820s. From *American Farmer* (August 11, 1820).

plows changed utterly. "Patent plows" became fixtures on American farms, particularly in the North and Midwest (figure 22.1). They permitted more ground to be tilled with less labor and new soils to be "broken" with a speed and efficiency hitherto impossible. Just as significantly, the new plows came to stand for the American farmers' acceptance and even expectation of unlimited mechanical improvement.[4]

In the first half of the nineteenth century, every mechanical component of farming became objects of American inventive efforts. Machines for sowing and planting began with variations of the so-called seed drill that had been around at least since the seventeenth century, but by the 1820s more complicated devices began to emerge. Some machines mechanized the broadcasting of seed, formerly done exclusively by hand, while others measured out seed and inserted it methodically into the ground. The success of these machines in speeding up planting was modest, but their widespread acceptance by farmers was further evidence that mechanical improvement had caught the agricultural imagination. Similarly, harrowing and cultivating, traditionally the real backbreaking routine of farmers, necessary for soil preparation and for keeping down weeds, were the subjects of a great range of inventions, particularly between 1815 and 1840.[5]

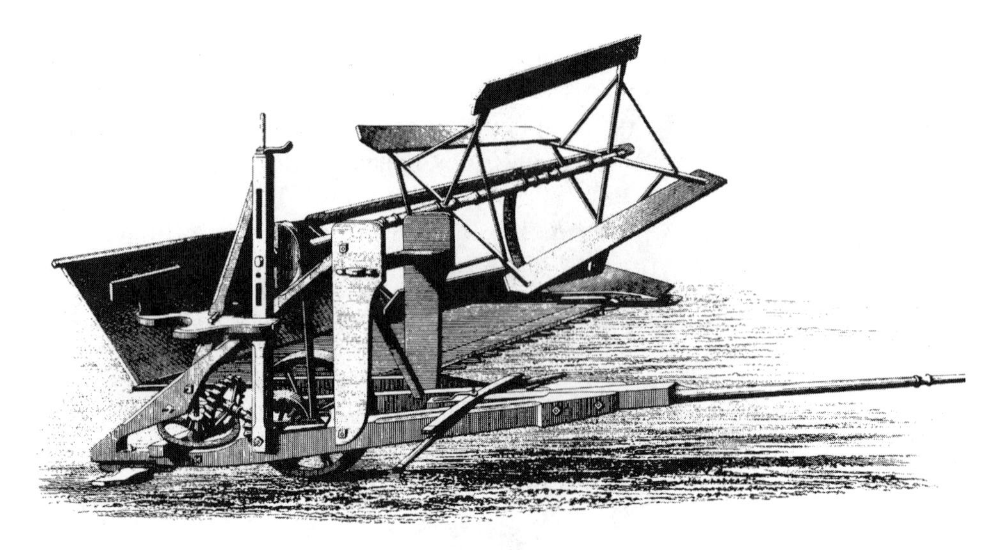

Figure 22.2
The McCormick reaper. From John C. Morton, *A Cyclopedia of Agriculture* (London, 1856).

The most spectacular and famous innovations, however, were in reaping and other harvest-related tasks. The harvest posed both the greatest allure and the greatest challenge for inventors. The time constraints of the harvest could be quite severe, and particularly in a country in which a shortage of farm labor was endemic, mechanization would always appear to be attractive. Unlike the prior tasks, the subjects of the harvest were living plants, and they had to be handled properly by tools or machines or else the loss in the farm product would be unacceptable. It is no wonder, then, that experiments in harvesting machinery, from mowers to reapers to the most complicated harvesters, binders, and combines of the later nineteenth century, should become the most famous of American agricultural inventions. In the 1830s two enterprising inventors, Cyrus McCormick in Virginia and Obed Hussey in Ohio, managed to solve the mechanical difficulties that had plagued dozens of earlier efforts. McCormick's reaper used a revolving reel to hold grain against the reciprocating cutter blades, and Hussey's machine incorporated a sawlike cutter bar that cut stalks far more effectively than McCormick's (figure 22.2). The result was initially two machines that were modest improvements on their predecessors, although their inventors trumpeted much more. Only when McCormick managed to get the rights to Hussey's cutter-bar mechanism about 1850 did there emerge a broadly satisfactory machine, one that would come to define the spread of American agriculture across the heartland. Through the 1850s and on into the period of the Civil War (when labor shortages on farms became acute), the reaper was the subject of continued improvement, which in turn encouraged its wider use throughout the grain growing regions of the United States (and beyond).[6]

Just as with most other categories of machinery, the patterns of improvement and adoption for agricultural machines were typically contingent on a wide variety of factors, often local in nature. Machines that both cut and threshed grain in a single operation, for example, were the subject of experiment as early as the simple reaper, and some measure of success was achieved as early as 1836, when Hiram Moore of Kalamazoo, Michigan, built a machine that was pulled by twelve horses and successfully reaped and threshed as it traveled across a wheatfield. By the 1840s there could be found "combines" that required sixteen horses and managed to harvest twenty-five acres a day. This machine, and similar models, found little favor in the East or Midwest, however, as farms were rarely large enough to accommodate such a large and bulky machine or to justify the high capital expense. In addition, the typically damp climate of the more easterly states did not suit the mechanical threshing action of the combine, thus much grain was wasted. Conditions were very different in California, however, and when the first large machines made their way to the California central valleys in the 1850s, they were adopted by farmers whose fields were huge by eastern standards and who found the machines perfect for their dry climate. The California machines required as many as forty horses, however, and the combine was not a completely satisfactory machine until newer, steadier, and more easily maneuvered power sources, such as steam and, later, internal-combustion powered tractors, became available. This connection between machines and power sources was, in fact, more general in the United States and in other parts of the world in which the technical and economic resources made mechanization a viable option for farmers after the mid-nineteenth century. Experiments and demonstrations of new and improved machines for every farm task became commonplace, but the widespread adoption of most of these awaited inexpensive and powerful self-propelled machines. By about 1905 the basic design for gasoline-powered tractors had emerged, and by 1915 tractors were widely available (figure 22.3). From roughly 25,000 tractors in that year, their usage exploded more than twenty-fold in the next decade and there were more than a million tractors on American farms by the mid-1930s. Some parts of Western Europe changed at about the same time, while much of the rest of the world saw little widespread change until the middle of the twentieth century.[7]

While nineteenth-century farmers, particularly in the United States, made use of improved machines and tools, these were not generally perceived as the key elements of agricultural improvement. New crops and breeds and new ways of treating soil continued to be, as they had been in previous centuries, the mainstays of efforts to improve the yields of farms and the work of farmers. Traditionally farmers kept their fields productive by a combination of crop rotations, fallowing, and manuring. These approaches added considerably to the labor required for farming—fallowing, for example, required constant tilling of the soil to keep down weeds, and manuring added obvious burdens in animal tending as well as the physical labor of collecting and spreading manure. It is no surprise, therefore, that in the labor-short United States

Figure 22.3
American steam tractor, 1890. Smithsonian Institution photograph.

such time-honored practices were avoided where possible. As long as land was cheap and easily accessible, it was easier simply to leave fields of declining yield and move on to fresher soils. As early as the eighteenth century in some areas, however, the obvious wastefulness of this drew comment from European observers, and by the end of the century the declines in soil fertility in many areas, both North and South, were matters of widespread concern and discussion. In the first decades of the nineteenth century, the public debate over "worn out soil" became perhaps the key impetus for agricultural improvement campaigns.

In 1813, the English chemist Humphry Davy published his *Agricultural Chemistry*, and the popularity of this work directed attention toward efforts to understand the basis of soil fertility and, indeed, of plant growth itself. For all the claims to lay the foundations of an agricultural science, however, all that actually appeared at this point were ad hoc theories that would support empirically derived practice and observation. Adding certain minerals to the soil, for example, such as calcium hydroxide—"lime"—clearly helped maintain fertility in some instances, but there was no sound explanation for why, much less was there any explanation for why manure seemed indispensable for the long-term maintenance of soils. Nonetheless, there emerged in the early 1800s a recycling system that helped to maintain the productivity of farms near the larger eastern cities: organic waste products from cities—ashes, bones, stable droppings, and even treated human waste—were collected and distributed to nearby farms, with minerals such as lime or gypsum supplementing. The result was often a short-term recovery of fertility, until some other fertilizer mixture was devised. This was because there was no informed theory of fertilizing—the materials used would supply one or two needed nutrients, but almost never a full complement.[8]

In the middle of the nineteenth century a number of elements converged to change the way soils were treated. The most spectacular of these was the discovery of the usefulness of Peruvian guano for restoring the productivity of worn-out fields. Both European (particularly British) and American farmers had experimented with the material from early in the century, but in the 1840s guano (mineralized bird droppings found in commercial quantities only on the west coast of South America and a few small islands) briefly transformed agricultural practice in some areas on both sides of the Atlantic. The readiness with which some farmers took to the material, which was never cheap due to its scarcity and the cost of shipping it thousands of miles, came from the urgency with which agriculturalists viewed soil exhaustion in some areas, especially in the middle Atlantic states, and the ease with which guano could be substituted for less effective traditional manures. Its period of real economic importance was short: American imports were 700 tons in 1845 and peaked less than a decade later at a little over 175,000 tons. The larger importance of guano was its role in transforming the farmer's sense of how soil was to be improved. After the supplies of natural guano diminished from a decade of mining, there were widespread efforts to make artificial guano, from minerals or, more successfully, from fish and animal by-products. Supplies of these substitutes, however, were always constrained by the cost of materials and manufacture.

The only other fertilizer of the mid-nineteenth century that experienced success similar to guano's was superphosphate, made by dissolving ground-up bone in sulfuric acid. A variety of superphosphate mixtures were made and promoted from the 1850s, many of them incorporating natural guano, ammonium sulfate, and other materials, which, at their best, yielded a well-balanced fertilizer, supplying both the phosphorous particularly wanting in eastern soils and the nitrogen needed for any kind of long-term fertilizer regimen. Unfortunately, the qualities of superphosphate mixtures varied greatly, which promoted government intervention and regulation to reduce fraud and assure farmers they were getting what they needed. By the 1870s and 1880s, however, there was emerging in the United States a system of artificial fertilizer production and use that came in the twentieth century to define the core means by which soil fertility was maintained. Such a system depended on an industrial infrastructure of recycling and, increasingly, mining and chemical processing, as well as specialized formulations for different crops. Just as importantly, however, the success of this system depended on a growing confidence in the science of agriculture, a confidence that emerged from a new system of knowledge production and dissemination pioneered by agricultural improvers.[9]

The linkage between agricultural improvement and science was promoted by the appearance in 1840 of Justus Liebig's *Organic Chemistry in Its Applications to Agriculture and Physiology*. Unlike Humphry Davy's work thirty years earlier, Liebig's book was firmly grounded in laboratory practice and in rapidly growing knowledge of organic chemistry. While there was still much speculation about the mechanisms of plant growth and nutrition, Liebig's effort finally gave agricultural chemistry a vocabulary and structure that could be built upon. Perhaps most impressive to his readers familiar with the problems of soil fertility, Liebig appreciated and explained to some extent the significance of minerals in the soil, and he held out the promise that further chemical research would provide farmers with ever more effective means of restoring and maintaining soil fertility. Within only months of its publication, English versions of the work appeared in both Europe and America, and agricultural journals were full of praises for Liebig and his ideas. Over the next few years the vision of "scientific agriculture" took hold in many quarters, although never without controversy. Liebig's allure lay in large part in the promise that he presented a theory that could actually be applied in the real world of farming. The promise was not, in fact, fully realized in the book, for there was actually much ambiguity, if not downright waffling, about the relative importance of, say, nitrogen versus minerals, but this was not as important as the message that chemistry was prepared to come to the aid of the farmer.[10]

In the 1840s and 1850s, the promise of agricultural science was touted in many quarters. Growing numbers of local agricultural societies turned their attention to promoting it. Older agricultural journals published countless articles on the subject, and newer ones were founded at a great rate to spread the word more widely. Perhaps most important, the precedent of a German academic chemist directing such careful attention to agriculture inspired and emboldened those eager to turn American colleges and universities towards service to agriculture and industry. The fact that Liebig held posts in prestigious German universities (first at Giessen, then at Munich) and worked from well-equipped laboratories was not lost on foreign observers. In fact, it was from this point that serious American students of chemistry came to believe that a stint in a German university laboratory was a necessary element of complete training in the science. Just as important as the training, however, was the model of laboratory-based research in the service of practical ends. This model became the dominant one for all technical improvement, particularly in industry, in the coming century, and its origins in agriculture are often forgotten or ignored. In the United States, agricultural science appeared in a variety of venues. In some of the established universities, such as Harvard and Yale, donors were persuaded to fund laboratories or endow professorships, although sustained support was hard to come by. Some of the agricultural societies that had been springing up around the country—there were more than nine hundred of them in the United States in 1858—devoted considerable effort to encouraging experimental work, and the almost sixty active journals circulating at this time eagerly reported both European and American experiments. Most significantly, perhaps, the societies and journals agitated at every level for more government involvement and investment in agricultural science and education.[11]

The use of federal resources for any direct assistance to education or productive activity was extremely limited in the years before the Civil War. Nonetheless, from the time that George Washington had voiced encouragement for the broader exchange of information among farmers across the country, the sentiment for finding the means for greater government assistance never was completely stilled. The most effective early assistance came from the first commissioner of patents, Henry Ellsworth, who persuaded Congress in 1839 to provide appropriations for the collection of agricultural information and statistics. The Patent Office went beyond collecting information, however; for the next two decades it was responsible for a range of activities, from the issuance of an annual report on the nation's farming to the distribution of thousands of packets of seeds throughout the country. Finally, in 1862, the government's commitment to agricultural improvement gained new status in the creation of the Department of Agriculture, the organization of which included provisions for scientific bureaus and data collection and dissemination. In that same year, the Congress—liberated from the conservative constraints formerly imposed by the now missing Southern congressmen—passed another piece of legislation that had an equally profound effect on agricultural improvement. Rep. Justin Morrill of Vermont had been pushing for years for federal funds in support of agricultural education, and the Morrill Act provided for grants of federal lands to the states for the express purpose of setting up "colleges for the benefit of agriculture and the mechanic arts." Morrill pointed to such institutions in Germany, France, Russia, and other countries, as well as fledgling efforts in a few states, most notably Michigan. He further made it clear that he expected the land-grant colleges to carry out and report on agricultural experiments, "such as can be made at thoroughly scientific institutions and such as will not be made elsewhere."[12]

Over the next decade or so, land-grant colleges were established throughout the United States, and in many cases money was provided to older (usually private) institutions to fund agricultural education. In the 1870s a number of states set up agricultural experiment stations, following German and English models, often attaching these to the land-grant colleges. These state-funded efforts promoted a wide range of experimental work, from the testing of new implements on local soils to further applications of agricultural chemistry and the collecting of seeds and hybrid livestock. The federal Department of Agriculture slowly expanded its own experimental efforts in this period, although with very limited staff and facilities. Finally, after years of agitation by the states and agricultural societies, the Hatch Act was passed in 1887 to provide for sustained federal funding of the experiment stations. With the secure source of funds, agricultural science—now a widely diversified field ranging from agronomy to veterinary science—thrived and expanded over the next decades. In the first decades of the twentieth century, the resources of the federal experiment stations also grew at a rapid pace. The combination of state-based stations and a federal research establishment allowed agricultural improvers to apply their efforts to a wide variety of local conditions and problems (soils, climates, plant and animal diseases, and the like) while at the same time developing an ever-expanding foundation of basic knowledge. Agriculture thus provided the most important model for a system of continuing, state-supported technological improvement in the twentieth century.[13]

In several other ways agricultural change established a larger pattern for technological change generally. The creation of the modern industrial civilization of the West took place against the background of a social and economic order that had for millennia been centered on farming. It should thus come as no surprise that the new order, appearing to revolve around urban life, machine-based manufacture, and an ever-expanding circle of techniques of communications, transport, and control, should find its earliest representations within the older, rural agricultural structures. In addition, the industrial and agricultural spheres were always intimately linked with one another, so the patterns of conduct, learning, and organization of one sphere naturally influenced the other. Trends that had been long developing in non-agricultural sectors of the economy inevitably showed up as farming itself took on more and more of the character of modern industry, with reliance on machinery, external sources of energy, scientific education and research, and the mechanisms of urban markets. The implications of this can best be seen in a seemingly small case, in which we can see the meanings of new paths and forms of technological improvement for the respective roles and relationships of men and women.

Cheesemaking is an ancient activity. In all societies that use animal milk (it's well to remember that many ethnic groups either cannot or do not use cow's milk), cheese has had a place. This is simply because milk spoils, and cheese is the most effective means of preserving milk nutrients over a period of time. A number of characteristics make cheese a particularly interesting technological artifact. One of these is simply the fact that this is a product that bears the stamp of its particular manufacture quite strongly. Cheese does not occur naturally, even though the processes that go into it are, of course, all organic. One of the glories of the product is its variety. There is perhaps no other foodstuff—not even wine or bread—that lends itself to such a great range of tastes and forms even within the same basic technical framework of curdled and cured milk. This variety is the product of a range of possible treatments and different kinds of milk, combined with many different bacteria, enzymes, and molds that enter into the cheese and give it much of its distinctive character. Both processing techniques and the nonmilk ingredients have historically another interesting feature—they tend to be intensely localized. The variety of cheese is traditionally—and still, to a degree not maintained in any other product (except, possibly, wine)—identified with place. This local character, so readily apparent in the case of cheese, is one of the distinctive features of most preindustrial technologies.

The local nature of cheese techniques and recipes, however, does mean that generalizations about cheesemaking must be made with care and circumspection. What is clearly true for one area may not be at all true for another. In particular, the social arrangements that govern this type of work, such as gender expectations, cannot be generalized from country to country, or even from region to region. In England before the nineteenth century the most important agricultural work for women was typically in the dairy. On just about every farm, women were in charge of making butter and cheese, at least for home use, and usually also with some additional for the market. The dairy was almost the exclusive province of women, from the proverbial "milkmaid" to the mistress of the household who supervised the final product and its marketing. Farm women were the traditional marketers of cheese, taking their products to nearby village and town market days and selling to local customers or to factors, who then dealt with city cheesemongers or their agents. Dairy work was hard work, typically starting with the first milking at 3:00 or 4:00 in the morning and going well into the evening. Cheese was particularly demanding, since timing was crucial—the first batch of cheese was typically started some hours after milking, after the milk had cooled, but before there was any danger of spoilage. It was customary for the women of the household to do all the work themselves: heating the milk in vats, preparing and adding rennet (typically made from calves' stomachs) to curdle the milk and set it, cutting up the curds and stirring at length, straining, salting, molding, pressing, turning, and so forth. All of this was guided by knowledge passed on from mother to daughter, from mistress to maid. There were numerous techniques for managing the rate of curdling, controlling the amount of liquid or whey in the new cheese, and manipulating the flavors of the result. This was women's knowledge, and its female nature was unchallenged until the eighteenth century.[14]

Like most preindustrial, home-based technologies, this activity is largely undocumented. Indeed, there is practically no written record of cheese making processes at all until the late sixteenth century, even though on the Continent there was a thriving interregional cheese trade by the late Middle Ages. In Britain, where such trade took much longer to become significant, the literature on cheese is scant until the late eighteenth century. Once a literature does begin to emerge, however, the role of women began to recede. At first this appears to have been in terms of managing the marketing of cheese, but the technical control soon slipped away as well. To some extent this was probably a matter of changing social and economic expectations: as larger farms became necessary for economic success, the more prosperous farm wives were more reluctant to devote themselves to onerous manual labor. But the changing form of technical knowledge was also important here. In the eighteenth century, cheesemaking, like all other agricultural occupations, began to be the subject of "scientific" study and discussion, done within the context of institutions, publications, and even language that was predominantly, if not exclusively, male. The search for "rational" methods of making cheese—like the making of almost everything else—was widely viewed, by men and women, as an appropriately masculine pursuit. Women were judged supreme in the realm of feeling, intuitive knowledge, and unruly organic processes. As cheese was moved into the realm of measurement, standards, precision, and, eventually, mechanical assistance, it was moved into a "defeminized" realm as well.[15]

More than social biases were at work here. The very means by which technical knowledge was seen amenable to analysis and improvement devalued the contributions closely associated with women's work and skill. In the depiction of the trades and crafts in the *Encyclopédie*, for example, the image of cheesemaking (as of almost everything else) placed attention on tools and readily describable techniques (with pictures showing men doing the work). The French attempted to create a "Natural History of Industry" (to adopt Charles Gillispie's phrase), in which artifacts and crafts were to be placed into categories and described with the rigor and uniformity of a naturalist's description of a species. The result was to channel improvement into specific approved paths—closing off those that did not fit the categorical scheme. At the time of the *Encyclopédie*, it is worth noting, there were only about thirty named varieties of cheese in France. This is not because there were only thirty kinds of cheese. Far from it. It was, instead, because *naming* of cheeses—reducing them to "types"—was still rudimentary. Today there are nearly four hundred named French cheeses: the process of categorizing has done its work.[16]

In mid-nineteenth-century America, factory cheesemaking began to displace home and farm industry, and the same process of "defeminization" took place. The ability to control milk temperatures, along with the introduction of new tools and techniques that made capital investment productive in cheesemaking, encouraged the creation of cooperatives and the centralization of cheese manufacture. Cheese factories appeared in New York State in the mid-1850s, and soon spread into the cheesemaking areas of the Midwest. As was typical in such industrialization, women were displaced widely from their previous positions of technical experts and managers. In the United States, while cheesemaking (like all dairy work) started out as women's work, the gendering of the knowledge itself was probably not as strong as in Europe, and so the passage of men into positions of authority in cheese making was probably smoother, although no less disruptive to the technical privileges of women. Just as in Britain earlier, the capacity of women to incorporate into their working the new elements of science and standardization was openly questioned. The pattern so evident in the changing personnel of cheesemaking repeated itself, although not always so overtly, in many areas of production where women once were intimately associated with valued knowledge.[17]

It is important to remember that the actual contributions of science to such occupations as cheesemaking in the nineteenth century were rather small. Very little understanding of the biological and chemical processes involved in a whole host of food transformations, from cheese to wine to beer and bread, went beyond the empirical knowledge that had been accumulating for centuries, at least until the 1870s or so. Just as in soil "science," however, these limitations hardly diminished the fervor of those dedicated to making agriculture into an enterprise guided by theory and improved by laboratory scientists rather than field practitioners. As in many areas of technology, improvement in agriculture and food production came more and more to be defined in terms of what could be accomplished through laboratory experiment, the collection and analysis of data, and the development and imposition of standards. The creation and spread of institutions of improvement—associations, journals, schools, and government bureaus—increased the scope and pace of technological change and made Europeans and Americans broadly aware of change, both actual and impending. But these institutions also served to channel improvement down particular paths, typically those responsive to measurement and markets, at the expense of other values, perhaps more aesthetic or less socially or environmentally disruptive. With limited tools and little workable theory, the scientific improvers of agriculture transformed humanity's most ancient occupation. In the last decades of the nineteenth century, the theory and tools that would carry this transformation through the twentieth century finally began to appear and to give it a scale and scope that would remake not only agriculture but the very experience of life itself.

Even the most "scientific" of cheese makers in the mid-nineteenth century lacked any understanding of what actually went on when milk was transformed into cheese. The same could be said for the makers of bread or beer or wine or a host of other ancient and valued commodities. Just as the analyzers of soil could measure the amounts of various chemicals in their samples and put forth generalizations about what was needed to restore fertility without a clue about why, so too the makers of food could provide recipes of ever-increasing exactness without any explanation for why some things worked and others didn't. It was not so much a workable theory that was lacking, as it turned out, as it was any comprehension of some of the most basic processes of organic transformation, and this ignorance, in turn, lay largely in the failure to recognize, or even see, the actual agents of change, microorganisms that everywhere influence human life and work. Bacteria and other microorganisms are the causes of fermentation, the keys to soil fertility, and the central agents of many diseases. One of the great scientific—and technological—transformations of the nineteenth century lay in the discovery of the actions of these organisms and the beginnings of means for controlling them.

Since the invention of the microscope in the seventeenth century and the work of Robert Hooke and Anton van Leeuwenhoek, the existence of bodies—apparently living organisms—of hitherto unimaginably small size was well known, but their character and action was little understood. In the 1820s and 1830s, improved microscopes, along with the growing interest in understanding the chemical reactions of organic substances, led to increased attention paid to the relation between microorganisms and chemical action. In 1837–38, several individuals, for example, published papers suggesting a relationship between fermentation—the conversion of starch or sugar into alcohol and carbon dioxide—was due to the action of living yeast feeding on sugar. This ran contrary to the received wisdom that yeast was not an organism but a complex chemical that assisted fermentation but did not cause it. No less a figure than Justus Liebig (along with Friedrich Wöhler) piled scorn on the idea of yeast as organism, thus the idea was put on the fringes of respectable science. The man who retrieved it and then made it a central element of his own science was a brilliant and ambitious French chemist, Louis Pasteur.[18]

In 1856 Pasteur was dean of the new faculty of sciences at the University of Lille, in the industrial region of northeast France, when he directed his attention to fermentation and the making of alcohol, an important industrial product of Lille. Pasteur had become comfortable with the microscope due to a long-standing interest in crystals, particularly the crystals of organic substances. When he turned his instrument on to the contents of the alcohol vats, he saw yeast bodies behaving very much like plants, producing buds that broke off to produce further bodies. In addition, he noticed that one of the alcohols produced in the vats was optically active—that is, that it polarized light in one direction. From earlier work, Pasteur had concluded that only living organisms produced chemicals that had this optical property (ordinary chemical reactions would not favor polarization in one direction over another, and so produced optically inactive mixtures). Through a series of remarkable experiments in the mid-1850s, Pasteur made a solid case for the fermenting action of living yeast. He went further, however, examining other types of fermentation, such as the production of lactic acid rather than alcohol by organisms that operated in the absence of air. We now refer to these as anaerobic bacteria, and the description of their action was one of the greatest steps in the creation of modern microbiology.[19]

Over the next several decades, Pasteur, rising quickly through the ranks of French science, transformed the entire vision of microbial life and, just as importantly, sought everywhere to exploit the technological and, eventually, medical implications of the new vision. He famously remarked, "There are no such things as pure and applied science—there are only science and the applications of science.'' The French wine industry attracted special attention, as did French beer making, in the aftermath of the humiliation of the Franco-Prussian War. He explained many of the techniques, some ancient and some new, for food preservation, contributing, of course, his own process for heat treatment to prevent bacterial action. He was acutely aware of the technological and industrial nature of these investigations, seeking, for example, to devise the equipment that would make large-scale Pasteurization safe and practical. Government authorities turned to Pasteur for assistance with both plant and animal diseases, and his success in isolating the bacterial causes of the diseases of the silkworm and, later, of anthrax, which annually killed hundreds of sheep and other livestock, gave impetus to efforts to apply bacteriology to human diseases. Pasteur was by no means alone in laying the foundations for the germ theory of disease or in demonstrating its significant applications in public health, disease prevention and treatment, and the larger understanding of microbial life. By the time that Pasteur's own campaigns against disease were attracting widespread attention in the 1870s and 1880s, the work of such men as the Hungarian Ignaz Semmelweis and Joseph Lister in England had established the efficacy of asepsis and antisepsis in promoting health in hospitals and clinics. Immunization had a long history, and had achieved a measure of respectability after the introduction of vaccination by Edward Jenner at the end of the eighteenth century. Such researchers as Robert Koch, in Germany, were working parallel to Pasteur, isolating the bacterial agents for such diseases as tuberculosis, typhoid, and diphtheria. It was Pasteur, however, with his great sense of showmanship and self-confidence, as displayed in his experiments on rabies, who made the germ theory the new orthodoxy for both the medical community and the general public. In addition, his arguments for the wider support of microbial research and applications helped to redefine improvement in medicine in a fashion that still dominates more than a century later.[20]

Until the nineteenth century, the practice of medicine was very much the art, rather than the science, of healing. As an art, it was little informed by the specifics of cause and effect and was much more shaped by human relationships. A physician spoke with his patient and constructed a diagnosis largely from the words and readily visible signs communicated by that patient. When, in 1819, René Théophile Hyacinthe Laennec reported on his use of an instrument he sometimes simply called "the cylinder," but often more grandly referred to as the "stethoscope" (from the Greek, "chest-viewer"), he set in motion a series of changes in medicine and in the relationships among doctor, patient, and instruments that completely remade our concepts of health, disease, and the human relationship with technology. What Laennec called "mediate auscultation" put the simplest imaginable device between a patient's chest and a physician's ear, initially a simple wooden tube, but this was the first time that the messages conveyed by an instrument were seen as the means for telling what was really going on in the body. Other simple instruments followed—the ophthalmoscope, which provided a clearer look into the eye, and the laryngoscope, which similarly aided the view of the throat. At the same time, in the middle decades of the nineteenth century, laboratory tests emerged as indispensable adjuncts to the physician's own probings and proddings (Richard Bright, to cite perhaps the earliest example, devised in 1827 the albumin test for the kidney disease that took his name), and microscopic examinations of fluids, tissues, and extracts became standard procedures by century's end. Anesthesia broadly expanded the possibilities of surgery and dentistry, while pain-reducing drugs, ranging from opiates to mild analgesics like aspirin, provided relief for a range of conditions. In truth, the ability of physicians to cure diseases was still quite limited at this point, although asepsis, sanitation and other public health measures, and the spreading use of a limited number of vaccines measurably improved the health of cities and the better-off portion of society. But there began to emerge a broader sense of the possibilities for medical improvement, at least as it could be defined by surer diagnosis, consistent description, and the alleviation of pain.[21]

In the next century, medical improvement was arguably the single most spectacular source of change in human life, at least for the citizens of the industrialized world. The curing and preventing of disease became such a central concern that some authors could speak of the "medicalization" of society, the orienting of institutions and values increasingly around the prolonging of human life and the granting of enormous authority to physicians, hospitals, and medical researchers. The two key facilitators of this phenomenon were the growing specialization of medical practice and the enormous expansion of medical technologies. The latter included not only new instruments for imaging, monitoring, and measurement but also a host of chemical, biological, and eventually genetic innovations that allowed physicians to target specific diseases and symptoms or to manipulate the body's own mechanisms in ways that enhanced defenses or, at least, alleviated suffering and extended life. Specialization and technology combined to transform not only the ways and means of physicians but also the expectations of patients and of society at large about the passages of life and the progress of death.[22]

The most visible early manifestation of the twentieth-century transformation of medicine was the spectacular appearance at the end of 1895 of Wilhelm Röntgen's mysterious X-rays. Within months of the German physicist's announcement of his discovery there appeared on the market a practical fluoroscope for viewing the rays and X-ray tubes that could be used with only a modicum of training (and were often used with none at all). So startling was the appearance of photographs showing the bones underneath human flesh that physicians began seeking uses for the new machine immediately, and within only a few years the X-ray machine had become a necessity for any up-to-date clinic or hospital. Other machines made their way between patient and physician. In 1860 the French physiologist Etienne-Jules Marey introduced a recording sphygomograph, which at first simply recorded the pattern of a patient's pulse. Over the next several decades, however, improved devices appeared that were capable of measuring blood pressure (the modern sphygmomanometer was introduced by Scipione Riva-Rocci in 1896). The spirograph appeared in the 1860s to record the patterns of chest movement while breathing. About the same time the first steps were taken to record the electrical impulses emitted by the action of the heart, but this too required decades of experimentation and improvement, until the Dutch physiologist Willem Einthoven introduced his string galvanometer type electrocardiograph in 1901. Somewhat simpler in terms of instrument, but complex in terms of interpretation was thermometry, the systematic recording of a patient's temperature. The knowledge that disease was often accompanied by heat (fever) or cold (chill) was, of course, ancient, but the nineteenth-century passion for measurement and system made the precise recording of body temperature and its fluctuations an article of faith for the physician, thanks in large part to the painstaking work of Carl Wunderlich, who sought to find the balance between scientifically informed methods and practical bedside procedures.[23]

In the twentieth century the use of these instruments became routine, and the institutional setting for ever more complex and more expensive technologies emerged in the modern hospital. Earlier, hospitals had been places of charity and civic protection, institutions less for marshalling medical expertise and curing people than for protecting the general population from the sick and lame who would otherwise interfere with normal life. Hospitals, however, turned out to be the perfect places for testing out new techniques and instruments and for training new generations of physicians in their uses. Over the course of the century, increasing complexity and specialization, along with the ever-growing problem of managing the costs of health care made the hospital the universal dispenser, the place where people were, in ordinary circumstances, born and where, most commonly, they went to die. That institutions so central to the basic cycle of life should also become the most thoroughly technological institutions that most people would ever encounter was a particularly striking testimony to the century's deep faith in technology and its improvement.[24]

23 Scale

Maurice Koechlin was, in 1884, a Swiss-trained engineer in his late twenties, working in the Paris engineering firm of G. Eiffel et Compagnie. In the late spring of that year, he began to join in conversations with a somewhat older colleague, Emile Nouguier, about possible projects for the centennial of the French Revolution, to be celebrated in 1889. On June 6, Koechlin sketched out in pencil the form of an iron tower, on a base of four latticework legs and three hundred meters in height, to be built in Paris for the occasion (figure 23.1). As his original sketch emphasized, such a structure would dwarf not only every other building in Paris, but any other structure built up to this time—he sketched in Notre Dame cathedral, the Statue of Liberty, the Arc de Triomphe, three famous monumental columns, and a six-story building, all on top of one another, simply to illustrate the ambitious height proposed for the tower.

A few days later Koechlin and Nouguier approached the architect Stephen Sauvestre to give their proposal a bit more dimension, and the architect drew in giant arches between each of the four legs and added platforms at two intervals partway up the tower. When the three men presented their proposal to their boss, bridge-builder Gustav Eiffel, a few days later, they called the monument "Gallia," a Latinized name for France. It's not certain what Eiffel's first reaction was, but within a space of a few months he had embraced the bold scheme, and over the next five years he was to pursue it to such a successful and impressive completion that his name will be forever associated with the result.[1]

Gustav Eiffel, by the mid-1880s, had already made a substantial reputation as master of a new structural idiom, the large metal arch. For more than a century, since the construction of the great iron bridge over the Severn at Coalbrookdale, iron structures had been used to proclaim the power of new materials and new engineering, but designs were largely imitative of older forms. In the hands of Eiffel and his engineers, iron members were turned into carefully calculated structures with great economy of materials and fabrication while providing the great strength and durability required by railroads. At the Paris Exhibition of 1867, Eiffel designed the

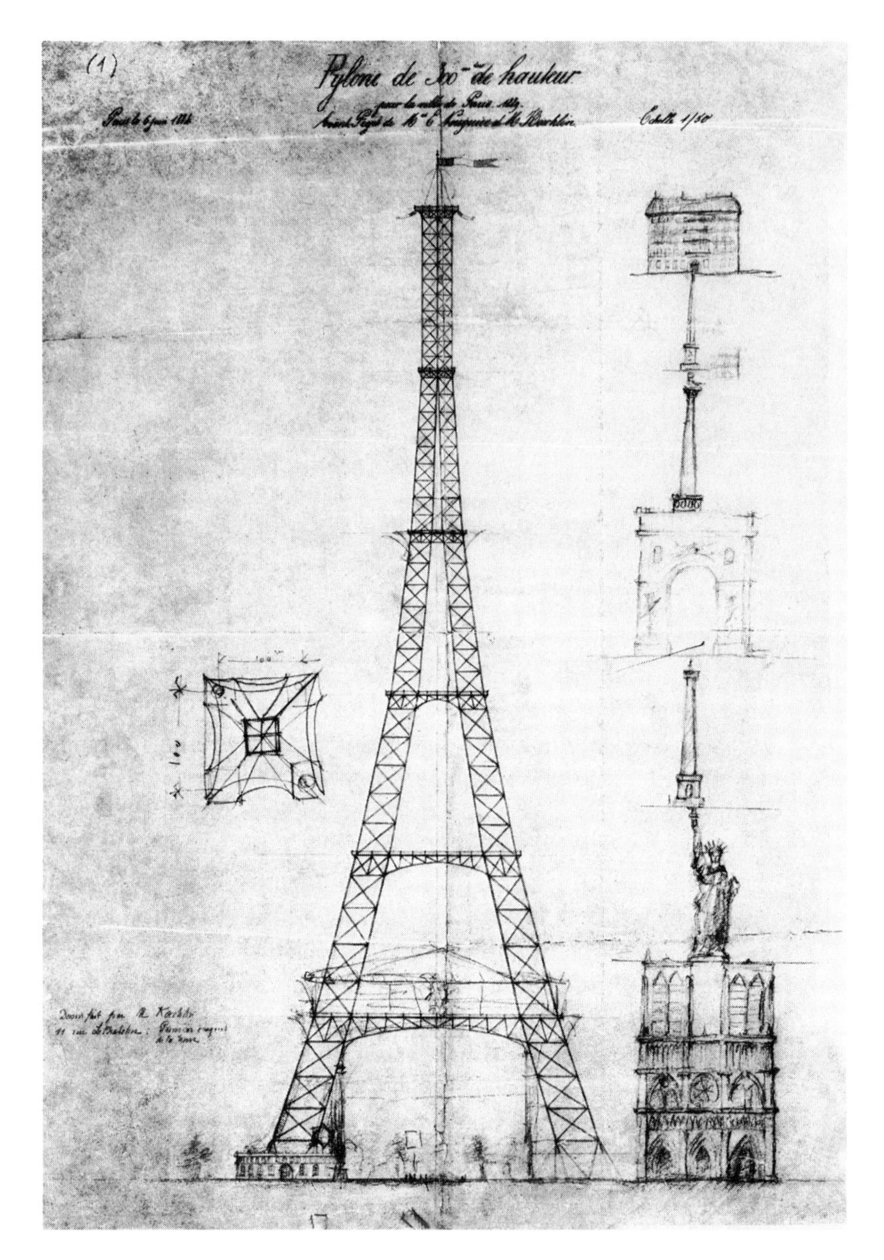

Figure 23.1
Maurice Koechlin's early sketch of the "300-meter tower" emphasized its height by juxtaposing such monuments as Notre Dame, Nelson's Column, and the Statue of Liberty. ETH Zürich, ETH-Bibliothek, Archive, Hs 1092.

great Machinery Hall that became a showpiece for engineering guided by structural theory. Eiffel proclaimed that the time had come "to create methods which opened up a sure path to progress, disengaged from all empiricism."[2]

Over the next several years Eiffel and his engineers built grand bridges at Oporto, in Portugal, and at Garabit, in southern France. In addition, the Eiffel firm turned out numerous less ambitious, but still important, structures, chiefly railroad bridges of iron trusses. In these projects, Eiffel took particular pride in showing the basic economy of the iron lattice structure, as it typically allowed lighter structures to be built more quickly and yet with substantial strength and safety. The great tower was an explicit attempt to apply the theory and lessons of the bridges to a monument, a structure whose primary notoriety lay in its sheer size. In many ways the Eiffel Tower epitomized the technological future, as the coming century was to be characterized by two of its most striking features: scale and system. The construction of the Eiffel Tower was carried out as a carefully coordinated campaign, in which each part was carefully calculated, based on both theory and economy, to play a specific role in a specific order (figure 23.2). Barely more than two years were required to build the world's tallest structure, contrasting markedly to its predecessor for that title, the Washington Monument—just a bit over half the height, which took some thirty-six years from start to completion (although, to be fair, only about ten years of that was actual construction time).

As much as Eiffel in his methods and techniques pointed towards the future, he also had one foot firmly planted in the nineteenth century. His tower, after all, was made of iron, rather than steel, the structural virtues of which had been demonstrated more than a decade before Eiffel began his tower, in the great St. Louis bridge of James B. Eads. Just a year after the tower's completion, the even larger steel bridge over the Firth of Forth in Scotland eclipsed the Brooklyn Bridge as the world's longest span (each of two spans are 1,710 feet). In the twentieth century steel spans of a variety of forms were to extend ever-further the scale of bridge construction. Gustav Lindenthal's Hell Gate Bridge over New York's East River (1916) showed the potential of the large steel arch (its arch span is 977.5 feet). Lindenthal's one-time assistant, Othmar Ammann, was to outdo his mentor with spectacular suspension bridges in steel. When his George Washington Bridge over the Hudson River opened in 1931, it was twice as long as any earlier suspension bridge, and the role of steel went beyond the cables and decking to incorporate the great towers as well. That same year also saw the opening of Ammann's own steel arch contribution, the 1,652 foot-span Bayonne Bridge, connecting Staten Island with the New Jersey mainland. Ammann's final masterpiece was the Verazzano Narrows Bridge, connecting Staten Island with Brooklyn across the entrance to New York Harbor. Its 4,260 foot span was the world's longest for many years after it opened in 1964. While the nineteenth century, largely through the impetus provided by the requirements of

Figure 23.2
The Eiffel Tower under construction, 1889. Courtesy Institution of Civil Engineers, London.

railroads and the capabilities provided by steam power and new iron and steel production techniques, saw great increases in the scale and extent of civil engineering structures, these were quickly surpassed by twentieth-century achievements in every area. The combination of new and more versatile materials (such as reinforced concrete as well as alloy steels and light metals), greater skills at calculating the predicted behaviors of structures under varying loads, new capacities for moving the earth through explosives and mechanization, and new techniques for managing the labor and logistics of large-scale works gave engineers in the new century the means to carry out projects that had hardly been dreams previously. One of the central themes of twentieth-century technology—more evident at the century's opening than at its close—was this new, gigantic scale of construction. Improvement came to be defined, for many at least, in terms of sheer size.[3]

No better example could be found of the new century's capabilities than a project that defeated the best engineers of the previous one, including Eiffel himself. In 1879, the French promoter Ferdinand de Lesseps, still basking in the glow of his triumphant organization of the building of the canal at Suez, Egypt, a decade before, organized a company to construct a sea level canal to connect the Atlantic and Pacific Oceans across the Isthmus of Panama. The dream of such a canal was an old one, and de Lesseps clearly felt that a canal half the length of Suez, even if over obviously more difficult terrain and under harsh conditions, was a manageable challenge. The tragedy of his effort—one in which hundreds of lives and the savings of thousands of Frenchmen were lost—was one of the most humbling technical and commercial defeats of the nineteenth century. A part of the same effort to recover past glories and proclaim technical prowess that had motivated Eiffel's tower, the French Panama Canal attempt reflected instead the limitations of poorly coordinated technical projects. The Suez effort which was the company's model was, in fact, striking in the primitive elements of its technology—the work was done by almost a million and a half Egyptians in forced labor, at a cost of 125,000 lives.

When it took over the effort in 1904, the United States government determined to face the obstacles of working in Panama forthrightly, with every technical resource available to a modern state. Modern explosives (sixty million tons of dynamite were consumed), the largest mechanized earth movers ever built, a dam that created the largest manmade body of water yet formed, and locks (a total of six pairs) that dwarfed anything ever attempted before were all combined in a military construction project that took a decade to complete (figure 23.3). Just as important as the gigantic scale of construction was the capacity to overcome less obvious but crucial obstacles, such as the tropical disease that had made every other effort so costly in human lives. The U.S. Army's William Crawford Gorgas designed and managed a campaign against the disease-carrying mosquitoes of the isthmian jungles that finally made the area safe for large-scale work. The canal opened less than fifteen years into the new

Figure 23.3
The Gatun Locks of the Panama Canal, just prior to opening, 1913. Library of Congress photograph.

century, but no other construction project was to surpass it in combined scale and strategic significance.[4]

One of the key elements of the Panama construction was the giant earthen dam, 105 feet high, one and a half miles long, and half a mile wide at the base, built to convert a large segment of the Chagres River into Gatun Lake (at the time the world's largest artificial body of water, at 164 square miles in area)—the heart of the canal and the great reservoir of fresh water that made its operation possible. The construction of giant dams, of a great variety of forms, materials, and functions, was among the twentieth century's most widespread and enduring civil engineering efforts. In earlier centuries, dams were often part of the systems required to support water mills or canals, and their use for flood control and water supply systems increased substantially during the nineteenth century.

As urban areas grew in size, the need to protect low-lying regions from occasionally rampaging rivers made dams strategically important structures. Similarly, the ever-larger water systems required by large cities required the creation of reservoirs of unprecedented size, utilizing sometimes ambitious dams. For all of this effort, however, dam construction was long guided exclusively by traditional methods and empirical data. Most efforts were thus conservative in form and material, utilizing massive amounts of earth or stone to assure dam integrity. Even so, failures did occur, sometimes with spectacular and tragic results, as at Johnstown, in southwestern Pennsylvania, when an earthen dam's failure after a lengthy period of rain cost more than two thousand lives on May 31, 1889. At just about this time, however, engineers began to apply new mathematical and physical tools to the problem of calculating the forces and stresses at work in large structures. The Scottish engineer W. J. M. Rankine, for example, published in his *Manual of Applied Mechanics* (1858) the rules that could be used to understand a large dam's requirements well enough so that the ratio between the width of a dam at the base and the final height could be reduced from three or four to one to about one to one. Such an apparently esoteric theoretical achievement had enormous implications, for it transformed many large dam projects from impossibly expensive and time-consuming efforts into very real achievements.

There could be no better illustration of how the twentieth-century dam differed from those of the past than the great dam built on the Colorado River at Black Canyon, between Nevada and Arizona, in the first half of the 1930s. The Colorado, one of North America's longest rivers, with a basin that reaches all the way north to Wyoming and flowing some 1,400 miles before emptying into the Gulf of California, is a classic example of what hydrologists call an "exotic" river. This means that the Colorado, like, say, the Nile, empties a large catchment area while flowing through drylands otherwise untouched by the rains feeding the river. Most such rivers experience great changes in their course and through the seasons, and the Colorado's floods were legendary to early western travelers. Increased settlement in southern California (especially in the Salton basin, which developers called the Imperial Valley) depended totally on harnessing the irrigation potential of the Colorado while protecting low-lying lands from flooding. Early attempts at irrigation canals from the river turned disastrous when floods turned out to be impossible to control.

The idea of a giant dam across one of the great canyons of the river to create a huge reservoir and to manage the river's flow was, appropriately enough, the brainchild of a man familiar with great works, having served as chief hydrologist for the Panama Canal project. Arthur Powell Davis became the director of the U.S. Reclamation Service (later the Bureau of Reclamation) after leaving Panama, and he wasted little time in proposing the great dam. A new wrinkle in the calculus of dams had been added since the nineteenth century, and Davis used this possibility for the generation of enormous amounts of electricity to overcome doubts about the economic feasibility of the Colorado project. More than a decade of political action was required before the states and the federal government could all agree on how this work could be managed, but the Boulder Canyon Project Act was passed in 1928, and in 1931 work began on the largest structure ever attempted at that time. After two years of preliminary work, which included building giant diversion tunnels to carry the river's water away from the dam site, the pouring of three million cubic yards of concrete began in June, 1933 and continued nonstop, night and day, for two years. The dam was completed and turned over to the federal government on March 1, 1936, two years ahead of schedule, and by that autumn it was supplying electricity to the small city of Las Vegas, Nevada, about thirty-five miles away (figure 23.4).[5]

The multifaceted nature of Hoover Dam, the Lake Meade reservoir that it created, and the electric power system that it supplied well represented the key technological tendencies of the first half of the twentieth century. Scale and system came to represent both methods and goals for engineers throughout the Western world and, increasingly, in those parts of the world motivated to "catch up" to the dominant Western technological order. The promise (and, increasingly, the compulsion) of large-scale electric power generation increasingly converted some of the most formidable and intimidating forces in nature into resources that had to be, according to ideologies ranging from freewheeling American capitalism to emerging Soviet Communism, harnessed by the most massive technological efforts. Even before the end of the nineteenth century, no less a natural wonder than Niagara Falls had been converted into a giant source of electric power, despite engineering challenges that only a generation earlier had seemed insurmountable. Throughout the entire twentieth century, few technologies took on such an international character as the harnessing of great rivers for hydroelectric power. Such efforts required not only large-scale dam construction (and related hydrological works) but also the development of technologies of electrical generation, transmission, and distribution that took on the same qualities of scale as the dams themselves. The first long-distance electric power transmissions were put in place in Europe soon after the Edison system first appeared. In 1882, the young German engineer, Oskar von Miller, organized an electrical exhibition in Munich, and installed a transmission line from the Alpine village of Miesbach, thirty-five miles away. Less than a decade later, von Miller sponsored the construction of a 25,000-volt transmission line between Frankfurt and a hydroelectric generator at Lauffen, on the Neckar River more than a hundred miles away. These systems were merely demonstrations, but they provoked widespread interest among engineers, financiers, and politicians. Long-distance large-scale electric power transmission was to be one of the formative technologies of the new century.[6]

Figure 23.4
Hoover (Boulder) Dam. Courtesy Institution of Civil Engineers.

The most vivid early demonstration of the close linkage between electricity and the new, larger scale of technology was the great power station that opened at Niagara Falls, New York, in 1895. The tremendous size and volume of water at Niagara Falls had long attracted industrialists looking for power, but harnessing such large amounts of power in one place posed such difficulties until the late nineteenth century that little was built. By the 1880s, however, the aesthetic values of the Falls had evoked protective measures in the form of public parks on both the American and the Canadian sides, presenting even greater challenges to those who would harness the falls for industrial purposes. Nonetheless, engineers and entrepreneurs were not deterred, and gigantic schemes, using canals and tunnels, began to attract serious investors by 1886. When the first engineering plans began to emerge a few years later, the proposals considered included the transmission of power from Niagara Falls to the city of Buffalo, some twenty miles distant, by compressed air or even by belts and pulleys. By this time, however, the attractions of electrical transmission were sufficiently apparent that the decision was made to build the largest power generating plant and transmission system yet conceived.

The solution of the specific problems this scheme presented set much of the basic pattern of large electric power systems for the coming century. The use of alternating current generators, high-voltage transmission lines, transformer and control systems for managing and distributing usable voltages, and the promotion of a variety of applications besides light—including motors, chemical extraction, and large electric furnaces—were all pioneered by the Niagara system. Each of the generators in the Niagara power plant was rated at 5,000 horsepower, dwarfing all earlier designs, and soon after the station first began operations in August 1895, ten of these generators were producing power, not only for the lights and streetcars of Buffalo but for new industries drawn close to the falls specifically to use the cheap electricity. In fact, the chemical and aluminum plants that soon proliferated around the power station took a much greater proportion of the generated power than Buffalo ever did. It was this fostering, seemingly overnight, of great new industries by large-scale electric power generation, that was to be the most important of the images that Niagara Falls was to provide to the new century.[7]

As important as was the model of the giant power station surrounded by thriving new industries, this was not, in fact, how electric power exercised its primary influence on the emerging twentieth century world. Electricity was a technology not of concentration, but of distribution. In the first decades after Niagara's opening, power systems not only of great size but also of great extent appeared throughout the Western world. The first electrical systems, such as that which Thomas Edison had built in lower Manhattan in the early 1880s, served typically only a square mile or two. By the 1890s, the early experiments in alternating current transmission and hydropower generation had demonstrated the possibilities of much larger service areas and these began to appear where the economics of generation and distribution fostered them. But it was not until the years after 1900 that the outlines of the full-scale electric system began to emerge. In these systems, multiple generating plants, serving not single towns or cities but entire urban regions and beyond, were interconnected into technical and economic entities of astonishing complexity. The development of such systems depended on a host of innovations in everything from large turbogenerators to complex financing schemes to bring together the capital from a host of smaller enterprises. The enormous costs of building and operating electric power systems, along with the sheer complexity of making a range of both suppliers and users of power work together smoothly, compelled engineers and managers to devise novel means for coordinating large systems.

The technical and managerial implications of this were made apparent in systems like that in greater Chicago. Samuel Insull became Thomas Edison's personal secretary as a young English immigrant in the early 1880s and moved quickly up the ranks of the Edison enterprises as they grew larger and more complex. When the Edison electrical manufacturing companies merged with other firms in 1892 to become the General Electric Company, however, Insull struck off in a new direction, becoming the head of the Chicago Edison Company, which was one of about twenty companies that supplied the fast-growing city with electric power. In less than two decades Insull transformed the company into Commonwealth Edison, widely regarded in 1910 as the largest electric utility in the world. Insull, who never claimed engineering expertise, accomplished this by taking advantage of the best advice on the fast-moving technical front of power generation, control, and management and by exercising political, managerial and financial skills that displayed daring and imagination. Electric power, as a public utility, always developed in a climate strongly shaped by politics, and only organizers who recognized and used this fact could succeed in the long run.

Insull recognized the economic advantages of large scale in generation, distribution, and marketing, and mobilized the capital and expertise required to use these advantages. The generators that he ordered for the Fisk Street station when it was built in 1903 used the largest steam turbines ever built. These 5,000-kw units, however, were soon dwarfed by 12,000-kw units, and in the older plant across the Chicago River at Quarry Street Insull installed six 14,000-kw units by 1910. As the heart of his growing system, the Fisk and Quarry Street stations required ever more sophisticated control technologies, especially since the power was required by consumers in a variety of forms, from the household current of 110 V and 60 cycles to the 25-cycle AC current used by large industrial motors and the 600 V DC current used in great quantities by the city's streetcars and elevated railways.

In addition, the requirements of consumers varied widely through the hours of the day and the seasons of the year and this required that all utilities practice "load management." Since the greatest expense of a power company lay in the construction of its generating capacity and distribution system, and the extent of these was determined by peak usage (or "load"), the most economic returns came when these were used as close to their maximum as possible. Insull understood this very well and he carefully measured variations in load and encouraged the kind of consumption that would even out loads (the transit systems were good examples) and maximize return on investment. He also experimented with a variety of rate schedules to encourage usage at off-peak times and promoted the use of electric motors and furnaces in industry—their daytime usage balanced a load otherwise heavily weighted toward evening lighting. Insull was by no means the sole pioneer in the organization and management of large power systems; the problems that he faced in Chicago were dealt with around the world, first in the larger cities and then, as the twentieth century went on, in more and more rural areas. Systems grew ever larger in scale, surpassing by the 1920s the urban system represented by Insull's Commonwealth Edison and developing into regional and even national systems. The consequence of this continued development was that electric power systems became the measure, not only in the West but around the world, of technological modernity throughout the first half of the twentieth century.[8]

The most radical demonstration of the large electric system's meaning for the understanding of improvement in the twentieth century was in its wedding to communist ideology in the first decades of the Soviet Union. In 1920, the leader of the Russian revolution and the head of its struggling communist government, Vladimir I. Lenin, declared to the Congress of Soviets that "Communism is Soviet power plus the electrification of the whole country." In the years before the revolution that broke out in 1917, Lenin studied not only the writings of Karl Marx and other philosophers of communism and socialism, but also the sources of the new emerging power and wealth of the United States. He was impressed, for example, by the writings of Frederick Taylor and the promise of efficiency gained by the scientific study of movement and work. He was even more impressed, however, by the conclusion that electrical technology represented an important new stage in industrial development, a stage that fit well into Marx's belief that the unfolding of the history of class struggle was often marked by the appearance and application of new technologies. Lenin himself remarked on how electricity could make factories safer and healthier, could liberate peasants and workers from drudgery and filth, and would make it possible for millions to extend the hours available for study and leisure, especially in the cold and dark of Russian winters.

The Electrification Plan that Lenin presented to his colleagues in 1920 was directed toward the realization of the promise of electricity for pulling a backward state sharply and firmly into the twentieth century. In this context it was natural that large-scale systems centered on large-scale generating plants should have the greatest appeal. During the 1920s, when the so-called New Economic Policy relaxed some of the reins of state control on the economy, progress towards large-scale electrification was steady but slow. Only after the death of Lenin and the rise of Stalin late in the decade did the pace change, and the construction of a giant hydroelectric station became one of the centerpieces of the first Five Year Plan. The gigantic Dnieprostroi project, built with the assistance of one of America's most accomplished dam builders, Hugh Cooper (who had designed the large hydroelectric station at Muscle Shoals, Alabama—the first element of what was to become the Tennessee Valley Authority system), began operation in 1932. Just as at Niagara Falls, Dnieprostroi became the center of a great industrial complex, surrounded by chemical, iron, and aluminum factories. Just as on the Tennessee and Colorado Rivers, the work on the Dnieper became the key to controlling great floods, and promoting agricultural development. And just as with the American projects, that on the Dnieper became a model for projects to come, not only in the Soviet Union but throughout the world, especially in the non-Western world, where the key to modernization was frequently seen to be in the construction of great dams on great rivers and the delivery of vast amounts of power where before there had been little or none.[9]

Through much of the twentieth century, very large-scale technologies represented to many, both in the West and outside it, the key instruments of modernity and progress. Great dams and electric power projects and systems continued to be particularly prominent representations of this. In the United States, the 1930s saw a continuation of the gigantic dreams embodied in Hoover Dam. As that dam was nearing completion, Franklin Roosevelt's New Deal embraced the gospel of big power as a means of deliverance from a host of troubles, ranging from rural poverty to capricious flooding. In the South, the Tennessee Valley Authority was finally organized after more than a decade of often acrimonious debate, and the government-directed delivery of electricity over a very large area became permanent policy. At Grand Coulee, on the Columbia River in Washington, an even more massive concrete structure than any on the Colorado was completed in the early 1940s, providing yet another demonstration of how cheap electricity could transform an entire region. Within months, eastern Washington became home to an enormous industrial complex serving the needs of the American effort to make an atomic bomb (the same thing happened using TVA power at Oak Ridge, Tennessee).

Overseas, the lessons of both the Americans and the Soviets were taken in even greater earnest, especially in the later years of the century. In Egypt the great High Dam at Aswan changed utterly the flow and uses of the Nile River beginning in the 1960s. At about the same time, on the other side of Africa the great dam on the Volta River at Akosombo, Ghana, produced the world's largest artificial lake and brought enormous amounts of electricity to a land that had nearly none before. At Itaipú, on the Paraná River, between Paraguay and Brazil, a huge set of dams completed in 1991 generated almost all of Paraguay's electricity and about a quarter of Brazil's. As the twenty-first century began, yet one more huge project, this the Three Gorges Dam on China's Yangtze River, promised to outdo all others in both environmental effects as well as power generation. By this time, however, the enormous and often unpredictable environmental consequences of damming large rivers and creating enormous reservoirs had reduced the allure of giant projects in the developed world, and even in places like China, pushing itself headlong toward industrial modernity, the environmental costs stirred anxiety and opposition. The giant dam, however, maintained its capacity through the century to represent a distinctive and very alluring form of technological improvement.[10]

The particular allure that big technologies had in the twentieth century went far beyond dams and electric power projects. The Soviet Union's fascination with large-scale electrification, for example, was matched by equal enthusiasm for other giant technologies, from factories to canals to subway systems. During the 1930s, especially, the pursuit of what were sometimes called "hero projects" or "superindustrialization" fit perfectly with the political and philosophical agendas of Stalinism. Such technologies promised highly visible indications of the progress of the Soviet state towards overtaking the West. The Moscow subway system, for example, was begun as a great showpiece for socialism in 1932 and its first line opened in May 1935. The lavish fittings and design of the system as well as its rapid construction required the diversion of resources from a host of other enterprises. Even more representative of Stalinist gigantism was the enormous iron and steel factory raised on the barren steppes east of the southern Urals near some small hills known as the "Magnetic Mountain." The richness of these hills in iron ore was obvious (as the name implied), but the environment was inhospitable, fuel was scarce, and there were no markets within hundreds of miles. But this became a target of Stalin's first Five Year Plan, and in hardly more than three years a city of almost a quarter-million arose along with an iron and steelmaking facility that dwarfed anything else. By the end of the 1930s the mills at Magnitogorsk were producing more than three million tons of iron and steel each year, and the new city became the indispensable source of supplies for the war effort to come. In later years the wastefulness of the effort to build and operate Magnitogorsk would be pointed to as hallmarks of the dysfunctional economics of the Soviet economy, but at the time, for much of the world, the great complex was yet another indicator that the future belonged to technologies of scale.[11]

Just as at the great dam on the Dnieper, the builders of Magnitogorsk turned to American engineers to show them just how to build on a large scale. Americans, indeed, provided the model and set the standard for large scale technologies in almost every direction. The great Soviet iron and steel concern, for example, could be said to have its earliest model in the great Edgar Thomson Steel works that Andrew Carnegie had constructed on the banks of the Monongahela in Pittsburgh in 1875. Under the direction of a brilliant metallurgical engineer, Alexander L. Holley, the new mill was designed from the beginning to rationalize steel production at every stage, thus allowing the scale of production to be increased well beyond previous experience. The layout of furnaces, the organization of transport systems in the mill, and the entire materials-handling capabilities were designed to facilitate ever-increasing production. It helped that the mill could rely on a steady demand for its products from the expanding railroads. The larger scale of production at Edgar Thomson gave Carnegie the means to lower his unit costs to such a level that competitors found their profits squeezed unmercifully. There could hardly have been a better or more dramatic illustration of the economic rationale for manufacture on a giant scale. Once steel, widely recognized by the late nineteenth century as a core industry of industrial culture, had made this dramatic shift into large-scale production, other industries followed.[12]

Another dramatic example of the possibilities of large-scale technologies, not only to transform great industries but to be the foundation for great fortunes, came from an industry of far smaller importance at the time, but one soon to be central not only to the world's economies but to its geopolitics as well—petroleum. Unlike iron and steel, which had been important for centuries, petroleum was a very new industry in the late nineteenth century. The 1859 construction of the world's first oil well at Titusville, in northwestern Pennsylvania, set off a boom of sorts, but it was a modest one in scope. The primary uses of the liquid hydrocarbon products found in Titusville and elsewhere were for illumination (in the form of the kerosene fraction) and for lubrication (in heavier oil fractions). From the outset, however, this product had to be treated to separate the useful portions from other parts (which at this time included the lightest fraction, which we know as gasoline). Initially refining was a batch process, in which quantities of petroleum were put into stills, where heat was applied to distill off the desired fractions, which were then collected in separate chambers and cooled. Different sources of petroleum yielded different proportions of useful fractions as well as different qualities, particularly in terms of freedom from impurities (like sulfur). Serious experimentation in refining was limited until the end of the nineteenth century, but the rising importance of new petroleum-based products, such as asphalt for pavements and gasoline for the rapidly growing automobile industry, spurred the development of methods for continuous refining. During the last two decades of the nineteenth century, much of the petroleum business in the United States was increasingly concentrated in the hands of John D. Rockefeller's Standard Oil Trust, and by 1900 the Trust had about 80 percent of the country's refining capacity. In an effort to increase profits by lowering costs, Standard Oil refiners devised the first large-scale continuous refining techniques. With the spread of these techniques, petroleum refining began to achieve a true industrial scale.[13]

The rapid expansion of the market for petroleum, however, along with the emergence of a radically altered competitive environment in the wake of the federal government's breakup of the Standard Oil monopoly in 1911, pushed ever greater innovation in the industry. Between the years 1911 and 1919 the number of refineries in the United States multiplied seven times. This paled, however, in comparison to the growth in gasoline supply, which rose from 7.4 million gallons to 351.5 million gallons in less than one decade. The explosive growth of the automobile industry, along with the surge in demand from World War I (when aviation and marine fuels came to be an additional factor), pushed the industry at a pace few could have anticipated at the century's beginning. New refining technologies permitted this growth and continued through the first half of the century to increase the scale of production as well as the size of refineries. In the final years of the Standard Oil monopoly, the Trust's head of refining operations was a PhD chemist, William M. Burton, who put together a highly trained research team to tackle the problem of increasing refinery output of desired products. Barely slowed by the Trust's breakup, Burton and his team, now working for Standard Oil of Indiana, devised a process known as thermal pressure cracking. The result was a spectacular increase in the yield of usable gasoline from a given amount of petroleum stock, as the process converted some of the less useful fractions into valuable gasoline. The savings in crude oil were enormous as were reductions in the cost of fuel; the cost of making a gallon of gasoline fell from the pre-cracking amount in 1913 of 18.7 cents per gallon to less than 10 cents per gallon by 1916. Just as important for the future, the introduction of cracking spurred further experimentation and improvement over the next several decades. In the years after World War I, the petroleum industry became truly the petrochemicals industry, as refining technology permitted a host of improvements in product yields for many different materials. Improved internal combustion engine technology raised requirements in fuel quality, which the petroleum industry met. The size and efficiency of thermal cracking units continued to increase, so that by the late 1930s it was not uncommon for refineries to have units with daily productions exceeding twenty thousand barrels. At this time, just prior to the outbreak of war in Europe, another innovation, catalytic cracking, further increased the efficiency and speed of refining. The process pioneered by Eugene Houdry increased the complexity of the refining process considerably, and the scale of the capital investment for modern units far outran that of older technologies. But the output of the Houdry process, soon followed by others, especially in the highest quality fuels—now, importantly, including large amounts of aviation fuel—was so impressive that investment in plants increased rapidly even during the war years. Petroleum refining came to represent, in highly visible (and, often, olfactory) ways, the scale of materials conversion typical of the mid-twentieth century.[14]

The primary reason, of course, that petroleum became one of the most crucial and most widely traded commodities of the twentieth century was the rise of the automobile industry. While the invention and development of the automobile itself is a key part of the modern history of mechanical engineering and transportation, the manufacture of the automobile was one of the central dramas in the dramatic spread of large-scale mass production. The translation of manufacturing approaches that had been pioneered in the nineteenth century for firearms, clocks, sewing machines, and typewriters into the production of a device as large and complicated as the automobile had implications not only for the history of production itself but also for the very place of the machine in modern culture. The mass manufacture of the automobile became the fundamental model for applying the technologies of the large factory and the assembly line to all the products of everyday life. The success of automobile production also confirmed, more than anything else, the central place of the United States in molding the form of the twentieth-century world. Societies seeking their own routes to the industrial future, even as diverse as Weimar Germany and Soviet Russia, would embrace "Fordismus" or "Fordizm" as embodying the American gospel. Success was seen in terms of thinking big, particularly in terms of production.[15]

While the origins of automobile mass production lie in the century of prior experience with the division of labor, interchangeable parts, and mechanized manufacture, the stage was set more immediately by a product that ushered in the automobile age in more ways than one—the bicycle. The first two-wheeled bicycle-like vehicles appeared in the first decades of the nineteenth century in the form of the "hobby horse." Pedals were applied to the front wheels of this device in the 1860s, and gradual improvements were made over the next decade, with the adoption of solid rubber tires, an enlarged front wheel, and ball and roller bearings to allow wheels and pedals to move smoothly. By the late 1870s, the high-wheeled bicycle began to climb in popularity, at least as a sporting device for young men. The modest boom in these vehicles, however, paled in comparison to the craze that met the introduction in the 1880s of the "safety" bicycle—the machine with two wheels of equal size, a chain drive, and eventually pneumatic tires. By the mid-1890s, when the craze reached its peak, more than three hundred companies in the United States produced over a million bicycles annually. The popularity of the bicycle introduced a culture of individual transportation which the rising automobile industry, once the cost of individual units had come down, made its own. This culture incorporated a number of elements important for the future, such as a widespread agitation for better (paved) roads and an acceptance of independent female mobility on a par with that of men.

Just as importantly, the bicycle craze supported the development of mechanisms, tools, manufacturing methods, and marketing approaches that all proved to be enormously valuable for the mass production of automobiles. The ball bearings that made cycle wheels move smoothly and reliably proved indispensable for a host of future machines. New stamping and welding techniques speeded up the manufacture of a host of critical parts. The most successful American bicycle maker, Albert A. Pope, pioneered the production of a series of models, each fancier and more costly than the next, that would tempt the bicycle purchaser to keep moving his or her sights higher as their aspirations for mobility or fashion (or both) were stoked by experience. By the time that automobile manufacture began in earnest in the America in the first decade of the twentieth century, the bicycle experience had provided a substantial foundation on which to build.[16]

The application of an engine to a road vehicle was an obvious idea, occurring to steam engine experimenters as early as the eighteenth century. The size and weight of steam engines and boilers discouraged much effort in this direction, however. The popularity of the railroad, along with the poor state of most road surfaces continued to reduce interest in motorized vehicles for most of the nineteenth century. The development of practical electric motors and internal combustion engines in the last decades of the century revived interest, and a range of designs emerged. The electric motor turned out to be a very popular option for streetcars, where power could be readily delivered via trolley wires. The difficulty of providing reliable and long-lasting sources of power for smaller independently operated vehicles, however, ultimately made the electric car a less attractive option than the gasoline-powered one. In 1885, Karl Benz, a mechanical engineer in Mannheim, Germany, installed an engine he had built on Nicolaus Otto's design in a three-wheeled vehicle he made for the purpose and proved its practicality on nearby roads. Benz sought to commercialize his idea, but without much success for several years. At almost the same time, Gottfried Daimler, the chief engineer in the firm that had been selling the "Silent Otto" engine since the 1870s, put a high-speed version that he had devised on to a bicycle, and soon afterward he put a 1.5 hp engine onto a carriage. The product ran well enough to prove its technical practicality, and in 1890 Daimler organized the first firm for the production of "horseless carriages." By 1893, both Daimler and Benz were producing motor cars for sale, although the numbers were quite small until the end of the decade.

The availability of the Daimler engine inspired others, like Emile Lavassor in Paris. The firm of Panhard & Lavassor made woodworking machinery, and this equipped Lavassor to explore the possibilities of the new German engine with an established machine shop at his disposal. His design of a vehicle to take advantage of the engine provided the most important model for all early automobile makers, incorporating a front-mounted engine driving the rear wheels through a shaft and differential and controlled through a friction clutch and change-speed gearbox. Equally importantly, Panhard & Lavassor became the first successful commercial manufacturers of cars, quickly turning to making their own engines (under license from Daimler) and promoting their product through increasingly more publicized road races. An equally adept exploiter of the road races was Armand Peugeot, who was France's most successful bicycle manufacturer. He began making automobiles in 1891, at first with very conservative designs, but by the mid-1890s he was making sophisticated models, and it was a Peugeot that won the first widely publicized auto race in 1895, a round-trip between Paris and Bordeaux (732 miles). By the late 1890s, the automobile was an emerging technology receiving enormous public attention, even if the vehicles themselves were still only rich enthusiasts' machines.[17]

At first the American experience with the automobile was little different from the European. The first car built for sale in the United States was the product of a bicycle manufacturer, the Duryea Brothers of Springfield, Massachusetts, in 1893. Within three years, dozens of American mechanics were experimenting with the new technology, particularly in the Midwest. Often the pioneers were bicycle or wagon and carriage manufacturers whose initial approach went little beyond adapting their familiar vehicles to the novelty of the internal combustion engine (as well as to, in decreasing numbers, small steam engines and electric motors). By the time that the U.S. Census Bureau took official notice of the new industry in 1899, thirty American firms and shops had set up in the automobile business, turning out a total of about 2,500 vehicles. New England firms dominated the briefly robust business of making electric or steam cars, but quickly the center of gravity for internal combustion (and, hence, in time for the whole industry) moved to Michigan, Indiana, and Ohio. Here an established tradition of carriage making was wedded to a newer one for supplying farms with gasoline engines for stationary farm machinery. Even more important, however, was simply the appearance of strong-willed and imaginative entrepreneurs, like Alexander C. Winton, Ransom E. Olds, Henry M. Leland, and Henry Ford. As the twentieth century began, a number of social and economic factors began to distinguish the American automobile industry from the European. The higher average incomes of Americans, the absence of substantial regulation or taxation on automobiles, and the greater distances confronted by rural Americans in their everyday lives made the United States more fertile ground for the new technology's popularization. Most important, perhaps, was the fact that by this time Americans expected new technologies to be widely accessible. An observer in 1901 commented "American manufacturers have set about to produce machines in quantity, so that the price can be reduced thereby and the public at large can have the benefit of machines which are not extravagant in price, and which can be taken care of by the ordinary individual." The idea of the car for the common man was not an uncommon vision.[18]

Making this vision a reality, on the other hand, posed difficulties. Before the automobile in America was a decade old, manufacturers were attempting to make cars that appealed to a broad market. In 1901, for example, Ransom Olds introduced a three-horsepower lightweight auto with a curved dashboard—the first

"Oldsmobile"—for sale at $650. This was still well out of reach of ordinary working-class Americans, but accessible to the growing upper-middle class. For a few years Olds was quite successful, making between 2,500 and 5,500 cars per year—more than one-third of total U.S. production in 1903. The Oldsmobiles were not bad cars—they proved themselves in a number of well-publicized cross-country trials, including one trip in 1903 that crossed the continent—but they were also not impressive in performance, durability, or technical sophistication. They were essentially horse carriages to which motors had been added, and this was true for almost all of the early efforts at inexpensive automobiles. In the same year that Olds introduced his car, the Daimler firm built its first "Mercedes," and this car served to show what a twentieth-century automobile could be: it had a 35-horsepower four-cylinder engine, a honeycomb radiator, an accelerator pedal connected to a modern carburetor, a chassis of steel, and could run over fifty miles per hour. But the Mercedes was, from the beginning and always, an expensive car, fewer than half-a-dozen emerging from the factory each day, even at the time the Daimler factory in Stuttgart employed about 1,700 workers in 1909. The two streams of improvement for the automobile—quantity production on the one hand and high performance on the other—remained separate, both in conception and in fact, until they were joined by Henry Ford, beginning in 1908.[19]

Like many who entered the automobile world in its first years, Ford started as an enthusiast, as a largely self-trained mechanic who found excitement in all the latest technologies of the day. He built his first vehicle, a so-called quadricycle, in 1896, and was so stimulated by the very experience of building and showing off the car that he leaped into the business quickly, persuading less mechanically-talented but better-heeled Detroiters that he knew how to build the best automobile yet. For several years Ford then showed his mechanical skills in high-performance vehicles, demonstrating their virtues by racing them himself. Technologically exciting as this may have been, it did not lead to commercial success. Finally, in 1903, Ford and a new backer formed the Ford Motor Co. and began to turn out a series of moderately priced (around $1,000) models which solidified Ford's reputation for well-made and well-designed automobiles. In 1906 Ford announced new goals, going in a direction recently pioneered by Henry Leland of the Cadillac Motor Company (one of Ford's former firms). Leland, who began as a machine-tool maker (and who became the founder of General Motors), took advantage of advances in machining tools and techniques to make a car with carefully fabricated interchangeable parts. His Model B Cadillac was a modest (though not inexpensive) car, but it showed how carefully planned production could make it possible to rationalize automobile manufacture. Ford's model N was intended "to revolutionize automobile construction." He announced that "I have solved the problem of cheap as well as simple automobile construction . . . a serviceable machine can be constructed at a price within the reach of many." The $600 Model N had a four-cylinder 15-horsepower engine, and it quickly became the most popular car in America. If Ford had followed the pattern established by others, he would have stopped development at this point, perhaps turning his focus to higher-performance models that Model N buyers could be induced to "step up" to.[20]

Ford's special contribution was his insistence in combining the two recognized paths of improvement—high performance and affordability—and understanding that doing this ultimately required a new form of production. All of the elements of Ford's system of mass production had existed for quite some time earlier. Standardized parts and interchangeability had been pioneered a century before, largely by armsmakers. The importance of precision had been recognized even earlier, most notably by clock makers. By the end of the nineteenth century, mass manufacture was the reality in a host of industries, as we have seen, but putting together all the elements for true mass production of a complex, widely-desired product began with Ford's Model T. The beginning was not sudden. Ford introduced the Model T in 1908 essentially as an improvement of the successful Model N. This was an exceptionally well-designed and well-made car. Its four-cylinder, 20-horsepower engine and other technical advances, from improved suspension to ease of repair, made the Model T's owners confident. Vanadium alloy steel proved lighter and tougher than cheaper alternatives, while innovations in casting engine parts and machining kept production costs down. For $825, one could not get anything close to it, and the car instantly found a heavy demand—fifteen thousand were on order before a single one was delivered.

The market for the Model T easily outstripped Ford's ability to produce it, and this would have been motive enough for reengineering the car's manufacture. But the Model T's price also represented a challenge to Ford and his vision of a car for the ordinary man. At his plant on Piquette Avenue in Detroit, Ford organized production as efficiently as the building and equipment allowed, and 13,840 cars were produced in the first full year of manufacture, 1909. On the first day of 1910, the Ford Company opened a completely new factory, in the Detroit suburb of Highland Park. This is where Henry Ford finally wrought the revolution he had been promising (figure 23.5).[21]

From the beginning, even before Ford and his colleagues had revealed the full outlines of their revolution, Highland Park was a remarkable statement of the new scale of production that Ford had in mind. Albert Kahn, the architect for the new works, built a "daylight factory," in which glass occupied almost 75 percent of the entire wall area, giving the main building astonishing light and openness. Eight other buildings made up the works and others were soon added, filling up the site's sixty acres completely. Power was provided from a giant electric generating plant, which Ford intended as one of the showpiece spaces of his works. A huge machine shop covered

Figure 23.5
Model T Fords as completed on the Highland Park assembly line, 1917. Library of Congress photograph.

a space 840 feet by 140 feet, connected to the four-story, 865-foot-long assembly building by a glass-enclosed craneway, in which materials distribution was coordinated. The key to the manufacturing, even before the assembly line was born, was relying as much as possible on specialized machines, each designed for specific tasks of molding, stamping, planing, drilling, and the like. Astonishingly complex machines could now be made by machine tool fabricators, and the Ford engineers took full advantage of this to reduce labor—particularly skilled labor—to a minimum. At the same time, the reliance on massive numbers of machines allowed the speed of production—throughput—to be increased enormously. The result was an astonishing rise in output. In 1911, the first full year of the Highland Park plant's production, 53,488 Model T's were made—about four times 1909's impressive output. Nonetheless, the Ford engineers—it was very much a team effort at this point—were dismayed at what they saw as substantial sources of inefficiency and waste in their system, and over the next two years they strove to eliminate each of these as they spotted and confronted them. There were many elements in the resulting accomplishment, but the central one—recognized by engineers, businessmen, workers, and the public alike—was the systematic, coordinated, controlled movement of each of the elements of the car in a rational system—the assembly line.[22]

The system of production that emerged at Highland Park and that was to be elaborated still further in a huge facility the Ford Company started building on the banks of the River Rouge in 1917 was so dramatic in the impression it made on observers that it was termed "Fordism," and spread around the world as not simply a method of manufacture but as a philosophy of economic life. The key concepts of the assembly line were simple: move the materials and parts of the product past stationery machines and workers, each of which was dedicated to simple specific tasks. If the product or subproduct was complex, then many different machines and tasks would be required, but the principle was the same: the work, not the worker, moved. Ford himself claimed later that the inspiration for this came from observing the "disassembly" of hogs on a meat production line—hog carcasses would be hung on conveyors and moved past cutters, each of whom had a specific cut to make or piece to remove. Other engineers pointed to canning lines, which by this time were highly mechanized operations in which cans were sent down conveyors to be filled, sealed, labeled, and packed in a continuous sequence. A large range of devices were available to move parts and assemblies, whether by overhead cranes, conveyor belts, or gravity chutes. The most difficult aspect of engineering the assembly line was fitting in the workers and determining how to pace them. Work would now be done strictly at the pace set by the machine; no initiative or option was to be left to the individual at the line.

The resulting assembly line system was not, apparently, the result of some sort of grand overarching plan to reform manufacture. It instead grew out of the Ford engineers' effort to reduce perceived inefficiencies—their pursuit of step-by-step improvement of the production process. In April 1913 the section at Highland Park dedicated to fabricating the flywheel magnetos for the Model T engine was reorganized around a moving assembly line. In the original system, a workman would spend about twenty minutes assembling a single magneto; the moving line reduced average assembly time to just over thirteen minutes; a year later the simple act of raising the working line eight inches to reduce strain on the workers reduced assembly time to seven minutes. Reducing labor, increasing speed, and trimming waste were the core goals of the new production engineering. The assembly line approach, which centered on keeping the right tools and machines at hand, reducing unnecessary motions to a minimum, and moving the product at a controlled, brisk pace through the work process gave the Ford engineers, and in short order industrialists everywhere, the means to achieve these goals. The result was manufacture on a scale hitherto unimaginable.[23]

The numbers of automobiles that poured from the Ford assembly plants told their own story: In 1913, the cost of the Model T Touring car (the more expensive of two levels) was reduced to $600 and in the first twelve months of the assembly line's operation (1913–14), Highland Park produced almost 250,000 automobiles. But this was only the beginning. In 1916, before World War I could disrupt production and sales, the touring car was reduced to $360 and the year's production was over 738,000 units. This was about half of all cars produced in the United States. By 1920, fully half of all the cars in the world (and three-fifths of those in the U.S.) were Model T Fords. By 1927, when Ford finally stopped making the Model T, more than fifteen million had been sold and the price had broken below $300. As demand increased in the early years, Ford expanded production as fast as he could, using the assembly line and all the advantages of large-scale manufacture. As this scale continued to increase, Ford stoked demand even further by reducing his prices. The basic principles of mass production and mass consumption were never better captured or displayed than by Henry Ford's achievement.

So too did Ford exemplify better than any other the twentieth century's lesson of scale applied to production. The Highland Park factory complex was constantly expanded, but before the decade was out it had reached its limits. In its place was built the huge River Rouge complex in Dearborn, Ford's hometown. On two thousand acres Ford brought together every element of production he could and he made it the hub of an even larger system that stretched from almost a million acres of forest in northern Michigan to Kentucky coal mines and Brazilian rubber plantations. As much as possible, Ford concentrated the conversion of these and other raw materials into useful products at River Rouge, integrating blast furnaces, steel mills, glass factories, leathermaking, and clothmaking into the machine shops and assembly lines. Although numerous changes, both visible and invisible, were made to the Model T over the years, Ford focused improvement on the process of production rather than the vehicle. This was the source of spectacular and world-changing success at the beginning, although by the mid-twenties the market had changed and become much more complex.[24]

Other automobile makers, most notably William Durant, William Knudson, and Alfred P. Sloan at General Motors, recognized both the extent and the limits of Ford's achievement, and set out to redefine automobile improvement. The Model T had introduced legions to mobility, but it prepared the ground for a different kind of demand, for greater performance, for more style, for variety and distinction. Technical advances like electric starting and higher compression engines began as optional luxuries for the upscale auto buyer but were turned into necessities through familiarity and marketing. Other elements, such as a wide choice of colors (Ford finally departed from black for the Model T, but only in the model's last years), body styles, and engine sizes came gradually to assume significance to the mass consumer. From the early 1920s units of General Motors began turning out annual models, although it was not until the thirties that the practice was made regular. "Sloanism" began to displace "Fordism" in the automobile business: in Alfred Sloan's own words, "a car for every purse and purpose." Sloan claimed this was "constant upgrading of product," but most observers saw it as the use of cosmetic changes and heavy marketing to make last year's models seem old-fashioned and obsolete compared to newer ones. William Knudson, whom Sloan brought in to make his Chevrolet division the mass-market leader, adapted the techniques he had used as a Ford engineer to the new philosophy, relying on the same level of mechanization and assembly organization as Ford, but using flexible factory layouts and machine tools, allowing for model changes to be put into place in a few weeks, compared to the six months that it took the Ford company to make the change from the Model T to the Model A in 1927. In World War I, Ford showed both the strengths and the limitations of his production system in transforming his plants to the manufacture of boats and aircraft—great quantities could be turned out, but too much time was required to engineer the production lines. In the Second World War, by contrast, the new more flexible systems of production pioneered by Knudson and Sloan were able to make the conversions to wartime needs much more readily. "Flexible mass production" became the key to mid-twentieth-century prosperity and power.[25]

No other single experience in the twentieth century emphasized the new role of bigness in every aspect of life, including technology, as World War II. As Europe lurched towards war in the late 1930s, the economies of the West were recovering very slowly from the Great Depression. Fitful increases in demand for such consumer goods as automobiles and the new necessity, refrigerators, were not enough to sustain a broad recovery. In Western Europe, only the acceleration of military preparedness at the decade's close moved the economy forward, and in the United States even this was put off as long as possible. After the German invasion of Poland in 1939, the French and British armed forces began to turn to the United States for supplies, and after his reelection in 1940, Franklin Roosevelt's Lend-Lease program facilitated the industrial recovery fueled by allied purchases. The United States itself only slowly moved its own economy to a war footing until after Pearl Harbor. From 1942 on, however, the capacity of the new American systems of production for yielding undreamed-of quantities of materiel and weapons was displayed spectacularly. While conceivably much of this might have been accomplished by the thousands of small businesses that were the traditional sources of production for most of the economy, this was not the pattern favored by either the new technologies or the politics of modern war. It was estimated that in 1940 about 175,000 firms were responsible for

70 percent of U.S. manufacturing, and the remaining 30 percent was concentrated in one hundred large companies. By the middle of 1943, total manufacturing output had been doubled due to war needs, but the ratio of producers had been reversed: 70 percent of this much larger output was the product of the one hundred largest firms. The war effort favored and fed huge factories for everything from aircraft to warships, and the enormous material requirements for all of this—great quantities of aluminum, steel, plastics, petroleum, and the like—were similarly met by ever larger factories and mills. The models of vertical integration, as at Ford's River Rouge, and of flexible mass production, as at General Motors, were applied on a scale that, in some ways, was national in scope. The War Production Board, with its powers for allocating materials and directing the output of factories, acted as a form of higher-level production management for the entire country. Both established large manufacturers and entrepreneurial smaller ones turned to large-scale war production. Ford, for example, built an enormous assembly plant (Willow Run) in Ypsilanti, Mich. to make thousands of B-24 bombers; Dodge constructed a factory in Chicago for aircraft engines that covered 6.5 million square feet—almost 50 percent of the auto industry's total capacity before the war. New entrepreneurial figures made bigness work for both the war effort and for their own fortunes. Henry J. Kaiser, for example, one of the primary contractors for the Boulder and Grand Coulee Dam projects, became the country's largest shipbuilder almost overnight. In ten shipyards, Kaiser's firms turned out ships at an astonishing rate, reducing the fabrication time required for the basic merchant vessel of the war, the Liberty ship, from an average of 355 days in 1941 to fifty-six days a year later, at the same time increasing output from one million tons aggregate to eight million tons. Kaiser's was perhaps the most famous of the scaling up efforts, but simply represented a phenomenon that could be seen across the economy.[26]

The greatest single example of large-scale production in the service of war, however, was not, at the time, visible at all, at least not to the public. In out-of-the-way corners of the country—the high desert of New Mexico, the hills of eastern Tennessee, and the barren drylands of the Columbia River basin in eastern Washington—there were built gigantic industrial establishments, practically overnight, for the most massive productive effort of all, the making of the first atomic bombs. The quantities produced, of course, were not large, but the go-for-broke effort to produce uranium and plutonium and to design and test an explosive device of a type completely unknown before generated some of the most massive production facilities ever constructed in America—plants in which size was measured not in acres, but in hundreds of square miles. The greatest drama of the bomb production effort was centered around the site of a former boys' school on a high mesa in central New Mexico, Los Alamos. There, in splendid isolation from any population centers, the U.S. Army constructed a laboratory complex for the physicists, chemists, and engineers designing and building the first atomic weapon. Less famous, but not less crucial to the bomb effort, were the facilities constructed elsewhere for manufacturing the material that would actually fuel the bombs. Never before had such resources been put into making such difficult-to-produce materials. The urgency of the effort—spurred on by the very real fear that Nazi Germany was on its own path to producing the superweapon—meant that making bomb materials had to move on several different fronts at once. The two candidate materials for effective nuclear fission were a rare isotope of the heavy metal uranium, so-called U-235 (from the number of neutrons, as opposed to ordinary U-238), and the even rarer artificial element plutonium, which had been made for the first time in tiny laboratory quantities only in early 1941. Maximizing the chances of success in making a bomb—when failure could be a truly catastrophic defeat—required making substances hitherto available in fractions of a gram in kilogram quantities, and doing so as quickly as possible.[27]

The acquisition of almost 60,000 acres of Tennessee woodland and worn-out farmland in September 1942 was one of the first major moves of the Manhattan Project (or, more strictly speaking, the Manhattan District of the U.S. Army Corps of Engineers). There the Project constructed the "Clinton Engineer Works," which was a giant facility to separate the fissionable U-235 from its more ordinary parent U-238. Two completely different techniques would be used, gaseous diffusion and electromagnetic separation, since no one yet knew which purely laboratory processes could be successfully scaled up to produce quantities sufficient for bombs. Great amounts of electricity could be supplied to the facility, more commonly known by the small town built to house workers on one edge of the reservation, Oak Ridge, thanks to the proximity of TVA dams. But many other problems of scaling up would have to confronted and overcome—some of them requiring ingenious expediencies, like borrowing fourteen thousand tons of pure silver from the U.S. Treasury for magnet windings, to save on critical copper. Twenty thousand construction workers built the huge separation plant (the main diffusion building was almost half a mile long) in only a matter of months, and almost five thousand employees working for the operating contractor, the Tennessee Eastman Company, labored around the clock for the next two years to produce enough fissionable uranium for the experimental bomb (figure 23.6). Even more astounding than the Oak Ridge facility was that built in a desolate area along the Columbia River in eastern Washington. About a half-million acres was acquired in this isolated region, chosen for its proximity to the clean water of the Columbia and to the electricity of the great dams at Grand Coulee and Bonneville and its distance from much of anything else, since the manufacture of plutonium was seen as not only a formidable engineering challenge but a potentially very hazardous one, generating more radioactivity than anyone had ever encountered.

Figure 23.6
The town and production facilities of Oak Ridge, Tennessee, 1945. Record Group 434-OR, National Archives and Records Administration, College Park, Md.

Into this wilderness about 50,000 construction workers (at the peak) were sent to build eventually more than 500 structures, 158 miles of railroad track, 386 miles of roads, and hundreds more miles of fencing, power lines, and other infrastructure elements—all at a cost approaching a quarter-billion dollars. The core reason for this huge establishment was speed and the sense that failure was not an option. Just as the manufacturers of everything from oil to automobiles had demonstrated, speed and scale went hand in hand, hence the enormous size of Oak Ridge, Hanford, and the entire atom bomb enterprise. This large scale, and the great expense attached to it, had another consequence. As the elements of diffusion plants of Oak Ridge and the reactors at Hanford were being put together, the physicists began to notice something disturbing to a number of them: rather than machines and structures built for a short-term war effort, answers to a great emergency, what they saw clearly had purposes extending into the future. The scale and cost were so great that they could only be justified in the long run if they did not simply answer the current emergency but laid the ground for the future, for a world in which atomic weapons were accepted weapons of war. In this way, the ultimate triumph of technological scale lay in the shaping of a future in which big technology—along with big science, big government, and enlarged institutions of every stripe to serve their purposes—came to seem not only natural but also necessary.[28]

24 The Corruption of Improvement

Late in the afternoon of April 26, 1937, a lone airplane was heard in the sky approaching the small Basque town of Guernica, in northern Spain. The bells of the church of Santa Maria rang out a warning, for the town of Durango, less than twenty miles away, had been bombed a few weeks before, many civilians had been killed, and the townspeople were wary. Spain had been wracked by civil war for many months, and although most of the fighting had been in the south of the country and then around Madrid, the northerners of the Basque country had seen the war come to them that spring. Still, despite the warnings and the growing ferocity of the civil war, nothing prepared the townspeople for what followed. At first the airplane, a German Heinkel 51, simply circled around the town, the pilot choosing to climb steeply to avoid possible ground fire. Encountering nothing, it circled back, coming directly over the heart of the city, and dropped several 550-pound bombs in the area around the railroad station. The plane then cruised away to the south and the people of Guernica came out to help the wounded and survey the damage. Less than a half-hour later, however, three more Heinkels appeared, accompanied by a half-dozen Messerschmidt fighters. Some 10,000 pounds of bombs were let loose on the heart of the town, and incendiaries scattered about, causing fires everywhere. Just a few minutes later, almost two dozen Junkers-52 bombers were in the air over Guernica—all in all, about 100,000 pounds of high-explosive, shrapnel, and incendiaries were dropped on the town. After several waves of Junkers had completed their work, a half-dozen Messerschmidts swooped down over the town, machine-gunning everyone they could catch in their sights. A few minutes later, the Heinkels returned and dropped more bombs. By nightfall, Guernica was in complete ruins and almost everything not already destroyed was aflame. The number of dead was never known, but the population was devastated; British and American war correspondents who made their way to the scene spoke of utter destruction.[1]

In the scale of violence that warfare visited on the world, and especially on Europe, in the twentieth century, the destruction of Guernica does not loom especially large. A generation earlier, many more thousands were killed in just a few hours on the Western Front of World War I than perished in Guernica's fires and bombs. Only a few years afterward, the armies and air forces of World War II, from the German *Blitzkrieg* of Poland to the atomic bombs exploded over Hiroshima and Nagasaki, would wreak destruction dwarfing anything seen in the Spanish Civil War or elsewhere in history. But in the effort to understand just how the technology of violence was utterly transformed in the twentieth century, Guernica is worthy of special notice. The large-scale violence of the twentieth century was typically waged more for psychological effect on observers than for the actual elimination of enemies or threats, and the bombing of this town, a place of no strategic significance, but the historic heart of Basque national traditions and aspirations, made sense (if it made any sense at all) only in psychological terms. The new communications technologies of the twentieth century supported this logic—the civil war in Spain was covered by an extraordinary host of journalists from noncombatant countries, and these availed themselves not only of the older technologies of telegraphy and the print media, but also the modern ones of radio and aviation. The news of Guernica's suffering spread rapidly throughout Europe and North America, as well as in Spain itself. While war journalism was almost a century old by this time, the speed and the reach of the news of the atrocities of the Spanish Civil War was unprecedented, and it helped to establish the pattern for military reporting in the wars to come. War had been captured by writers and then photographers through much of the preceding century, and now war was to be recorded and reported by radio and motion pictures, providing an immediacy in both time and experience that was uniquely modern.

Yet one other aspect of the bombing of Guernica made it an event of greater interest both at the time and to historians than it might seem to merit: the pilots of all of the aircraft in the air over the town were not Spanish, but German. This was heatedly denied by the Nationalist Spanish, who were the only ones to possibly gain by the destruction of the largely Republican Basques, and by the Nazi leaders in Berlin, who were signatories to an international agreement to stay out of the Spanish conflict. Nonetheless, despite the denials, it was clear to most observers at the time as well as to historians afterward that the bombing was a German exercise—in fact, a trial run by the leaders of the Luftwaffe for the terror-bombing tactics they had been developing for several years and which they were to use to great effect in the war that began in 1939. This was neither the first nor the last intrusion of the German "Condor Legion" into the fighting, but it became the most famous single instance of the manner in which the fascists in Germany and Italy used the Spanish conflict as a testing ground for the new technologies of warfare. So complex and expensive had modern war become that weapons, materiel, and tactics had to be tried out thoroughly so that the complete systems of battle, on land, sea, and air, could be observed and improved. The improvement of violence became a large-scale enterprise itself, with enormous resources given over to it for the long term.

The improvement of military capabilities through the systematic marshalling of expertise and resources was an old idea by the twentieth century. The French Revolutionary governments in the 1790s pressed the *savants* into government service and established institutions like the Conservertoire des Arts et Mètiers to make readily available the fruits of advanced technical and scientific knowledge for the purposes of the state. Through the course of the nineteenth century a variety of models was devised to expedite the development of improved weapons and other useful military technologies, most typically consisting of the formation of institutions with government sponsorship that would foster inventive activity to serve state needs. During the Civil War, for example, the U.S. Congress chartered the National Academy of Sciences, although there is no evidence that the scientists meeting in Washington had much effect. World War I was more obviously a conflict with advanced scientific and technological dimensions, but the institutional innovations were still rudimentary. The American examples are again instructive: a group of well-known inventors, chaired by Thomas Edison, was constituted as the Naval Consulting Board in 1916. Their job was to solicit and screen proposals from inventors and aspiring inventors around the country, and to direct the most promising to firms that could build and test prototypes. While not without some value, this effort produced little in the end—it was a model that looked backward to an already disappearing era of independent inventors rather than forward to the new age of organized research. The more constructive model was provided by the National Research Council, formed under the auspices of the National Academy of Sciences in 1916, and charged with organizing the means to solve specific problems brought to it by government agencies.

In some areas, the government took even more deliberate steps to provide the means for improvement. The National Advisory Committee for Aeronautics (NACA) was formed more than two years before the United States entered the European war, but advancement in aviation technology was a national priority and required special measures. The purely advisory nature of the agency gave way in 1920 to a more active role, when wind tunnels were constructed and a laboratory was outfitted on the Virginia coast. In the decades to come, the NACA was to take the lead in applying the most advanced aerodynamics to the improvement of every aspect of aviation. The British followed much the same pattern, both in creating general instruments for research (the Department of Scientific and Industrial Research was set up in 1916) and in the special pursuit of aeronautical research (the Royal Aircraft Establishment at Farnborough was created in 1918, from the Royal Balloon Factory). The First World War illustrated the increasingly crucial nature of aerial warfare, but it also brought home the lesson that the technological edge that was indispensable to success in the air could not be left to the whims of independently generated research, but would have to be actively pursued by those wishing to control it.[2]

Not only governments came to understand that technological opportunities could be best exploited by those who created them themselves. Capitalists, too, came to see improvement as an instrument for corporate growth that needed to be shaped and controlled to serve specific ends. Out of the workshops and testing rooms of the nineteenth-century inventor and manufacturer there emerged laboratories of a new kind, later to be dubbed as sites for "research and development," which became the twentieth century's principal sites for technical innovation. Around the turn of the century a number of American corporations established permanent laboratories, at first largely extensions of the shops that had become common for testing materials and instruments. In the chemical and electrical industries these laboratories, at least in a few major companies, came to be increasingly devoted to the pursuit of new technologies, of inventions that would insure continued corporate success through the management of improvement itself. By the beginning of the twentieth century, for example, the testing laboratory of the Standard Oil Company of Indiana had become a research laboratory for improving the efficiency of refineries. In 1902, the DuPont Company, primarily then an explosives manufacturer, set up a laboratory headed by a PhD-holding chemist, at first primarily to standardize and make more efficient company procedures and formulae, but within a few years Eastern Laboratory staff was devoting attention to possible new chemical products. At about the same time, America's premier electrical manufacturer, the General Electric Company (heir to Thomas Edison's first electric light enterprises) undertook a similar transformation of its testing and standards departments into centers of innovation. The General Electric Research Laboratory was formally organized in Schenectady, New York, in 1900, under the leadership of a young chemistry professor from MIT, Willis R. Whitney. The new laboratory was explicitly charged with improving the company's products (primarily in lighting), and with giving the firm the means of protecting or extending its control over a number of markets, either through improved products or through management of the patents that resulted from corporate research.[3]

The emerging laboratory model explicitly combined industrial interests and needs with the pursuit of the most advanced science, typically led by individuals with advanced academic training. This philosophy was behind the formation in 1911 of the "Research Branch" within the Engineering Department of the Western Electric Company, which was the manufacturing arm of AT&T, the holder and operator of the Bell telephone patents and the company in charge of all but a small fraction of America's telephone lines. The new operation devoted itself to improvement through "investigation covering the fundamental principles" of telephony and associated technologies, and it, in time, under the name of the Bell Telephone Laboratories (familiarly, "Bell Labs"), became the flagship institution for American industrial research. Initially Bell Labs was formed largely to protect the AT&T monopoly as the company's patents on the telephone system expired, but its mission soon grew well beyond this. This was spurred in part by the nature of the perceived challenges. The technical problems of creating a truly national telephone network, with full long-distance capability, were formidable, and the company learned that the most advanced understanding of electrical phenomena would be needed to solve them. In addition, the possible challenge posed by the appearance of "wireless" telephony—that is, radio—spurred intensive investigation of that technology as well. The resources of the telephone monopoly were applied to creating the largest and best funded research effort in the country (about eight times the size of that at General Electric, for example, in the mid-1920s). The growth of the larger laboratories in this period was fostered by the conviction that scientific research, carried out without specific products in mind but instead to extend knowledge of relevant physical phenomena, was the key to long term growth (and industrial control). As the century wore on, researchers at AT&T, GE, and elsewhere garnered not only hundreds of key patents for important inventions, but also Nobel Prizes and memberships in the most elevated scientific circles. The marriage between advanced science and "high technology" was a close and enduring relationship.[4]

Despite spectacular products of scientific research from the industrial laboratories, including nylon (from DuPont in the late 1930s) and the transistor (from Bell Labs in 1947–48), the most important results of the marriage between cutting-edge science and technology came not from industry but from government-sponsored research. The experiences of World War I and the modest steps taken by a number of governments to set up laboratories, such as those of the NACA and others attached directly to the military services (the Naval Research Laboratory, for example), primed governments to seek out advanced research much more readily at the onset of an even greater war in 1939. The products of government funded and directed research in the 1940s were extraordinary in both their range and their long-term importance. Radar, rocketry, new medicines, synthetic materials, digital computers, and atomic weapons helped to determine the war's course, and reshaped almost every sphere of society in the decades after. All of these, pursued with varying vigor and success by the Allies and the Axis powers alike, depended fundamentally on scientists and engineers working in close cooperation in structured, disciplined research environments.

In the United States, these environments took several different forms, each of which was to make their mark on the postwar world. In 1940, alarmed by the wars in Europe and Asia and by a growing sense of unpreparedness, the Roosevelt administration began to reshape the relations between industrial and academic research capabilities and the government. A National Defense Research Committee was formed to create a contract system to foster widespread weapons research efforts. An even stronger agency, the Office of Scientific Research and Development (OSRD), emerged the following year under the leadership of Vannevar Bush, a former MIT electrical engineering professor. Bush and his colleagues, almost all of whom were drawn from academia, were convinced that only thoroughly coordinated and well funded cooperative projects, bringing together academic scientists with engineers from government and industry, could meet the military emergency they were certain was looming. The most famous outcome of their agitation and efforts was the Manhattan Project, the effort that resulted in the atomic bombs. The project turned out to be so large and ambitious, as the facilities at Hanford and Oak Ridge exemplified, that it was spun off from OSRD jurisdiction, but other equally crucial efforts remained under the eye of Bush.

The development of radar technology was the project of many laboratories in both Britain and America. As early as the 1920s, physicists at the Naval Research Laboratory in Washington, D.C., observed that radio waves were reflected off ships passing by transmitters, but it was not until the mid-1930s that the electronics for making sense of the reflected signals began to appear. At Britain's National Physical Laboratory, Robert Watson-Watt was able to devise equipment for detecting aircraft and determining their location by timing the reflected radio signal (hence the term "radar," for *RA*dio *D*etection *A*nd *R*anging), and by 1939 radar stations were in place in southern England to track German bombers and fighters. Meanwhile, the U.S. Navy had not neglected its early start entirely, and a few ships were getting equipment by 1938. By 1941 the Navy was able to put radar aboard aircraft, giving American night fighters an enormous advantage. A key to the extraordinary effectiveness of the Allied radar systems was the development of powerful microwave generators, such as the British cavity magnetron and the American klystron. These gave radar a small size, range, and precision that earlier systems lacked. The most spectacular success was the creation of the proximity fuse, which was a radar set small and tough enough to fit into artillery shells. The complexity of the technical difficulties involved in designing and manufacturing such devices by the thousands was daunting, and solving these required enormous amounts of money and expertise—by war's end, it is estimated that a quarter of the entire electronics industry of the United States had been put to work making them. At the same time, radar research transformed universities like Johns Hopkins and MIT into large military contractors, with implications for the structure and character of American academic research that reached far into the future.[5]

As was the case with radar, a number of the most important technical successes of the war years came from the intensive development of technologies that had begun to emerge in the decades before. While rockets of some sort had been used ever since the invention of gunpowder, and the Congreve rocket entered the British arsenal in the early nineteenth century (its "red glare" was memorialized in the *Star Spangled Banner*), modern rocketry began in the 1920s with the liquid-fueled devices of Robert H. Goddard, a physicist at Clark University. By the mid-1930s British, Russian, and

Figure 24.1
V-2 rockets on their mobile launch platforms, Peenemünde, Germany, 1945. Courtesy Archives Division, National Air and Space Museum, Smithsonian Institution.

German experiments had begun to make serious progress. After 1940, the Germans made the greatest efforts in this connection, with spectacular success. The German V-2 rocket had a range of three hundred miles and could carry a warhead of over two thousand pounds (figure 24.1). When sent on their way to England from the launching grounds in north Germany, the V-2s reached supersonic speeds and terrified the target populations, which felt quite defenseless. Fortunately for the Allies, this weapon was not ready until near the war's end, and so was of little strategic significance. The larger importance of the V-2 was the foundation that it provided for the large-scale rocket and missile programs of the two postwar superpowers. The Russians and Americans rushed into defeated Germany to snatch away not only the remaining rockets but, more importantly, the scientists and engineers who made them. Both the military and the civil rocket and space programs of the 1950s and beyond were built on these wartime foundations.[6]

Not all wartime research went into spectacular weapons. Among the most important infrastructure achievements was the large-scale development of synthetic rubber.

The vagaries of rubber supply, depending as it did on tropical plantations, largely under the control of a few colonial powers, encouraged experiments into rubber chemistry and synthesis from the beginning of the century, when the growing importance of the automobile began to make rubber an indispensable commodity. All early efforts yielded materials that were both far less durable and much more expensive than natural rubber. Some of these products, such as polymerized vinyl chloride, became important in their own right, often substituting for rubber or other materials. It was not until the early 1930s that a commercially successful synthetic rubber, dubbed Neoprene by the DuPont Company, came to market. German chemical firms came up with their own rubber substitutes, and these products captured modest niche markets, based on some superior properties. The onset of world war in 1939, however, changed the economic and strategic equations, and developing synthetic rubbers became a matter of national priority. The U.S. government organized a massive effort to free the country from dependence on imports, cutting through the tangle of patents, marketing agreements, and shortages of raw stock. By war's end, more than fifty rubber plants were built, operated by several dozen different manufacturers under the direction of the Rubber Reserve Company. A variety of different "rubbers" emerged from this effort, which became a model for transforming the outlets for a valuable natural material into a whole range of markets for the products of chemical synthesis.

Other chemistry related projects had arguably an even greater effect on the war effort as well as on postwar life. The first antibiotic drugs were found in the years before the war, but it was wartime that pushed the transformation of these agents into widely available medical tools. The German biochemist G. J. P. Domagk first observed the antibacterial action of the dye Prontosil in the early 1930s, but it was not until 1936 that the first of the derived sulfanilamide drugs became available to physicians. An entire family of sulfa drugs followed, and during the war the widespread use of these agents drastically cut the infections that typically ran rampant in military encampments, particularly in tropical zones. An even more remarkable antibacterial agent was discovered in the form of penicillin, derived from a common mold. British biochemist Alexander Fleming stumbled on the antibacterial mold action in the late 1920s, but could not see any effective means for applying it. Through the 1930s, penicillin was of little more than theoretical interest, but the looming problems of war spurred the engagement of both government and corporate laboratories to solving the production problem in earnest. In Britain, Howard Florey and Ernst Chain provided the first evidence in 1940 that penicillin could be produced by a batch process, and the American pharmaceutical firm Pfizer, among others, commenced to invest heavily in novel production techniques that eventually turned out millions of doses of this powerful antibiotic. Equally importantly, the techniques used to discover how to mass-produce penicillin, requiring the careful controlled growth of large amounts of microbial agents and purifying the products from them, established a model for what later became known as "biotechnology." In the decades after World War II, the once solid barriers that divided the techniques and processes of the inorganic from the organic world were repeatedly breached, so that by the century's end the realm of the living was seen as every bit as amenable to technical manipulation as that of the mechanical and the electrical.

It would be deeply misleading to conclude that the collaborative, laboratory-based model of improvement was the only important one to affect profoundly the technologies of the twentieth century. Other venues for technological development emerged and demonstrated their significance during the war. The urgencies of World War II revealed the new capacities of industry to adapt and expand technical capabilities in ways that were often less spectacular than the novel devices that emerged from the laboratories, but which were arguably every bit as significant for the ways in which technology reshaped daily life in the postwar world. The massive expansion of production in everything from vacuum tubes to aircraft revealed new dimensions of the industrial world that were exploited in the decades following to create a material prosperity in the West that had hardly been imaginable before (figure 24.2). To some degree, this expansion was the straightforward product of investment in the infrastructure of industry. Aluminum production, for example, increased more than sevenfold during the war years, as the struggling aircraft industry of the 1930s was transformed into a gigantic manufacturing enterprise. Plastics, too, began transformation from a small, specialized class of materials into a large-scale substitute for almost every other fabricating substance. Where, in 1939, the United States had produced 213 million pounds of synthetic resin, by war's end annual output was 818 million pounds. This figure tripled again in the six years following the war, and materials that were once scorned as merely ersatz substitutes for the "real thing" were now regarded as key commodities. These changes in technology and industry involved the solution of myriad problems of scaling up production, saving skilled labor, economizing scarce materials, and organizing distribution efficiently. Solutions were often unspectacular, but the scale and range of problems successfully mastered in war production were far beyond anything ever confronted before. Entirely new approaches to problem solving and management emerged, some of them formalized in the form of such fields as "operations research," and others simply integrated into factory or corporate organization. The postwar economic boom was to a large degree the result of these innovations, even if they were less visible than the novel products that came from the laboratories.[7]

The pursuit of improvement, guided by the prevailing belief that there were scientific and technical solutions for almost all human problems, found other outlets during this bloodiest of conflicts. To some degree, these outlets were straightforward extensions of problems that were easily put into a technological framework, such as

Figure 24.2
Aircraft assembly plant with B-24 Bombers under construction. Library of Congress Photograph (Office of War Information photograph).

military tactics. Beyond this, however, there were other realms of improvement that were explored in the first decades of the twentieth century and then in the context of fascism and war that provided the opportunity for criminality on a scale hitherto never experienced. When the notions of improvement, along with the faith that there were no boundaries on human capacity to pursue it, were applied to the human species itself, the consequences were horrific, and the culture of improvement itself was profoundly changed.

That war is a great field for experiment is an old truth, and that technology is a key element in such experimentation is nearly as old, as the giant wooden horse at the gates of Troy exemplifies. Of course, World War II was a vast theater of new weapons, but it was more than that. The entire philosophy behind the use of weapons and of armed force in general was open to experiment and improvement. The German experiments at Guernica and elsewhere during the Spanish Civil War suggested how much this willingness to explore new directions had permeated modern combat. More than anything else, as Guernica demonstrated, it was aerial warfare that invited new approaches, not only to combat but to killing and destruction in general. The conflicts of the 1930s, from the Italians in Ethiopia to the Japanese in Manchuria and the combatants in Spain, all were laboratory exercises studied by the powers, great and small, for clues about the nature of modern war. Perhaps the clearest lessons drawn were the importance of communications and logistics and the frightening potential of aircraft for terrorizing noncombatants. These lessons were rapidly applied, particularly by the Axis powers, once war began in September 1939.

Mobility and overwhelming force were the keys to German war plans, and the rapid conquest of Poland—Warsaw surrendered in less than four weeks—suggested that the Nazi army had the means to make this work. After a lapse of many months, the strategy of blitzkrieg was put to work again with enormous success, as Denmark and Norway were conquered in a few weeks, followed by a campaign from May 10 to June 19, 1940, that conquered the Netherlands, Belgium, and France. The battle with Britain that ensued had far different results for the Germans, and thus began a series of experiments in aerial warfare. The Battle of Britain lasted from midsummer to late fall of 1940. It was largely combat among fighter aircraft, with the most decisive element being simply the ability of the two sides to keep up with enormous losses through maintaining aircraft production and pilot training. Up to the end of October, the Royal Air Force lost just under 800 fighters while the Luftwaffe nearly 1,400. In the following months, the Germans turned their attention to a bombing campaign: up to May 1941 German aircraft dropped over 20,000 tons of explosives and 80,000 incendiary devices on cities all over Britain. About 42,000 were killed in the campaign, and thousands more wounded, but neither British morale nor industrial capacity were much diminished. This was a war of a very new type—with both offense and defense relying on novel machines and tactics—but it was not a very effective war from the standpoint of would-be invaders.[8]

Making the newly mechanized and aerial warfare more effective was the subject of much effort in World War II. Probably the most profoundly important and contentious attempt was what became known as "strategic bombing." At first, the use of bombs from airplanes was thought of in terms similar to the use of artillery—a means of delivering shot or explosives to support troops in battle, what might be referred to as "tactical bombing." The episode at Guernica illustrated how the theory of aerial warfare moved beyond this to incorporate targeting civilians—a tactic not unknown to the artillery as well. But the scale on which bombing was directed at targets behind enemy lines in the Second World War was far beyond anything seen earlier. Some theorists as early as 1921 had speculated that a properly equipped and used air force could be sufficient, when used against industry and population centers, to win wars with little additional effort. At the very least, such promoters of air power as William ("Billy") Mitchell argued, control of the air could be decisive in giving land and sea forces the winning edge, bypassing the kind of stalemate that was so demoralizing in World War I. The initial experience of the German blitzkrieg supported this doctrine, but the Battle of Britain showed how limited, in fact, aerial war could be in cowering an enemy. Nonetheless, Allied strategists pursued strategic bombing as the primary means for overcoming Germany. The British, in particular, were explicit about their use of area bombing to destroy both cities and war industries and to demoralize populations. American air strategy tended to be more oriented around "precision" bombing, but, even with improved bombsights and heavier bombers, this was not particularly effective.[9]

The efforts to improve the results of strategic bombing created some of the most horrific images of the world war. In late July 1943, for example, British and American bombers visited the skies over Hamburg, Germany in a series of waves, the British at night and the Americans by day. On July 24, 27, and 29 and August 2, almost eight hundred British bombers rained explosives and, even more devastating as it turned out, incendiaries down on the city. The raid of July 27, in particular, produced a result never before seen: the bombs in the center of the city set off fires that merged into one huge conflagration that probably reached 1,000 degrees Celsius. The firestorm eventually covered six square miles of the city, sucking all the oxygen out of buildings and shelters. The raids in Hamburg are estimated to have killed as many as fifty thousand people. Other German cities were targeted, but only one experienced anything like Hamburg: On the night of February 13, 1945, as Germany began reeling from direct attacks by Allied armies on both fronts, nearly eight hundred British bombers began raids over the hitherto untouched city of Dresden, a city of palaces and churches, known as the jewel of the Elbe and crowded with refugees fleeing the Russian advance to the east. The firestorm there was perhaps twice as devastating as

Hamburg's. Almost 90 percent of the structures in the city's center were destroyed, over eleven squares miles burned, and the death toll is unknown even today, but it probably exceeded 100,000. Less than a month later, an American raid with a smaller number of bombers but even more intent on burning created a similar firestorm in the Japanese capital, Tokyo. Again, perhaps 100,000 perished and the city's center was essentially destroyed to the ground. The casualties of these "conventional" raids exceeded those of the atomic bombings in August, and the improvement of strategic bombing had reached its climax.[10]

The sheer destructiveness of World War II was unparalleled in history, and the role of constantly improving technologies and techniques in this was evident to all observers. The fact that the war ended in August 1945 on the heels of the demonstrations, at Hiroshima and Nagasaki, of the most horrific invention ever conceived was widely noted, and it gave the war's close a particularly technological cast. The announcement of the atomic bomb, in which President Truman declared, "the force from which the sun draws its power has been loosed on those who brought war to the Far East," left no doubt that the new weapon was the product of scientists. Images were conjured up that drew on both ancient and more modern mythologies and fears. In the first years of the "atomic age," large numbers of people believed it possible or even likely that atomic physics and its uses would bring great destruction to humanity, either through weapons or through accidental unleashing of new forces. The very language adopted to describe the powers of war became apocalyptic. As more was learned about the human experience at Hiroshima, through war correspondent John Hershey's famous account of that name as well as through interviews with the physicists and engineers, the more it seemed to many that the world had crossed some kind of crucial threshold. Manhattan Project director J. Robert Oppenheimer's remark that the scientists had "known sin" captured the idea that science and technology had taken a step with consequences that were uncontrollable.[11]

The atomic bombs themselves, as the ensuing years were to demonstrate horrifically, opened up new avenues for improvement that appeared to have no end but Armageddon. Almost two years before the Trinity test of the first bomb in July 1945, Edward Teller, a Manhattan Project physicist and Hungarian refugee, turned his primary attention to the design of an even greater weapon, one that would release energy through the fusion of atoms of heavy hydrogen (Deuterium). The result, according to Teller's calculations, would be a bomb with the potential explosive power of not thousands of tons of TNT, like the fission bomb then being developed, but of millions of tons—a thousand times larger. Such a weapon would require enormous heat to trigger the reaction (at least 40 million degrees C), and the fission bomb then being built provided for the first time the possibility of such heat. Little time was lost after the atomic bomb was completed in carrying out key calculations for the design of what Teller called the "Super," although political and economic considerations slowed development for several years, until the Soviet Union demonstrated in August, 1949 that the American monopoly on nuclear weapons was over. On November 1, 1952, a test device called "Mike" was exploded on the island of Elugelab, in the mid-Pacific atoll of Eniwetok. The "yield" was 10.4 megatons of TNT, about a thousand times more than the bomb that destroyed Hiroshima. As the century reached its midpoint, the two great superpowers embarked on a nuclear arms race—a race of improvement of gigantic forces of destruction which culminated in the accumulation of arsenals of thousands of weapons, each many times the power of the Hiroshima and Nagasaki bombs.[12]

Improvement on this scale provoked an unusual response. In over a decade of nuclear tests, the United States and the USSR exploded hundreds of devices, largely in the atmosphere, but increasingly underground and underwater as well. Finally, in 1963, responding to widespread condemnation of the weapons testing programs, the Americans and the Soviets, joined by the British (but not by the other two nuclear powers at that point, France and China) signed a treaty to ban tests in the atmosphere, outer space, or under water. About ten years later, these same powers agreed to test no weapons larger than 150 kilotons in size (about ten times the Hiroshima bomb, but considerably less than high-yield hydrogen weapons). In 1996, the United Nations adopted a Comprehensive Test Ban Treaty, to forbid all testing of nuclear weapons. This, however, was not ratified by the United States or by newer nuclear powers such as India or Pakistan. While treaties to reduce the size, cost, and power of armaments have considerable precedent, particularly in the twentieth century, the drive to halt not only the spread (a nonproliferation treaty came into force in 1970) but the improvement of nuclear weapons was the most prominent indicator that some areas of improvement were seen as simply too dangerous to pursue.

As awful, in the literal as well as the common sense, as atomic weapons were, they were not the source of the greatest shock to the culture of improvement in the twentieth century. The effort to extend to the human race itself the techniques of improvement that the industrializing Europeans and Americans celebrated so widely in the nineteenth century came to be closely linked to the twentieth century's greatest horror. The Nazi effort to exterminate entire peoples, most especially the Jews, had many roots, but one of them was eugenics, the idea that humanity could be improved by selective breeding. In the first decades of the twentieth century, eugenics attracted a wide and influential following throughout much of the west, especially in the most industrialized countries, such as Britain, Germany, and the United States. The origins of eugenics lay directly in the culture of improvement, a product of scientific ideas and technical means, directed towards an end that seemed to many observers to be a logical extension of the material progress that marked the West through much of the nineteenth century. The term, and much of the early philosophy, of eugenics was the product of Francis Galton, an English polymath who left his stamp on developing sciences of anthropology, meteorology, and statistics. Above all, Galton was a counter, a man who sought to understand the numerical patterns and relationships that he felt underlay every aspect of nature and of human life. In the 1860s, he turned his attention to heredity and to understanding both the mechanism of inheritance and its consequences for society.

Very little was known in the nineteenth century about how organisms actually passed on their characteristics from parents to offspring, and this subject was to be largely shrouded in mystery and conjecture until the middle of the next century. But the basic rules of breeding both plants and animals had been known for millennia, and Charles Darwin had given the processes of "selection" new prominence by suggesting how the competition for survival in nature could act as a selection mechanism, much as a farmer acted in improving the stock of sheep or corn. It was against this background that Galton considered the problem of how favorable traits among humans, such as "genius," could be promoted in successive generations. There was still much debate about whether the characteristics that an individual acquired during life could be passed on to offspring. If they could, then the improvement of an individual could, to some degree, be retained in successive generations. But no one could demonstrate the inheritance of acquired characteristics, and Galton chose instead to focus on improvement through the selection of the "best" parents. His adoption of the language of improvement was explicit very early in his work. In 1865 he wrote, "If a twentieth part of the cost and pains were spent in measures for the improvement of the human race that is spent on the improvement of the breed of horses and cattle, what a galaxy of genius might we not create!" He later dubbed such measures of improvement "eugenics," from the Greek for "well-born." By the end of the nineteenth century, a wide range of followers had adopted Galton's goals, in the belief that scientifically guided human improvement was the next logical aim of modern science.[13]

But just what constituted improvement in this case? And what characteristics of humans were to be selected out? Such questions clearly went to the very heart of how humans identified themselves and their qualities. That many people a century ago came to believe that such questions could be answered and then acted on by scientific and technical means is a measure of the enormous influence of the belief in technological improvement in Western thought. Such characteristics as genius, loyalty, creativity, persistence, and self-discipline were widely perceived as largely products of inheritance—of "breeding"—and thus to be cultivated and promoted in the same way that the sweetness of corn or the wooliness of sheep might be improved. Equally disturbing in hindsight is that most of these characteristics were widely seen, at least among Europeans and Americans, as already selected out through the categories of race. For most northern Europeans and Americans in the early twentieth century, race was an unproblematic concept, although boundaries and definitions might get difficult to mark on occasion. And the belief that a wide range of human traits, from musical abilities to the willingness to do physical work, were linked in some way to racial qualities was also quite common, as was at least some fuzzy notion of relative superiority. What eugenics inserted into this racist landscape was the notion that scientific tools were at hand for promoting the prevalence of desired qualities and diminishing undesirable ones. Eugenics was not necessarily a racist program, but in the United States, England, and Germany it typically took on racist coloration, as race and ethnic origin were often used as a kind of shorthand for identifying the most and least desirable strains. Again, the analogies to animal and plant breeds easily led champions of the new science to oversimplify the challenges and ambiguities of defining human improvement.

There were, to most observers, two essential means for eugenic action: promoting breeding among the carriers of the best qualities and discouraging or preventing breeding among "undesirables." The first approach, sometimes referred to as "positive eugenics," generally took benign forms, seeking ways of encouraging certain types or groups to have children and to raise them according to accepted standards. This naturally acquired its own elitist, and often racist, tone, as even positive eugenics was seen as promoting "purity" among certain groups and avoiding "pollution" from undesirables. The second approach, called "negative eugenics," was more controversial, and, in fact, led to moral disaster. The most extreme form was genocide—the systematic effort to eliminate entire groups of individuals on the basis of their belonging to some ethnically or racially defined class. While genocide in the twentieth century had numerous roots in each of the cases in which it occurred, and it was never simply the result of eugenic thinking, the logical linkage between the one and the other was so apparent to most observers that eugenics itself became, after the horrors of the Holocaust in particular, widely abhorrent.

In light of the later image of eugenics, the widespread influence that it exercised early in the century is striking. There is perhaps no better illustration of the pervasive nature of the culture of improvement, guided now by a deep faith in science, throughout the industrialized West and beyond. Eugenics appealed to men and women across a wide political spectrum and from diverse backgrounds and interests. The goals could be framed in very general terms—essentially to create better human beings over time—or in more specific, self-interested terms: to promote the welfare and status of particular human beings, often but not necessarily defined in racial or ethnic terms. In the United States, the early eugenics movement found supporters and sponsors in the most respectable quarters of society and eventually made great inroads in the scientific and political institutions of the country.

The American Breeders Association was formed in 1903, largely through efforts by the U.S. Department of Agriculture to organize animal breeders and plant hybridizers. The recent discovery of the work of Gregor Mendel, whose laws of inheritance appeared to give breeders important new tools, energized those seeking to put breeding on a scientific foundation. Charles Davenport, an experimental biologist out of Harvard and Chicago, was particularly struck by the potential for applying the Mendelian rules to human betterment, and he approached the Breeders Association at its first meeting to urge that it adopt human as well as animal and plant improvement as a core mission. He had already persuaded the Carnegie Institution to fund a "Station for the Experimental Study of Evolution" at Cold Spring Harbor, New York. Within a few years Davenport was able to use the Breeders Association and his laboratory to promote a wide-ranging eugenics agenda. Studies were made of the inheritability of a wide range of traits, from insanity to epilepsy to criminality, as well as corrective approaches. The Association established, for example, a "Committee on Sterilization and Other Means of Eliminating Defective Germ Plasm." One of Davenport's core ideas, picked up by many although not all in the eugenics movement, was that undesirable inheritable traits were evidence of inferior or defective hereditary material (the "germ plasm"). Davenport believed that the only way to diminish the influence of this defective material in the larger population was to identify its sources and carriers and to prevent these individuals from passing it on to another generation. At his Cold Spring Harbor laboratory he established a "Eugenics Record Office" to collect data about human genetics and to educate the public about the need for eugenic action. In his 1911 book *Heredity in Relation to Eugenics*, a work that was to be cited in over a third of high school biology texts in the two decades after World War I, Davenport gave voice to the crudest type of heriditarian doctrine, warning of impending social chaos due to the influence of immigrant populations on the United States. Americans would become, if no action were taken, "darker in pigmentation, smaller in stature, more mercurial, more attached to music and art, more given to crimes of larceny, kidnapping, assault, murder, rape, and sex-immorality and less given to burglary, drunkenness, and vagrancy." In these sorts of terms, Davenport tied eugenics to the most blatant sorts of prejudices about human types and ethnicity.[14]

The broadening appeal of eugenics was heralded by the convening of the First National Conference on Race Betterment at Battle Creek, Michigan, in 1914. Here many of the key issues that really lay at the heart of eugenic thinking were aired. The great social worker Jacob Riis spoke of the limits of inheritance and the importance of social and physical environments in shaping human lives and characters. The great African American educator Booker T. Washington spoke eloquently and with some irony of the challenges met and overcome by black Americans, once again arguing that environment trumped any qualities to be linked to inheritance, especially when the latter was spoken of in broad racial categories. But these figures were largely drowned out by speech after speech extolling the importance of "biological strength" for the well-being of the nation and the race, however "race" might be

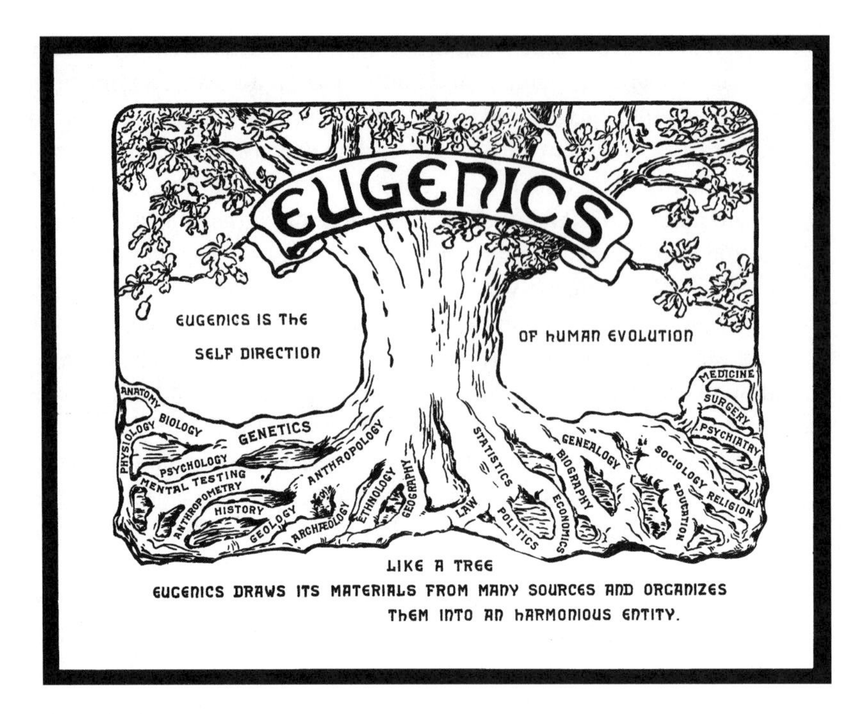

Figure 24.3
The Second International Congress of Eugenics, 1921, used this logo to illustrate the many roots claimed for the field. Library, American Philosophical Society.

defined (figure 24.3). The central and triumphant idea was that some people were inferior to others and that the breeding of such people was a negative force that had to be controlled if human betterment was to be a reality. In the next year's follow-up conference, the crux of the matter was clearly stated by Paul Popenoe, who would go on to co-author *Applied Eugenics*, the most popular text of the movement. Popenoe told the conference, "Science knows no way to make good breeding stock out of bad, and the future of the race is determined by the kind of children which are born and survive to become parents in each generation." In a few years, Popenoe became one of the most vocal champions of the legal instruments for applying eugenic principles in America, the laws that were enacted in thirty states that called upon authorities to seek out "defectives" and, if they could not be persuaded to take measures on their own, to sterilize them so that their "germ plasm" would go no further.[15]

As early as the 1890s, American physicians working in state institutions began to castrate patients to prevent their having offspring. Throughout the country every state had institutions such as the "Kansas Home for the Feebleminded," where the superintendent sterilized fifty-eight children before he was stopped by a public outcry. The idea, however, that the state could and should sterilize the "defectives" in its care began to catch on, and was debated and promoted in medical associations, reform groups, and state legislatures. In Indiana, the chief physician of the Indiana Reformatory in Jeffersonville, Harry Clay Sharp, began to sterilize inmates in 1899 and over the next few years operated on dozens of men, in a quick operation with no anesthetic. He went public with his crusade, urging for authority to institutions to "render every male sterile who passes its portals, whether it be an almshouse, insane asylum, institute for the feeble minded, reformatory, or prison." He admitted that his own authority was not clear-cut, but rallied members of his legislature to remedy this quickly. He was supported by eminent scientists, such as the president of Indiana University (and later the first president of Stanford), David Starr Jordan. In 1902, Jordan wrote, "The pauper is the victim of heredity, but neither Nature nor Society recognizes that as an excuse for his existence."[16]

Jordan's words hint at how far some eugenicists were beginning to urge action, as works began to appear that went beyond sterilization campaigns and urged more direct efforts to weed out undesirables and defectives from society, including institutionalized killing. For example, W. Duncan McKim's innocently titled 1901 book, *Heredity and Human Progress*, began by lamenting "the inefficiency of the measures which we bring to bear against the weakness and depravity of our race." His solution rested on "the gentle removal from this life of such idiotic, imbecile, and otherwise grossly defective persons as are now *dependent for maintenance upon the State*, and of such criminals as commit the most heinous crimes, or show . . . that they are hopelessly incorrigible." In this class he went on to include habitual drunkards, epileptics, and "nocturnal house-breakers." While McKim's ideas were extreme, they were not so far-fetched in the context of the times as to be dismissed out of hand. And other milder notions, such as widespread legally mandated sterilization, gained popularity in state after state. From Indiana's law of 1907, finally sanctioning Dr. Sharp's long-standing practice, to California's, passed a couple of years later and used to force sterilization on thousands of state wards over the next three decades, to twenty-eight more states, most Americans by the 1930s lived under mandatory sterilization laws.[17]

From their beginning, these laws were looked upon by many as constitutionally suspect, and they were widely challenged. A number of state courts threw out them out, but the eugenicists fought back by seeking a test in the U.S. Supreme Court. Virginia passed its sterilization law in 1924, and eighteen-year-old Carrie Buck immediately fell in its clutches. Her mother, Emma, a widow with no visible means of support, had been declared defective four years before and sent to the Colony for Epileptics and Feebleminded near Lynchburg. Carrie lived her childhood in foster care, but at seventeen she got herself pregnant and was thrown out of her home and reported to the local judge as feebleminded. After a short hearing she too was committed, and after she delivered her baby girl, she was sent to the Colony. A few months after her arrival, a review board ruled that she was "the probable potential parent of socially inadequate offspring," and ordered her to be sterilized. The novelty of the Virginia law combined with the desire of the eugenicists for a test case gave rise to a suit that made its way eventually to the Supreme Court as *Buck v. Bell* and to one of the most notorious judicial rulings of the twentieth century. On May 2, 1927, Justice Oliver Wendell Holmes, already revered for his stout defense of individual liberties, wrote for the court (there was one dissent) in a typically carefully worded decision. Therein he included some particularly striking language that gave the eugenicists their most important encouragement:

> It is better for the world, if instead of waiting to execute degenerate offspring for crime, or to let them starve for their imbecility, society can prevent those who are manifestly unfit from continuing their kind. The principle that sustains compulsory vaccination is broad enough to cover cutting the Fallopian tubes. Three generations of imbeciles is enough.[18]

The *Buck v. Bell* decision had an influence that is hard to overstate. Even in states that had sterilization laws, many government officials held off in using them while constitutional questions were still open. But after 1927, these restraints were widely removed. It is estimated that between 1907 and 1940, almost 36,000 men and women were forcibly sterilized by law in the United States, but more than 80 percent of these were in the years after the Supreme Court ruling.

Just as significantly, the new sanctions for the movement in the United States energized the eugenics movement overseas, particularly in Europe. Just as in the United States, eugenics held out particular appeal to educated elites worried about "racial hygiene," to use the term most popular in Germany. In Scandinavia the American example seemed particularly compelling, and sterilization programs were undertaken in Denmark, Norway, and Sweden. Between 1929, when Denmark passed Europe's first sterilization law, on the American model, and the mid-1970s, when the Scandinavian program became largely voluntary, more than 100,000 operations were performed on the "unfit," the vast majority of them women. No other area of the world took up the American model with such enthusiasm. In many countries, including most Catholic ones, religious objections stopped most eugenic proposals cold. In Britain, eugenics never lost its Galtonian flavor, with its strong emphasis on positive programs for encouraging children from favored groups rather than turning to the negative eugenics of sterilization. Only in Germany did the negative program find even more fertile fields in which to grow, with results that were profoundly disastrous.[19]

Eugenics in Germany—more commonly called "race hygiene" (*Rassenhygiene*)—originated much as it had in other countries, achieving its first popularity among the same sorts of groups that supported it elsewhere. Race hygiene appealed to a broad spectrum of the educated middle classes—those whose faith in science and technical improvement was a foundation of nineteenth-century life. Just as in the United States, the movement was not restricted to a narrow political base or ideology, but reached across lines of party, economic interests, and racial attitudes. The term racial hygiene does suggest, however, some differences in emphasis. A key concern among the German eugenicists was avoiding the "pollution" of the race. What was meant by "race" in this context, however, differed greatly among proponents. For some, racial hygiene was a matter of maintaining the best qualities of human beings through the generations, regardless of ethnic or racial identity, while for others it specifically referred to upholding a standard of "purity" for certain groups, preventing the mixing of "inferior" types with "superior" ones, and eliminating potential sources of deficiencies or defects. As the term "hygiene" implied, German eugenics had a strongly medical bent; to a large degree eugenic measures were a logical extension of those adopted for public health in the nineteenth century. One of the movement's leaders, Alfred Ploetz, for example, wrote that his observation of social disorder led him to believe that the source of much misery was "the low quality of human beings," and that "for this reason, I must direct my efforts *not merely toward preserving the race but also toward improving it*." In the social and political turmoil that shaped much of German experience in the first decades of the twentieth century, these sentiments were recast in terms of the German nation itself, and this easily took on a racial coloration that had not been present earlier.[20]

German intellectuals, politicians, and citizens in general were no more or less focused on the idea of race than were most other Europeans in the early twentieth century. In Britain, the idea of an "Anglo-Saxon" race was part of the common language, and similar kinds of terms were widely found elsewhere, particularly in the United States. In America, the readily visible wide distribution of peoples of different national origins made race, defined most often in terms of color, an apparently unproblematic way of talking about people and differences. In Germany, racial categories were a bit less obvious, hence debates arose about such terms as "Germanic," "Nordic," and "Aryan," and their definition was just as likely to be in contrast to identified "others," such as Slavs, Gypsies, or Jews. Just as significantly, the relevance of these categories to racial hygiene was much disputed. More readily accepted was the importance of identifying the "unfit," those who should be discouraged from reproducing. Just as in the United States, a consensus grew in the 1920s that some social and economic problems might be alleviated by reducing the population of disabled, feebleminded, and other categories of individuals who were often wards of the state. Anxiety about the German social and political order combined with other political concerns to shift the rhetorical emphasis from positive eugenics to negative. The example set by American sterilization laws had considerable influence, but no action was taken until the Nazi seizure of power in 1933.[21]

"National Socialism is nothing but applied biology," said Rudolf Hess, one of Adolf Hitler's most prominent deputies. The ideology and political regime of the German National Socialists was both more and less than this, but there was a degree to which many Nazi policies had some roots in a biological conception of the world and of human progress. The Nazi penchant for racial labeling, their abhorrence of racial mixing, and their development of programs for weeding out the "unfit" of all varieties betray the eugenic element of the ideology. What distinguished them from others was the readiness with which they wedded these approaches to other agendas, fed largely by ancient prejudices, class anxieties, and an extreme nationalism. The Nazis were also distinguished by the scale on which they were willing to embark on the programs for racial hygiene they adopted. It would not be accurate to term the Nazi murder of more than six million people simply because of the categories into which they fell a eugenic program in any rational meaning of that term. But by the same token, no exact boundary can be drawn between eugenics, with its scientific and technical foundations, and the "Final Solution" and other killing programs undertaken by the Nazis between 1933 and 1945. Within a few months of taking power, the Nazi government began passing laws with heriditarian bases. The Law on Preventing Hereditarily Ill Progeny, passed in July 1933, established an extensive system for mandating sterilization of those with mental and physical deficiencies. The Law Against Dangerous Habitual Criminals extended the reach of sterilization to the "criminal classes" a few months later. In 1935 the Law for the Protection of Heredity prohibited the feeble-minded from marrying healthy individuals. That same year, the so-called Nuremberg Laws put an explicitly anti-Semitic stamp on these efforts at "protecting heredity," forbidding the marriage of Jews and "citizens of German blood."[22]

During the twelve years the Nazis held power, their "applied biology" led to the murder of millions of individuals. The sterilization program was pursued on a scale that dwarfed anything seen elsewhere and within just a few years thousands of individuals were sterilized, typically on the flimsiest of pretexts. By the end of the 1930s, as Germany shifted to a wartime footing, the Nazi version of eugenics changed significantly. In mid-1939, Hitler authorized a program for killing handicapped children. Under the direction of his personal chancellery, a system of killing centers for eliminating children with a variety of conditions, including all deformities and mental conditions, was set up. More than twenty "euthanasia wards" operated throughout Germany, using a variety of techniques, from starvation to lethal injections. Within a few months of setting this in motion, the program was extended to handicapped adults, on a even larger scale. This required new killing methods, eventually leading to the building of centers for gassing "patients." By mid-1941, the extermination program was extended to the Jews in occupied countries, especially the Soviet Union and Poland, and then to Jews throughout Europe, including Germany itself.

The Nazis constantly sought improved methods for their killing, and they took every opportunity to use the victims in the concentration camps and killing centers for experiments directed at improving everything from submarine rescue to eye surgery. While great efforts were made to hide the enormity of this effort from the world, including the German people themselves, the Nazi rhetoric was consistently the rhetoric of racial hygiene and improvement, a rhetoric that never looked very different from that of the eugenicists of the previous half-century.[23]

The end of the Nazi regime in 1945, the exposure of the death camps, and the long series of war crimes trials that followed cemented the demise of eugenic rhetoric for the next half-century. Even before the outbreak of war in 1939, the publicity given to Nazi racial policies, particularly by Jewish refugees from German and occupied lands, brought eugenics into considerable disrepute. In addition, serious questions raised by geneticists and anthropologists cast much doubt on the science. Through much of the 1930s, the racial cast of much of eugenics, at least as it was talked about in the United States and Britain, gave pause to scientists and laymen alike. The failure of American eugenicists, in particular, to disavow Nazi policy and practice alienated much of the community that had accepted, through much of the 1920s, the claims of the eugenicists to speak for a new science of human improvement. In the second half of the twentieth century, a new genetic science emerged, strongly rooted not only in the kind of statistical observation that had been at the foundation of the work of Galton and Mendel, but now also based on new understandings of the physics and chemistry of heredity. The discovery in 1953 by Francis Crick and James Watson of the structure of deoxyribonucleic acid (DNA) and of the mechanism of replication that allowed it to pass on genetic codes sent genetic science down the path of molecular biology, and thus sent the dreams of improvement into those realms as well, with consequences that are still to be worked out in the twenty-first century.

25 Networks

Of all the technologies that trace their beginnings to the Second World War, arguably none had such a pervasive effect on later twentieth-century life than the electronic digital computer. Like the technologies already mentioned, this too had origins deeply rooted in the earlier decades of the century, but the complexities and costs of moving prototypes and ideas into the world of real artifacts and systems were so great that the agency of war—of a war for national survival—was required to make sufficient intellectual and industrial resources available to turn complicated ideas into working realities. The idea of making machines that would perform arithmetic was almost as old as geared machines themselves. The astonishing clockwork of Giovanni di Dondi, built in the mid-fourteenth century, for example, was in some ways a programmed counting and calculating machine, and in the following centuries plans for such machines were not uncommon. By the late seventeenth century, mathematicians and mechanics had already collaborated on more general-purpose arithmetic machines, but they were little more than esoteric curiosities.

In the nineteenth century, the combination of the increased capabilities of machinists, familiarity with expanded applications of external power sources (such as water and steam), and the vast expansion of both scientific and commercial computation (in everything from astronomical to actuarial tables) fostered an increased interest in mechanical calculating. The most famous product of this was the work of the English mathematician and scientific promoter Charles Babbage. Struck by the growing need for the construction of mathematical tables, Babbage envisioned machinery that would take the labor of computing tables from men and women and apply external power sources to it, much in the fashion that textile machines transformed spinning and weaving. Babbage encountered enormous frustrations in his efforts, for the task proved more complicated than the technology of his day could encompass and more expensive than even his considerable resources would provide for.

While twentieth-century computers owed little to Babbage except perhaps for the historical resonance that his vision provided the pioneers, the efforts that fostered their development arose from some of the same impetus that drove him—the pressing need to perform enormous numbers of mathematical calculations. Adding machines were commercially available by the middle of the nineteenth century, and by the 1880s they were making headway in the offices of insurance companies, banks, and government agencies. The spread of typewriters and cash registers at about the same time made office machinery a growing technical sector by century's end, and by the 1920s some of this machinery was electrified, establishing, finally, Babbage's vision of power-driven calculation. Perhaps the most complicated of these devices were sorting and tabulating machines. In 1889, an American engineer, Herman Hollerith, devised a sorting and counting machine using punched cards, of the kind that had long been used in fancy looms (and were derived in turn from music machines). His immediate goal was to supply the U.S. Census Bureau with the means to handle the 1890 Census. The tabulation of returns from the previous census took seven years to complete, and the larger census impending threatened to overwhelm all available capabilities. Hollerith's machines succeeded, using more than sixty-two million cards, one for each enumerated person, to prepare the census report in only two and a half years. This success spurred him to organize the Tabulating Machine Company in 1896, which became one of the foundations for the International Business Machines Company (IBM) in 1924.[1]

By the late 1930s, machines for calculating, tabulating, recording, and sorting numbers were common in businesses and in some laboratories. In addition, some technical problems spurred the creation of even more complex machines. The growing size and complexity of electric power systems, for example, required the calculation of ever more complex models to determine connections among stations, users, and circuits. Hand calculations increasingly proved inadequate for solving the complex mathematics of these systems, so mechanical devices that could mimic mathematical relationships were designed and built and by the 1930s were in widespread use. These "analog computers" were descendents of a technological tradition that could be traced back to eighteenth-century planetary models and the tide predictor constructed by William Thomson (Lord Kelvin) in the 1870s. In the years before the First World War, in the midst of the naval arms race that consumed the great powers, other analog computing devices were put to work aboard warships for controlling long-range guns. The enormous challenges of calculating the trajectory of guns firing over several miles of sea from rapidly moving platforms to rapidly moving targets encouraged intensive development of devices that would not only make complicated calculations quickly, but also would transmit this information rapidly to control mechanisms. The efforts of electrical engineers to build on this analog computing tradition gave rise to the first general-purpose analog computers. Vannevar Bush, while still teaching electrical engineering at MIT, built a "differential analyzer" in the early 1930s that could handle a variety of engineering problems requiring the solution of differential equations. His machine was copied and used in a number of laboratories in both the United States and overseas in the years before World War II, and was arguably one of the most important sources of the idea of the general purpose programmable computing machine.

While mechanized calculating and computing had found users and enthusiasts in the first decades of the twentieth century, markets were really quite narrow. Bookkeepers and actuaries relied on mechanical tabulating machines. Scientists and engineers found uses for analyzers. Government bureaus made ever greater use of record-keeping machinery (the U.S. Social Security Administration began work in 1935 by using more than four hundred machines from IBM to keep track of accounts from more than twenty-six million workers). Making more complicated and expensive machines, however, could only be justified by the urgent needs of war. In particular, two areas emerged in which prodigious feats of mathematical calculation were at the heart of the pursuit of military victory: codebreaking and ballistics. The codebreaking remained enshrouded in secrecy for many decades, so its influence on the further development of computers was indirect (some of the key figures in British cryptography, such as Alan Turing, did emerge as major voices in computing). But the need to calculate the increasingly complicated firing tables required by the rapidly improving guns used on both land and sea presented urgent problems that eventually sprang loose the resources required to take the calculating and tabulating machine into vastly richer territory.

One of the reasons that these apparently specialized machines turned out to have such broad applications is that a number of them were promoted and designed by individuals with agendas that were different from those of their military sponsors. The machines that emerged were therefore often designed to handle a range of problems. The first to come to the public's attention, even in the midst of wartime secrecy, was proposed by a young Harvard physicist, Howard Aiken. In the late 1930s, still at work on his doctoral dissertation, Aiken sought to devise a machine that would solve nonlinear differential equations. The analog computers that were used by engineers to handle differential equations could not manage nonlinear ones, so Aiken began thinking in terms of a high-speed digital calculating machine. Undaunted by skepticism among the academics, Aiken approached industry figures, finally coming to IBM's chief engineer. By early 1938 design work began for a programmable automatic calculating machine, and IBM began committing resources to the project. In the next couple of years the project came to be part of the firm's military-support efforts, and while Aiken had little to do with it at this point, there was enough momentum to sustain it to completion. In January 1943 the so-called Mark I solved its first problem, and it was eventually put to work at Harvard calculating tables for the Navy's Bureau of Ships. The Mark I, however, while arguably the world's first fully automatic programmable computing machine, was also what computer historians Martin Campbell-Kelly and William Aspray call a "technological dead end." It was an electromechanical marvel, weighing five tons and operating off of a 5-hp electric motor turning a fifty-foot shaft which was connected to a gigantic array of gears (there were about three-quarters of a million parts all told) to carry out some three additions per second. The great virtue of the Mark I was that it could operate automatically for days at a time, once its program (on punched paper tape) was in place. This was a considerable advance on earlier calculating machines, but its reliance on electromechanical systems had inherent limits in terms of speed and reliability. Aiken's was one of a number of machines from the pre-electronic era that showed both the promise and the problems of electromechanical computing. In Germany, Konrad Zuse designed and built a series of such machines during this period, and at Bell Labs George Stibitz designed and built a "Complex Calculator," first of a number of machines that his company provided to military clients. Such machines served to illustrate the limitations of mechanical technology as they also advertised the promise of rapid computing machinery.[2]

By the late 1930s electrical engineers had built up considerable expertise in the use of electronic vacuum tubes in a variety of circuits, beginning of course with those needed for radio but extending to a range of other uses from detectors to counters. At least one builder of computing machinery, John Atanasoff at Iowa State University, experimented with using vacuum tubes to make a faster machine, and his device incorporated a number of important features, such as binary computation and electronic logical units and memories. While Atanasoff's computer was never quite finished, his efforts still attracted the attention of a physics college teacher, John W. Mauchly, who was on the verge of reeducating himself to become an engineer. In mid-1941, Mauchly visited Ames, Iowa, and studied the Atanasoff machine for several days. A few weeks later, he enrolled in a course of study at the Moore (Engineering) School at the University of Pennsylvania, and after a few weeks further he talked himself into a place on a faculty depleted by war-induced shortages. He quickly ran into a young electrical engineering instructor, J. Presper Eckert, and won him over to the idea of making an all-electronic high-speed computing machine. The operational advantages of such a machine over earlier devices were clear—the possibilities of true high-speed computing were vastly greater than with electromechanical devices, and, properly designed, an electronic machine promised great flexibility in the way it could be reconfigured for a variety of tasks.

It was also clear, however, that an electronic computer would pose design and fabrication challenges of the first order, as well as costs beyond anything ever spent for a calculating machine. Only the most urgent needs were likely to spring loose these kinds of funds, but these in fact showed themselves in short order. Not far from the Moore School in Philadelphia the U.S. Army's Ballistics Research Laboratory in Aberdeen, Maryland, was responsible for testing the key armaments in the American arsenal and providing the information needed to make the guns most effective. Particularly important in an era in which artillery was steadily increasing in range and power were the so-called firing tables, which allowed gunners to make corrections in aim and elevation for the various conditions and ranges they were asked to deal with. These tables were based on the calculation of several thousand trajectories for each type of gun; each trajectory typically required some twelve hours of expert work at a desk calculator. After the United States entered World War II, the problems associated with table production began to appear increasingly acute. The engineers at Aberdeen acquired a differential analyzer, which sped up work somewhat, but problems in maintaining precision argued for using digital rather than analog approaches, and the analyzer was still too slow to clear an increasing backlog of computations. In early 1943, officers at Aberdeen learned of the proposals of Mauchly and Eckert, and in a few months funding was found to push the project ahead.[3]

The Eckert-Mauchly machine, dubbed the Electronic Numerical Integrator and Computer, or ENIAC, was a monster (figure 25.1). It was by far the most complicated electronic device ever built, distinguished most by the large number of vacuum

Figure 25.1
The ENIAC computer filled an entire room and required enormous cooling equipment to maintain the thousands of vacuum tubes. U.S. Army photograph.

tubes—some 18,000. In addition, 70,000 resisters, 10,000 capacitors, 6,000 switches, and 1,500 relays equipped a machine that filled a 1,500-square-foot room and produced as much heat as a thousand 150W light bulbs (and thus required enormous air conditioning facilities). The ENIAC took much more time and money to build than had been projected, and it was completed too late to contribute to the war effort, taking on its first problem in November 1945, more than three months after the Japanese surrender. But it was still deemed a great success, in part simply because solving its engineering and fabrication problems demonstrated the larger possibilities of electronic computing. These larger possibilities were perhaps most apparent to a brilliant mathematician who was an adviser to a host of wartime projects, John von Neumann. Von Neumann learned of the ENIAC more than a year before its completion and began consulting with its designers. He is given credit for a central feature of its final design, the idea of the stored-program computer. In such a machine, both the data and the instructions for how this is to be used are stored in a memory, each retrieved and used when required, and fed back into memory for further use as necessary. This notion is still at the heart of the digital computer, giving the ENIAC and subsequent machines a flexibility and scope that far exceeded any previous device.[4]

The success of the ENIAC inspired a number of efforts to extend the technology. Von Neumann himself was able to persuade the Institute for Advanced Study, his academic home in Princeton, New Jersey to support design and construction of a computer to demonstrate better the mathematical importance of the new technology. In Britain, Maurice Wilkes at Cambridge University studied both Aiken's Mark I and the ENIAC and returned home to build the EDSAC, which simplified a number of the earlier systems and was vastly easier to program than Eckert and Mauchly's machine. The Pennsylvania pioneers proceeded to explore the commercial possibilities of the new technology amid widespread skepticism about the utility and cost-effectiveness of the electronic computer outside the environment of military procurement. The ENIAC, a very difficult and expensive machine to program and maintain, nonetheless found enormous demand for its services, and this suggested that if better computers were built, uses would found for them. Eckert and Mauchly found customers willing to back their efforts to make a reliable general-purpose machine, and in 1951 the UNIVAC computer was introduced to the public with great fanfare—the first machine being delivered to the U.S. Census Bureau. The first six UNIVAC computers, costing about a million dollars apiece, were delivered to U.S. government agencies, largely for defense work and research. In 1954, however, the first private customer, the General Electric Company, took delivery of a machine, and a number of large industrial corporations, insurance companies, and utilities soon followed suit. The uses of these computers were often mundane, constituting little more than the automating of familiar payroll, billing, and inventory tasks, but corporate as well as government investment was clearly based on the belief that the electronic digital computer had future capabilities that would help reshape the postwar world.[5]

This belief in a computerized future, whatever that might mean, was a public phenomenon. Even while the UNIVAC was still exclusively a government machine, it was introduced to the public in a dramatic fashion. On the evening of November 4, 1952, as the American people awaited the results of the closely contested presidential election between Adlai Stevenson and Dwight D. Eisenhower, those with access to the newest medium for delivering such news, television, were treated to a view of the UNIVAC (actually, what viewers saw was a mock-up, conforming to popular ideas of what an "electronic brain" should look like) at work. For weeks before the election, the computer's engineers had been gathering data from the previous two presidential contests and working out algorithms to process early election night returns to predict the final results, state by state. The CBS television network, which had bought into the idea of using the UNIVAC that night to distinguish their coverage from that of competitors, introduced the machine as a herald of an electronic age. When, at 10:00 that evening, the broadcasters reported UNIVAC's first prediction, they called the election a toss-up, with each candidate winning twenty-four states. An hour later, however, the numbers clearly indicated an Eisenhower landslide, and the computer engineers in considerable embarrassment confessed that UNIVAC had, much earlier in the evening, with only three million votes reported, predicted the Republican sweep, but they had not trusted their machine sufficiently. In one very public episode, some of the larger implications of an age in which information was better analyzed by machinery than by people were laid bare to a wondering public.[6]

While the basic form and principles of the computer were in place by the early 1950s, the machine itself underwent continual transformation in the coming decades. There have been, in a sense, few more ripe avenues for improvement in modern technology than the computer, although few understood this at the beginning. In 1956, Howard Aiken, whose Mark I device had been far superseded by the UNIVAC and its imitators, declared, "If it should ever turn out that the basic logics of a machine designed for the numerical solution of differential equations coincide with the logics of a machine intended to make bills for a department store, I would regard this as the most amazing coincidence that I have ever encountered." And yet, already at this point, it was clear to more prescient observers that to an extent this was precisely the case. Improvement on two distinct but intertwined fronts were responsible for this, as both hardware (the machines themselves) and software (the programs of instructions that made the machines do useful things) developed rapidly. The most important changes in hardware involved the form of memory and the electronic components. The ability of the computer to keep elements of its calculations and its programming in a memory, to be used, changed, and altered as appropriate, was crucial to its working, and larger and faster memories greatly enhanced functionality. The relatively slow and expensive memory systems of earlier machines—vacuum tubes, mercury delay lines, and magnetic tape—were replaced by magnetic drums and then magnetic cores, webs of wire and ferrite (iron) washers that were expensive but very effective and rapid memory devices. They were particularly valued for providing "random access" to memory; each item in memory was given an "address," which could then be accessed immediately when a program needed it. To make this technology practical, however, required a strong financial push from military customers. Customers like the U.S. Air Force wanted to be able to rely on computers for their radar-based warning systems. This required both speed and reliability that earlier machines lacked, but the military could provide the money to allow industry to work out the basic problems with the new technologies. This pattern of development, with the military providing the essential capital to overcome technical difficulties, repeated itself often through the period of the Cold War, and it gave American industry important advantages (as well as some competitive disadvantages when it came to consumer goods).[7]

The crucial role of military support in early computer development is even more apparent in the case of electronic components. The vacuum tubes at the heart of early computers were sources of heat, energy consumption, and unreliability. Enormous efforts were required to check and recheck every tube of the thousands that went into a machine like the UNIVAC, and even then continued working could not be guaranteed for any extended period. The path out of this problem was through replacing vacuum tubes with solid-state devices, that is, components that would perform the same functions as tubes but did so within carefully engineered crystals. The advantages of the solid-state diode, or rectifier, were known to early radio experimenters, whose "crystal sets" used a piece of semiconducting material, such as lead sulfide (galena), to receive radio transmissions. These devices also illustrated the disadvantages, as such crystals were notoriously fickle in their action, and they could not amplify signals, like a triode vacuum tube.

The development of devices that transcended these limitations was the work of physicists, engineers, and chemists at Bell Telephone Laboratories. Since the 1930s the telephone giant had been interested in finding a replacement for vacuum tubes in the far-flung components of its network, where the maintenance of electronic devices was becoming increasingly burdensome. In that decade physical theory finally began to provide some tools for experimenters to work with, and wartime researches pushed both knowledge and technique forward. Toward the end of 1947, three Bell Labs researchers, John Bardeen, William Brattain, and William Shockley, demonstrated the action of electrical contacts pressed against a small piece of germanium crystal—in the right configuration the device performed the function of a three-element vacuum tube. The invention was dubbed the transistor, and modest publicity accompanied its announcement. This "point-contact transistor" was itself a fickle device, extremely difficult to fabricate consistently or to predict behavior. But like most key inventions, it too could be improved, and in a few years there was a more reliable version with a different construction, the junction transistor. This worked by making large single crystals of semiconducting material, such as germanium, and creating very limited areas in the crystal that were "doped" with impurities that altered the electronic qualities of the material. Manipulating the doped areas allowed the construction of devices that acted as three-element amplifiers with far greater reliability and predictability than the point contact devices. Bell Labs announced this invention in 1951, and it finally became possible to "transistorize" a whole host of devices, from hearing aids (the first commercial applications) to radios to computers.[8]

For years, transistors were much more expensive to make and to use than vacuum tubes. Only applications that put very large premiums on low power consumption, quick response, and small size could make use of the new technology. While there were a few consumer items that fit into that category, much more important for the transistor's development were military applications. For the Air Force, for example, it was worth paying hundreds of times the cost of conventional devices for components that would allow a missile system to function reliably and quickly in a very small space. With this kind of support, the technology of semiconductor manufacture advanced rapidly in the United States, even if products themselves changed slowly. The first transistor radios appeared in 1954, but they were expensive novelty items. By the 1960s, however, transistors and other semiconductor devices began to be commonplace and prices fell to the point that they could displace tubes almost everywhere. For most of the public, this was largely evident in entertainment devices like radios, televisions, and phonographs, but of much larger significance was the transformation of the digital computer and other complex electronic systems.[9]

The transistor was so significant for the development of computer technology that the new computers were called "second-generation" machines. The first general-purpose computer to use transistors was a machine built for the supersecret National Security Agency between 1956 and 1958, and commercially available machines followed quickly. The replacement of vacuum tubes with transistors began the transformation of the computer from a huge room-sized device serviced by large air conditioning systems into desk-sized devices that could fit into the corner of an ordinary office. Equally important, the small semiconductors could be jammed together into much smaller spaces, requiring a fraction of the wires of earlier machines, and they were far more reliable than tubes, thus increasing the dependency of users on their computer systems. At the same time, a key invention in memory storage, the magnetic disk, provided computers with true random access memories that were reliable and fast. Perhaps more than any other addition to the basic processing units, the disk memory gave the computer the key feature that was to make it the preferred means of data storage. In the 1960s the computer was still an expensive, formidably complicated machine, but falling costs, better software (an entirely novel and truly significant realm of improvement), and greater familiarity inched the machine toward the ubiquity that characterized it at the end of the twentieth century.[10]

The discrete transistor, attached to circuit boards and connected with other components, such as resistors, capacitors, and other electronic elements, was an important invention, but it was not in fact a revolutionary one. An improvement of the transistor, however, appeared about a decade after the original device, and this—the integrated circuit—was the key to making electronics, and accompanying control, data storage and retrieval, and processing capabilities, the most important continuing source of novelty in the second half of the century. At first, the integrated circuit was indeed conceived as little more than an improvement in the means by which transistors were fabricated and connected. As the advanced computers and control systems required by such military applications as missiles grew particularly demanding, especially in terms of electronics reliability and complexity, the advantages of new means of making components that were uniform and sturdy, even under extreme conditions, were widely apparent. Military markets suggested to innovators that costs need not be a hurdle if stringent technical requirements could be met. At one of the pioneering transistor manufacturers, Texas Instruments Corporation, in Dallas, engineer Jack Kilby decided to experiment with extreme miniaturization by building all the components of a circuit on a single piece of semiconductor, such as silicon or germanium. In the summer of 1958, Kilby succeeded in fabricating an oscillator circuit on a single half-inch piece of germanium, and he soon applied for a patent for "miniaturized electronic circuits." A few months later, Robert Noyce of the Fairchild Semiconductor Company in California devised the means to accomplish Kilby's feat more easily, using a new kind of transistor known as "planar," because its elements were on the flat surface of a semiconductor crystal wafer. Noyce's "unitary circuit" was sufficiently distinctive to merit its own patent, and the approach he outlined became the basic principle for the manufacturing processes that would make the integrated circuit the most dramatically "improvable" technology of the modern age.[11]

The integrated circuit (IC) was, like almost all novel electronic devices, expensive and difficult to manufacture at first. Again, repeating a decades-old pattern, the development of integrated circuits into reliable and economical products owed a great deal to military demand and military funding. The small size and tiny energy requirements of these microscopic circuits had enormous appeal to the designers of missiles, satellites, and high-performance aircraft, and their premium price was of little concern. Through the early 1960s, this military-driven demand allowed manufacturers—both established electronics firms such as RCA or Fairchild and new ones such as Intel—to refine the technology, increase the density of circuitry, and, above all, reduce unit costs. In 1968, fifty times more ICs were produced than five years before—some 250 million—while prices came down to less than a tenth of their earlier level. The lower cost and increasing capabilities of ICs sparked development of other markets. At first, these were largely directed at industrial and business customers; integrated circuits were used to simplify the manufacture of complicated machinery and control devices. More common consumer electronic items, such as radios and televisions, simply did not involve the complexity or number of electronic elements that the new miniature devices incorporated, so the significance of "microelectronics" for ordinary life was not at all evident in its first decades.[12]

This changed in the 1970s. At the beginning of the decade, a new level of complexity and utility appeared with the invention of the microprocessor—the "computer on a chip." A number of elements came together to lead integrated circuit manufacturers to design and market devices that combined all of the primary elements of the computer—memory, processing instructions, and arithmetic functions—into a single piece of semiconductor. At first this was seen as a way to simplify the engineering of advanced handheld calculators, which were beginning to find markets in the late 1960s. Engineers at the Intel Corporation proposed to put all of the complex functions of a calculator on a single IC, which could be combined with three other ICs—for random access memory, for permanent program instructions (read-only memory), and for input and output functions—to make a versatile, compact calculating device. The Intel 4004 chip appeared in 1971, and it set off a series of innovations and improvements that still drives much of modern computing. The most spectacular result, the personal computer, was by no means a necessary outcome of these electronics innovations. This owed a great deal to the enthusiasm of individuals who were fascinated by expanding computer capabilities but were frustrated at the limitations imposed by the cost and complexity of the computer as it had developed up to this point. Computers expanded in capability and shrank in size throughout the 1960s; by the decade's end "minicomputers" had displaced in many applications the large machines that were still the primary heirs of the UNIVAC and IBM machines of the 1950s. But minicomputers still cost many thousands of dollars and required great skill and experience to program and even to make use of for scientific calculations, record keeping, accounting, and the other applications with which computers were identified. Access to such machines was typically restricted to those with demonstrable need and the skills to make use of them. The availability of the microprocessor, however, changed all that.[13]

One of the remarkable characteristics of technological improvement over the past century or so has been the role played by enthusiasts and hobbyists. The irony of this is that it has typically been in technological realms of great complexity that such individuals have had their greatest influence. Early in the twentieth century, the automobile was as complicated a technology as most individuals were ever to encounter, and yet its engine (and related systems) quickly became the subject of endless tinkering—the garage was transformed from simply a housing for vehicles into a venue for creative mechanics. In the years around World War I, wireless telegraphy was transformed into "radio" in large part due to the enthusiastic ingenuity of young men (and a very few women) who found the new technology remarkably accessible. After World War II, the flood of surplus electronic components gave further impetus to an entirely new generation of hobbyists who turned these components into everything from door openers to loud speakers. Not a few of these were inspired to make their mark even more deeply at the intersections of technology and commerce, perhaps only in a radio and TV repair shop but frequently in more creative ways. The personal computer that emerged in the mid-1970s owed a great deal to the same impulses that had worked on earlier generations, using earlier devices and components. It is no coincidence at all that the first personal computers were strictly hobby machines, doing little that could be called "useful," but inspiring and provoking sustained efforts to make something "cool" that no one else had made before—at least no one else nearby. The announcement of the microprocessor in 1971 was followed within only a few years by impromptu kits and manuals and clubs that beckoned the hobbyists (not yet "hackers") to see what they could make the new chips do. This spurred, of course, not only the wielders of soldering irons and pliers, but also the pencil-wielding writers of software, and yet another realm of technical improvement opened up to masses of enthusiasts. It should be added that the role of hobbyists in twentieth-century technologies has also tended to reinforce the gender identities of the technologies; for a range of social reasons, such hobbies were always more accessible and attractive to young men than to young women.[14]

In 1974, electronics hobby magazines began to announce the availability of kits built around microprocessors, typically the Intel 8008, that would allow enthusiasts to build computers at home. The resulting machines did not do much, since programming was through flipping switches, there was no way of storing instructions, and output was initially little more than blinking lights. But from this beginning, from kits costing $400 or $500, further machines came from small shops and entrepreneurial firms. Tape recorders were adapted to store programs, a new form of a very accessible programming language, BASIC, became available, and keyboards and cathode-ray monitors were adapted for input and output. Finally, in 1977, the personal computer appeared in a form we would recognize today. The hobby-oriented firm Radio Shack offered a $400 machine in its stores, and, of even greater importance, two computer enthusiasts living near Palo Alto, California, offered for sale the Apple II, a machine that worked with disk drives and even a color monitor. For a couple of years the Apple II sold reasonably well, but still largely as a hobbyist machine for users interested in programming or in devising games. In late 1979, however, a couple of skilled computer programmers introduced a product called VisiCalc for the Apple II. This was arguably the true beginning of the personal computer age, for this spreadsheet application showed how these small machines could do things that even mainframe computers were not set up to do, and do them for anyone patient enough to read an instruction manual. This signaled to some of the larger computer makers, most notably IBM, that the personal computer was not going to stay in hobby shops and enthusiasts clubs. The introduction of the IBM PC in 1981 cemented this conclusion, making the personal computer a respectable purchase for a vast range of users, from office workers to factory managers, engineers, and college professors. As the hardware began a decades-long process of expanding capabilities and complexity, so too did the software grow quickly from its hobby roots through spreadsheets, word processors, database managers, and, eventually graphics-oriented programs. These latter were given a special boost by the introduction in 1984 of the Apple Macintosh computer, with its easily mastered graphical interface (complete with mouse, icons, and all the paraphernalia now taken for granted in computers).[15]

By the 1990s, the computer in all its manifestations had become a universal feature of technical life in the West. Mainframes and large systems controlled everything from air traffic to pension checks, mid-sized computers were used in businesses and factories to manage every conceivable process, and personal computers kept falling in cost and size and increasing in capability. While the economics and technical achievements of microelectronics were central to making the computer so accessible, the machine's universal character owed more to the creation of something largely unanticipated by the computer's pioneers: networks. At the end of the twentieth century, the electronic digital computer became as much a communications device as an information processor and in this way joined itself to another of the century's key theaters of improvement.

Since the introduction of the electromagnetic telegraph, communications systems of two basic modes emerged and spread, first in the West and then rapidly through the entire world. The first mode was point-to-point; the original telegraph was the classic form of this, and its elaboration and extension through the telephone made instant communication between individuals on a global scale one of the defining technologies of modernity. The second mode was broadcasting, and its emergence and spread was arguably a much more surprising phenomenon of the twentieth century, although no less universal in its reach or effect. These systems of instantaneous communications were augmented significantly by a series of technologies for capturing information, both aural and visual: the phonograph, cinematography, and a host of mechanical, magnetic and electro-optical recording technologies that continue to evolve. Together these developments made "media" the most visible and socially and culturally disruptive technologies of the past century.

The media technologies recorded and/or transmitted three distinct products: words, images, and sound. Each of these were the subject of remarkable innovations in the nineteenth century. Telegraphy and telephony made the transmission of words nearly instantaneous, and wireless freed this transmission from the constraints of specific pre-determined routes and points. Photography and, at century's end, cinematography, provided radically new means for capturing images, and new printing technologies, like lithography and its simpler and cheaper heirs, made the proliferation of images a hallmark of Victorian life. And the phonograph, which Thomas Edison called his most "surprising" invention, promised the means to make sound as recordable as text, although the improvement of sound recording was to prove a century-long struggle of surprising complexity and difficulty. The new and sometimes surprising means by which words, images, and sound were combined, stored, and moved about in the twentieth century—the new "media," in other words—generally arose through new understandings of chemical and physical phenomena, from photosensitive chemicals to the quantum electronics that made the laser possible. In a pattern that became particularly important for both culture and commerce, the most advanced products of science and engineering came regularly to be seen as opportunities for making new media. As we have seen, radio was the product of the scientific theories and discoveries of Maxwell, Hertz, and their followers. Time and again in the next hundred years this pattern of exploiting science was pursued to generate new capabilities. It is important to note, however, that the particular ways in which these new capabilities were developed and exploited were not determined by science at all, but by the complex combination of personal ambitions, social currents, markets, and politics that lay at the foundation of all important technological change.

Sometimes the scientific bases for new media were not, in fact, complex or esoteric. Making pictures move was an idea that predated the nineteenth century, but it was not until the 1820s that experimenters attempted to apply the "persistence of vision" (the tendency of a viewer to see an image briefly after it has been removed from sight) to a device to portraying motion. A series of scientific toys demonstrated the phenomenon, the most successful being the zootrope of William George Horner, which was widely popular in the 1860s. This was an open drum with a series of slots; a strip of drawings of successive positions was placed inside the drum, below the level of the slots, and a viewer looking at the spinning drum through these would see the images merge successively to depict motion. Using a series of photographs to make moving images was a logical next step, but it required solving some difficult technical problems. In the 1870s, the advent of faster photographic processes permitted a French physician, Étienne-Jules Marey, and an English immigrant to America, Eadweard Muybridge (who thought his original name of Edward Muggeridge was too "ordinary") independently to devise techniques for taking sequential photographs of animals and people in motion. Muybridge was the greater showman, and by 1880 he was giving presentations of his successive pictures of moving animals and people in through a projecting device (an adaptation of the "magic lantern" that had been used to project still images, like later slide projectors) that he called a "zoopraxiscope."[16]

By the end of the 1880s, numerous experimenters were attempting to make devices to capture and depict motion. Among these was Thomas Edison, whose projects in his West Orange, New Jersey, laboratory included efforts improve the phonograph. In the fall of 1888, he filed a claim for a device "which does for the eye what the phonograph does for the ear." Some months before, Muybridge had demonstrated his zoopraxiscope for Edison, but Edison's own approach was quite different. He adapted his cylinder phonograph to carry a long spiral of tiny images, which could then be viewed in moving sequence by turning a crank and peering through a magnifying eyepiece. This "kinetoscope" was not a high priority for Edison, but he put one of his more energetic assistants, W. K. L. Dickson, on to the project and over some months a device emerged. The most crucial improvement came when Edison saw a demonstration in Paris of Marey's use of a new photographic medium for taking sequential photographs. Marey adapted the long strips of photosensitive paper (and later celluloid film) that the photography entrepreneur George Eastman introduced for the amateur system he called the "Kodak" and designed a camera capable of taking sixty frames per second. Thus is was that the kinetoscope that Edison introduced to the public in 1894 used reels of film and a shutter like device adapted from his telegraph days. Edison's vision of the new technology, however, was quickly overtaken by others, who saw motion pictures not as a personal experience to be viewed through peepholes but as public, theatrical ones, depending on projected images. In 1895 a host of inventors introduced projectors, with the "cinematograph" of Louis Lumière being the most successful (and the source of the name of the new medium: cinematography).[17]

The rapid development of motion pictures in the twentieth century was a combination of technical and cultural innovations. This was not simply a new medium, but it set out new expectations of what mass communications could do, expectations that helped to shape the direction of all of the new media that followed. Most important, perhaps, motion pictures shaped the idea of the mass experience. The idea that someone with a message, whether it be art, news, propaganda, or instruction, could package and deliver it in a form that would be seen (and after the 1920s, heard) in a carefully controlled manner by a limitless audience was a notion that would reshape the very idea of community. Subsequent media, from radio to television to the internet, became above all means for creating and shaping this kind of mass experience. Different technologies and different social and political contexts gave each medium a distinctive character (or, better, characters, as these varied from place to place and over time). The particular markets and uses of the new media technologies proved to be remarkably resistant to successful prediction, as did their social consequences.

Cinema, for example, emerged as one of the most distinctive and successful art forms of the new century. As it continued to be the subject of improvement—with sound, color, and a variety of enhancements (some of which, like three-dimensionality, continued to elude developers), cinema resisted repeated forecasts of impending obsolescence and remained a remarkably resilient shaper of popular culture and imagery. Just as significantly for our story, the emergence and continuing development of motion pictures established a pattern of constant change that linked technology and popular culture through the twentieth-century.

The other primary pattern-setting medium of the twentieth century, radio, followed different paths from cinema, and from any prior medium, for that matter. The "wireless" that emerged out of advanced physical experiments at the end of the nineteenth century found rapid application as a communications tool, but two decades of innovation and experimentation preceded the emergence of radio's most significant form, broadcasting. In the nineteenth century, tentative steps had been taken to apply new communications technologies to the broadcast idea. The stock tickers that became so important for financial markets in the 1860s and 1870s were essentially broadcast devices using telegraphy. Among the Bell Telephone Company's earliest demonstrations were the broadcasts of concerts and other events through its network. But such applications were no more than peripheral, whereas the broadcasting model became the core of radio's use. Marconi's efforts in the first years of the century, and those of the competitors that followed, were built around the point-to-point model of wireless telegraphy and then telephony. Broadcasting was, to an extent, a by-product—initially a deficiency—of the technology itself. A radio signal, once transmitted, cannot be restricted to a single receiver (not without considerable manipulation, the kind that now makes the cellular telephone system workable). Perhaps the most remarkable thing about the emergence of radio broadcasting was the conversion of this deficiency into an asset—indeed, into the primary virtue of the medium.

A series of technical innovations in the first two decades of the twentieth century brought the crude wireless telegraphy of Marconi to an increasingly reliable and powerful form of radio telephony that found widespread application during World War I. A true technology of "electronics" (although that term was yet to come) emerged, based on the ability to apply currents to control, amplify, and modulate the typically weak electronic signals of wireless. The core technology was that of the two- and three-element vacuum tube. The "diode" was the result of observations that had been made in the first years of experiments with Thomas Edison's vacuum light bulb. About twenty years after Edison observed the apparent passage of current across the vacuum in his bulb, British electrical engineer Alexander Fleming applied the effect to construct a "valve," converting the oscillations of a received radio signal into a much more easily registered direct current. A few years later, in 1907, an

American experimenter, Lee DeForest, patented a three-element tube that he called an audion. This had the capability of amplifying signals, although it was some years before it could be made with sufficient reliability. The audion, also called a triode, could be used as an oscillator as well; that is, it could produce high frequency current shifts required to produced radio waves of usable frequency. In the decade before World War I, an astonishing variety of devices, systems, and apparatus was invented throughout Europe and America, largely to increase the range and precision of wireless telegraph signals.

Initially Marconi's vision of wireless as primarily used where land telegraph lines were not available (most importantly, at sea) governed the directions of the technology and its applications. In short order, however, the sheer range of experimentation and ingenious technical approaches outran this limited idea. On Christmas Eve 1906 a brilliant American experimenter, Reginald Fessenden, startled others in the field with the transmission, not of the usual Morse code, but a concert, complete with himself on the fiddle, Christmas carols, and a phonograph playing Handel. This was conceded a bit of a stunt, but it made the point that there were no technical hurdles in the way of transmitting words and music as easily as dots and dashes, and within a few years radiotelephony had begun to find uses at sea and in hastily set up land stations as well. By 1914 and the outbreak of general war in Europe, radio communications had become a valued asset both at sea and for rapidly moving forces on land. The war reshaped the emerging technology in many ways. The strategic importance of radio communications brought governments, particularly military interests, actively into both the promotion and the regulation of its development. Wartime applications exposed vastly more people to both the devices and their possible uses. And the fractures and tensions of war reshaped the commercial environment profoundly, as older monopolies and national systems were broken up or rearranged for strategic purposes. In the United States, for example, the wide control that Marconi (British) interests had over radio patents and infrastructure was deemed as unacceptable, and the federal government promoted the creation of a new company, the Radio Corporation of America (RCA), to assist in the pooling of patents and maintaining an American stamp on the emerging systems.[18]

The rapid emergence of radio broadcasting in the years after World War I was the product of a combination of technical developments, popular enthusiasm, commercial interests, and government concerns. The companies with interests in radio were largely concerned at first either with selling equipment or with operating and controlling communications networks. In the United States these companies strongly urged government interests to remove themselves from the arena, and this created a range of opportunities for enthusiasts and entrepreneurs. In 1920 Frank Conrad, an engineer for the Westinghouse Company, which made radio receiving equipment, began transmitting regular phonograph "concerts" from a Pittsburgh Department store

Figure 25.2
Broadcasting controls at the first radio station, KDKA, Pittsburgh. Library of American Broadcasting, University of Maryland.

(figure 25.2). Sales of both receivers and phonograph records soared, and the manufacturers noticed. The radio "boom" that followed was dramatic, as hundreds of licenses for broadcasting were issued over the next few months, and as many as a hundred thousand receivers were sold in 1922 alone, and five times that many the following year.

The boom continued through the 1920s. At first almost all broadcasting was done by independent local stations, but by 1930 about 30 percent of stations were parts of national networks, such as the National Broadcasting Company, set up by RCA as a "public service." Very quickly, the primary financing for broadcast stations in the United States came from air time used for advertising, establishing a tension between the "service" and the "commercial" aspects of the new medium that persists to this day. In other countries, such as Great Britain, alternative models emerged. There, the monopoly over radio that was early asserted by the Post Office led to the formation of the British Broadcasting Company in 1922. Funded largely through license fees paid by all radio set owners, the public (but ostensibly independent) broadcaster represented a much narrower but service-oriented model for the new medium. This model, often reshaped to make radio into a political voice, was dominant for most of the world through the century.[19]

Even before radio was a widely available technology, the possibility of adding images to the transmission of sound was bandied about. In the 1870s some experimenters began speculating on how the phenomenon of photoelectricity—the capacity of some materials to respond to light by generating weak voltages—was explored as a possible means for capturing and transmitting pictures. A host of technical problems stymied the first efforts, especially the difficulty of scanning moving images in such a way that they could be successfully reconstructed at the receiving end (still images were less of a problem, and early versions of facsimile transmission were available by World War I). Both mechanical and electronic scanning techniques were the subject of extensive experimentation. Paul Nipkow, a young German engineer, proposed a design based on a spinning disk in 1884, but it was not until another twenty years had passed that a working system could be demonstrated. Boris Rosing in Russia constructed an improved mechanical scanning system in 1908 and other experimenters made their own stabs at the idea, most notably John Loglie Baird in England in the mid-1920s. Baird's system, like that of C. Francis Jenkins in the United States, produced extremely low resolution pictures and represented an unlikely path to success. A more promising direction was set out in the electronic scanning systems pioneered in the 1920s by Philo T. Farnsworth, working in California, and Vladimir Zworykin, a Russian immigrant engineer at Westinghouse. When RCA hired Zworykin in 1930, the substantial corporate commitment required to bring out a complete television system was launched.[20]

If anything, however, the American pursuit of television was more tentative than the European. Britain's BBC gave that country a single strong institution with a powerful commitment to exploiting all the potential of broadcasting, and the company threw its support toward the development of a native-born television system. Baird's mechanical scanning system became the basis of the first broadcasts in the summer of 1930. A German, all-electric system was demonstrated the following year in Berlin. Rival television technologies received enormous public attention, and by the mid-1930s experimental broadcasts were also being made in the Netherlands and France. American television was in this period largely in the hands of RCA engineers, who began test broadcasts in 1936. The public debut of television in the United States was the broadcast of President Roosevelt opening the New York World's Fair in April 1939. By this time the electronic systems had clearly won out over mechanical ones, but different standards emerged in different countries, reflecting different decisions about quality, economy, and national styles. The expansion of television was interrupted by the outbreak of World War II later that year: German and French broadcasts continued sporadically through the first war years, but British and American efforts were suspended entirely. Even when government restrictions were lifted in 1946, it was not immediately clear what direction television development would take. In the United States, for example, uncertainty was so great that the Federal Communications Commission froze the allocation of broadcast licenses between 1945 and 1949, hurting the independent developers at the expense of established broadcasters like NBC and CBS. When restraints were removed, however, a television boom reminiscent of the radio boom of the early 1920s followed: only eight thousand sets were receiving broadcasts in 1946; this expanded to one million in 1949 and in only two more years was over ten million. By 1960, 90 percent of American homes had sets, and television was widely acknowledged as the single greatest shaper of American popular culture.[21]

Over the years, television was the subject of a great range of technical improvements. Color broadcasting was introduced in the 1950s, although it took more than a decade to move from the fringes to the mainstream of broadcasting. Recording technologies were introduced early in the United States as the difficulty of relying on live telecasts in a nation that spread over four time zones (not including Alaska and Hawaii) gave studios extra incentives to push nascent videotape capabilities. When these became available for homes in the 1970s, the television began its transformation from a simple receiving instrument into a versatile entertainment center, a transformation that expanded with the spread of cable transmission systems and video games in the 1980s and 1990s. The development of international satellite relay and transmission capabilities through the 1960s and beyond gave television a global dimension that did more than perhaps any other single technology in making the culture of the late twentieth century boundaryless. Even from its earliest form as a conveyer of grainy black-and-white images at best only ten or twelve inches across, television displayed a remarkable and largely unanticipated capacity for transforming the idea of community. While radio had made information and entertainment immediately available to the masses, television's visual dimensions made everything from football championships to the landing of men on the moon into global experiences. The technical developments that made this possible were all the products of organized corporate efforts, abetted by government laboratories and regulators that sought to juggle an often confusing mix of commercial, social, political, and national interests in giving the technology its modern form. As the product of intertwined government and corporate interests, television also was the model for much of technological change in the second half of the twentieth century—often very much in the public eye, but never, seemingly, subject to any kind of plan, management, or even logic.

One other feature of television helped to shape the media innovations to come. More than any other technology, television conveyed the power of the network, of the web of connections among nodes that people intimately associated with the real value of the medium. The intensely corporate roots of the technology, based on efforts of RCA, the BBC, and other nation-straddling entities, gave television a network-oriented character that always dominated. Even with the proliferation of low-powered UHF stations or niche cable channels, the true power of television was always seen to lie in its networks, both substantive (NBC, BBC) and virtual (the collectivity of *Seinfeld* or *Star Trek* viewers, for example). The power of the local, of the technology that operated purely at the level characterized by where an individual was or where he or she could be in a short time, was further diminished by television and by the sense it conveyed that only the global mattered any more. This is one of the reasons that the Internet, the tying together of computers throughout the world into an apparently boundless network of networks, found such a receptive audience at century's end. Nothing seemed more natural than to create a means for linking together a "world wide web" of computers and users, in a system that combined the autonomy of individuals with the intimate connectedness of virtual communities.

The beginnings of the Internet were more prosaic. The tensions of the Cold War and the nuclear arms race between the United States and the Soviet Union created considerable anxiety about the integrity of military command and control systems in the event of large-scale war. By 1960 these concerns had led planners in the U.S. Department of Defense to seek out the means to connect systems in redundant ways, so that the breakdown of one key link would not bring down the entire command structure. Already computer systems had become crucial elements, particularly for air defense, so the linking together of such systems was seen as a much needed security measure. The techniques for doing this, however, were not easy or straightforward, for in many ways "sharing" a computer ran against the grain for those fortunate enough to have access to these still large and expensive machines. But a series of innovations, including time-sharing technologies and an approach called "packet switching" that allowed computers to communicate efficiently, opened the door by the late-1960s to the creation of reliable and far-reaching networks of defense computers. The Advanced Research Projects Agency in the Pentagon sponsored the ARPANET, linking the four largest of the military-funded computer centers around the United States in 1969. At about the same time, "local area networks," or LANs, were being constructed in large businesses and on university campuses. By the mid-1970s, a variety of specialized networks (for computer science, seismology, and the like) and of proprietary networks (such as the IBM-based BITNET) emerged from the effort to link computer resources and computer users into larger and larger systems. In this unstructured environment, a range of uses sprang up, including electronic mail and bulletin boards—uses never envisioned by the original network designers (much less by those footing the bills). Innovation picked up in the 1980s, especially as a host of techniques emerged to make communications easier and more reliable—the domain name addressing system and the transmission control program (TCP) protocol, for example. By this time the expansion of computer networking took on a momentum that was clearly beyond the vision of any single user or agency.[22]

In the 1980s the networking of computers began its transition from a largely military and scientific affair into the modern Internet. Strong academic interests, principally from computer scientists and engineers, but also joined by scientists and scholars from many other fields, helped guide the erstwhile military system into broader arenas. The U.S. military recognized the extent to which its creation had grown beyond its core concerns, and the National Science Foundation took over support for the growing system's core elements. In the second half of the 1980s, the growth of the system became explosive, going from about 2,000 connected computers in late 1985 to almost 30,000 connections only two years later and more than 160,000 two years after that. Growth continued at an accelerating pace, and by the mid-1990s, the U.S. government eased itself out of the responsibility for the Internet, turning administration over to private entities.

In this same period, a new application (or set of applications) emerged that was to give the Internet its most distinctive character at the end of the twentieth century. The World Wide Web was in large part the brainchild of a computer scientist, Tim Berners-Lee, at the primary European nuclear research facility, CERN, in Geneva, Switzerland. Seeking a means to make it easier for computer users with a wide range of competences and using a variety of systems to exchange information, Berners-Lee devised a graphics-friendly system of presenting information and then embedding "hyperlinks" within it, allowing users to follow a great range of possible paths in pursuing connections. The result would be, in Berners-Lee's words, "a pool of human knowledge." Berners-Lee and his colleagues designed the language (HTML), addressing system (URLs), and browsers that made the Web a readily accessible system for users with little computer knowledge and experience as well as old hands. CERN began distributing the technology in 1991 and a few other laboratories around the world began making use of it. When one of these, the University of Illinois's National Center for Supercomputing Applications, began to distribute an improved browser program, Mosaic, in 1993, the Web took off, becoming in a few years the defining technology of the 1990s.[23]

Like all truly significant technologies, the Internet and Web were distinguished by the degree to which they were perceived as platforms for improvement. The last decade of the twentieth century saw the pattern of explosive experimentation that had accompanied the launching of other key technologies, from the railroad to radio, in earlier decades repeated once again, along with the economic and technical uncertainties that accompanied those booms. Giant new economic entities emerged, built less around machines or even networks and more around software and applications: Microsoft, AOL, Yahoo!, Amazon, and Google were the creative responses to and corporate results of the opportunities for improvement sought out by the winners (at least in the short term) in the race to exploit the remarkable confluence of computing and networking. There was no shortage of losers in this race either, as the disappointed investors in the "dot-com bust" found out at century's end. But sorting out other losers and winners was a bit more complicated, as the social and cultural dislocations that often accompany technological change occurred on a truly global scale. Worries of a "digital divide," separating those groups with access to the Internet and the opportunities it represented and those with little or no access took hold on both local and international levels. The displacement of workers—a common effect of new technologies—took place in a new pattern, one that found men and women in California's Silicon Valley, for example, resentful of new opportunities for the residents in Bangalore, in southern India. The linkages that constituted the 'Net and the Web bound people and nations together in ways that were unprecedented, but that also resurrected anew the age old questions of what really constituted improvement, and for whom.

26 Improvement's End

On December 17, 1972, Eugene A. Cernan became the last man on the moon. Educated as an electrical engineer and trained as a naval jet pilot, Cernan was a particularly well experienced astronaut. His trip on Apollo 17 was the third one he had taken into space. He was well aware, as he left the surface of the moon that day, that there was no one scheduled to follow him for the foreseeable future, and so he felt compelled to devise some pithy remarks for the occasion. Like his fellow astronaut Neil Armstrong upon first stepping out on to the moon's surface some forty-one months before, Cernan seemed a bit overwhelmed by the occasion. Armstrong's "small step for man, giant leap for mankind" at least had a memorable cadence to it, while Cernan's "America's challenge of today has forged man's destiny of tomorrow" sounded too much like a politician's platitude. Perhaps more to the point was what Cernan said to his moon-lander companion a few hours later as they sent their vessel off the surface back toward Earth, "Okay, Jack, let's get this mutha outta here"—the last words on the moon.[1]

The Apollo missions to the moon were as much politics as technology, and by the end of that last, sixth, moon landing, the political effort had exhausted itself, even as the technology was seen as triumphant. In the middle decades of the twentieth century the improvement of technologies of all kinds appeared to be an imperative—political, social, economic, even cultural—throughout the West, particularly in the United States. The Second World War had sown the seeds of an astonishing range of technological crops, from widely available antibiotics and pesticides to bombs, missiles, and jet aircraft that promised to transform utterly large-scale warfare. Few agendas seemed clearer both to politicians and to the public at large than the pursuit of technological promise. Great possibilities were opened up by the war and expanded enormously by the institutional offspring of war: the giant research and development laboratories that so effectively combined the material and organizational resources of the state with the initiative and creativity of industrial corporations. Even in its heyday, however, this agenda made some thoughtful observers uneasy.

In his 1961 Farewell Address to the American people, President Dwight D. Eisenhower referred to the beliefs that fostered this particular form of improvement:

> Crises there will continue to be. In meeting them, whether foreign or domestic, great or small, there is a recurring temptation to feel that some spectacular and costly action could become the miraculous solution to all current difficulties. A huge increase in newer elements of our defense; development of unrealistic programs to cure every ill in agriculture; a dramatic expansion in basic and applied research—these and many other possibilities, each possibly promising in itself, may be suggested as the only way to the road we wish to travel.[2]

The departing president went on to speak of the dangers of the "military-industrial complex" that this faith had spawned, and to caution that the solitary inventor and the independent university researcher were fast disappearing as "a government contract becomes virtually a substitute for intellectual curiosity." He was quick to add mention of "the equal and opposite danger that public policy could itself become the captive of a scientific-technological elite." Never before had the shapers of technology been seen to wield such influence over the affairs of the state as well as of everyday life. The fruits of this power were, in the mid-twentieth century, widely regarded as well worth the dangers.

Just a few months after Eisenhower sounded his warnings, the new president, John F. Kennedy, gave voice to the strongest faith in the capacities for technological improvement. In a May 1961 message to Congress, he declared, "I believe this nation should commit itself to achieving the goal, before this decade is out, of landing a man on the moon and returning him safely to earth. No single space project in this period will be more impressive to mankind, or more important for the long-range exploration of space; and none will be so difficult or expensive to accomplish. . . . But in a very real sense, it will not be one man going to the moon—if we make this judgment affirmatively, it will be an entire nation."[3] The boldness of Kennedy's declaration is easily overlooked in retrospect, but it must be remembered that he spoke little more than three years after the United States had put its first spacecraft into earth orbit: the tiny Explorer I, launched on January 31, 1958. The twenty-pound Explorer had been launched only after a number of earlier attempts to orbit a satellite had ended in very public failures and also, even more importantly for the kind of political concerns Kennedy referred to, after the Soviet Union had successfully launched two Sputnik satellites, the first on October 4, 1957, and the second only a month later (figure 26.1). This latter, Sputnik 2, carried the dog Laika and a host of instruments, representing a payload of more than a thousand pounds. Like the United States, the Soviet Union built its space program on research documents, rockets, and scientists and engineers taken from Germany at the end of World War II. And like the United States, the Soviets had by the mid-1950s converted these resources into a growing arsenal of missiles, most of them capable of carrying nuclear warheads hundreds or

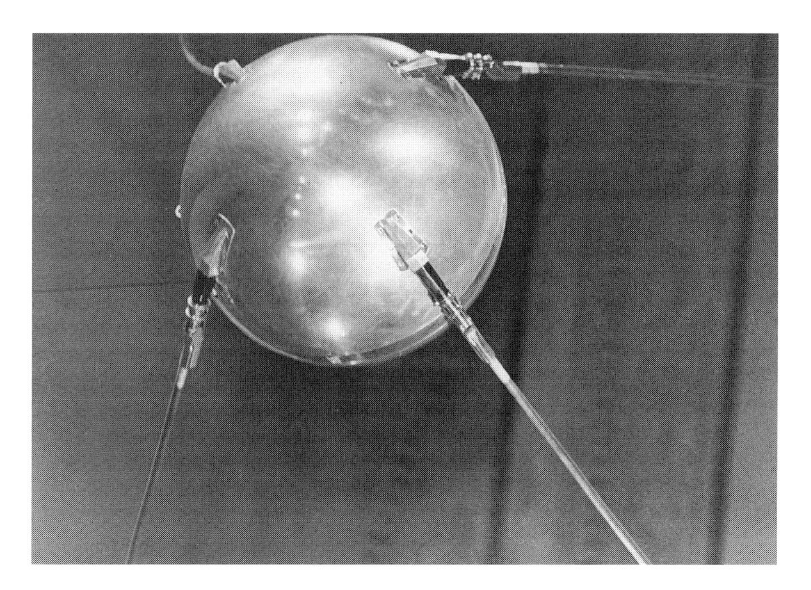

Figure 26.1
Replica of the first artificial satellite to orbit the earth, the Soviet Union's Sputnik I. Courtesy Archives Division, National Air and Space Museum, Smithsonian Institution.

even thousands of miles to a target. The space programs of both countries were built directly on these missiles and their successors and thus never lost something of a military flavor, despite their avowed civilian aims and direction.

The mixed military and civilian character of the mid-twentieth-century space efforts was perhaps simply the most glaring example of a hybridization that characterized much of the large-scale technological efforts of midcentury, at least in the superpowers that found themselves engaged in a Cold War of epic dimensions. In every field, from aviation and rocketry to electronics to medicine, the pressure of defense needs combined with unprecedented resources to give a military coloring to all the advanced technologies of the day. Some technical efforts were so large in scale and fraught with danger that private, nonmilitary initiative was almost entirely suppressed. This was true not only in rocketry but also in the development of nuclear energy. In the United States, civilian agencies such as the National Aeronautics and Space Administration (NASA) and the Atomic Energy Commission (AEC) might be given statutory responsibility, but the primary resources were still in large part military. In other areas, such as electronics and computing, civilian and corporate efforts were more independent and varied, but the largest source of funds for advanced work long continued to be from the armed services or other security agencies. This support sometimes pushed technologies in directions and at speeds that would have been unlikely, if not in fact unthinkable, otherwise. The rapid development of solid-state electronics, for example, was unquestionably due to military interests. In some countries where such interests had little influence, as in Japan, this actually led to concentration on consumer-oriented applications that provided long-term market advantages compared to, say, American developments, where defense-funded work was a surer and more lucrative entrepreneurial lure.

President Kennedy had been right—the American space program, with its focus on Project Apollo and the moon, was indeed difficult and expensive. Year by year, the NASA engineers and the contractors who became wedded to government business pushed their technology to its limits. Astronauts were recruited with great fanfare: the first American in space was Alan B. Shephard, who flew a suborbital flight just three weeks before Kennedy's bold declaration of American intentions. American celebrations over this feat were dampened somewhat by the Soviet orbiting the previous month of Yuri Gagarin, who spent almost two hours in flight, compared with Shephard's fifteen minutes. And so it went for several years, with the space programs of the two superpowers being used as obvious propaganda displays, and with the Soviets generally taking the lead, largely due to their ability to lift much larger payloads into space and their willingness to take risks, made easier by the secrecy that surrounded their efforts. The American space program, on the other hand, had of political necessity to take place in the glare of public scrutiny. The manned programs were the spectaculars, but alongside them emerged other more practical efforts in space. The first weather satellite was put in orbit in 1960, and a few months later a "passive" communications satellite—the 100-foot-diameter Echo—provided the first space-based assistance to earth communications. These were followed by more sophisticated orbiting instruments, gradually transforming everything from geophysical and astronomical research to electronic communications and all varieties of navigation and mapping. In the mid-1960s unmanned probes were successfully sent by both nations to Venus and the moon, but as the end of the decade neared the Americans continued their steady progress towards a moon landing, while the Soviet Union steered away from that objective, focusing more on developing the capacity for staying in orbit for many weeks in large craft—the first versions of "space stations." By the time that Neil Armstrong set foot on the moon on July 20, 1969, human lunar exploration had been ceded to the Americans. It was, in fact, widely perceived as just the sort of popular technological triumph that Kennedy had projected.

In the midst of triumph, however, discordant voices were heard. The president at the time was Kennedy's old nemesis, Richard M. Nixon, and he had come to office precisely six months before, riding a wave of dismay over a frustrating war overseas (in Vietnam) and over continuing civil unrest at home, fed by both antiwar sentiment and the continuing struggle for the civil and economic rights of African Americans. Not only were social and political priorities not as clear-cut as they had been to Kennedy, but questions were being asked about the scientific and technical establishment that had been the foundation of the space program's commitments. With greater efforts to confront directly some of the social ills that 1960s activism had highlighted, such as poverty and racial inequality, and with the growing costs of the Vietnam War, pressures were put on many government initiatives, and the efforts to sustain them over time in fact led to economic difficulties in the following decade, as the Apollo program wound down to a close. Other aspects of growing social and political discord had even more explicit technological implications, none more so than the rising and broadening concern over the health of the air, water, and earth.

The year of Apollo 11 would witness the flowering of the growing environmental movement. Congress passed the Environmental Policy Act, which established an Environmental Protection Agency with far-ranging powers to enforce both old and new laws restricting pollution of water and air and the release of toxic chemicals, such as pesticides and herbicides, into the environment. A few months later, in April 1970, popular protestors, politicians, and journalists marked the first "Earth Day," in which "teach-ins"—discussion groups and meetings of an argumentative character typically associated with the anti–Vietnam War movement—were organized throughout the United States to raise citizen concern and to instruct on everything from recycling waste to controlling engine and power plant emissions and halting the use of pesticides. The movement that emerged at this time, in both its legal and its popular manifestations, was a departure from the traditional concerns that had moved the friends of nature for about a century previous. Ever since the boisterous growth of industry, settlement, and the railroads in the nineteenth century, some Americans had expressed concern about the costs exacted in changes to the landscape and the natural world. National parks and forests were established, some control exerted over the spread of cities and industries, and just occasionally a dam was not constructed, an animal not hunted, or a forest not cut down. But the movement that gained force and political muscle in the late 1960s had much broader concerns, many of which were responses, either directly or indirectly, to technology. Environmentalists questioned the costs associated with a wide range of technologies, from the internal combustion engine to nuclear power to plastics. The hitherto little-challenged definitions of improvement in everything from energy to materials were now subject to new and intense scrutiny. "Environmental Impact Statements" were now required for almost everything that the government was engaged in, from building roads to raising dams, and the resulting assessments frequently cast such work in a new, suspect light.

The questioning of technology and of once common notions of progress grew quite widespread in the midst of the social and political unrest that marked the end of the 1960s. One example, from the popular 1971 book by biologist Barry Commoner, *The Closing Circle*, can stand for the great volume of critical comment:

> We live in a time that is dominated by enormous technical power and extreme human need. The power is painfully self-evident in the megawattage of power plants, and in the megatonnage of nuclear bombs. The human need is evident in the sheer numbers of people now and soon to be living, in the deterioration of their habitat, the earth, and in the tragic world-wide epidemic of hunger and want. The gap between brute power and human needs continues to grow, as the power fattens on the same faulty technology that intensifies the need.[4]

Commoner's essential message was not new, for criticisms of large-scale technologies had been common for at least a century, but the political responses this message received in the early 1970s was, in fact, a new development. Particularly since the end of World War II, the impressive achievements of technology along with the rapidly expanding scale of Western economies had reduced the influence of technological critics to the fringe of public life. Even the warnings of respected, conservative observers such as President Eisenhower found little response in the larger society. The kind of technical bravado voiced in Kennedy's moon challenge was much more typical of popular opinion. But this changed in many ways as the turbulent sixties drew to a close.

The new skepticism of technological improvement found expression in numerous ways. Projects that only a few years before had seemed to be logical extensions of important technologies came to be viewed very differently. The promise of nuclear energy, for example—touted in the 1950s as the solution to a host of impending problems—began to be challenged and even widely spurned. The novel synthetic materials that seemed to earlier generations to be "wonder substances" came to be the butt of jokes and scorn (even as they continued to find more uses). The sleek superhighways of the Interstate Highway System were no longer universally greeted as ties to bind together different regions as well as city and countryside, but increasingly came to be seen as dividers of communities, sources of pollution, and creators of sprawl. Even projects that had originally seemed to be straightforward extensions of rapidly developing technologies ran into doubts, controversy, and obstruction. One of the best examples came in a realm that experienced spectacular and much-lauded growth in the 1940s and 1950s—jet-powered aviation.

The jet was the product of a host of independent experiments in the 1930s. Piston-driven propeller aircraft developed rapidly during that decade as the flimsy multi-winged planes of the World War I era were transformed into all-metal "monoplanes" capable of carrying sufficient cargo or passengers to become real agents of commerce and transport. The technical foundations of this transformation were broad, ranging from the development of new lightweight aluminum alloys to new understanding of aerodynamic shapes and new engines. Engine development alone had been dramatic since the Wright Brothers. Where their first production aircraft, in 1909, had a four-cylinder engine generating 30 hp, the famous Liberty engine used on American Army Air Corps planes ten years later had twelve cylinders putting out 400 hp. The most successful airplane of the 1930s, the Douglas DC-3, had, 1,000-hp engines, weighing barely more than the Liberty engines of fifteen years before. The economies of the DC-3 were so great that it was adopted by all major airlines, although not to the exclusion of others. Alternatives included Lockheed's Electra, which was a bit smaller but could cruise at over 200 mph. Advances in both military and civilian airplanes appeared in Europe as well, with German and British manufacturers having the greatest success. By the end of the 1930s, aircraft design and manufacture, centered on continuous improvement of piston-driven propeller engines, had become among the most remarkably successful technologies of the day. But some engineers saw inherent limits approaching in the capacity of piston/propeller craft to be improved and these are the ones that sought out an alternative in turbojets.[5]

The essential foundations of turbojet technology came from the development of turbine power plants in the late nineteenth century. At first using water and then applying steam, turbines had become the mainstay of electric power generation by the 1920s. Experiments were made with alternative forms of turbines, such as gas and internal combustion turbine engines, but these ran into technical difficulties in mechanisms, materials, and heat transfer. The engines that could be made to work failed any tests of economy, and so experiments were severely limited. By the mid-1930s, however, a number of aeronautical engineers had concluded that the limits of piston engine aircraft were being approached, particularly as the desire grew to increase speeds and altitudes. In England, a Royal Air Force engineer, Frank Whittle, came up with the idea of a turbojet engine, based on his belief that future flight technology would need to go to higher altitudes and greater speeds. Since propellers lose their efficiency in the thin air of high altitudes, a different kind of power plant was required, and through the 1930s Whittle was able to get just enough support to demonstrate the practicability of the turbojet idea. As World War II broke over Britain, the key problems in the Whittle jet were worked out and the Gloster Meteor jet entered service in very limited numbers in July 1944. The chief German champion of the jet was physicist Hans von Ohain, who was able to persuade German aircraft manufacturer Ernst Heinkel to back development of a turbojet engine. Just a few days before Hitler's armies invaded Poland, a Heinkel He178, powered by a von Ohain designed engine, flew the first fully jet-powered flight. Other German manufacturers also took up the challenge of jet-powered flight, so that by World War II, the idea was a largely proven one. The war diverted resources toward the massive production of more conventional propeller-driven aircraft so that actual combat jet aircraft saw service only in the war's last few months, but this was enough to make it clear to aircraft engineers that a substantial part of the future lay with jets.[6]

Jets rapidly took over military aviation in the years after 1945. The transition to jets began slowly, but by 1951 the U.S. military had largely abandoned piston aircraft. The change in the civilian sector took more time. The first jet airliner was introduced by the British De Havilland firm with its Comet in 1952. The Comet, however, demonstrated some of the hurdles faced in moving the jet into civil aviation, as the need to combine commercial economy with the stringent technical requirements of the high-flying high-speed aircraft posed challenges that had not been fully anticipated. After a number of highly publicized crashes, the Comet was removed from service and the American aircraft manufacturers were able to take advantage of the experience afforded them by large military orders. It was still not until 1958 that the Boeing 707 jet went into regular service, to be followed the next year by the Douglas DC-8. The power plants and other elements of these large four-engine jets were based on prototypes developed for military tankers and bombers. The development costs for these were naturally paid for by the military customers who ordered them, and this sustained a basic pattern of government support for key aircraft improvement. At the same time, the complexity of the aircraft and the technical demands made on engineering, materials, and construction increased to a level difficult for private industry to manage without government subsidy. In this way, the aviation industry was the archetype of "big technology" in the mid-twentieth century.[7]

Nothing was more exciting to the public witnessing the unfolding of the "jet age" than news of aircraft breaking the sound barrier. Flying faster than sound had hardly been a dream a generation earlier, but even before the war ended experimenters began figuring out how it could be done. In the fall of the 1947 a rocket-powered plane, the X-1, flown by test pilot Chuck Yeager, flew faster than Mach 1—the speed of sound (about 760 mph at sea level)—and the public was thrilled by the news. A Douglas Skyrocket, with a rocket engine, flew Mach 2 in 1953, and within a few years fighter aircraft used both by navies and air forces were flying at twice the speed of sound and more. It was the most logical thing in the world for the public (along with the airline industry and the government) to assume that civilian supersonic flight would soon follow, and as early as 1958, just as the first civilian jet airliners took flight, aviation writers were charting the likely advent of such service within a few years. Even more enthusiastic than American observers were members of the British aviation establishment, who came to see the building of a successful supersonic transport, or SST, as a promising escape route from the doldrums their industry had entered due to the failure of the Comet and other difficulties. Even before 1960, the British and the Americans saw themselves as rivals in producing the first successful SST.[8]

From the start it was clear to most observers that designing an aircraft as expensive and pathbreaking as the SST could only be done with considerable government assistance. In the United States, the Federal Aviation Administration was instructed to put together a sustained program, relying on NASA for advanced research, the Defense Department for managerial expertise, and on industry for actual prototype planes (to be built at government expense). In November 1962, the British and French governments announced a joint project, under the name of "Concorde," to design and build a Mach 2.2 aircraft. By the summer of 1963, the American government had announced a similar goal of its own, centered on an even larger and faster plane. By this date, however, questions began to be raised about both the economics of such a project and the environmental effects of the commercial fleets of the world operating SSTs.

Through the mid- and late-1960s SST designers and promoters struggled with the challenges of making a supersonic airliner that could carry enough passengers at a sufficiently moderate operating cost to actually be profitable. Fuel costs, in particular, began to loom as an ever-larger consideration as signs appeared that the era of cheap oil was nearing an end. The environmental questions loomed even larger for the skeptics. The high altitudes required to make supersonic flight feasible put the aircraft into a portion of the atmosphere that seemed particularly sensitive to the kinds of emissions that were unavoidable for the giant engines of the craft. Worry began to be voiced about lasting damage that large numbers of SSTs might wreak in the earth's protective layer of ozone, although the science behind the concern was very uncertain. More immediate was worry about the sonic booms that such a large aircraft would make as it traveled at supersonic speed. Sonic booms are the felt effects of the shock waves that are created as a body moving faster than sound collides with the compressed air in front of it. These shock waves expand as they travel, and it was estimated that an SST traveling between 60,000 and 70,000 feet above the earth would produce a boom that would affect a sixty-mile wide path beneath the plane. An airliner crossing the United States at supersonic speeds, therefore, would generate a shock that would affect thousands, if not millions, of people. By the late 1960s, action groups were protesting loudly at the prospect of SST-generated sonic booms, and political pressure was being brought to bear on the government.[9]

Despite the questions raised and the political protests, as late as 1970 the construction of an American SST appeared to most observers as an inevitable, if not in fact a natural, event. On the last day of 1968, the Soviet Union's TU-144 supersonic transport flew for the first time, and Soviet authorities announced service over the country's vast distances would be available in three years. A little more than two months later, a prototype of the Anglo-French Concorde flew a short flight over the English Channel (figure 26.2). A bit more than seven years afterward, the Concorde entered service across the Atlantic, although accompanied by frenzied protests from environmentalists. By that time, however, the American SST project was dead. The combination of insoluble economic problems and the environmental issues finally came to a head in March 1971, when both houses of the U.S. Congress voted to stop all funding for the development of the SST. What had only a few years before seemed to be an inevitable next step in technological improvement, one that an ambitious world

Figure 26.2
A technical triumph, but an economic failure: the Anglo-French Concorde supersonic transport. British Aerospace photograph.

power like the United States would necessarily take a leading role in, came to be seen instead as an unsupportable assault on fiscal and environmental responsibility. The American supersonic transport was never built, and in the first years of the twenty-first century even the Concorde—always a financial basket case—ceased flying. Talk remained of alternatives, such as "aerospace planes" that could take passengers from Washington to Tokyo in two hours, but no one seriously advanced the illusion that such technologies were "natural next steps." Civilian aviation instead revolved around larger aircraft (such as the Boeing 747, which began flying in the early 1970s), more economical aircraft (with much emphasis on fuel economy), and more environmentally acceptable craft (with special attention given to engine noise). The complex means by which modern societies determined the character of improvement had no better illustration than the experience of civilian jet aircraft in the late twentieth century.[10]

The beginning of the 1970s saw a remarkable confluence of forces that collectively cast into doubt the once ascendant culture of improvement. The cessation of manned lunar exploration, the grinding to a halt of the expansion of nuclear power, the increasing challenge to technological projects, both small and large, on the grounds of environmental damage or hazards, all gave evidence of widespread questioning in the West of the consensus on what constituted improvement. The new doubts affected even the boldest new technologies, such as the efforts to exploit molecular biology. During the 1950s and 1960s, biologists and chemists built an extraordinary research tradition from the model of DNA and its genetic coding announced by James Watson and Francis Crick in 1953. Watson and Crick's "double helix" became the basis for a rapidly expanding understanding of the mechanism of inheritance in all living things—the mechanism that, in fact, made any particular organism just what it was and could be. During these decades this new knowledge, expanded largely at government expense, became the center of a new discipline, molecular biology. At first this was a discipline with no practical applications, no technological outcomes, but by the mid-1960s change was on the horizon. Geneticist Hermann Muller proposed the possibilities at a seminar in 1963 of "genetic surgery" that would not only treat inherited diseases but that could also improve such traits as "intelligence, the native strength of social predispositions, general vigor, and longevity." The term "genetic engineering" was introduced a couple of years later to refer to such manipulations. One scientist predicted confidently in 1966 that "eventually . . . man will have the power to alter, specifically and consciously, his very genes."[11]

In 1968 molecular biologist Joshua Lederberg made one of the first serious proposals for a practical mechanism to make such possibilities real, suggesting ways to put new combinations of DNA into a bacterium in such a way that the organism would reproduce the new combination on its own. In 1971 researchers at Stanford University carried out the first successful experiments in "gene splicing," using viruses as agents for inserting portions of genes from two different organisms into a third. The bacterium that seemed most useful for these experiments was *Escherichia coli*—universally referred to as *E. coli*—since it was by far the best known and most used organism for molecular genetics. It is also, however, a familiar occupant of the human gut (and that of other mammals and larger animals). The prospect of reengineering an organism that was commonly transmitted from animal to animal and to humans raised alarms that evoked an unusual response.[12]

Within a few months of the Stanford experiments, there was widespread discussion in the scientific community of "biohazards" attached to genetic research. The techniques pioneered at Stanford turned out to be remarkably easy and effective, and genetic material was being moved around between organisms with unanticipated ease. While the real hazards attached to this kind of research were very difficult to pin down, concern was sufficiently widespread to move the scientific community to action. In 1974 a committee of the U.S. National Academy of Sciences recommended an international meeting to discuss what policies should guide scientists to protect the public and in February 1975 about 150 molecular biologists and geneticists met at the Asilomar Conference Center in Pacific Grove, California. There issued from this meeting a remarkable consensus that, while genetic engineering and recombinant DNA research were enormously promising fields, caution had to be exercised. A yearlong moratorium on risky research was declared while guidelines were developed for "safe" laboratories. This halt in some areas of development was widely observed, while government agencies, in the United States and elsewhere, developed mechanisms for monitoring research and certifying safe conditions. The possible hazards of research had been encountered before in the twentieth century—radioactive materials and X-rays caused great harm to some early researchers and users and medical researchers had long taken precautions with contagious substances. The Asilomar moratorium, however, marked a departure, for it was based on possibilities that were barely imaginable. It also raised specters in the public mind that continued to color the perceptions of genetic engineering.[13]

The controversy in the scientific community over regulating recombinant DNA and related research alerted the wider public to not only the debates over dangers but also the promises and some of the ethical worries that were emerging. After about sixteen months the Asilomar moratorium was ended, government agencies issued safety regulations for research, and the door was opened not only to scientific research but to commercial exploitation. The first effort to patent a genetically engineering organism was an application in 1972 from General Electric for an oil-eating bacterium. The proposition of patenting such a creation was so radical that it took a Supreme Court decision (*Diamond v. Chakrabarty*, 1980) to determine that it was indeed an invention within the terms of U.S. patent law. Further examples of genetic manipulation followed. By 1976 a commercially driven project to use genetic engineering to synthesize the key hormone insulin (used to treat diabetes) was under way, and new companies with names like Genentech and Biogen were being created to push the technology forward. By the early 1980s, genetic engineering was out of the realm of speculation and becoming a part of technology and commerce, although the hyperbole attached to many efforts often seemed to outstrip the real accomplishments.[14]

The sense of danger that was attached to genetic engineering, however, did not go away. The introduction of genetically modified foods, for example—plants genetically altered to be more disease or insect resistant or less fertilizer dependent—sparked heated debates about safety in the food supply and possible ramifications for agricultural practice as well as human (and animal) health. In Europe, in particular, popular movements arose to resist the new plants and governments tightened their regulation of what could be sold as food and how it was to be labeled. Concerns went beyond safety to include environmental, social, and economic implications—some modified plants were engineered to limit their capacity for self-propagation with the intent, it seemed, to make farmers dependent on large corporations for their seed supplies. There was also widespread concern about genetic diversity: Would engineered plants that resisted pests, for example, crowd out traditional ones more vulnerable to natural processes, and eventually lead to their disappearance? Would some of them escape their cultivated environment and invade other settings, becoming noxious weeds resistant to usual control techniques? The debates sparked by these and similar concerns raged into the twenty-first century.

The questions raised about genetically modified organisms became particularly difficult, of course, when the subject organism was human. From the beginning of recombinant DNA research, the human gene was the ultimate target of interest. Indeed, even in the first days of genetics, in the early twentieth century, researchers aimed at a more precise understanding of the mechanisms of human heredity, mapping characteristics on chromosomes to understand sex linkages and other genetic relationships. When the tools of recombinant DNA research emerged, they were quickly turned to further efforts to map more precisely the genetic sequences of human beings. By the late 1980s, proposals for a "Human Genome Project" came from several quarters, and throughout the 1990s the project was pursued with both government and private resources. By the end of the decade, a fairly complete "draft sequence" was available—setting out the basic elements of the genetic code to be found on all human chromosomes (the proportion of each individual's code that is distinctive is a tiny fraction, so that the human genome can be generalized). Continued research aimed at understanding how the genetic sequences actually functioned (it turned out, for example, that only about 2 percent of the sequences actually coded for proteins, raising enormous questions about the functions of the bulk of the code). Perhaps most striking was the effort made to examine, throughout the course of the project and beyond, the ethical and social implications of scientific knowledge in this area and of possible applications. Genetic therapies began to emerge, sparking great controversy about this interference with the basic construction of a human being. Even more controversial were questions raised about the possibilities of genetic screening and even genetic "design" of unborn humans. Knowledge of the "normal" human genome is easily applicable to mechanisms for identifying the "abnormal" ("inferior?"). Beyond this, the goal—so clearly expressed by the eugenicists a century before—of "improving" human beings clearly looms large for many. Except that now, genetic engineering avoids the problem of sifting through nature's variations for the desired and less-desired individuals and presents the potential for designing improvements—disease resistance, intelligence, strength and agility, even beauty—directly through genetic manipulation.

On February 22, 1997, Dr. Ian Wilmot of the Roslin Institute, a research center outside Edinburgh, Scotland, announced that his research team had engineered the birth of a sheep named Dolly by transferring the genetic material from the mammary gland of a ewe into a fertilized egg cell. In so doing, the Roslin team had, for the first time, created an individual with an exact copy of the genetic material of another mammal—a process called cloning. The result was a scientific and public uproar. Wilmot's process succeeded after hundreds of failed attempts, and many were skeptical of its real utility, but within months other clonings, and even clones of clones, were reported. The Roslin group disavowed any interest in the cloning of humans, but it was widely obvious that the rapidly improving techniques could indeed be applied to any mammalian species, including our own. The debates over the morality of attempts to clone humans were strident and are ongoing. These ran headlong into the faith, central to the culture of improvement, that, in John von Neumann's words, "Technological possibilities are irresistible to man."[15]

In the twenty-first century, technological possibilities are everywhere. Spacecraft have landed on the surface of Mars and even a moon of Saturn. Wireless transmission of messages, music, and the like potentially tie individuals ever more tightly into networks of information and entertainment that have no boundaries. Pharmaceuticals appear weekly with new properties and new promises to alleviate suffering, cure diseases, and prolong life. Entire libraries are put into electronic form and made accessible with astonishing immediacy and ease. Genetic knowledge is translated into new agricultural, pharmaceutical, and medical capabilities at a pace that sometimes seems bewilderingly rapid. Some things change more slowly—problems in energy production, transport, and materials processing resist quick solution—but even these are typically seen as matters of time and effort rather than intractable difficulties. The culture of improvement—once largely a distinctive characteristic of the West, of Europe and European America—is now a worldwide set of beliefs and expectations. The belief in technological improvement, however, does not and cannot extend to shared beliefs about culture, values, or the best way for humans to live. The dramatic power of technology and the powerful promise of its unending improvement have led to the misperception of more widely shared values in the world. This has, in turn, led to serious errors of judgment, policy, and understanding.

At the beginning of the last century, the belief was widespread that technological improvement, so heady in its speed and extent, was the inevitable harbinger of moral and social progress. Liberate human beings, so the thinking went, from the toil and drudgery associated with winning their daily bread, clothing and sheltering themselves, and otherwise providing the material elements of a decent and comfortable life, and the chief forces that lead to violence, degradation, and inequity will be lessened if not eliminated. This secular, materialist version of the long-abiding Western faith in progress—widely thought to be rooted in Jewish and Christian millenarian belief—has had enormous significance in the modern age. As any number of scholars have pointed out, progress is not a simple idea and it has many different manifestations and expressions, not all of them consistent with one another. In the course of the twentieth century, the widespread faith in progress was badly battered by historical events. World War I, in particular, undermined the more naïve expectations of the Europeans and Americans about the inevitably uplifting effects of scientific and technological achievement. In the decades between the world wars, numerous commentators reflected on the limitations of technology, on the uncertainty of a future in which science could be expected to be a constant source of unpredictable change and in which technology appeared inevitably to bring greater and greater power, without greater tools for judgment or self-control.[16]

The larger cultural significance of the West's creation of a world of undirected and apparently accelerating change evoked reflections about the meaning of science and technology for human happiness and well-being. In 1927, the English bishop of Ripon, E. A. Burroughs, asked "Is Scientific Advance Impeding Human Welfare?" and he floated the suggestion that scientific and technical research should perhaps be brought to a halt for a while, at least until people could be more comfortable with possible developments. Scientists themselves, while not countenancing the idea of a moratorium on research, acknowledged the sense of unease. In 1923, to give just one example, the English physiologist J. B. S. Haldane mused in a lecture at Cambridge University:

> Science is as yet in its infancy, and we can foretell little of the future except that the thing that has not been is the thing that shall be; that no beliefs, no values, no institutions are safe.... The future will be no primrose path. It will have its own problems. Some will be the secular problems of the past, giant flowers of evil blossoming at last to their own destruction. Others will be wholly new. Whether in the end man will survive his accessions of power we cannot tell. But the problem is no new one. It is the old paradox of freedom re-enacted with mankind for actor and the earth for stage.[17]

In the years between the world wars, science and technology became "problems" to an extent never so widely recognized. The critic Lewis Mumford published his extended essay on "Technics and Civilization" in 1934, decrying the moral implications of industrial technology, for which he adopted the term from his teacher, the Scottish philosopher Patrick Geddes, of "paleotechnics." Ultimately, however, Mumford saw possible redemption in the spread of newer technologies—a "neotechnics"—that would displace iron and coal and steam with aluminum and plastics, electricity and central power systems. Over the course of his long life as a public intellectual (he died in 1990, at the age of ninety-four), Mumford returned to examine what he came to call the "megamachine" of modern technology, with deepening pessimism about the capacity of Western culture to save itself and its moral center from the gigantism and anonymity that large-scale modern technical systems appeared to cultivate and thrive on.[18]

In the years after World War II, Mumford's pessimistic voice was joined by many other observers. Critics of technology came from both the right and the left of the classic political spectrum; technology was criticized as subversive of traditional order and as a necessary prop of traditional wielders of power. The apparent "autonomy" of technology was critiqued as a challenge to human independence and choice, and the manipulability of technology was held up as further evidence of the capacity of the powerful to reinforce their power. One of the most prominent critics was French philosopher and theologian Jacques Ellul, whose damning critique of "La technique" was published in 1954. In a later English-language article he wrote, "Technique has become the new and specific *milieu* in which man is required to exist.... It is artificial, autonomous, self-determining, and independent of all human intervention." The idea that technology (the distinction from Ellul's "technique" is not always clear) had somehow become "independent of all human intervention" was echoed by a number of critics at midcentury, although others dismissed it as illogical or, at least, unsupported by history. These general critiques of technology, however, were lonely voices in the larger culture. Even among intellectuals, the allure of technology, particularly of new technologies just over the horizon, was often irresistible. One observer of this parodied Karl Marx's description of religion's place among the masses by calling technology the "opiate of the intellectuals," and, despite the critics, this status hardly changed in the course of the late twentieth century. That is to say, technology and technological solutions remained, for most people in all sectors of society, a source of expected change, and typically, of improvement in everything from waging war to entertainment to prolonging life.[19]

That the relationship between technologies and human choices, between technical capabilities and moral consequences, is a complicated one, not subject to simple correlations, is a message reinforced time and again in contemporary experience. Usually the reminders, for those willing to take note, are small and circumscribed—personal and sometimes subtle indicators that the possession of a new device, the installation of a new system, or the use of a new power are no guarantees of happiness or security or prosperity. Occasionally, however, the reminders are on a larger scale: the destruction of a space vehicle, the unforeseen side-effects of a new drug, or the compromising of a promising new form of communications by junk and pornography serve to raise questions, at least momentarily, about the balance sheet of much touted progress. Once in a long while, even, events can call into question assumptions about shared values and shared technologies in a multilateral world. The use of civilian airliners, for example, as giant weapons of destruction, as occurred on September 11, 2001, necessarily challenged widespread assumptions about the moral and technical boundaries around the use of modern technologies. Many in the West looked on with horror not only at the acts of destruction themselves but also at the applause for these actions in parts of the world that had not seemed, to many in the West, to be so different. Jumbo jets, guided by fanatics, were, it could hardly be denied, improved weapons of terror—more effective than anything ever tried before. At one level, no doubt, the jubilation in some Middle Eastern bazaars at the pictures of famous American buildings in ruins was a celebration of a particular kind of improvement, as well as of a violent challenge to the technology-centered hegemony of the West. Lord Haldane's premonitions of several generations before, his visions of continued struggle over freedom and power in a world with ever-increasing scientific and technical capabilities, seemed more apposite that ever before.

In many ways, this entire history of the West's millennium of technology has been about freedom and power and the uneasy relation between the two. Technology and the pursuit of improvement are ultimate expressions of freedom, of the capacity of humans to reject the limitations of their past and their experience, to transcend the boundaries of their biological capacities and their social traditions. At the same time, however, the pursuit of improvement is the pursuit of and the struggle over power. Technology can, indeed, be defined as a pursuit of power over nature. Francis Bacon, almost four centuries ago, conjured up the vision of power with a startling explicitness as he called on individuals and states to devote themselves to commanding nature. There are other kinds of power involved in the culture of improvement, however, and the social and political meaning of these are much more ambiguous than power over natural forces. At every step along the way in this history there have been debates, sometimes quiet, often violent, about improvement. Who should define it? Who should benefit from it? Who must pay the inevitable costs?

These debates will—and should—continue. But their capacity to enlighten the choices that must be made, to enlighten our view of the future's possibilities and dangers rather than to obscure it, depends on a clearer understanding of the history and values that have shaped technology. Such understanding begins with humility and skepticism, but it deepens only with a confident sense that our technologies are the product of human desires and human capabilities, ultimately answerable to our own wills.

Notes

Chapter 1

1. One of the best expressions of the importance of gradual improvement is the remark of the French historian, Fernand Braudel: "In a way, everything is technology; not only man's most strenuous endeavors, but also his patient and monotonous efforts to make a mark on the external world; not only the rapid changes we are a little too ready to label revolutions... but also the slow improvements in processes and tools, and those innumerable actions which may have no immediate innovating significance but which are the fruit of accumulated knowledge—the sailor rigging his boat, the miner digging a gallery, the peasant behind the plow or the smith at the anvil." Braudel, *Structures of Everyday Life*, (New York: Harper & Row, 1981), 1:334, as quoted in Douglas Harper, *Working Knowledge; Skill and Community in a Small Shop* (Chicago: University of Chicago Press, 1987), 18. Harper himself, in his chronicle of a modern machine repair shop, remarks, "Technology itself was worked into new forms with every object made."

2. A good recent example of a discussion of the comparative questions is David S. Landes, *The Wealth and Poverty of Nations; Why Some Are So Rich and Some So Poor* (New York: Norton, 1998). A distinguished alternative contribution that attempts to draw together the great range of factors involved is Eric Jones, *The European Miracle; Environments, Economies, and Geopolitics in the History of Europe and Asia* (New York: Cambridge University Press, 2003).

3. In a remarkable passage of his *Civilization and its Discontents* (New York: Norton, 1961 [orig. 1930], 34–35), Sigmund Freud addressed the problematic nature of technology's contribution to human happiness: "Men are proud of [scientific and technological] achievements, and have a right to be. But they seem to have observed that this newly-won power over space and time, this subjugation of the forces of nature, which is the fulfillment of a longing that goes back thousands of years, has not increased the amount of pleasurable satisfaction which they may expect from life and has not made them feel happier." After regarding the joy one gets from hearing a child's voice even from a great distance or learning of a friend's successful journey, he reflects, "If there had been no railway to conquer distances, my child would never have left his native town and I should need no telephone to hear his voice; if traveling across the ocean by ship had not been introduced, my friend would not have embarked on his sea-voyage and I should not need a cable to relieve my anxiety about him."

4. Joel Mokyr's *The Lever of Riches; Technological Creativity and Economic Progress* (New York: Oxford University Press, 1990) is a good example of a treatment of technology that talks largely in progressive terms and that uses the distinctions discussed here.

5. Perhaps the most complete example of the evolutionary treatment of technology is George Basalla, *The Evolution of Technology* (New York: Cambridge University Press, 1988). Mokyr also devotes a chapter to the subject.

Chapter 2

1. As reprinted in Jeremy deQuesnay Adams, *Patterns of Medieval Society* (Englewood Cliffs, N.J.: Prentice-Hall, 1969), 291.

2. The following discussion of medieval agriculture owes much to the work of Lynn White, Jr., especially his *Medieval Technology and Social Change* (New York: Oxford University Press, 1964). Some of White's claims for the importance of the agricultural changes he describes have been much disputed, but this debate does not alter the essentials of his descriptions of the changes themselves, which are based on a century of prior scholarship.

3. The fullest statement of Pirenne's argument was in his *Mohammed and Charlemagne* (New York: Barnes and Noble, 1939).

4. See especially Jacques Le Goff, "Labor, Techniques, and Craftsmen in the Value Systems of the Early Middle Ages," in *Time, Work, & Culture in the Middle Ages* (Chicago: University of Chicago Press, 1982).

5. George Ovitt Jr., *The Restoration of Perfection; Labor and Technology in Medieval Culture* (New Brunswick, N.J.: Rutgers University Press, 1987), 143–150; Henri Pirenne, *Economic and Social History of Medieval Europe* (New York: Harcourt, Brace, 1937), 68.

6. See esp. Bruce M. S. Campbell, "Economic Rent and the Intensification of English Agriculture, 1086–1350," in Grenville Astill and John Langdon, eds., *Medieval Farming and Technology* (Leiden: Brill, 1997), 225–226.

7. The criticisms of White's thesis are summarized in Kelly DeVries, *Medieval Military Technology* (Peterborough, Ontario: Broadview Press, 1992), 95–122.

8. Charles Gladitz, *Horse Breeding in the Medieval World* (Dublin: Four Courts Press, 1997), esp. 154–161; see also John Langdon, *Horses, Oxen and Technological Innnovation: The Use of Draught Animals in English Farming from 1066 to 1500* (Cambridge: Cambridge University Press, 1986).

Chapter 3

1. Thomas Roscoe, *Lives of the Kings of England* (Philadelphia: Lea and Blanchard, 1846), 284; Thomas B. Macaulay, *The History of England from the Accession of James II* (New York: Harper & Bros., 1851), 1:10.

2. White's analysis is in his *Medieval Technology and Social Change* (New York: Oxford University Press, 1964), chapter 3.

3. The primary source on early water wheels, unless otherwise noted, is Terry S. Reynolds, *Stronger Than a Hundred Men: A History of the Vertical Water Wheel* (Baltimore: Johns Hopkins University Press, 1983), esp. chapters 1 and 2.

4. As quoted in ibid., 17–18.

5. The developments at Toulouse are described ibid., 65–66.

6. Ibid., esp. 55–64.

7. On windmills, see Richard Holt, *The Mills of Medieval England* (Oxford: Basil Blackwell, 1988), esp. chapter 2.

8. The technical details come largely from ibid., chapter 8.

9. Reynolds, 71–74.

Chapter 4

1. Abbot Suger, *Abbot Suger and the Abbey Church of St.-Denis and Its Art Treasures*, ed. Erwin Panovsky (Princeton, N.J.: Princeton University Press, 1946), 43–45.

2. Anne Freemantle, *Age of Faith* (New York: Time, 1965), 122.

3. Suger, 47.

4. L. R. Shelby, "The Role of the Master Mason in Mediaeval English Building," *Speculum* 39 (July 1964): 387–403.

5. Quoted in Ian Sutton, *Western Architecture from Ancient Greece to the Present* (New York: Thames and Hudson, 1999), 88.

6. This story is told in David Jacobs, *Master Builders of the Middle Ages* (New York: American Heritage, 1969), 52–53.

7. Robert Mark, "Technological Innovation in High Gothic Architecture," in Elizabeth B. Smith and Michael Wolfe, *Technology and Resource Use in Medieval Europe: Cathedrals, Mills, and Mines* (Brookfield, Vt.: Ashgate, 1997), 11–14.

8. Mark, 14–16.

9. Stephen Murray, *Beauvais Cathedral, Architecture of Transcendence* (Princeton, N.J.: Princeton University Press, 1989), esp. chapter 2.

10. Much has been written on this event. The sources used here include Murray, esp. chapter 6; Robert Mark, *Experiments in Gothic Structure* (Cambridge, Mass.: MIT Press, 1982), esp. chapter 5; Maury I. Wolfe and Robert Mark, "The Collapse of the Vaults of Beauvais Cathedral in 1284," *Speculum* 51 (July 1976): 462–476; Mario Salvadori, *Why Buildings Stand Up* (New York: Norton, 1980), chapter 12.

11. Murray, chapter 8; Salvadori, 222–224.

12. Shelby, 399; Sidney Toy, *A History of Fortification from 3000 B.C. to A.D. 1700* (London: Heinemann, 1966), esp. chapters 8–13.

13. Marjorie Nice Boyer, "The Bridgebuilding Brotherhoods," *Speculum* 39 (1964): 635–650; H. Shirley Smith, *The World's Great Bridges* (New York: Harper & Bros., 1953), 34–38.

Chapter 5

1. Theophilus Presbyter, *On Divers Arts*, trans. John G. Hawthorne and Cyril Stanley Smith (New York: Dover, 1979), 95.

2. Ibid., 11–12.

3. See, for example, Henri Pirenne, *Economic and Social History of Medieval Europe* (New York: Harcourt, Brace, and World, 1937), 140–143.

4. T. K. Derry and Trevor I. Williams, *A Short History of Technology* (New York: Oxford University Press, 1961), 96–97; Pamela Long, *Technology, Society, and Culture in Late Medieval and Renaissance Europe, 1300–1600* (Washington, D.C.: American Historical Association, 2000), 13–14; John H. Munro, "Textile Technology," in J. R. Strayer, ed., *Dictionary of the Middle Ages* (New York: Chas. Scribner's Sons, 1988), 11:693–711.

5. Besides Pirenne, see Gerald A. J. Hodgett, *A Social and Economic History of Medieval Europe* (New York: Harper & Row, 1974), 137–156.

6. Derry and Williams, 96.

7. Dard Hunter, *Papermaking; the History and Technique of an Ancient Craft* (New York: Knopf, 1943), frontispiece, 109–114.

8. R. F. Tylecote, *The Early History of Metallurgy in Europe* (New York: Longman, 1987), esp. chapters 5 and 7.

9. The size of blooms showed a great deal of variation over times and places. The average size, to give one example, of a bloom from a British hearth in the classical or early medieval period was about sixteen pounds. H. R. Schubert, *History of the British Iron and Steel Industry from c. 450 B.C. to A.D. 1775* (London: Routledge & Kegan Paul, 1957), chapters 3 and 4.

10. Bartholomew the Englishman, *De Proprietatibus Rerum*, quoted in Schubert, 94.

11. Schubert, chapter 7; H. W. Koch, *Medieval Warfare* (London: Bison Books, 1982), 66–73.

12. See Tylecote, chapter 9; Theodore A. Wertime, *The Coming of the Age of Steel* (Chicago: University of Chicago Press, 1962), chapter 2.

13. Bert S. Hall, *Weapons and Warfare in Renaissance Europe: Gunpowder, Technology, and Tactics* (Baltimore: Johns Hopkins University Press, 1997), chapter 2.

14. Roger Bacon, *Epistola de secretis operibus artiis et naturae*, as quoted in Frances and Joseph Gies, *Cathedral, Forge, and Waterwheel* (New York: HarperCollins, 1994), 206.

15. Hall describes these issues in some detail in his chapter 3.

16. Ibid., 59–60.

17. In addition to ibid., chapter 3, see also Kelly DeVries, *Medieval Military Technology* (Peterborough, Ont.: Broadview Press, 1992), 150–159.

Chapter 6

1. Reconstructed from Edward Rosen, "The Invention of Eyeglasses," *Journal of the History of Medicine* 11 (January 1956): 29 and 35.

2. Quoted in Alberto Manguel, *A History of Reading* (New York: Viking, 1996), 293–294.

3. Besides Rosen, see Vincent Ilardi, "Renaissance Florence: The Optical Capital of the World," *Journal of European Economic History* 22, no. 3 (1993): 507–541. Ilardi states that Rosen's conclusions about the origins of spectacles, despite the age of his article, are still "generally accepted." For an argument for the accidental nature of the invention of spectacles, see Vasco Ronchi, *The Nature of Light: An Historical Survey* (London: Heinemann, 1970; orig. Italian edit., 1939), 69–72.

4. See Lewis Mumford's perceptive comments on clear glass and its significance in *Technics and Civilization* (New York: Harcourt, Brace and Co., 1934), 124–131.

5. W. Patrick McCray, *Glassmaking in Renaissance Venice: The Fragile Craft* (Aldershot: Ashgate, 1999), esp. chapter 3; Theophilus (see note 1, chapter 5), 52–57.

6. McCray, chapter 3, esp. 43–49.

7. McCray, 96–107.

8. On silk in Venice, see Luca Molà, *The Silk Industry of Renaissance Venice* (Baltimore: Johns Hopkins University Press, 2000), esp. chapters 1, 7, and 8.

9. Frederic C. Lane, *Venice, a Maritime Republic* (Baltimore: Johns Hopkins University Press, 1973), chapter 5; William H. McNeill, *Venice, the Hinge of Europe 1081–1797* (Chicago: University of Chicago Press, 1974), chapter 2; the classic source is Frederic C. Lane, *Venetian Ships and Shipbuilders of the Renaissance* (Baltimore: Johns Hopkins University Press, 1934).

10. H. W. Longfellow trans. from Canto xxi of *Inferno*, quoted in Lane, *Venice, a Maritime Republic*, 163.

11. Lane, *Venetian Ships*, chapters 8 and 9; see also Robert C. Davis, *Shipbuilders of the Venetian Arsenal* (Baltimore: Johns Hopkins University Press, 1991), esp. chapter 1, for a discussion of the later, larger Arsenale.

12. The latest broad popular survey in a large literature on clocks is David S. Landes, *Revolution in Time: Clocks and the Making of the Modern World* (2nd ed.) (Cambridge, Mass.: Harvard University Press, 2000). See esp. chapter 3.

13. H. W. Longfellow translation from Canto xxiv of *Paradiso*, quoted in Ernest L. Edwardes, *Weight-driven Chamber Clocks of the Middle Ages and Renaissance* (Altringham: John Sherratt and Son, 1965), 20.

14. Among many discussions of the escapement, see that of Carlo Cipolla, *Clocks and Culture, 1300–1700* (New York: Norton, 1978), 111–112. A better illustration is in Abbott Payson Usher, *A History of Mechanical Inventions* (rev. ed.) (New York: Dover, 1988; from the Harvard University Press edit. of 1954), 200.

15. Silvio A. Bedini and Francis R. Maddison, *Mechanical Universe, the Astrarium of Giovanni de' Dondi*, published as *Trans. of the American Philosophical Society*, n.s. vol. 56, pt. 5, 1966; among discussions of de' Dondi, see Lynn White, *Medieval Technology and Social Change* and Cipolla, *Clocks and Culture*.

16. Bedini and Maddison, 31–33.

17. Paolo Galluzzi, *Renaissance Engineers from Brunelleschi to Leonardo da Vinci* (Florence: Giunti, 1996), introduction and sec. I; see also, for the classic (and somewhat dated) discussion of Renaissance engineering, William Barclay Parson, *Engineers and Engineering in the Renaissance* (Cambridge, Mass.: MIT Press, 1968, from the 1939 edit.), chapter XXXV.

18. Frank D. Prager and Gustina Scaglia, *Brunelleschi: Studies of His Technology and Inventions* (Cambridge, Mass.: MIT Press, 1970), 85–88. Cited in Eugene S. Ferguson, *Engineering and the Mind's Eye* (Cambridge, Mass.: MIT Press, 1992), 4–5.

19. Ferguson, 77–82.

20. As quoted in Molà, 187.

21. On early patents, see esp. Pamela O. Long, "Invention, Authorship, 'Intellectual Property,' and the Origins of Patents: Notes Toward a Conceptual History," *Technology & Culture* 32 (October 1991): 846–884; the quotation from Brunelleschi is in Prager and Scaglia, 111–123.

Chapter 7

1. Francis Bacon, *Novum Organum* (1620) (New York: Colonial Press, 1900), 366.

2. There are many important sources for the history of printing. The discussion here draws from Elizabeth L. Eisenstein, *The Printing Press as an Agent of Change: Communications and Cultural Transformations in Early-Modern Europe* (New York: Cambridge University Press, 1979); Eisenstein, *The Printing Revolution in Early Modern Europe* (New York: Cambridge University Press, 1983); S. H. Steinberg, *Five Hundred Years of Printing* (New Castle, Del.: Oak Knoll Press, 1996, rev. ed.); Lucien Febvre and Henri-Jean Martin, *The Coming of the Book: The Impact of Printing 1450–1800* (London: NLB, 1976, trans. of *L'Apparition du Livre*); Michael Clapham, "Printing," in Chas. Singer et al., *A History of Technology* (Oxford: Clarendon Press, 1957), 3:377–411; Abbott Payson Usher, *A History of Mechanical Inventions* (New York: Dover, 1988), 238–257.

3. See Seinberg, 18; Febvre and Martin, 167–182.

4. William Caxton, *The Recuyell of the Histories of Troye*, as quoted in Seán Jennett, *Pioneers in Printing* (London: Routledge & Kegan Paul, 1958), 29–30.

5. Jennett, 28–46.

6. Eisenstein (1979, 1983) is the authoritative source on the effects of printing. Her reference to Vespucci is from *The Printing Press*, 1:120.

7. Ibid., 2:482.

8. This is the argument of Samuel Y. Edgerton, Jr., *The Heritage of Giotto's Geometry* (Ithaca, N.Y.: Cornell University Press, 1991), 145–147. See also Leonard C. Bruno, *The Tradition of Technology: Landmarks of Western Technology in the Collections of the Library of Congress* (Washington, D.C.: Library of Congress, 1995), 20–21, 75.

9. Edgerton, 131–139; Ladislao Reti, "Francesco di Giorgio Martini's Treatise on Engineering and Its Plagiarists," *Technology & Culture* 4, no. 3 (Summer 1963): 287–298.

10. In addition to Edgerton, see Pamela O. Long, "Power, Patronage, and the Authorship of *Ars*: From Mechanical Know-How to Mechanical Knowledge in the Last Scribal Age, *ISIS* 88, no. 1 (March 1997): 1–41; Robert Zwijnenberg, *The Writings and Drawings of Leonardo da Vinci* (Cambridge: Cambridge University Press, 1999), esp. 35–46; Kim H. Veltman, *Studies on Leonardo da Vinci: Linear Perspective and the Visual Dimensions of Science and Art* (Munich: Deutscher Kunstverlag, 1986).

11. Marcus Popplow, "Protection and Promotion: Privileges for Inventions and Books of Machines in the Early Modern Period," *History of Technology* 20 (1998): 103–124; Alex Keller, "Mathematical Technologies and the Growth of the Idea of Technical Progress in the Sixteenth Century," in Allen G. Debus, ed., *Science, Medicine, and Society in the Renaissance; Essays to Honor Walter G. Pagel* (London: Heinemann, 1972), 1:11–27.

12. Edgerton, 185–192; A. G. Keller, *A Theatre of Machines* (London: Chapman and Hall, 1964); William B. Parsons, *Engineers and Engineering in the Renaissance* (Cambridge, Mass.: MIT Press, 1968), 108–118.

Chapter 8

1. Paracelsus (Theophrastus Bombastus von Hohenheim), *Labyrinthus Medicorum Errantium* (1587), as quoted in Friedrich Klemm, *A History of Western Technology* (Ames: Iowa State University Press, 1991), 145.

2. See the introductory matter on Paracelsus in *Four Treatises together with Selected Writings by Paracelsus* (Birmingham, Ala.: The Classics of Medicine Library, 1988).

3. The best source on mining and metalworking in the medieval and early modern period is still John U. Nef, *The Conquest of the Material World* (Chicago: University of Chicago Press, 1964). See chapter 1 for the medieval period.

4. *Bergwerk- und Probierbüchlein* (New York: American Institute of Mining and Metallurgical Engineers, 1949), translation by Anneliese Grünhaldt Sisco and Cyril Stanley Smith; the quotation is from 39. While published anonymously, the *Bergbüchlein* is credited by scholars to a physician in the eastern German city of Freiberg, Ulrich Rülein von Kalbe, known as "Calbus." The author of the *Probierbüchlein* is simply unknown.

5. Vannoccio Biringuccio, *Pirotechnia* (Cambridge, Mass.: MIT Press, 1966, from the 1959 edition of the AIME), translated from the Italian by Cyril Stanley Smith and Martha Teach Gnudi; quotations are from 440 and 444.

6. The editions and translations of Biringuccio are discussed in the Smith and Gnudi edition, xvi–xxiii.

7. Modern understanding of Agricola is based on the fine translation and editorial work of Herbert Hoover and his wife Lou Henry Hoover, who published their annotated edition of *De Re Metallica* in 1912. This was made widely accessible in a Dover edition (New York, 1950). There have been many studies of Agricola and his contributions, such as that of Bern Dibner, *Agricola on Metals* (Norwalk, Conn.: Burndy Library, 1958), but all begin with the Hoovers' work. For a broader view of Agricola, see Owen Hannaway, "Georgius Agricola as Humanist," *Journal of the History of Ideas* 53, no. 4 (October–December 1992): 553–560.

8. On the subject of openness and technical knowledge, see Pamela O. Long, *Openness, Secrecy, Authorship: Technical Arts and the Culture of Knowledge from Antiquity to the Renaissance* (Baltimore: Johns Hopkins University Press, 2001).

9. The quotations are from the Hoover edition, preface.

10. On the sixteenth-century economy and technology, see Nef, *The Conquest of the Material World.* On France in particular, see Henry Heller, *Labour, Science and Technology in France, 1500–1620* (Cambridge: Cambridge University Press, 1996).

11. The literature on ceramics is extensive. See, for the Italian story, Richard A. Goldthwaite, "The Economic and Social World of Italian Renaissance Maiolica," *Renaissance Quarterly* 42, no. 1 (Spring 1989): 1–32.

12. Gerhard Dohrn-van Rossum, *History of the Hour: Clocks and Modern Temporal Orders* (Chicago: University of Chicago Press, 1996), esp. chapter 5.

13. In addition to Dohrn-van Rossum, see David Landes, *Revolution in Time* (Cambridge, Mass.: Harvard University Press, 2000), esp. chapter 5.

14. William H. TeBrake, "Hydraulic Engineering in the Netherlands during the Middle Ages," in Paolo Squatriti, ed., *Working with Water in Medieval Europe* (Leiden: Brill, 2000); Richard L. Hills, *Power from Wind: A History of Windmill Technology* (Cambridge: Cambridge University Press, 1994), chapter 6.

15. Richard A. Goldthwaite places this change a couple of centuries earlier, in the Italian Renaissance: "If, as Burckhardt would have it, the Renaissance saw the development of the individual and the discovery of what he called 'the full, whole nature of man,' this happened largely because man attached himself in a dynamic and creative way to things, to material possessions; and with the discovery of things modern civilization was born, for man embarked on the adventure of creating that dynamic world of goods in which he has found his characteristic identity." *Wealth and the Demand for Art in Italy, 1300–1600* (Baltimore: Johns Hopkins University Press, 1993), 255. Despite Goldthwaite's arguments, however, I believe the Dutch society at the turn of the seventeenth century represented a much more broadly based culture of possession, suggesting the urban middle-class basis of the consumption of goods that was to sustain widespread technological change in the coming centuries.

16. Michael North, "Art and Commerce in the Dutch Republic," in Karel Davids and Jan Lucassen, eds., *A Miracle Mirrored: The Dutch Republic in European Perspective* (Cambridge: Cambridge University Press, 1995), 285; Jonathan Israel, *The Dutch Republic: Its Rise, Greatness, and Fall, 1477–1806* (Oxford: Clarendon Press, 1995); Simon Schama, *The Embarrassment of Riches: An Interpretation of Dutch Culture in the Golden Age* (New York: Knopf, 1987), 350–363; the classic source on the tulip mania is N. W. Posthumus, "The Tulip Mania in Holland in the Years 1636 and 1637," *Journal of Economic and Business History* 1, no. 3 (May 1929): 434–466.

Chapter 9

1. From *The Good Wife's Fore-cast* in W. Chappell and J. Ebsworth, eds., *Roxburghe Ballads* (1866–99), as quoted in Sarah Mendelson and Patricia Crawford, *Women in Early Modern England 1550–1720* (Oxford: Clarendon Press, 1998), 328.

2. This discussion relies entirely on Judith M. Bennett, *Ale, Beer, and Brewsters in England; Women's Work in a Changing World, 1300–1600* (New York: Oxford University Press, 1996). The quotation is from 88–89.

3. See Robert B. Shoemaker, *Gender in English Society, 1650–1850: The Emergence of Separate Spheres?* (New York: Longmans, 1998), 182–183. See also Bonnie S. Anderson and Judith P. Zinsser, *A History of Their Own; Women in Europe from Prehistory to the Present* (New York: Harper & Row, 1989), 1:417–422.

4. Anderson and Zinsser, 1:407–410; Shoemaker, 161–162.

5. Jordan Goodman and Katrina Honeyman, *Gainful Pursuits; the Making of Industrial Europe 1600–1914* (London: Edward Arnold, 1988), 110–112; Deborah Simonton, *A History of European Women's Work: 1700 to the Present* (London: Routledge, 1998), 77–78; Merry E. Wiesner, *Women and Gender in Early Modern Europe* (Cambridge: Cambridge University Press, 1993), 103–104.

6. The best-known exponents of the Marxist position in English were the English scientist J. D. Bernal, whose *Science in History* went through many editions after its first appearance in the early 1950s, and Robert K. Merton, a sociologist whose *Science, Technology, and Society in Seventeenth Century England* appeared in 1938 and influenced generations of historians. Merton was not a Marxist, but he was strongly influenced by materialist discussions of science, most notably the work of Russian Boris Hessen, who wrote in the early 1930s of the "social and economic roots" of Newton's science.

7. A still useful treatment of the scientific revolution is A. Rupert Hall, *The Scientific Revolution, 1500–1800: The Formation of the Modern Scientific Attitude* (Boston: Beacon Press, 1966). Hall was one of the most vocal opponents of Bernal and other Marxist interpreters of science.

8. The literature on Galileo is enormous, and much of the discussion here is based on general sources. Two recent works that enlarge the perspectives here are John Henry, *The Scientific Revolution and the Origins of Modern Science* (New York: St. Martin's Press, 1997) and H. Floris Cohen, *The Scientific Revolution: A Historiographical Inquiry* (Chicago: University of Chicago Press, 1994).

9. On Galileo and technology, see Lynn White Jr., "Pumps and Pendula: Galileo and Technology," originally published in 1966 and reprinted in his *Medieval Religion and Technology* (cited in Cohen, 346–347). A much more recent treatment of this subject can be found in articles in a theme issue of *Science in Context* 13, nos. 3–4 (2000), particularly Wolfgang Lefèvre, "Galileo Engineer: Art and Modern Science" (281–297) and Mario Biagnoli, "Replication or Monopoly? The Economics of Invention and Discovery in Galileo's Observations of 1610" (547–590).

10. See Marie Boas Hall, "Salomon's House Emergent: The Early Royal Society and Cooperative Research," in Harry Woolf, ed., *The Analytic Spirit: Essays in the History of Science in Honor of Henry Guerlac* (Ithaca, N.Y.: Cornell University Press, 1981): 177–194; M. B. Hall, "Science in the Early Royal Society," in Maurice Crosland, ed., *The Emergence of Science in Western Europe* (London: Macmillan, 1975), 57–77.

11. *Novum Organum*, aphorism 36; see Francis Bacon, *Selected Philosophical Works*, ed. Rose-Mary Sargent (Indianapolis: Hackett, 1999), xix–xxiii, Marie Boas Hall, "Royal Society of London," in Wilbur Applebaum, ed., *Encyclopedia of the Scientific Revolution* (New York: Garland Press, 2000): 582–585; see also the article by Peter Dear, "Experience and Experiment," 219–223. Much has been written about the role of Puritanism; see, for example, Charles Webster, *The Great Instauration: Science, Medicine and Reform 1626–1660* (New York: Holmes & Meier, 1976) and, for the later period, Larry Stewart, *The Rise of Public Science; Rhetoric, Technology, and Natural Philosophy in Newtonian Britain, 1660–1750* (Cambridge: Cambridge University Press, 1992).

12. The literature on Newton is vast, and is barely touched upon here. For the clockwork metaphor, see Otto Mayr, *Authority, Liberty & Automatic Machinery in Early Modern Europe* (Baltimore: Johns Hopkins University Press, 1986), esp. chapter 3.

13. Mayr, 12–15; Landes, *Revolution in Time*, 128–135.

Chapter 10

1. The Case of Monopolies, Darcy v. Allin is reported as 11 Co Rep 84, 74 ER 1131, in Phillip Griffith, "Patent Law: A History of the Patent System," as posted to http://www.austlii.edu.au/ (Australasian Legal Information Institute), "Intellectual Property Online," 16 February 2000.

2. For the politics of monopolies, see Charles Wilson, *England's Apprenticeship 1603–1763*, 2nd ed. (New York: Longman, 1984), 100–103.

3. The first Defoe quotation from *An Essay Upon Projects* (1697), the second from *A Plan of the English Commerce* (1728), as quoted in Christine MacLeod, *Inventing the Industrial Revolution: The English Patent System, 1660–1800* (New York: Cambridge University Press, 1988), 267 and 208. See MacLeod, 201–222.

4. Ibid.

5. Ibid., 209–219.

6. John Locke, *Second Treatise of Government* (1690); John Weaver, "Laying Claim to Property Rights: Concepts of Waste and Improvements on British and American Frontiers, 1650–1900," paper delivered at the conference "Property Rights in the Colonial Imagination and Experience," University of Victoria, B.C., 22–24 February 2001; Laura Brace, "Husbanding the Earth and Hedging out the Poor," paper delivered to conference "Land and Freedom; 18th Annual Conference of the Australia and New Zealand Law and History Society," July 2000; John Broad, "The Verneys as Enclosing Landlords, 1600–1800," in John Chartres and David Hey, eds., *English Rural Society, 1500–1800* (Cambridge: Cambridge University Press, 1990), 28.

7. In addition to items above, see G. E. Mingay, *The Agricultural Revolution; Changes in Agriculture 1650–1880 (Documents in Economic History)* (London: Adam & Charles Black, 1977); B. A. Holderness, *Pre-Industrial England, Economy and Society 1500–1700* (London: J.M. Dent & Sons, 1976), chapter 3; J. D. Chambers and G. E. Mingay, *The Agricultural Revolution 1750–1880* (New York: Schocken Books, 1966). Details on drainage of the Fens can be found in Richard L. Hills, *Power from Wind* (Cambridge: Cambridge University Press, 1994), 133–145.

8. See Brace; Mingay (1977).

9. Wilson, *England's Apprenticeship 1603–1763*, 42–46.

10. Holderness, *Pre-Industrial England*, 141–142; T. K. Derry and Trevor L. Williams, *A Short History of Technology* (New York: Oxford University Press, 1961), 436–438.

11. Robert Payne, *The Canal Builders* (New York: Macmillan, 1959), 95–104.

12. Roland Turner and Steven Goulden, eds., *Great Engineers and Pioneers in Technology* (New York: St. Martin's Press, 1981), 1:219–220; W. H. G. Armytage, *The Rise of the Technocrats: A Social History* (London: Routledge and Kegan Paul, 1965), 374–375, n. 11; Hans Straub, *A History of Civil Engineering* (London: Leonard Hill, 1952), 118–120.

13. Straub, 120–126; R. J. Forbes, "Roads to c. 1900," in Charles Singer et al., eds., *A History of Technology* (Oxford: Clarendon Press, 1957), 4:520–547; S. B. Hamilton, "Building and Civil Engineering Construction," in Singer et al., 4:442–443; Frederick B. Artz, *The Development of Technical Education in France, 1500–1850* (Cambridge, Mass.: MIT Press, 1966), 81–83.

14. Philip S. Bagwell, *The Transport Revolution from 1770* (New York: Barnes & Noble Books, 1974), 35–38; H. J. Dyos and D. H. Aldcroft, *British Transport: An Economic Survey from the Seventeenth Century to the Twentieth* (Leicester: Leicester University Press, 1969), 29–35; Eric Pauson, *Transport and Economy: The Turnpike Roads of Eighteenth Century Britain* (New York: Academic Press, 1977), 65–86.

15. The emergence of the hero engineers was marked by the appearance of a small but popular literature touting their virtues and contributions. The most famous and useful of this was the work of Samuel Smiles, as represented in his *Lives of the Engineers* (1861); see Samuel Smiles, *Selections from Lives of the Engineers* (Cambridge, Mass.: MIT Press, 1966).

16. Karl-Gustaf Hildebrand, *Swedish Iron in the Seventeenth and Eighteenth Centuries* (Stockholm: Jernkontoret, 1992).

17. "Christopher Polhem" in Lance Day and Ian McNeil, eds., *Biographical Dictionary of the History of Technology* (New York: Routledge, 1996), 564–565; Svante Lindqvist, *Technology on Trial: The Introduc-*

tion of Steam Power Technology into Sweden, 1715–1736 (Uppsala: Almqvist & Wiksell, 1984), 23–33; Samuel E. Brig, "A Contribution to the Biography of Christopher Polhem," in *Christopher Polhem, the Father of Swedish Technology*, William A. Johnson, trans. (Hartford: Trinity College, 1963), 15–40.

18. Hildebrand, 25–43; Charles K. Hyde, *Technological Change and the British Iron Industry, 1700–1870* (Princeton, N.J.: Princeton University Press, 1977), 7–22; Arthur Raistrick *Dynasty of Ironfounders; the Darbys and Coalbrookdale* (New York: Longmans, Green, and Co., 1953), 17–29.

19. Raistrick, 30–46.

20. Hyde, 53–76; Hildebrand, 39–42. Swedish imports continued to be roughly between 15,000 and 20,000 tons annually in this period; Russian imports grew from about 5,000 tons to between 20,000 and 25,000, and the total domestic production grew from 30,000 to 90,000 tons.

21. John U. Nef, *The Rise of the British Coal Industry* (London: George Routledge & Sons, 1932), 1:19–22, 1:190–223.

Chapter 11

1. Thomas Savery, *The Miners Friend* (London: S. Crouch, 1702; reprint 1979); for the context, see A. F. C. Wallace, *The Social Context of Innovation* (Princeton, N.J.: Princeton University Press, 1982), chapter 1.

2. H. W. Dickinson, *A Short History of the Steam Engine* (New York: Reprints of Economic Classics, 1966), 18–22; "An Account of Mr. Savery's Engine," *Philosophical Transactions of the Royal Society* (June 1699): 228.

3. John Harris, *Lexicon Technicum* (London: for Dan. Brown et al., 1704), v. 1, not paginated.

4. L. T. C. Rolt and J. S. Allen, *The Steam Engines of Thomas Newcomen* (New York: Science History, 1977); Svante Lindqvist, *Technology on Trial: The Introduction of Steam Power Technology into Sweden, 1715–1736* (Stockholm: Almqvist & Wiksell, 1984); Richard L. Hills, *Power from Steam: A History of the Stationary Steam Engine* (Cambridge: Cambridge University Press, 1989); John Kanefsky and John Robey, "Steam Engines in 18th-Century Britain: A Quantitative Assessment," *Technology & Culture* 21, no. 2 (April 1980): 161–186.

5. Rolt and Allen, 107–116; Hills, 31–37; John Harris, "Recent Research on the Newcomen Engine and Historical Studies," *Transactions of the Newcomen Society* 50 (1978–79): 175–180.

6. On popular science in Britain, see Larry Stewart, *The Rise of Public Science* (Cambridge: Cambridge University Press, 1992) and Margaret Jacob, *Scientific Culture and the Making of the Industrial West* (New York: Oxford University Press, 1997). The problem of longitude is discussed thoroughly in Dava Sobel, *Longitude* (New York: Walker, 1995); the edition used here is Dava Sobel and William J. H. Andrewes, *The Illustrated Longitude* (New York: Walker, 1998).

7. Sobel and Andrewes report (143) that Harrison calculated the loss of 5.1 secs. for a duration of 81 days, not 62 (the period from Nov. 18, when he set out from Portsmouth, to Jan. 19, when his ship arrived in Jamaica). The eighty-one-day figure appears in Humphrey Quill, *John Harrison: The Man Who Found Longitude* (London: John Baker, 1966), 105, citing a 1763 pamphlet of Harrison, but it appears to be a miscount.

8. "The organ is, together with the clock, the most complex of all mechanical instruments developed before the Industrial Revolution," *Grove's New Dictionary of Music and Musicians* (online edition, 2000). I owe the connection to organ making to Lindqvist (1984), 310–312.

9. H. W. Dickinson, *James Watt: Craftsman and Engineer* (Cambridge: for Babcock & Wilcox, 1935), chapter 2.

10. Besides Dickinson, this account follows R. J. Law, *James Watt and the Separate Condenser* (London: HMSO, 1969); Hills, chapter 4; Dickinson, chapter 5; Eric Robinson and A. E. Musson, *James Watt and the Steam Revolution* (London: Adams and Dart, 1969).

11. Watt to Smeaton quoted in Dickinson, 74.

12. The figures are from Hills, chapter 5.

13. The quotation is from Hills, 69.

14. Figures for steam engine numbers are from Kanefsky and Robey.

Chapter 12

1. The quotation is from *Thursday's Journal* (6 August 1719), as quoted in Beverly Lemire, *Fashion's Favourite: the Cotton Trade and the Consumer in Britain, 1660–1800* (Oxford: Pasold Research Fund/ Oxford University Press, 1991), 37; see Lemire for the scene in Spitalfield Market.

2. Lemire, 34–42; Natalie Rothstein, "The Calico Campaign of 1719–1721," *East London Papers* 7, no. 1 (July 1964): 3–21.

3. Lemire, 12–29; on the promotion of silk fashions, including annual changes, see Carlo Poni, "Fashion as Flexible Production: the Strategies of the Lyon Silk Merchants in the Eighteenth Century," in Charles Sabel and Jonathan Zeitlin, eds., *World of Possibilities; Flexibility and Mass Production in Western Industrialization* (Cambridge: Cambridge University Press, 1997), 37–74.

4. Henry Martin, *The Advantages of the East India Trade to England, Considered* (1697–98), in John Smith, ed., *Chronicum rusticum-commerciale* (1747), as quoted in Lemire, 30; John E. Wills Jr., "European Consumption and Asian Production in the Seventeenth and Eighteenth Centuries," in John Brewer and Roy Porter, eds., *Consumption and the World of Goods* (London: Routledge, 1993), 133–147.

5. The statistics are reported in Lemire, 54.

6. The debates over the extent and meaning of the Industrial Revolution are found in a vast literature. Some recent summaries include Patrick K. O'Brien, "Modern Conceptions of the Industrial Revolution," in Patrick K. O'Brien and Roland Quinalt, eds., *The Industrial Revolution and British Society* (Cambridge: Cambridge University Press, 1993), 1–30; Joel Mokyr, "The New Economic History and the Industrial Revolution," in Mokyr, ed., *The British Industrial Revolution: An Economic Perspective*, (2nd ed.) (Boulder, Colo.: Westview Press, 1999), 1–127; and David Landes, "The Fable of the Dead Horse; or, The Industrial Revolution Revisited," in Mokyr, 128–159.

7. The varieties of cloth is one of the key lessons in Eric Kerridge, *Textiles Manufactures in Early Modern England* (Manchester: Manchester University Press, 1985). See also the articles in J. Geraint Jenkins, ed., *The Wool Textile Industry in Great Britain* (London: Routledge & Kegan Paul, 1972). On the New Draperies, see D. C. Coleman, "An Innovation and Its Diffusion: The 'New Draperies,'" *Economic History Review*, 2nd ser., 22, no. 3 (December 1969): 417–429, reprinted in S. D. Chapman, ed., *The Textile Industries* (London: I. B. Tauris, 1997), 2:613–626.

8. W. English, *The Textile Industry* (London: Longmans, 1969), 11–20, 41–44; Kerridge, 133–137; Abbott Payson Usher, *A History of Mechanical Inventions* (New York: Dover, 1954, rev. ed.), 277–281.

9. Kerridge, 172; English, 27–34; Usher, 285–287.

10. English, 18, for the riots against frames.

11. Ibid., 21–26; Kerridge, 169–171; Luca Molà, *The Silk Industry of Renaissance Venice* (Baltimore: Johns Hopkins University Press, 2000), 190–197; S. Ciriacono, "Silk Manufacturing in France and Italy in the XVIIth Century: Two Models Compared," in Chapman, *The Textile Industries*, 3:1–29.

12. Kerridge, 169–170; C. Aspin and S. D. Chapman, *James Hargreaves and the Spinning Jenny* (Helmshore, Lanc.: Helmshore Local History Society, 1964), 15; the patent is that issued to Richard Dereham and Richard Haines.

13. Aspin and Chapman, 9–28; Richard L. Hills, *Power in the Industrial Revolution* (Manchester: Manchester University Press, 1970), 55–59; English, 45–51.

14. Richard Hills, "Textiles and Clothing," in Ian McNeil, ed., *An Encyclopaedia of the History of Technology* (London: Routledge, 1990), 828–829.

15. English, 55–64.

16. Hills, "Textiles and Clothing," 830; Karen Fisk, "Richard Arkwright: Cotton King or Spin Doctor?" *History Today* (March 1, 1998); Hills, *Power*, 73–88.

17. Hills, 116–120; Hector C. Cameron, *Samuel Crompton* (London: Batchworth Press, 1951), 57–66.

18. Quotation from P. Gaskell, *The Manufacturing Population of England* (1833), quoted in Hills, 124; Cameron, 64–70 and Hills, 116–122.

19. Figures are from Hills, 129–130.

20. Hills, 208–220; Edward Baines, *History of the Cotton Manufacture in Great Britain* (London: Frank Cass & Co., 1966, from the 1835 edition), 229–231; English, 89–94.

21. E. P. Thompson, *The Making of the English Working Class* (Harmondsworth: Penguin Books, 1968), 297–346; Kirkpatrick Sale, *Rebels Against the Future: The Luddites and Their War on the Industrial Revolution* (Boston: Addison-Wesley, 1995); Adrian Randall, *Before the Luddites; Custom, Community and Machinery in the English Woolen Industry, 1776–1809* (Cambridge: Cambridge University Press, 1991).

22. The quotation is from David S. Landes, *The Unbound Prometheus; Technological Change and Industrial Development in Western Europe from 1750 to the Present* (Cambridge: Cambridge University Press, 1969), 1.

Chapter 13

1. Denis Diderot, entry for "Art," vol. 1 of *Encyclopédie: ou Dictionnaire raisonné des sciences, des arts et des métiers* (Paris; Briasson, 1751) in John Hope Mason, *The Irresistible Diderot* (London: Quartet Books, 1982), 91.

2. On Diderot generally, see P. N. Furbank, *Diderot: A Critical Biography* (New York: Knopf, 1992), esp. chapter 4, and Arthur M. Wilson, *Diderot* (New York: Oxford University Press, 1972).

3. On the origins of the *Encyclopédie*, see Furbank; see also Wilson, chapter 6.

4. Denis Diderot, entry for "Art," vol. 1 of *Encyclopédie*, as trans. in *Encyclopedia, Selections*, trans. Nelly S. Hoyt and Thomas Cassirer (Indianapolis: Bobbs-Merrill, 1965), 14–15.

5. John Harris, "Law, Espionage, and the Transfer of Technology from Eighteenth Century Britain," in Robert Fox, ed., *Technological Change: Methods and Themes in the History of Technology* (Amsterdam: Harwood, 1996), 134–135. *Rameau's Nephew and Other Works*, trans. Jacques Barzun and Ralph H. Bowen (Indianapolis: Bobbs-Merrill, 1964), 304.

6. Denis Diderot, entry for "Encyclopedia," vol. 5 of *Encyclopédie* (1755), as quoted in *Rameau's Nephew and Other Works*, 283 and 293.

7. Ibid., 304–305; on the culture of control, see Meriam R. Levin, ed., *Cultures of Control* (Amsterdam: Harwood, 2000), esp. articles by Levin, Rosalind Williams, and Daryl M. Hafter.

8. On the technical illustrations of the *Encyclopédie*, see Denis Diderot, *A Diderot Pictorial Encyclopedia of Trades and Industry*, ed. Charles C. Gillispie (New York: Dover, 1959), esp. the editor's introduction, xi–xxvi; the pinmaking plates are reproduced on vol. 1, plates 184–186.

9. Robert Darnton, *The Business of Enlightenment: A Publishing History of the Encyclopédie, 1775–1800* (Cambridge, Mass.: Harvard University Press, 1979), 33–37.

10. The primary source for the remarks on porcelain here is W. D. Kingery, "The Development of European Porcelain," in W. D. Kingery, ed., *High-Technology Ceramics: Past, Present, and Future* (Westerville, Ohio: American Ceramic Society, 1986), 153–180.

11. The primary source for the Staffordshire potteries and the work of the Wedgwoods is Archibald Clow and Nan Clow, *The Chemical Revolution: A Contribution to Social Technology* (London: Batchworth Press, 1952), 293–319.

12. In addition to Clow, see *Josiah Wedgwood. "The Arts and Sciences United,"* (Barlaston, England: Jos. Wedgwood & Sons Ltd., 1978) (the quotations are from pages 16 and 12, respectively).

13. The marketing techniques of Wedgwood are the subject of Neil McKendrick, "Josiah Wedgwood and the Commercialization of the Potteries," in Neil McKendrick et al., *The Birth of a Consumer Society* (Bloomington: Indiana University Press, 1985), 99–145.

14. Cissie Fairchilds, "The Production of Marketing of Populuxe Goods in Eighteenth-Century Paris," in John Brewer and Roy Porter, eds., *Consumption and the World of Goods* (New York: Routledge, 1993), esp. table 11.1, p. 230. The earlier British shift is also discussed in this volume, particularly in Carole Shammas, "Changes in English and Anglo-American Consumption from 1550 to 1800," 177–205, and Lorna Weatherill, "The Meaning of Consumer Behavior in Late Seventeenth- and Early Eighteenth-Century England," 206–227.

15. Sources on Birmingham include Eric Hopkins, *The Rise of the Manufacturing Town: Birmingham and the Industrial Revolution* (Stroud, England: Sutton, 1998), esp. chapters 1 and 3; Maxine Berg, *The Age of Manufactures: Industry, Innovation, and Work in Britain 1700–1820* (Totowa, N.J.: Barnes & Noble Books, 1985), esp. chapter 12; and John Styles, "Manufacturing, Consumption, and Design in Eighteenth-Century Britain," in Brewer and Porter, 527–554.

16. Quoted in Berg, 287; the verse is from James Bisset's *Poetic Survey Around Birmingham* (1800), as quoted in Hopkins, 33.

17. W. K. V. Gale, *The British Iron & Steel Industry: A Technical History* (Newton Abbot, England: David & Charles, 1967), 35–36.

18. Gale, 43–47; Charles K. Hyde, *Technological Change and the British Iron Industry, 1700–1870* (Princeton, N.J.: Princeton University Press, 1977), 76–92.

19. Quotation from Young: Barrie Trinder, ed., *'The Most Extraordinary District in the World' Ironbridge and Coalbrookdale* (Chichester, England: Phillimore & Co., 1988), 36; Arthur Raistrick, *Dynasty of Iron Founders: the Darbys and Coalbrookdale* (London: Longmans, Green and Co., 1953), 193–200.

Chapter 14

1. B. Franklin to Sir Joseph Banks, July 27, 1783, in *The Writings of Benjamin Franklin*, ed. Albert Henry Smith (New York: Haskell House, 1970), 9:73–75.

2. Charles C. Gillispie, *The Montgolfier Brothers and the Invention of Aviation, 1783–1784* (Princeton, N.J.: Princeton University Press, 1983), 3–7; Christa Jungnickel and Russell McCormmach, *Cavendish: The Experimental Life* (Cranbury, N.J.: Bucknell University Press, 1999), 206–207, 360–361; Tiberius Cavallo, *The History and Practice of Aerostation* (London: for the author, 1785), 30–42; Archibald Clow and Nan Clow, *The Chemical Revolution: A Contribution to Social Technology* (London: Batchworth Press, 1952), 151–155.

3. L. T. C. Rolt, *The Aeronauts: A History of Ballooning, 1783–1903* (London: Longmans, 1966), 31–35; Gillispie, 27–33.

4. B. Franklin to Sir Joseph Banks, Aug. 30, 1783, in *The Writings of Benjamin Franklin*, 9:79–85; Sir Joseph Banks to B. Franklin, Sept. 13, 1783, in *Writings*, 9:85.

5. B. Franklin to Sir Joseph Banks, Nov. 21, 1783, in *The Writings of Benjamin Franklin*, 9:113–118; Rolt, 44–48; Gillispie, 44–56.

6. I. Bernard Cohen, *Benjamin Franklin's Science* (Cambridge, Mass.: Harvard University Press, 1990), 198.

7. The best account of the Montgolfiers' thinking is in Gillispie, esp. 15–16.

8. B. Franklin to Sir Joseph Banks, Dec. 1, 1783, in *The Writings of Benjamin Franklin*, 9:119–122; B. Franklin to Henry Laurens, Dec. 6, 1783, in *Writings*, 9:122–123; Rolt, 50–54, 86–88; Gillispie, 56–64; Carl Van Doren, *Benjamin Franklin* (New York: Viking Press, 1938), 700–702.

9. Joseph Priestley, *The History and Present State of Electricity* (London: various pub., 1767), iv.

10. Bern Dibner, *Early Electrical Machines* (Norwalk, Conn.: Burndy Library, 1957), 7–28; J. L. Heilbron, *Electricity in the 17th and 18th Centuries* (Berkeley: University of California Press, 1979), 169–179, 229–260, 309–318; Herbert W. Mayer, *A History of Electricity and Magnetism* (Norwalk, Conn.: Burndy Library, 1972), 11–19; see also William Hackmann, *Electricity from Glass: The History of the Frictional Electrical Machine, 1600–1850* (Alphen aan den Rijn, Netherlands: Sijthoff & Noordhoff, 1978); Michael Brian Schiffer, *Draw the Lightning Down: Benjamin Franklin and Electrical Technology in the Age of the Enlightenment* (Berkeley: University of California Press, 2003).

11. Van Doren, 156–164; Heilbron, 324–352.

12. Mayer, 34–41; Heilbron, 491–494.

13. Mayer, 41–42.

14. Ibid., 42–44; Jungnickel and McCormmach, 479–485.

15. J. R. Partington, *A History of Chemistry* (London: Macmillan & Co., 1962), 3:496–509; Maurice Daumas, ed., *A History of Invention & Technology*, vol. 3, *The Expansion of Mechanization, 1725–1860* (New York: Crown, 1979), 556–557; Clow and Clow, 172–194.

16. Clow and Clow, 130–150; Daumas, 558–562.

17. W. A. Campbell, "The Alkali Industry," in Colin A. Russell, ed., *Chemistry, Society and Environment: A New History of the British Chemical Industry* (Cambridge: Royal Society of Chemistry, 2000), 75–82; Robert P. Multhauf, *Neptune's Gift: A History of Common Salt* (Baltimore: Johns Hopkins University Press, 1978), 139–143; Clow and Clow, 91–129.

18. Campbell, 82–98.

19. Cooper and Jefferson quoted in Robert Friedel, "Defining Chemistry: Origins of the Heroic Chemist," in Seymour H. Mauskopf, ed., *Chemical Sciences in the Modern World* (Philadelphia: University of Pennsylvania Press, 1993), 220–221.

Chapter 15

1. Oliver Evans, "On the Origin of Steam Boats and Steam Waggons," *Niles Register* (1812), as quoted in Caroline E. MacGill et al., *History of Transportation in the United States Before 1860* (Washington, D.C.: Carnegie Institution of Washington, 1917), 308.

2. Eugene S. Ferguson, *Oliver Evans: Inventive Genius of the American Industrial Revolution* (Greenville, Del.: Hagley Museum, 1980), 11–32; Greville Bathe and Dorothy Bathe, *Oliver Evans: A Chronicle of Early American Engineering* (New York: Arno Press, 1972), 1–48.

3. Ferguson, 35–39.

4. Ibid., 39–41; Bathe and Bathe, 108–112; James Thomas Flexner, *Steamboats Come True* (Boston: Little, Brown, 1978), 3–4, 267–268.

5. Flexner, 169–170, 277–279.

6. Ibid., 262–269; Cynthia Owen Philip, *Robert Fulton, a Biography* (New York: Franklin Watts, 1985), 3–27.

7. Robert Fulton, *A Treatise on the Improvement of Canal Navigation*... (London: I. & J. Taylor, 1796), x.

8. Philip, 144–148; Flexner, 286–293.

9. Philip, 198–205; Flexner, 322–327.

10. Flexner, 341–345; Philip, 270–275. See also Charles W. Dahlinger, "The *New Orleans*...," *Pittsburg Legal Journal* 59, no. 42 (1911): 570–591.

11. Flexner, 348–354.

12. Thomas Tredgold, *A Practical Treatise on Rail-Roads and Carriages* (London: Josiah Taylor, 1825), 3–15; Philip S. Bagwell, *The Transport Revolution from 1770* (New York: Barnes and Noble, 1974), 88–90.

13. Bagwell, 38–49.

14. L. T. C. Rolt, *The Cornish Giant* (London: Lutterworth Press, 1960).

15. Harold Perkin, *The Age of the Railway* (Newton Abbot, England: David & Charles, 1971), 71–73; Bagwell, 91.

16. Perkin, 78–88; Bagwell, 91–92; Jack Simmons and Gordon Biddle, eds., *The Oxford Companion to British Railway History from 1603 to the 1990s* (New York: Oxford University Press, 1997), articles, "Liverpool and Manchester Railway" and "Rainhill trials."

17. John F. Stover, *History of the Baltimore and Ohio Railroad* (W. Lafayette, Ind.: Purdue University Press, 1987), 10–20; MacGill, 397–401.

18. Maurice Daumas, *A History of Technology & Invention* (New York: Crown, 1979), 367–368; David P. Billington, *The Innovators: The Engineering Pioneers Who Made America Modern* (New York: John Wiley & Sons, 1996), 108–110.

19. Daumas, 346–349, 360–367.

20. Charles Hadfield, *Atmospheric Railways: A Victorian Venture in Silent Speed* (New York: Augustus M. Kelley, 1968), chapters 1–5.

Chapter 16

1. *The Diary of Joseph Farington*, ed. Kathryn Cave (New Haven, Conn.: Yale University Press, 1983), 12:4209 (5 Sept. 1812); this is discussed in Paul Johnson, *The Birth of the Modern: World Society, 1815–1830* (New York: HarperCollins, 1991), 165. Farington made an earlier reference to the telegraph in his diaries, reporting in September 1798 that the news of Admiral Nelson's victory in the Battle of the Nile had been relayed from Dover to London by telegraph. *Diary*, 3:1059 (20 Sept. 1798).

2. Gerald Holzmann and Björn Pehrson, *The Early History of Data Networks* (Los Alamitos, Calif.: IEEE Computer Society Press, 1995), 47–59.

3. Ibid., 59–64.

4. Ibid., 71–126.

5. Quoted in C. Mackenzie Jarvis, "The Origin and Development of the Electric Telegraph," *Journal of the I.E.E.* (March 1956), 132–135, reprinted in George Shiers, ed., *The Electric Telegraph: An Historical Anthology* (New York: Arno Press, 1977).

6. Jarvis, 130–138.

7. On Oersted: P. M. Harmon, *Energy, Force, and Matter: The Conceptual Development of Nineteenth-Century Physics* (New York: Cambridge University Press, 1982), 30–33.

8. W. James King, "The Development of Electrical Technology in the 19th Century: The Telegraph," *Contributions from the Museum of History and Technology*, Bull. 228, paper 29 (1962), 277–286; as reprinted in Shiers, *The Electric Telegraph*.

9. King, 286–290.

10. Ibid., 294–298.

11. Ibid., 295–298; Lewis Coe, *The Telegraph: A History of Morse's Invention and Its Predecessors in the United States* (Jefferson, N.C.: McFarland & Co., 1993), 28–31; Thomas Gray, "The Inventors of the Telegraph and Telephone," *Smithsonian Annual Reports, 1892* (Washington, D.C.: Superintendent of Documents, 1893), 649–650, as reprinted in Shiers, *The Electric Telegraph*; personal communication from Paul Israel, July 2001.

12. Tom Standage, *The Victorian Internet* (New York: Basic Books, 1999), 48–58.

13. Paul Israel, *From Machine Shop to Industrial Laboratory: Telegraphy and the Changing Context of American Invention, 1830–1920* (Baltimore: Johns Hopkins University Press, 1992), chapter 2.

14. See Israel, *From Machine Shop to Industrial Laboratory*, for a general discussion of this.

15. The best available biography of Edison is Paul Israel, *Edison: A Life of Invention* (New York: John Wiley & Sons, 1998).

16. The story of Edison's electric light work is told in Robert Friedel and Paul Israel, *Edison's Electric Light: Biography of an Invention* (New Brunswick, N.J.: Rutgers University Press, 1986); additional materials can be found in Israel, *Edison*, chapters 10–12.

17. The best source on Bell is Robert V. Bruce, *Bell: Alexander Graham Bell and the Conquest of Solitude* (Boston: Little, Brown, 1973), esp. chapters 11–19.

18. Brian Winston, *Media Technology and Society: A History, from the Telegraph to the Internet* (New York: Routledge, 1998), chapter 13; see also Claude S. Fischer, *America Calling: A Social History of the Telephone to 1940* (Berkeley: University of California Press, 1992).

Chapter 17

1. On the 1851 Exhibition, see John R. Davis, *The Great Exhibition* (Stroud, England: Sutton, 1999) and Yvonne Ffrench, *The Great Exhibition* (London: Harvill Press, 1950).

2. David A. Hounshell, *From the American System to Mass Production, 1800–1932* (Baltimore: Johns Hopkins University Press, 1984), chapter 1.

3. The story of the Gribeauval reforms and subsequent implications for the history of mass production is told in Ken Alder, *Engineering the Revolution; Arms and Enlightenment in France, 1763–1815* (Princeton, N.J.: Princeton University Press, 1997). See also Merritt Roe Smith, "Army Ordnance and the 'American System of Manufacturing' 1815–1861," in Merritt Roe Smith, ed., *Military Enterprise and Technological Change* (Cambridge, Mass.: MIT Press, 1987), 46–49.

4. Carolyn C. Cooper, "The Portsmouth System of Manufactures," *Technology & Culture* 25 (1984): 182–223; see also Hounshell, chapter 1.

5. Hounshell, chapter 1.

6. Smith, "Army Ordnance," 40–64.

7. Hounshell, 35–39.

8. L. T. C. Rolt, *A Short History of Machine Tools* (Cambridge, Mass.: MIT Press, 1965), 83–86.

9. Ibid., 92–113.

10. Ibid., 113–121; W. H. G. Armytage, *A Social History of Engineering* (Cambridge, Mass.: MIT Press, 1961), 126–131.

11. Derrick Beckett, *Telford's Britain* (Newton Abbot, England: David & Charles, 1987).

12. L. T. C. Rolt, *Victorian Engineering* (Harmondsworth: Penguin Books, 1974), 27–31.

13. Ibid., 88–95; Armytage, 135–136; see also Isambard Brunel, *The Life of Isambard Kingdom Brunel, Civil Engineer* (Rutherford, N.J.: Fairleigh Dickinson University Press, 1971).

14. On Eads, see Henry Petroski, *Engineers of Dreams: Great Bridge Builders and the Spanning of America* (New York: Knopf, 1995), chapter 2; on Roebling, the best source is still David McCullough, *The Great Bridge: The Epic Story of the Building of the Brooklyn Bridge* (New York: Simon & Schuster, 1972).

Chapter 18

1. Howard T. James, "Why Photography Wasn't Invented Earlier," in *Pioneers of Photography* (Springfield, Va.: SPSE, 1987), 12.

2. Robert Hirsch, *Seizing the Light: A History of Photography* (New York: McGraw Hill, 2000), 12.

3. Helmut Gernsheim and Alison Gernsheim, *L. J. M. Daguerre: The History of the Diorama and the Daguerreotype* (New York: Dover, 1968), 48–70.

4. Gernsheim and Gernsheim, 69–84.

5. Beaumont Newhall, *The History of Photography* (New York: Museum of Modern Art, 1982), 18–21.

6. Newhall, 59; Robert Friedel, *Pioneer Plastic: The Making and Selling of Celluloid* (Madison: University of Wisconsin Press, 1983), 4–5.

7. Friedel, *Pioneer Plastic*, 6–12, 41–49.

8. Ibid., 41–66.

9. Anthony S. Travis, *The Rainbow Makers: The Origins of the Synthetic Dyestuffs Industry in Western Europe* (Bethlehem, Pa.: Lehigh University Press, 1993), 31–55.

10. Ibid., 67–84.

11. Ibid.; Fred Aftalion, *A History of the International Chemical Industry* (Philadelphia: University of Pennsylvania Press, 1991), 49.

12. Aftalion, 49–53.

13. Ibid., 72–79.

14. The production data are from Thomas J. Misa, *A Nation of Steel: The Making of Modern America, 1865–1925* (Baltimore: Johns Hopkins University Press, 1995), esp. 16. See also Robert B. Gordon, *American Iron, 1607–1900* (Baltimore: Johns Hopkins University Press, 1996).

15. Misa, 5–11.

16. Misa, 16, 40–41; Carl Condit, "Buildings and Construction, 1880–1900," in M. Kranzberg and C. Pursell, eds., *Technology in Western Civilization* (New York: Oxford University Press, 1967), 1:603–619; Nicholas Faith, *The World the Railroads Made* (New York: Carroll & Graf, 1991), 128–134.

17. Misa, esp. chapters 5 and 6.

18. On aluminum, see Robert Friedel, "'A New Metal!' Aluminum in Its Nineteenth Century Context," in *Aluminum by Design: Jewelry to Jets* (New York: Abrams, 2000); see also Eric Schatzberg, "Symbolic Culture and Technological Change: The Cultural History of Aluminum as an Industrial Material," *Enterprise and Society*, 4 (2003): 226–271.

Chapter 19

1. The story of Bushnell's Turtle and the morning of September 7 is told in many places, although usually with variations. The most reliable account is in Alex Roland, *Underwater Warfare in the Age of Sail* (Bloomington: Indiana University Press, 1977), chapters 5 and 6. The account here also draws on George Pararas-Carayannis, "Turtle: A Revolutionary Submarine," *Sea Frontiers* 22, no. 4 (July–August, 1976): 234, and "David Bushnell's American Turtle," Connecticut River Museum, Essex, Connecticut.

2. For the literature of imaginary warfare, see I. F. Clarke, *Voices Prophesying War 1763–1984* (London: Oxford University Press, 1966); "The Reign of George VI" is discussed on 4–5.

3. In addition to Roland's chapters on Fulton (7, 8, and 9), see also Cynthia Owen Philip, *Robert Fulton: A Biography* (New York: Franklin Watts, 1985), chapters 5–11.

4. The draft letter to Pitt is quoted by Roland, 106–107; the letter to Grenville (2 Sept. 1806) is quoted in Jeremy Black, *Western Warfare, 1775–1882* (Bloomington: Indiana University Press, 2001), 182. Fulton's attitudes are also examined by H. Bruce Franklin, *War Stars: The Superweapon and the American Imagination* (New York: Oxford University Press, 1988), chapter 1.

5. The quotation from Fulton is in Franklin, 13–14, from Edison, in Franklin, 76. See esp. the first three chapters of Franklin for a discussion of the idea of the superweapon in American history.

6. David Gates, *Warfare in the Nineteenth Century* (New York: Palgrave, 2001), 4–5; Michael Howard, *War in European History* (Oxford: Oxford University Press, 1977), 76–84.

7. Brodie and Brodie, 131–139; Gates, 75–77; Geoffrey Wawro, *Warfare and Society in Europe, 1792–1914* (New York: Routledge, 2000), 102, 137–138; on the machine gun, see also John Ellis, *The Social History of the Machine Gun* (Baltimore: Johns Hopkins University Press, 1986).

8. Brodie and Brodie, 139–140; Gates, 5–8.

9. Gates, 81–82; John Allwood, *The Great Exhibitions* (London: Studio Vista, 1977), 45.

10. On the rapidity of obsolescence, see Brian Bond, *War and Society in Europe, 1870–1970* (Montreal: McGill-Queen's University Press, 1998), 43–44; on the reaction to the Japanese victory, see Michael Adas, *Machines as the Measure of Men: Science, Technology, and Ideologies of Western Dominance* (Ithaca, N.Y.: Cornell University Press, 1989), 360–361.

11. Brodie and Brodie, 155; Black, 71–72; Philip, 328–329.

12. Black, 74–77.

13. On the *Monitor*, see David A. Mindell, *War, Technology, and Experience Aboard the USS Monitor* (Baltimore: Johns Hopkins University Press, 2000), esp. chapter 2 on the ship's design and chapter 5 on the Battle of Hampton Roads.

14. Mindell, 113–116; Bond, 43–44; Edward W. Byrn, *The Progress of Invention in the Nineteenth Century* (New York: Russell & Russell, 1970), 399–401; Brodie, 160–162; John B. Hattendorf, "Sea Warfare," in Charles Townshend, ed., *The Oxford Illustrated History of Modern War* (New York: Oxford University Press, 1997), 221. The American satirist Ambrose Bierce commented on the arms race between shells and armor in his 1899 short story, "The Ingenious Patriot," in which an inventor peddles a series of constantly improving shells and armor plate to a worried ruler, only in the end to be sentenced to death for "ingenuity" (*Fantastic Fables*, 1899).

15. Discussion of the British-German naval race and the *Dreadnought* are largely drawn from Wawro, chapter 7, and Lawrence Sondhaus, *Naval Warfare 1815–1914* (New York: Routledge, 2001), chapter 8.

16. Wawro, 181–182; Brodie and Brodie, 188–189; Peter Padfield, *Battleship* (Edinburgh: Birlinn, 2000), chapter 16; Sondhaus, 200–205. On fire control, see Jon T. Sumida, *In Defence of Naval Supremacy* (London: Routledge, 1993).

17. John Terraine, *White Heat: The New Warfare 1914–18* (London: Sidgwick & Jackson, 1982), 33–35.

18. Spencer Tucker, "The First World War," in Jeremy Black, ed., *European Warfare, 1815–2000* (New York: Palgrave, 2001), 87–90; Terraine, 95–97.

19. Terraine, 155–161.

Chapter 20

1. [George S. Emmerson], *Engineering Education: A Social History* (Newton Abbott, England: David & Charles, 1973), 91–93; "John Anderson," in *The Dictionary of National Biography* (New York: Macmillan, 1908), 1:383–384.

2. "George Birkbeck," in *The Dictionary of National Biography*, 2:542–543.

3. Carl Bode, *The American Lyceum: Town Meeting of the Mind* (New York: Oxford University Press, 1956), 3–6.

4. Bode, 7–15; Emmerson, 136; Edward W. Stevens, *The Grammar of the Machine: Technical Literacy and Early Industrial Expansion in the United States* (New Haven, Conn.: Yale University Press, 1995), 105–132.

5. On the Franklin Institute, see Bruce Sinclair, *Philadelphia's Philosopher Mechanics: A History of the Franklin Institute, 1824–1865* (Baltimore: Johns Hopkins University Press, 1974).

6. Emmerson, 29–33, 40–42, 77–85.

7. Charles R. Richards, *The Industrial Museum* (New York: Macmillan, 1925), 7–8; Charles R. Day, *Education for the Industrial World: The Écoles des Arts et Métiers and the Rise of French Industrial Engineering* (Cambridge, Mass.: MIT Press, 1987), 18–21.

8. Ulrich Pfammatter, *The Making of the Modern Architect and Engineer* (Basel: Birkhäuser, 2000), 103–122; Day, 12–16; see also John H. Weiss, *The Making of Technological Man: The Social Origins of French Engineering Education* (Cambridge, Mass.: MIT Press, 1982).

9. Emmerson, 43–49, 85–90, 193–194.

10. The terms "shop culture" and "school culture" were coined by Monte Calvert in *The Mechanical Engineer in America, 1830–1910: Professional Cultures in Conflict* (Baltimore: Johns Hopkins University Press, 1967).

11. Emmerson, 140–141; Daniel H. Calhoun, *The American Civil Engineer: Origins and Conflict* (Cambridge, Mass.: The Technology Press, 1960), 36–40; Lawrence P. Grayson, *The Making of an Engineer* (New York: John Wiley & Sons, 1993), 17–23 (the quotation by Adams is on page 18).

12. Grayson, 27–28 (quotation from Rensselaer is on page 28); Emmerson, 141–146; Stevens, 148–169.

13. Grayson, 28–30.

14. Ibid., 39–44; Emmerson, 146–152; James Gregory McGivern, *First Hundred Years of Engineering Education in the United States (1807–1907)* (Spokane, Wash.: Gonzaga University Press, 1960), 91–93.

15. Lawrence Veysey, *The Emergence of the American University* (Chicago: University of Chicago Press, 1965), 81–86; John S. Brubacher and Willis Rudy, *Higher Education in Transition*, 4th ed. (New Brunswick, N.J.: Transaction, 1997), 158–164.

16. Grayson, 40–45; Emmerson, 169–171; "Robert Henry Thurston," in *Dictionary of American Biography* (New York: Charles Scribner's Sons, 1964), IX:518–520.

17. Calhoun, 182–189; Garth Watson, *The Smeatonians: The Society of Civil Engineers* (London: Thomas Telford, 1989), 43; Emmerson, 253–255, 259–264; Bruce Sinclair, *A Centennial History of the American Society of Mechanical Engineers, 1880–1980* (Toronto: University of Toronto Press, 1980), 22–27; [R. Angus Buchanan], *The Engineers; a History of the Engineering Profession in Britain, 1750–1914* (London: Jessica Kingsley, 1989), 75–81.

18. Terry S. Reynolds, "Defining Professional Boundaries: Chemical Engineering in the Early 20th Century," *Technology and Culture* 27, no. 4 (October 1986): 707–712; see also the introductory essays by Reynolds in Terry S. Reynolds, ed., *The Engineer in America* (Chicago: University of Chicago Press, 1991).

19. Ruth Schwartz Cowan, *A Social History of American Technology* (New York: Oxford University Press, 1997), 111.

Chapter 21

1. J. S. Mill, *Principles of Political Economy with Some of their Applications to Social Philosophy*, 2 vols. (London: 1849, 2nd ed.), 2:24, as quoted in Delores Greenberg, "Energy, Power, and Perceptions of Social Change in the Early Nineteenth Century," *American Historical Review* 93 (June 1990): 711.

2. Sadi Carnot, *Réflexions sur la puissance motrice du feu et sur les machines propres à développer cette puissance*, as quoted in D. S. L. Cardwell, *From Watt to Clausius: The Rise of Thermodynamics in the Early Industrial Age* (Ithaca, N.Y.: Cornell University Press, 1971), 192.

3. Ibid., 196.

4. Ibid., 220–221; Mikael Hård, *Machines Are Frozen Spirit: The Scientification of Refrigeration and Brewing in the 19th Century* (Boulder, Colo.: Westview Press, 1994), 31–33; D. S. L. Cardwell, *Wheels, Clocks, and Rockets: A History of Technology* (New York: Norton, 1995), 239–243.

5. Hård, 33; P. M. Harmon, *Energy, Force, and Matter: The Conceptual Development of Nineteenth-Century Physics* (New York: Cambridge University Press, 1982), 49–58.

6. See, for example, Cardwell, *Wheels*, 313–314.

7. Horst O. Hardenberg, *The Antiquity of the Internal Combustion Engine, 1509–1688* (Warrendale, Pa.: Society of Automotive Engineers, 1993); C. Lyle Cummins, *Internal Fire* (Warrendale, Pa.: Society of Automotive Engineers, 1989), chapters 1–3.

8. Cummins, chapters 5–6; Aubrey F. Burstall, *A History of Mechanical Engineering* (Cambridge, Mass.: MIT Press, 1965), 332–333.

9. Cummins, 106–112; Cardwell, *Wheels*, 340–342.

10. Cummins, 112–146.

11. Cummins, 134–184.

12. W. H. G. Armytage, *A Social History of Engineering* (Cambridge, Mass.: MIT Press, 1961), 191; Hård, esp. chapters 7 and 8; Donald E. Thomas Jr., *Diesel: Technology & Society in Industrial Germany* (Tuscaloosa: University of Alabama Press, 1987), esp. chapters 3 and 4.

13. W. Garrett Scaife, *From Galaxies to Turbines: Science, Technology, and the Parsons Family* (Philadelphia: Institute of Physics, 2000), 151–153; Burstall, 339–340.

14. Norman Smith, *Man and Water: A History of Hydro-Technology* (New York: Charles Scribner's Sons, 1975), chapters 12 and 13; Hunter Rouse and Simon Ince, *A History of Hydraulics* (New York: Dover, 1963), 113–123.

15. Smith, 171–180.

16. Ibid., 180–185; Rouse and Ince, chapters 11 and 12.

17. John D. Anderson Jr., *A History of Aerodynamics and Its Impact on Flying Machines* (Cambridge: Cambridge University Press, 1997), chapters 2 and 3.

18. Anderson, 55–83.

19. Ibid., 115–188.

20. The key sources on the Wright Brothers' invention are ibid., chapter 5; Peter Jakab, *Visions of a Flying Machine* (Washington, D.C.: Smithsonian Institution Press, 1990); Charles H. Gibbs-Smith, *Aviation: A Historical Survey* (London: HMSO, 1970); and Tom Crouch, *The Bishop's Boys* (New York: Norton, 1981).

21. Anderson, 244–296.

22. Olivier Darrigol, *Electrodynamics from Ampère to Einstein* (Oxford: Oxford University Press, 2000), 4–12.

23. Ibid., 16–41; the quotation from Faraday appears on 24.

24. Ibid., 137–176; Harmon, 72–98.

25. Hugh G. J. Aitken, *Syntony and Spark—the Origins of Radio* (New York: John Wiley & Sons, 1976), 48–79.

26. Ibid., 80–115.

27. Ibid., 179–188.

28. Ibid., 188–229.

29. The further development of radio is discussed in W. Rupert Maclaurin, *Invention and Innovation in the Radio Industry* (New York: Macmillan, 1949; reprinted Arno Press, 1971) and Hugh G. J. Aitken, *The Continuous Wave: Technology and American Radio, 1900–1932* (Princeton, N.J.: Princeton University Press, 1985).

30. This argument in a much more limited form is found in Hård, esp. 14–18 and 230–231, 239–241; see also James R. Beniger, *The Control Revolution: Technology and Economic Orgins of the Information Society* (Cambridge, Mass.: Harvard University Press, 1986), 166–168.

31. Lester F. Ward, *Applied Sociology: A Treatise on the Conscious Improvement of Society by Society* (Boston: Ginn & Co., 1906); Beniger, 294–298; the Brandeis quotation is from Frank B. Gilbreth, *Primer on Scientific Management* (Easton, Pa.: Hive, 1985), 3; Christopher Lasch, *The True and Only Heaven: Progress and Its Critics* (New York: Norton, 1991), 139–147.

Chapter 22

1. *New England Farmer*, 1 (January 4, 1823), 180, quoted in Peter D. McClelland, *Sowing Modernity: America's First Agricultural Revolution* (Ithaca, N.Y.: Cornell University Press, 1997), 217–218.

2. Washington (Dec. 7, 1796) quoted in A. C. True, *A History of Agricultural Experimentation and Research in the United States, 1607–1925* (Washington, D.C.: U.S. Government Printing Office, 1937; reprinted NY: Arno Press, 1980), 19.

3. R. Douglas Hurt, *American Agriculture: A Brief History* (Ames: Iowa State University Press, 1994), 43–47.

4. McClelland, 52–63.

5. Ibid., 64–128.

6. Ibid., 129–164; John T. Schlebecker, *Whereby We Thrive: A History of American Farming, 1607–1972* (Ames: Iowa State University Press, 1975), 114–115; Hurt, 142–144.

7. Hurt, 144–146; Schlebecker, 116–118; Willard W. Cochrane, *The Development of American Agriculture: A Historical Analysis*, 2nd ed. (Minneapolis: University of Minnesota Press, 1993), 196–199.

8. Margaret W. Rossiter, *The Emergence of Agricultural Science: Justus Liebig and the Americans* (New Haven, Conn.: Yale University Press, 1975); Richard A. Wines, *Fertilizer in America: From Waste Recycling to Resource Exploitation* (Philadelphia: Temple University Press, 1985), 6–21.

9. Wines, passim.

10. Rossiter, chapters 2 and 3.

11. Margaret W. Rossiter, "The Organization of Agricultural Improvement in the United States, 1785–1865," in Alexandra Oleson and Sanborn C. Brown, eds., *The Pursuit of Knowledge in the Early American Republic* (Baltimore: Johns Hopkins University Press, 1976), 290–293.

12. True, 22–46; A. C. True, *A History of Agricultural Education in the United States, 1785–1925* (Washington, D.C.: U.S. Government Printing Office, 1929; reprinted Arno Press, 1980), 95–111.

13. True (*Experimentation*), 118–207; True (*Education*), 119–129.

14. Ivy Pinchbeck, *Women Workers and the Industrial Revolution, 1750–1850* (London: Frank Cass, 1930), 10–13.

15. Robert B. Shoemaker, *Gender in English Society, 1650–1850: The Emergence of Separate Spheres?* (New York: Longman, 1998), 156.

16. C. C. Gillispie, "The Natural History of Industry," in A. E. Musson, ed., *Science, Technology, and Economic Growth in the Eighteenth Century* (London: Methuen, 1972); John V. Pickstone, "Bodies, Fields, and Factories: Technologies and Understandings in the Age of Revolutions," in Robert Fox, ed., *Technological Change: Methods and Themes in the History of Technology* (Amsterdam: Harwood, 1996), 51–61; Bruno Battistotti et al., *Cheese: A Guide to the World of Cheese and Cheesemaking* (New York: Facts on File, 1984), 17.

17. Sally McMurray, *Transforming Rural Life: Dairying Families and Agricultural Change, 1820–1885* (Baltimore: Johns Hopkins University Press, 1995), 168.

18. Raymond W. Beck, *A Chronology of Microbiology in Historical Context* (Washington, D.C.: ASM Press, 2000), 65–67.

19. Hubert A. Lechevalier and Morris Solotorovsky, *Three Centuries of Microbiology* (New York: McGraw-Hill, 1965), 15–29; Howard Gest, *The World of Microbes* (Madison, Wis.: Science Tech Publications, 1987): 14–18; Robert Reid, *Microbes and Men* (n.p.: Saturday Review Press, 1975), 41–46; Gerald L. Geison, *The Private Science of Louis Pasteur* (Princeton, N.J.: Princeton University Press, 1995), 90–109.

20. Quotation in René Dubos, *Pasteur and Modern Science* (Garden City, N.Y.: Doubleday, 1960), 69; Lechevalier and Solotorovsky, 41–119.

21. Stanley Joel Reiser, *Medicine and the Reign of Technology* (Cambridge: Cambridge University Press, 1978), chapters 2, 3, 4, and 6.

22. See, for example, Ivan Illich, *Medical Nemesis: The Expropriation of Health* (New York: Pantheon Books, 1976).

23. Reiser, chapter 5.

24. Joel D. Howell, *Technology in the Hospital: Transforming Patient Care in the Early Twentieth Century* (Baltimore: Johns Hopkins University Press, 1995).

Chapter 23

1. Henri Loyette, *Gustave Eiffel* (New York: Rizzoli, 1985), 111–114; David P. Billington, *The Tower and the Bridge* (New York: Basic Books, 91–94).

2. Quotation in Billington, 65; see ibid., chapters 4 and 6; Loyette, 114–150, Neil Parkyn, ed., *The Seventy Wonders of the Modern World* (New York: Thames & Hudson, 2002), 171–178.

3. Billington, chapter 8.

4. Tom F. Peters, *Building the Nineteenth Century* (Cambridge, Mass.: MIT Press, 1996), chapter 7; David McCullough, *The Path Between the Seas: The Creation of the Panama Canal 1870–1914* (New York: Simon & Schuster, 1977).

5. On the Hoover (Boulder) Dam: James Tobin, *Great Projects* (New York: Free Press, 2001), chapter 2; Henry Petroski, *Remaking the World: Adventures in Engineering* (New York: Knopf, 1997), 184–193.

6. Thomas P. Hughes, *Networks of Power: Electrification in Western Society, 1880–1930* (Baltimore: Johns Hopkins University Press, 1983), 131–139.

7. Robert Belfield, "The Niagara System: The Evolution of an Electric Power Complex at Niagara Falls, 1883–1896," *Proceedings of the IEEE* 64, no. 9 (1976): 1344–1350.

8. Hughes, chapter 8.

9. Anne D. Rassweiler, *The Generation of Power: The History of Dneprostroi* (New York: Oxford University Press, 1988), esp. chapter 1; Thomas P. Hughes, *American Genesis* (New York: Viking, 1989), 250–268; Jonathan Coopersmith, *The Electrification of Russia, 1880–1926* (Ithaca, N.Y.: Cornell University Press, 1992); Harold Dorn, "Hugh Lincoln Cooper and the First Détente," *Technology & Culture* 20, no. 2 (1979): 322–347; Paul R. Josephson, "'Projects of the Century' in Soviet History: Large-Scale Technologies from Lenin to Gorbachev," *Technology & Culture* 36, no. 3 (1995): 519–559.

10. Paul C. Pitzer, *Grand Coulee: Harnessing a Dream* (Pullman: Washington State University Press, 1994); Erwin C. Hargrove, *Prisoners of Myth: The Leadership of the Tennessee Valley Authority, 1933–1990* (Princeton, N.J.: Princeton University Press, 1994); Tom Little, *High Dam at Aswan: The Subjugation of the Nile* (New York: John Day, 1965); James Moxon, *Volta: Man's Greatest Lake* (London: Andre Deutsch, 1984); Shiu-Hung Luk and Joseph Whitney, eds., *Megaproject: A Case Study of China's Three Gorges Project* (Armonk, N.Y.: M. E. Sharpe, 1993).

11. Josephson, 530–538; Stephen Kotkin, *Magnetic Mountain: Stalinism as a Civilization* (Berkeley: University of California Press, 1995), 42–71; Hughes, *American Genesis*, 278–284.

12. Alfred D. Chandler Jr., *The Visible Hand: The Managerial Revolution in American Business* (Cambridge, Mass.: Harvard University Press, 1977), 259–269; Thomas J. Misa, *A Nation of Steel: The Making of Modern America, 1865–1925* (Baltimore: Johns Hopkins University Press, 1995), 21–28.

13. Chandler, 253–256; Harold F. Williamson et al., *The American Petroleum Industry: The Age of Energy, 1899–1959* (Evanston, Ill.: Northwestern University Press, 1963), 110–132.

14. Williamson et al., 132–150; 603–635.

15. Hughes, *American Genesis*, 269–294.

16. David A. Hounshell, *From the American System to Mass Production, 1800–1932* (Baltimore: Johns Hopkins University Press, 1984), 189–208.

17. T. P. Newcomb and R. T. Spurr, *A Technical History of the Motor Car* (New York: Adam Hilger, 1989), 15–28; James J. Flink, *The Automobile Age* (Cambridge, Mass.: MIT Press, 1988), 10–18.

18. Flink, 22–40; the quotation, taken from J. A. Kingman, "Automobile Making in America," *American Monthly Review of Reviews* (Sept. 1901): 302, is on 40.

19. Flink, 25–35; Newcomb and Spurr, 33–37.

20. Flink, 36–39 (the quotation is on 37); John B. Rae, *The American Automobile Industry* (Boston: Twayne, 1984), 31–34; Hounshell, 217–218; Newcomb and Spurr, 32.

21. Flink, 37–39; Hounshell, 222–227; Rae, 35–37.

22. Hounshell, 222–237; Lindy Biggs, *The Rational Factory: Architecture, Technology, and Work in America's Age of Mass Production* (Baltimore: Johns Hopkins University Press, 1996), 101–107; Flink, 40–49.

23. Hounshell, 229–256; Henry Ford (with Samuel Crowther), *My Life and Work* (Garden City, N.Y.: Doubleday, 1922), 79–85; Biggs, 107–117.

24. Ford, 145; Flink, 36–38; Rae, 37–39; Biggs, 106–107, 150–152; Hounshell, 267–275.

25. Hounshell, 263–267; Hughes, 219; Flink, 234–235; Rae, 50–53, 66–67.

26. John M. Blum, *V Was for Victory: Politics and American Culture During World War II* (New York: Harcourt, Brace, Jovanovich, 1976), 110–124; Robert Friedel, "Scarcity and Promise: Materials and American Domestic Culture During World War II," in Donald Albrecht, ed., *World War II and the American Dream* (Cambridge, Mass.: MIT Press, 1995), 42–89; Joel Davidson, "Building for War, Preparing for Peace: World War II and the Military-Industrial Complex," in Albrecht, 184–229; Ruth S. Cowan, *A Social History of American Technology* (New York: Oxford University Press, 1997), 257.

27. Richard Rhodes, *The Making of the Atomic Bomb* (New York: Simon & Schuster, 1986), chapters 14 and 15.

28. Rhodes, 486–500; Michele Stenehjem Gerber, *On the Home Front: The Cold War Legacy of the Hanford Nuclear Site* (Lincoln: University of Nebraska Press, 1992), 23–26, 31–45.

Chapter 24

1. The bombing of Guernica was for many years a matter of some controversy, as the victorious Nationalists of Francisco Franco sought to deny it, or at least to deny any responsibility for it. Scholarship in recent decades, however, has removed all credible doubt about the destruction or the perpetrators. The sources used here are Gordon Thomas and Max Morgan Watts, *Guernica: The Crucible of World War II* (New York: Stein and Day, 1975); Herbert R. Southworth, *Guernica! Guernica! A Study of Journalism,*

Diplomacy, Propaganda, and History (Berkeley: University of California Press, 1977), and Paul Preston, *The Spanish Civil War, 1936–39* (London: Weidenfeld and Nicolson, 1986).

2. Roger E. Bilstein, *Flight in America: From the Wrights to the Astronauts* (Baltimore: Johns Hopkins University Press, 1987), 31, 70–71.

3. Leonard S. Reich, *The Making of American Industrial Research: Science and Business at GE and Bell, 1876–1926* (Cambridge: Cambridge University Press, 1985), chapters 4 and 10.

4. David A. Hounshell, "The Evolution of Industrial Research in the United States," in Richard S. Rosenbloom and William S. Spencer, eds., *Engines of Innovation: U.S. Industrial Research at the End of an Era* (Boston: Harvard Business School Press, 1996), 22–24; Reich, chapter 7.

5. Guy Hartcup, *The Effect of Science on the Second World War* (New York: St. Martin's Press, 2000), chapters 2 and 3; Tom Shachtman, *Terrors and Marvels: How Science and Technology Changed the Character and Outcome of World War II* (New York: William Morrow, 2002).

6. Shachtman, chapter 10.

7. Robert Friedel, "Scarcity and Promise," 42–89; on plastics, see Jeffrey Meikle, *American Plastic: A Cultural History* (New Brunswick, N.J.: Rutgers University Press, 1995), esp. 125–126; on operations research, see Hartcup, chapter 6.

8. R. A. C. Parker, *The Second World War: A Short History* (New York: Oxford University Press, 1997), esp. chapters 2 and 3; Jeremy Black, *Warfare in the Western World, 1882–1975* (Bloomington: Indiana University Press, 2002), chapter 4.

9. Black, 153–162; Parker, 151–156.

10. Parker, 156–172; Black, 162–164.

11. Spencer Weart, *Nuclear Fear: A History of Images* (Cambridge, Mass.: Harvard University Press, 1988), esp. chapter 6.

12. Richard Rhodes, *Dark Sun: The Making of the Hydrogen Bomb* (New York: Simon & Schuster, 1995), esp. chapter 13.

13. The quotation is from Francis Galton, "Hereditary Talent and Character," *Macmillan's Magazine*, June and August, 1865, as quoted in Ruth Schwartz Cowan, "Francis Galton's Statistical Ideas: The Influence of Eugenics," *Isis* 63, no. 219 (December 1972): 511; see also Daniel J. Kevles, *In the Name of Eugenics: Genetics and the Uses of Human Heredity* (New York: Knopf, 1985), chapter 1.

14. Davenport quoted in Steven Selden, *Inheriting Shame: The Story of Eugenics and Racism in America* (New York: Teachers College Press, 1999), 6; see also Edwin Black, *War Against the Weak: Eugenics and America's Campaign to Create a Master Race* (New York: Four Walls Press, 2003), 28–41; Kevles, 44–56.

15. Popenoe quoted in Selden, 11.

16. Black, chapter 5, the quotation from Jordan is on 65; Elof Axel Carlson, *The Unfit: A History of a Bad Idea* (Cold Spring Harbor, N.Y.: Cold Spring Harbor Laboratory Press, 2001), chapter 13.

17. W. Duncan McKim, *Heredity and Human Progress* (New York: G. P. Putnam's Sons, 1901), iii, 189–192.

18. The entire opinion by Holmes can be found in *Buck v. Bell* 274 U.S. 200 (1927); see Black, chapter 6.

19. Black, chapter 12.

20. Ploetz quoted in Sheila Faith Weiss, "The Race Hygiene Movement in Germany 1904–1945," in Mark B. Adams, ed., *The Wellborn Science: Eugenics in Germany, France, Brazil, and Russia* (New York: Oxford University Press, 1990), 15. Weiss is the primary source on the German eugenics movement.

21. In addition to Weiss, see Henry Friedlander, *The Origins of Nazi Genocide: From Euthanasia to the Final Solution* (Chapel Hill: University of North Carolina Press, 1995, esp. chapter 1; Stephan Kühl, *The Nazi Connection: Eugenics, American Racism, and German National Socialism* (New York: Oxford University Press, 1994), esp. chapter 6.

22. Kühl, esp. chapters 4 and 9; Friedlander, chapter 2; Black, chapter 14 (Hess quoted on 270).

23. See Friedlander for extended discussion of killing of the handicapped as well as the techniques of the "Final Solution." See also Black, chapter 15; Carlson, chapter 18.

Chapter 25

1. The best single source for the early history of computers is Martin Campbell-Kelly and William Aspray, *Computer: A History of the Information Machine* (New York: Basic Books, 1996). These paragraphs draw from chapters 1–3.

2. Ibid., 69–74; Harry Wulfurst, *Breakthrough to the Computer Age* (New York: Charles Scribner's Sons, 1982), 39–48; Steven Lubar, *Infoculture* (Boston: Houghton Mifflin, 1993), 301–304.

3. This discussion is based largely on Wulfurst, 7–36.

4. Campbell-Kelly and Aspray, 85–98; Wulfurst, 49–84.

5. Paul E. Ceruzzi, *A History of Modern Computing* (Cambridge, Mass.: MIT Press, 1999), 14–34.

6. Wulfurst, 161–171.

7. Ceruzzi, 49–53; Campbell-Kelly and Aspray, chapter 7.

8. Michael Riordan and Lillian Hoddeson, *Crystal Fire: The Birth of the Information Age* (New York: Norton, 1997), chapters 7–9.

9. Riordan and Hoddeson, chapter 10.

10. Ceruzzi, chapter 2.

11. Riordan and Hoddeson, chapter 12; Ceruzzi, chapter 6.

12. Lubar, 338–339.

13. Ceruzzi, chapter 7.

14. Ceruzzi, chapters 7 and 8.

15. Ceruzzi, chapter 7; Lubar, 342–346.

16. Rebecca Solnit, *River of Shadows: Eadweard Muybridge and the Technological Wild West* (New York: Viking, 2003), 200–203.

17. Paul Israel, *Edison: A Life of Invention* (New York: John Wiley & Sons, 1998), 292–296; Solnit, 228–231; Asa Briggs and Peter Burke, *A Social History of the Media* (Cambridge: Polity Press, 2002), 166–168.

18. Briggs and Burke, 154–157; the best source on this period of radio development is Hugh G. J. Aitken, *The Continuous Wave: Technology and American Radio, 1900–1932* (Princeton, N.J.: Princeton University Press, 1985).

19. Briggs and Burke, 157–163.

20. Lubar, 243–247; Briggs and Burke, 173–177.

21. Lubar, 247–257; Briggs and Burke, 176–179, 233–235.

22. The history of the Internet has been treated in a number of works, and a great deal of information is also available (appropriately enough) online. The most authoritative work on the origins is Janet Abbate, *Inventing the Internet* (Cambridge, Mass.: MIT Press, 1999). For online sources, the best place to begin is the site of the Internet Society: www.isoc.org/internet/history/.

23. Abbate, chapter 6.

Chapter 26

1. Eugene Cernan and Don Davis, *The Last Man on the Moon* (New York: St. Martin's Press, 1999), 337–338.

2. Dwight D. Eisenhower, "Farewell Address," January 17, 1961.

3. John Kennedy, "Address to Congress on Urgent National Needs," May 25, 1961, quoted in Eugene M. Emme, *A History of Space Flight* (New York: Holt, Rinehart & Winston, 1965), 179–180.

4. Barry Commoner, *The Closing Circle: Nature, Man & Technology* (New York: Knopf, 1971), 293.

5. On aircraft developments, see Roger E. Bilstein, *Flight in America from the Wrights to the Astronauts* (Baltimore: Johns Hopkins University Press, 1984), esp. chapters 3 and 4; Edward W. Constant III, *The Origins of the Turbojet Revolution* (Baltimore: Johns Hopkins University Press, 1980), chapter 5.

6. Constant, chapter 7.

7. Bilstein, 178–182, 218–230; Charles D. Bright, *The Jet-Makers: The Aerospace Industry from 1945 to 1972* (Lawrence: Regents Press of Kansas, 1978), 15–18, 85–93.

8. Bilstein, 182–184; Mel Horwitch, *Clipped Wings: The American SST Conflict* (Cambridge, Mass.: MIT Press, 1982), 7–18.

9. Horwitch covers SST difficulties and controversies in considerable detail; see esp. chapters 6, 9, 10, and 14. See also William A. Shurcliff, *S/S/T and Sonic Boom Handbook* (New York: Ballantine Books, 1970); this work not only gives the arguments regarding the SST's environmental effects, but also reproduces documents relating to economics and other issues.

10. Horwitch, esp. chapters 19 and 20.

11. Susan Wright, *Molecular Politics* (Chicago: University of Chicago Press, 1994), 123. The quotation was by the California Institute of Technology's Robert Sinsheimer.

12. Wright, 71–75; Clifford Grobstein, *A Double Image of the Double Helix: The Recombinant-DNA Debate* (San Francisco: W. H. Freeman, 1979), 12–18.

13. Wright, chapter 3; Grobstein, chapter 2.

14. Besides Wright and Grobstein, see also Thomas A. Shannon, *Genetic Engineering: A Documentary History* (Westport, Conn.: Greenwood Press, 1999), and Robert Bud, *The Uses of Life: A History of Biotechnology* (Cambridge: Cambridge University Press, 1993), chapters 8–9.

15. Von Neumann is quoted by Kirkpatrick Sale, "Ban Cloning? Not a Chance" (*New York Times*, March 7, 1997), excerpted in Shannon, 263. See Shannon, part VIII.

16. A good examination of the critiques of progress, particularly in the political context, is Christopher Lasch, *The True and Only Heaven: Progress and Its Critics* (New York: Norton, 1991).

17. J. B. S. Haldane, *Daedalus, or Science and the Future* (New York: E. P. Dutton, 1923).

18. Mumford's key works in this connection were *Technics and Civilization* (New York: Harcourt, Brace, 1934) and *The Myth of the Machine* (New York: Harcourt, Brace and World, 1967–70).

19. Jacques Ellul, "The Technological Order," *Technology & Culture* 3 (Fall 1962): 10, as quoted in M. R. Smith, "Technological Determinism in American Culture," in M. R. Smith and Leo Marx, eds., *Does Technology Drive History?* (Cambridge, Mass.: MIT Press, 1994), 30; John McDermott, "Technology: The Opiate of the Intellectuals," *New York Review of Books* 13, no. 2 (July 31, 1969): supplement.

Index